T0177159

Principles of Thermodynamics

In this introductory textbook, thermodynamics is presented as a natural extension of mechanics so that the laws and concepts learned in mechanics serve to get acquainted with the theory. The foundations of thermodynamics are presented in the first part. The second part covers a wide range of applications, which are of central importance in the fields of physics, chemistry and engineering, including calorimetry, phase transitions, heat engines and chemical reactions. In the third part, devoted to continuous media, Fourier and Fick's laws, diffusion equations and many transport effects are derived using a unified approach. Each chapter concludes with a selection of worked examples and several exercises to reinforce key concepts under discussion. A full solutions manual is available at the end of the book. It contains more than 150 problems based on contemporary issues faced by scientists and engineers that are solved in detail for undergraduate and graduate students.

Jean-Philippe Ansermet is a professor of physics at École Polytechnique Fédérale de Lausanne (EPFL), a fellow of the American Physical Society and a past president of the Swiss Physical Society. He coordinated the teaching of physics at EPFL for 12 years. His course on mechanics, taught for 25 years, was based on his textbook and a massive open online course (MOOC) that has generated over half a million views. For more than 15 years, he has taught thermodynamics to engineering and physics students. An expert in spintronics, he applies thermodynamics to analyse his pioneering experiments on giant magneto-resistance, or heat–driven spin torques and predict novel effects.

Sylvain D. Brechet completed his PhD studies in theoretical cosmology at the Cavendish Laboratory of the University of Cambridge as an Isaac Newton fellow. He is lecturer at the Institute of Physics at EPFL. He teaches mechanics, thermodynamics and electro-magnetism to first-year students. His current research focuses on theoretical modelling in condensed matter physics and more particularly in spintronics. Merging the fields of non-equilibrium thermodynamics, continuum mechanics and electromagnetism, he brought new insight to spintronics and fluid mechanics. In particular, he predicted in 2013 the existence of a fundamental irreversible thermodynamic effect now called the Magnetic Seebeck effect.

Principles of Thermodynamics

JEAN-PHILIPPE ANSERMET

École Polytechnique Fédérale de Lausanne

SYLVAIN D. BRECHET

École Polytechnique Fédérale de Lausanne

CAMBRIDGE
UNIVERSITY PRESS

CAMBRIDGE
UNIVERSITY PRESS

University Printing House, Cambridge CB2 8BS, United Kingdom

One Liberty Plaza, 20th Floor, New York, NY 10006, USA

477 Williamstown Road, Port Melbourne, VIC 3207, Australia

314–321, 3rd Floor, Plot 3, Splendor Forum, Jasola District Centre, New Delhi – 110025, India

79 Anson Road, #06–04/06, Singapore 079906

Cambridge University Press is part of the University of Cambridge.

It furthers the University's mission by disseminating knowledge in the pursuit of education, learning, and research at the highest international levels of excellence.

www.cambridge.org
Information on this title: www.cambridge.org/9781108426091
DOI: 10.1017/9781108620932

© Jean-Philippe Ansermet and Sylvain D. Brechet 2019

First published 2019

First edition in French published by Presses Polytechniques et Universitaires Romandes, 2016.

Printed in the United Kingdom by TJ International Ltd, Padstow Cornwall

A catalogue record for this publication is available from the British Library.

Library of Congress Cataloging-in-Publication Data
Names: Ansermet, Jean-Philippe, 1957- author. | Brechet, Sylvain D., 1981– author.
Title: Principles of thermodynamics / Jean-Philippe Ansermet
(École Polytechnique Fédérale de Lausanne), Sylvain D. Brechet
(École Polytechnique Fédérale de Lausanne).
Other titles: Thermodynamique. English
Description: Cambridge ; New York, NY : Cambridge University Press, 2018. |
Originally published in French: Thermodynamique (Lausanne : EPFL, 2013). |
Includes bibliographical references and index.
Identifiers: LCCN 2018030098 | ISBN 9781108426091 (hardback : alk. paper)
Subjects: LCSH: Thermodynamics–Textbooks. | Thermodynamics–Problems, exercises, etc.
Classification: LCC QC311.28 .A5713 2018 | DDC 536/.7–dc23
LC record available at https://lccn.loc.gov/2018030098

ISBN 978-1-108-42609-1 Hardback

Additional resources for this publication at www.cambridge.org/9781108426091

Contents

Preface

Thermodynamics is a theory which establishes the relationship between the physical quantities that characterise the macroscopic properties of a system. In this textbook, thermodynamics is presented as a physical theory which is based upon two fundamental laws pertaining to energy and entropy, which can be applied to many different systems in chemistry and physics, including transport phenomena. By asserting that energy and entropy are state functions, we eliminate the need to master the physical significance of differentials. Thus, thermodynamics becomes accessible to anyone with an elementary mathematical background. As the notion of entropy is introduced early on, it is readily possible to analyse out-of-equilibrium processes taking place in systems composed of simple blocks.

Students engaging with thermodynamics have the opportunity to discover a broad range of phenomena. However, they are faced with a challenge. Unlike Newtonian mechanics where forces are the cause of acceleration, the mathematical formalism of thermodynamics does not present an explicit link between cause and effect.

Nowadays, it is customary to introduce temperature by referring to molecular agitation and entropy by invoking Boltzmann's formula. However, in this book, the intrusion of notions of statistical physics are deliberately avoided. It is important to start off by teaching students the meaning of a physical theory and to show them clearly the very large preliminary conceptual work that establishes the notions and presuppositions of this theory. Punctual references to notions of statistical physics, which are not formally presented, give the impression that in science the results from another theoretical body of knowledge can be borrowed without precaution. By doing so, students might not perceive thermodynamics as a genuine scientific approach. It is clear that the introduction of entropy with a mathematical formula is somewhat reassuring. However, it is by performing calculations of entropy changes in simple thermal processes that students become familiar with this notion and not by contemplating a formula that is not used in the framework of thermodynamics.

This book is broken up into four parts. The first part of the book gathers the formal tools of thermodynamics, such as the thermodynamic potentials and Maxwell relations. The second part illustrates the thermodynamic approach with a few examples, such as phase transitions, heat engines and chemical reactions. The third part deals with continuous media, including a chapter that is devoted to interactions between electromagnetic fields and matter. A formal development of the thermodynamics of continuous media results in the description of numerous transport laws, such as the Fourier, Fick or Ohm laws and the Soret, Dufour or Seebeck effects.

At the end of each chapter, there are worked solutions that practically demonstrate what has been presented, and these are followed by several exercises. In the last part of the book, these exercises are presented with their solutions. Some exercises are inspired by physics auditorium demonstrations, some by research, for example: the melting point of nanoparticles, an osmotic power plant, a Kelvin probe, the so-called ZT coefficient of thermoelectric materials, thermogalvanic cells, ultramicroelectrodes or heat exchangers.

Thanks to the theory of irreversible phenomena which was elaborated in the period from approximately 1935 to 1965, thermodynamics has become an intelligible theory in which Newtonian mechanics and transport phenomena are presented in a unified approach. The book demonstrates that thermodynamics is applicable to many fields of science and engineering in today's modern world.

Acknowledgments

The authors are indebted to their mentor and friend Doctor François Reuse for the diligence with which he introduced them to the approach of his master, Professor Stückelberg. The authors were introduced to a school of thought through numerous discussions with the students of Professor Stückelberg like Professor Christian Gruber and Professor André Chatelain and with Professor Jean-Pierre Borel.

The authors gratefully acknowledge the stimulating discussions they had with the specialists whom they invited to contribute to a MOOC on thermodynamics: Chantal Maatouk and Marwan Brouche of the École Supérieure d'Ingénieurs de Beyrouth, Lebanon; Marthe Boyomo Onana, Paul-Salomon Ngohe-Ekam, Théophile Mbang and André Talla of the École Nationale Supérieure Polytechnique at Yaoundé and the Université de Yaoundé I, Cameroun; Etienne Robert of the École Polytechnique de Montréal, Canada; Miltiadis Papalexandris of the Université Catholique de Louvain, Belgium; and Michael Grätzel of the École Polytechnique Fédérale de Lausanne.

Graphic designer Claire-Lise Bandelier took great care in producing the figures according to the authors' wishes, in particular when making sketches of auditorium experiments. Professor Christian Gruber proofread the original French manuscript and made critical suggestions. Editor Evora Dupré secured the English translation and perfected the style of the text during the many meetings she held with the authors.

Finally, we express our sincere gratitude towards thousands of students and hundreds of tutors who took part in the course that led to this book. It is in the context of this large and vigilant audience that this book took shape.

Part I

FOUNDATIONS

Thermodynamic System and First Law

James Prescott Joule, 1818–1889

J. P. Joule was the owner of a brewery in England and worked as a self-educated scientist making major contributions to the development of thermodynamics. In 1840, he stated the law that bears his name on power dissipated by a current passing through a resistance. In 1843, there began a series of observations on the heat equivalent pertaining to mechanical work.

1.1 Historical Introduction

At the beginning of the nineteenth century, steam engines had been converting heat into work for about 150 years. Scientific investigations were under way to establish a quantitative equivalence of heat and work. With time, this concept became the law of energy conservation, which we will discuss in this chapter.

In 1839, Marc Séguin, nephew of the famous Mongolfier, published his 'Study on the influence of railways' [1]. It was clear for him that the condenser of a heat engine played a role that was equally important as that of the furnace [2]. Séguin assumed that the steam that caused the volume of the cylinder to increase performed work that was equal in

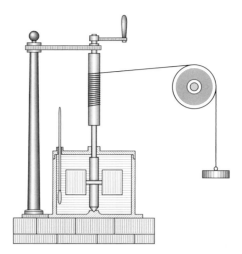

Figure 1.1 Joule's calorimeter: the fall of the weight at nearly constant velocity (because of the fluid viscosity) drives a stirring mechanism in the calorimeter. The change in potential energy of the weight determines the work performed on the liquid. The increase in energy of the liquid is deduced from its temperature rise.

magnitude to the heat lost by the steam in the process. He sought to estimate this heat loss by measuring the heat taken from the furnace and returned to the condenser.

A year later, Julius Robert von Mayer travelled to the tropics as a medical doctor. He observed that the colour difference between veinous and arterial blood was more pronounced there than at latitudes where the climate is colder. He attributed this difference to the heat released by the body. His thoughts on the human body as a heat engine led him to the idea of an equivalence between heat and work, which he later tested on inert matter.

In his 1842 treatise [3], he asked the following question: what is the change in temperature of a stone when it hits the ground after falling from a given height? James Prescott Joule succeeded in doing this measurement, thanks to his development of highly sensitive thermometers. Joule also observed that an electrical current dissipates heat, an effect that bears his name. In 1845, he published his fundamental work on energy conservation [4]. With a calorimeter (Fig. 1.1), he determined the equivalence between work (defined by masses going down in the gravitational potential) and heat (corresponding to the warming of the liquid stirred by the device). Before Joule established this link, it was customary to measure heat in *calories*. Joule established the conversion to the unit that bears his name, the ***joule*** $[J]$: 1 [cal] = 4.1855 [J].

These are units of energy. Joule and others observed in various circumstances that the heat provided to a system and the work performed on it are equal and opposite to each other for every process that brings the system back to its initial state [5]. Von Mayer, in his treatise of 1842, expressed the idea that every system is characterised by a quantity, energy, that can only be modified by an external action in the form of work or heat [6]. Hermann von Helmholtz, in 1847, gives energy conservation the status of a physical law [7].

1.2 Thermodynamic System

A ***thermodynamic system*** consists of matter contained in a region of space delineated by a closed surface, called an ***enclosure***, that separates the system from its ***environment*** or ***surroundings***. The system is assumed to be large, in the sense that the amount of substance it contains is typically counted in moles. A mole corresponds to the Avogadro number $6.02 \cdot 10^{23}$ that is used to count the number of elementary constituents of matter. This definition can be extended to a thermodynamic system that includes radiation and any other kind of physical field. In the first two parts of this book, we will discuss systems consisting of matter only. We will introduce electromagnetic fields in the third part.

We use the following terms to characterise how a system interacts with its environment. A system is said to be:

- ***open***, if its enclosure allows convective matter exchange with the environment
- ***closed***, if its enclosure does not allow convective matter exchange with the environment
- ***diathermal***, if its enclosure allows conductive heat exchange with the environment
- ***adiabatic***, if its enclosure does not allow conductive heat exchange with the environment
- ***isolated***, if its enclosure does not allow any interaction with the environment

A thermodynamic system can be decomposed in ***subsystems*** that can be considered as thermodynamic systems themselves. The separation between two subsystems is called a ***wall***. The enclosure between a system and its environment consists of one or several walls. We use the following terms to characterise a wall. A wall is said to be:

- ***fixed***, if it cannot move
- ***movable***, if it can move
- ***permeable***, if it allows convective matter exchange with the environment
- ***impermeable***, if it does not allow convective matter exchange with the environment
- ***diathermal***, if it allows conductive heat exchange with the environment
- ***adiabatic***, if it does not allow conductive heat exchange with the environment

1.3 State, Variables and State Functions

The ***state*** of a system is characterised by physical properties that are described by a set of ***state variables***. The state is entirely specified by the values of these state variables and it does not depend on the history of the system. The set of these state variables is written as,

$$\{X_1, X_2, X_3, X_4, X_5, \ldots\}$$

A ***state function*** is a physical property that depends only on the state of the system. Thus, a state function is expressed as,

$$F(X_1, X_2, X_3, X_4, X_5, \ldots)$$

1.3.1 Partial Derivatives of a Function

Let $f(x, y)$ be a function of two variables x and y. The **partial derivatives** with respect to the variables x and y are defined by,

$$\frac{\partial f(x, y)}{\partial x} \equiv \lim_{\Delta x \to 0} \frac{f(x + \Delta x, y) - f(x, y)}{\Delta x}$$
$$\frac{\partial f(x, y)}{\partial y} \equiv \lim_{\Delta y \to 0} \frac{f(x, y + \Delta y) - f(x, y)}{\Delta y} \tag{1.1}$$

Thus, in the calculation of a partial derivative with respect to one variable, the other variables are kept constant. As an example, let us consider the function $f(x, y) = x^2 + 3xy$. The partial derivatives of this function are,

$$\frac{\partial f(x, y)}{\partial x} = 2x + 3y \qquad \text{and} \qquad \frac{\partial f(x, y)}{\partial y} = 3x$$

1.3.2 Differential of a Function

Let $f(x, y)$ be a function of two variables x and y [8]. The variation of the function $f(x, y)$ from point (x, y) to point $(x + \Delta x, y + \Delta y)$ is written as,

$$\Delta f(x, y) = f(x + \Delta x, y + \Delta y) - f(x, y) \tag{1.2}$$

It can be recast as,

$$\begin{aligned} \Delta f(x, y) &= f(x + \Delta x, y + \Delta y) - f(x, y + \Delta y) + f(x, y + \Delta y) - f(x, y) \\ &= \frac{f(x + \Delta x, y + \Delta y) - f(x, y + \Delta y)}{\Delta x} \Delta x \\ &+ \frac{f(x, y + \Delta y) - f(x, y)}{\Delta y} \Delta y \end{aligned} \tag{1.3}$$

The **differential** $df(x, y)$ is defined as the infinitesimal limit of the variation $\Delta f(x, y)$,

$$df(x, y) \equiv \lim_{\Delta x \to 0} \lim_{\Delta y \to 0} \Delta f(x, y) \tag{1.4}$$

Taking into account the limit

$$\lim_{\Delta y \to 0} \left(f(x + \Delta x, y + \Delta y) - f(x, y + \Delta y) \right) = f(x + \Delta x, y) - f(x, y) \tag{1.5}$$

in equation (1.3), the differential (1.4) can be written as,

$$\begin{aligned} df(x, y) &= \lim_{\Delta x \to 0} \frac{f(x + \Delta x, y) - f(x, y)}{\Delta x} \Delta x \\ &+ \lim_{\Delta y \to 0} \frac{f(x, y + \Delta y) - f(x, y)}{\Delta y} \Delta y \end{aligned} \tag{1.6}$$

Using the definition (1.1) of the partial derivatives of a function, the differential (1.6) is reduced to,

$$df(x, y) = \frac{\partial f(x, y)}{\partial x} dx + \frac{\partial f(x, y)}{\partial y} dy \tag{1.7}$$

1.3.3 Time Derivative of a Function

Let $f(x,y)$ be a function of two time-dependent variables x and y. The time derivative of the function is obtained using the chain rule,

$$\dot{f}(x,y) = \frac{\partial f(x,y)}{\partial x}\dot{x} + \frac{\partial f(x,y)}{\partial y}\dot{y} \tag{1.8}$$

where

$$\dot{f}(x,y) \equiv \frac{df(x,y)}{dt} \qquad \dot{x} \equiv \frac{dx}{dt} \qquad \dot{y} \equiv \frac{dy}{dt}$$

1.4 Processes and Change of State

A thermodynamic system can interact with its environment through **processes** that change the state of the system. We distinguish three types of physical processes:

- mechanical processes
- thermal processes
- chemical processes

As we will see, thermodynamics allows us to provide a quantitative characterisation of the effects that processes have on the state of a system. Mechanical processes can lead to mechanical or thermal changes of the state. In the experiment of Joule's calorimeter (Fig. 1.1), work is performed on the system and its temperature changes. Thus, a mechanical process can lead to a thermal change of state. Likewise, thermal processes can lead to thermal or mechanical changes of the system state. In the experiment illustrated in Fig. 1.2, heat is provided to the system by hand contact. The pressure of the gas rises, causing a shift of the water levels in the U-shaped tube. Thus, a thermal process can lead to a mechanical change of state.

Figure 1.2   A vessel contains a fixed amount of gas. The liquid in the tube measures the changes of pressure when the gas is heated by the hand.

In the context of describing the types of processes that take place between a system and its environment, the term "chemical process" refers specifically to matter being exchanged between a system and its environment. This is not to be confused with chemical reactions among different substances. Such reactions occurring within the system will be explored in Chapter 8. The enclosure of the system should be defined so as to encompass the regions of space where chemical reactions take place.

1.5 Extensive and Intensive Quantities

A quantity is called *extensive* when it has the following property: its value for the whole system is equal to the sum of its values for every subsystem. The following are extensive quantities:

- mass
- momentum
- angular momentum
- energy
- volume

Sometimes we refer to extensive quantities divided by the volume, the mass or the number of moles of the system. They are called a volume density, a mass density or a molar density. In that case we speak of a *specific quantity* or a *reduced extensive quantity*. The following are densities:

- mass density
- momentum density
- angular momentum density
- energy density

A quantity is called *intensive* when it is conjugated to an extensive quantity, which means that it is defined as the partial derivative of the energy with respect to this extensive quantity. The following are intensive quantities:

- velocity
- pressure
- temperature

To determine if a quantity is extensive or intensive, it is useful to imagine what happens to this quantity when the size of the system doubles. If the quantity is extensive, its value doubles, but if it is intensive, its value does not change. It is also useful to clarify that *certain quantities are neither extensive nor intensive*. This is the case of the entropy production rate, which we will introduce in Chapter 2.

A system is called *homogeneous* when all the intensive scalar functions conjugated to the extensive scalar state variables do not depend on position. This means that they have the same value for every subsystem. A system is called *uniform* when the intensive vectorial functions conjugated to the extensive vectorial state variables do not depend on position. This means that they have the same norm and orientation in every subsystem.

1.6 First Law of Thermodynamics

Thermodynamics is based on two fundamental laws. Their justification is based on the empirical validity of their implications. In this chapter, we discuss the first law. We will present the second law in Chapter 2.

The *first law of thermodynamics* states that:

For every system, there is a scalar extensive state function called *energy* (E). **When the system is isolated, the energy is conserved**.

The energy conservation law is mathematically written as,

$$\dot{E} = 0 \quad \text{(isolated system)} \tag{1.9}$$

where $\dot{E} \equiv dE/dt$. This conservation is related to time homogeneity [9]. It implies that energy is defined up to a constant.

When the system interacts with its environment, the energy evolution results from the power of the processes exerted on the system. We distinguish four types of external processes and write [10]:

$$\dot{E} = P^{\text{ext}} + P_W + P_Q + P_C \quad \text{(open system)} \tag{1.10}$$

- P^{ext} represents the *power* associated with the *external forces* and *torques* that modify the *translational kinetic energy* of the centre of mass and the *rotational kinetic energy* around the centre of mass. These forces and torques do not modify the shape of the system.
- P_W represents the *mechanical power* associated with the work performed by the environment on the system that results in a deformation of the system without any change in its state of motion, in particular its kinetic energy.
- P_Q represents the *thermal power* associated with heat exchange with the environment through conduction.
- P_C represents the *chemical power* associated with matter exchange with the environment through convection.

Any physical process performing work is called a *mechanical action*. Any physical process in which heat is exchanged is called a *heat transfer*. A physical process in which matter is exchanged is called a *matter transfer* or *mass transfer*. When a heat transfer takes place through a matter transfer, it is called a heat transfer by *convection*. When a heat transfer occurs without matter transfer, it is called a heat transfer by *conduction*. In general, a matter transfer leads simultaneously to a mechanical action and to a heat transfer.

We have characterised systems according to ways in which they interact with their environment. We can now specify such characteristics in terms of the powers of the various processes considered here. Hence, a system is called:

- **rigid** if no work by deformation is possible, i.e. $P_W = 0$.
- **closed** if there is no matter transfer, i.e. $P_C = 0$, and **open** otherwise.
- **adiabatic** if there is no heat transfer, i.e. $P_Q = 0$ and **diathermal** in the opposite case.
- **isolated** if it is rigid, adiabatic and closed in the absence of external forces and torques, i.e. $P^{ext} = P_W = P_Q = P_C = 0$.

When a system is closed, the energy evolution equation (1.10) reduces to,

$$\dot{E} = P^{ext} + P_W + P_Q \quad \text{(closed system)} \tag{1.11}$$

The first law can be expanded on to include two other conservation laws that impose additional constraints on the possible states of the system. The first is related to the translational state of motion and the second to the rotational state of motion.

Concerning translations, we have the following conservation law:

For every system, there is a vectorial extensive state function called _momentum_ (P). When the system is isolated, the momentum is conserved.

The momentum conservation law is mathematically written as,

$$\dot{P} = 0 \quad \text{(isolated system)} \tag{1.12}$$

where $\dot{P} \equiv dP/dt$. This conservation law is related to the homogeneity of space [9]. It implies that the momentum is defined up to a constant.

In the case of a system interacting with the environment, the evolution of the momentum with respect to an inertial frame of reference is given by the centre-of-mass theorem [11],

$$\dot{P} = F^{ext} \tag{1.13}$$

where F^{ext} is the net external force exerted on the system. If the system undergoes a uniform translational motion, the momentum is constant. A net external force causes a departure from the state of uniform translation.

Concerning rotations, we have the following conservation law:

For every system, there is a vectorial extensive state function called _angular momentum_ (L). When the system is isolated, the angular momentum is conserved.

The angular momentum conservation law is mathematically written as,

$$\dot{L} = 0 \quad \text{(isolated system)} \tag{1.14}$$

where $\dot{L} \equiv dL/dt$. This conservation law is related to the isotropy of space. It implies that the angular momentum is defined up to a constant.

In the case of a system interacting with the environment, the evolution of the angular momentum with respect to an inertial frame of reference is given by the angular momentum theorem [11],

$$\dot{L} = M^{ext} \tag{1.15}$$

where M^{ext} is the net external torque exerted on the system. If the system undergoes a uniform rotational motion, the angular momentum is constant. A net external torque causes a departure from the state of uniform rotation.

1.7 Thermodynamics and Mechanics

The first law, expanded with the two mechanical conservation laws, links thermodynamics and mechanics. We will now illustrate this link by considering a particular system that is homogeneous, deformable, closed, diathermal, electrically neutral and cylindrical in shape. The centre of mass of the system is moving with a velocity v that is small compared to the speed of light in vacuum. The system rotates with an angular velocity ω around its axis of symmetry, which is a principal axis of inertia [12]. The motion of the system is such that the velocity v is collinear to the angular velocity ω.

We assume further that the deformation of the system is symmetric in the direction of the rotational axis and leaves the radius of the cylinder constant. It is a contraction or dilatation of the system that does not modify its translational kinetic energy. The deformation does not modify the rotational kinetic energy either since the moment of inertia with respect to the rotation axis is constant (Fig. 1.3). It modifies the total kinetic energy of the system since there is a symmetric motion of matter relative to the centre of mass. However, we consider here a deformation that is sufficiently slow for this kinetic energy variation to be negligible.

The translational motion of the system is characterised by its momentum P and the rotational motion by its angular momentum L. According to the first law, these quantities are extensive state functions that can be chosen as extensive state variables. A state variable is a trivial state function for which all other variables are kept constant. The state of the system at time t is determined by a set of $n + 3$ extensive state variables $\{P, L, X_0, X_1, \ldots, X_n\}$.

According to the first law, energy is a state function. Thus, it is a function of all the state variables of the system, i.e. $E = E(P, L, X_0, X_1, \ldots, X_n)$. The **momentum** is proportional to the velocity v,

$$P = M v \tag{1.16}$$

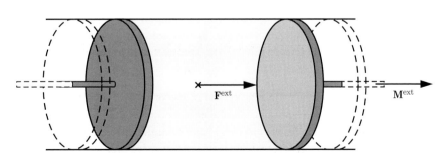

Figure 1.3 A net external force F^{ext} and a net external torque M^{ext} are exerted on a homogeneous system that is symmetrically deformed.

where the mass M is constant, i.e. $\dot{M} = 0$. Substituting the expression (1.16) into the centre of mass theorem (1.13), we obtain,

$$F^{\text{ext}} = M\dot{v} \tag{1.17}$$

The *velocity v* is the *intensive quantity conjugated* to the momentum P,

$$v = \frac{\partial E\left(P, L, X_0, X_1, \ldots, X_n\right)}{\partial P} \tag{1.18}$$

The *angular momentum L* is proportional to the angular velocity ω,

$$L = I\omega \tag{1.19}$$

where I is the moment of inertia with respect to the rotational axis. Since the system is cylindrical in shape with a constant radius R_0, the moment of inertia $I = \frac{1}{2} MR_0^2$ is constant, i.e. $\dot{I} = 0$. Substituting the expression (1.19) into the angular momentum theorem (1.15), we obtain,

$$M^{\text{ext}} = I\dot{\omega} \tag{1.20}$$

The *angular velocity ω* is the *intensive quantity conjugated* to the angular momentum L,

$$\omega = \frac{\partial E\left(P, L, X_0, X_1, \ldots, X_n\right)}{\partial L} \tag{1.21}$$

In order to satisfy the definitions, (1.18) and (1.21), taking into account relations (1.16) and (1.19), the energy of the system $E\left(P, L, X_0, X_1, \ldots, X_n\right)$ is given by [13],

$$E\left(P, L, X_0, X_1, \ldots, X_n\right) = \frac{1}{2} v \cdot P + \frac{1}{2} \omega \cdot L + U(X_0, X_1, \ldots, X_n) \tag{1.22}$$

The first term is the translational kinetic energy of the centre of mass and the second term the rotational kinetic energy around a principal axis of inertia. The state function $U \equiv U(X_0, X_1, \ldots, X_n)$ is called the *internal energy* of the system, because it is independent of its state of motion. If we wanted to identify the internal energy of a system which is more complex than the one examined here, we would have to split it into homogeneous subsystems (as in Chapter 3) and apply the present approach to each of the subsystems. If the system is too complex, then it has to be described by the continuum approach that will be developed in Chapter 10.

For the simple system treated here, we can determine the time derivative of its energy by taking the time derivative of expression (1.22). Applying the mathematical rule (1.8) for the time derivative of a function of several variables, and taking into account the definitions (1.18) and (1.21), we obtain:

$$\dot{E}\left(P, L, X_0, X_1, \ldots, X_n\right) = v \cdot \dot{P} + \omega \cdot \dot{L} + \dot{U}(X_0, X_1, \ldots, X_n) \tag{1.23}$$

Taking into account relations (1.16) and (1.19), the expression (1.23) of the time derivative of the energy becomes,

$$\dot{E} = M v \cdot \dot{v} + I \omega \cdot \dot{\omega} + \dot{U} \tag{1.24}$$

Using the laws of dynamics (1.17) and (1.20), the energy evolution equation (1.24) becomes,

$$\dot{E} = \boldsymbol{F}^{\text{ext}} \cdot \boldsymbol{v} + \boldsymbol{M}^{\text{ext}} \cdot \boldsymbol{\omega} + \dot{U} \tag{1.25}$$

Thus, we see that the power P^{ext} resulting from the action of the net external force $\boldsymbol{F}^{\text{ext}}$ and net external torque $\boldsymbol{M}^{\text{ext}}$ is written as,

$$P^{\text{ext}} = \boldsymbol{F}^{\text{ext}} \cdot \boldsymbol{v} + \boldsymbol{M}^{\text{ext}} \cdot \boldsymbol{\omega} \tag{1.26}$$

Therefore, the energy evolution equation (1.25) is reduced to,

$$\dot{E} = P^{\text{ext}} + \dot{U} \tag{1.27}$$

1.8 Internal Energy

Comparing the energy evolution equations (1.10) and (1.27) pertaining to the particular system described in the previous section, we obtain the evolution equation for the internal energy,

$$\dot{U} = P_W + P_Q + P_C \quad \text{(open system)} \tag{1.28}$$

In the case of a closed system, i.e. $P_C = 0$, the internal energy evolution equation (1.28) is reduced to,

$$\dot{U} = P_W + P_Q \quad \text{(closed system)} \tag{1.29}$$

In the case of an isolated system, i.e. $P_Q = P_W = 0$, the internal energy U is constant,

$$\dot{U} = 0 \quad \text{(isolated system)} \tag{1.30}$$

The internal energy evolution equations (1.28)–(1.29) and conservation equation (1.30) describe the thermodynamics of a system with respect to a frame of reference where it is at rest, i.e. a frame of reference where its kinetic energy is negligible.

For certain thermodynamic systems, there is no frame of reference where the kinetic energy vanishes. The time evolution of such systems is described by the first law (1.11) expressed in terms of the total energy E, as for instance in § 3.6.3 and § 3.6.4.

The terms on both sides of the evolution equation (1.28) can be multiplied by an infinitesimal time interval dt:

$$\dot{U}\,dt = P_W\,dt + P_Q\,dt + P_C\,dt \quad \text{(open system)} \tag{1.31}$$

In the case of a closed system, i.e. $P_C = 0$, equation (1.31) is reduced to,

$$\dot{U}\,dt = P_W\,dt + P_Q\,dt \quad \text{(closed system)} \tag{1.32}$$

The infinitesimal energy variation is given by,

$$dU = \dot{U}\,dt \tag{1.33}$$

The work performed on the system during an infinitesimal process is defined as,

$$\delta W \equiv P_W \, dt \tag{1.34}$$

The heat provided to the system during an infinitesimal process is defined as,

$$\delta Q \equiv P_Q \, dt \tag{1.35}$$

where the symbol Q comes from the German word "Quelle" that means source. We call

$$\delta C \equiv P_C \, dt \tag{1.36}$$

the **chemical work** performed on the system during an infinitesimal process. We use the expression chemical work to describe the energy variation resulting from the matter transfer between the system and its environment in order to conform to the most common terminology used in thermodynamics. However, we should keep in mind that a matter transfer can perform (mechanical) work on a system and it can also cause a heat transfer. The chemical work could also be called **convective chemical heat**.

Using the definitions (1.33)–(1.36) the expression for the infinitesimal internal energy variation (1.31) is written as,

$$dU = \delta W + \delta Q + \delta C \quad \text{(open system)} \tag{1.37}$$

In the case of a closed system, i.e. $P_C = 0$, the infinitesimal internal energy variation (1.37) is reduced to,

$$dU = \delta W + \delta Q \quad \text{(closed system)} \tag{1.38}$$

The internal energy U is a state function, which means that its variation depends only on the initial and final states. Thus, it is independent of the process that drives the system from the initial state to the final state. Hence, the infinitesimal variation of the internal energy is a differential, dU, as defined in (1.7).

By contrast, the work δW, the energy exchange δC caused by matter transfer and the heat transfer δQ provided by conduction depend on the process that took place during the infinitesimal time interval dt to bring the system from one state to another. Therefore, these quantities are not differentials. In order to distinguish these infinitesimal quantities from differentials, we use the symbol δ.

In order to obtain an expression for the internal energy variation, we now integrate relation (1.38). When integrating dU, we can use the fact that the variation of U depends only on the initial and final states of the system. This yields:

$$\Delta U_{if} = \int_{U_i}^{U_f} dU = U_f - U_i \tag{1.39}$$

When integrating δW, we get the **work** performed on the system,

$$W_{if} = \int_i^f \delta W = \int_{t_i}^{t_f} P_W \, dt \tag{1.40}$$

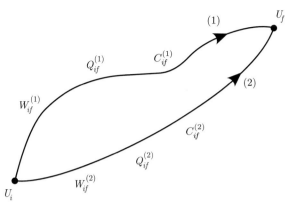

Processes (1) and (2) bring the system from the initial state i to the final state f. $W_{if}^{(1)}$ is different from $W_{if}^{(2)}$, $Q_{if}^{(1)}$ from $Q_{if}^{(2)}$ and $C_{if}^{(1)}$ from $C_{if}^{(2)}$, but their sum for a given process yields the same energy change ΔU_{if}.

When integrating δQ, we get the **heat** transferred to the system,

$$Q_{if} = \int_i^f \delta Q = \int_{t_i}^{t_f} P_Q \, dt \qquad (1.41)$$

And finally, when integrating δC, we get the **chemical work** performed on the system,

$$C_{if} = \int_i^f \delta C = \int_{t_i}^{t_f} P_C \, dt \qquad (1.42)$$

Using relations (1.39)–(1.42), the integral of the infinitesimal internal energy variation (1.37) from an initial state i to a final state f is written as,

$$\Delta U_{if} = W_{if} + Q_{if} + C_{if} \quad \text{(open system)} \qquad (1.43)$$

For a closed system, i.e. $P_C = 0$, the internal energy variation (1.43) is reduced to,

$$\Delta U_{if} = W_{if} + Q_{if} \quad \text{(closed system)} \qquad (1.44)$$

We illustrate (Fig. 1.4) the distinction between the variation of a state function that depends only on the initial and final states, and the integrals W_{if}, Q_{if} and C_{if} that depend on the process. Two different processes are drawn, labelled (1) and (2), that bring the system from an initial state i to a final state f. Since the work is not a state function, $W_{if}^{(1)}$ and $W_{if}^{(2)}$ can be different. Similarly, since the heat is not a state function, $Q_{if}^{(1)}$ and $Q_{if}^{(2)}$ can also be different. Likewise, since chemical work is not a state function, $C_{if}^{(1)}$ and $C_{if}^{(2)}$ can be different. Since the internal energy is a state function, its variation $\Delta U_{if} = U_f - U_i$ depends only on the initial and final states. Thus, it is independent of the process.

In order to express the evolution of the internal energy in terms of the state variables, we apply the rule (1.8) to the time derivative to the state function internal energy $U(X_0, X_1, \ldots, X_n)$,

$$\dot{U}(X_0, X_1, \ldots, X_n) = \sum_{i=0}^{n} \frac{\partial U(X_0, X_1, \ldots, X_n)}{\partial X_i} \dot{X}_i \qquad (1.45)$$

The **intensive quantities** $Y_i \equiv Y_i(X_0, X_1, \ldots, X_n)$, called the **conjugates** of the extensive state variables, are defined as

$$Y_i(X_0, X_1, \ldots, X_n) \equiv \frac{\partial U(X_0, X_1, \ldots, X_n)}{\partial X_i} \tag{1.46}$$

Thus, the time derivative (1.45) of the internal energy is reduced to,

$$\dot{U} = \sum_{i=0}^{n} Y_i \dot{X}_i \tag{1.47}$$

When the system is isolated (1.30), the internal energy is constant. Thus the term on the right-hand side of equation (1.47) vanishes but the state variables may still depend on time. In that case, the contribution to the internal energy resulting from the variation of a state variable X_j is compensated by the contribution to the variation resulting from the other state variables X_i, where $i \neq j$, in order to maintain constant the internal energy.

The time evolution of the state of the system is described by the time dependence of the state variables. Identifying the evolution equations (1.29) and (1.47), we can relate the physical processes to that time evolution in the following way,

$$\sum_{i=0}^{n} Y_i \dot{X}_i = P_W + P_Q + P_C \quad \text{(open system)} \tag{1.48}$$

When the system is closed, i.e. $P_C = 0$, equation (1.48) is reduced to,

$$\sum_{i=0}^{n} Y_i \dot{X}_i = P_W + P_Q \quad \text{(closed system)} \tag{1.49}$$

1.9 Damped Harmonic Oscillator

In the following example, borrowed from mechanics, we will show that a non-mechanical state variable is needed in order to satisfy the first law of thermodynamics. Let us consider a point mass M subjected to an elastic force and a viscous friction force in a fluid (Fig. 1.5).

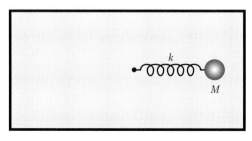

Figure 1.5 The system consists of a point mass attached to a spring oscillating in a viscous fluid.

The system, which is considered as isolated, consists of the damped harmonic oscillator and the fluid. The point mass is not isolated since it interacts with the spring and the fluid. The system is large enough for the point mass to be negligible with respect to the mass of the fluid. This implies that the kinetic energy of the fluid is negligible with respect to the kinetic energy of the point mass. It follows therefore that the fluid can be considered at rest in the centre-of-mass frame of reference. In mechanics, we consider that the point mass position and the momentum are the variables defining the state of the system. With this example, we wish to show that in thermodynamics a new state variable has to be added in order to define an energy that is a state function.

The system consisting of the point mass, the spring and the fluid is isolated. Thus, according to the first law (1.9), its total energy E is conserved. The total energy E is the sum of the kinetic energy and the internal energy U. Hence, the internal energy U has to increase to compensate the kinetic energy loss when the mass stops oscillating. Therefore, there has to be a non-mechanical contribution to U.

In order to take into account this non-mechanical contribution, we will introduce a new extensive variable X_0 that does not depend on the point mass position r and momentum P. Then the internal energy U is not only a function of r but of X_0 also. The total energy (1.22) is the sum of the kinetic energy and the internal energy $U(X_0, r)$ [14],

$$E(P, r, X_0) = \frac{P^2}{2M} + U(r, X_0) \tag{1.50}$$

The internal energy is the sum of the elastic potential energy $\Phi\left(r\right)$ and an additional term,

$$U(r, X_0) = \Phi\left(r\right) + U(\mathbf{0}, X_0) \tag{1.51}$$

The elastic potential energy $\Phi\left(r\right)$ is defined as,

$$\Phi\left(r\right) = \frac{1}{2} k r^2 \tag{1.52}$$

Since the total energy (1.50) is a state function and a constant of motion and since the velocity v is the intensive function conjugated to the momentum P (1.18), the time derivative of the energy is written as,

$$\dot{E} = v \cdot \dot{P} + \frac{\partial U(r, X_0)}{\partial r} \cdot \dot{r} + \frac{\partial U(r, X_0)}{\partial X_0} \dot{X_0} = 0 \tag{1.53}$$

Using relations (1.51) and (1.52), the partial derivative of the internal energy with respect to the position yields,

$$\frac{\partial U(r, X_0)}{\partial r} = \frac{d\Phi\left(r\right)}{dr} = k r \tag{1.54}$$

The velocity v of the point mass is $\dot{r}$. In light of definition (1.46), the conjugate to the state variable X_0 is the intensive state function Y_0 given by

$$Y_0 = \frac{\partial U(r, X_0)}{\partial X_0} = \frac{dU(\mathbf{0}, X_0)}{dX_0} \tag{1.55}$$

Thus, using the definitions (1.54) and (1.55), the evolution equation (1.53) becomes,

$$\left(\dot{P} + k r\right) \cdot v + Y_0 \dot{X_0} = 0 \tag{1.56}$$

The forces exerted on the point mass are the elastic force $\boldsymbol{F}^{\mathrm{el}}$ and the friction force $\boldsymbol{F}^{\mathrm{fr}}$,

$$\begin{aligned} \boldsymbol{F}^{\mathrm{el}} &= -k\,\boldsymbol{r} \\ \boldsymbol{F}^{\mathrm{fr}} &= -\lambda\,\boldsymbol{v} \end{aligned} \tag{1.57}$$

The elastic constant k and the the viscous friction coefficient λ are positive. The centre-of-mass theorem (1.13) reads [15],

$$\dot{\boldsymbol{P}} = \boldsymbol{F}^{\mathrm{el}} + \boldsymbol{F}^{\mathrm{fr}} = -k\,\boldsymbol{r} - \lambda\,\boldsymbol{v} \tag{1.58}$$

The elastic force $\boldsymbol{F}^{\mathrm{el}}$ and the friction force $\boldsymbol{F}^{\mathrm{fr}}$ are treated as external forces for the mechanical system, which consists of the point mass. However, these forces are considered internal forces for the thermodynamic system which consists of the point mass, the spring and the fluid.

We now establish an evolution equation for the new state variable X_0 by combining the centre-of-mass theorem (1.58) and the energy conservation (1.56). We find that,

$$\dot{X}_0 = \frac{\lambda\,\boldsymbol{v}^2}{Y_0} = -\frac{\boldsymbol{F}^{\mathrm{fr}} \cdot \boldsymbol{v}}{Y_0} \tag{1.59}$$

Thus, if Y_0 is positive then $\dot{X}_0$ is also positive, which means that X_0 increases over time. The term $\boldsymbol{F}^{\mathrm{fr}} \cdot \boldsymbol{v}$ corresponds to the power dissipated by viscous friction. Intuitively, we know that friction generates heat and leads to an increase in temperature.

We will see in the next chapter that the intensive state function Y_0 corresponds to the temperature T of the system and that the extensive state variable X_0 corresponds to the entropy S. Then, the evolution equation (1.59) becomes,

$$\dot{S} = \frac{\lambda\,\boldsymbol{v}^2}{T} = -\frac{\boldsymbol{F}^{\mathrm{fr}} \cdot \boldsymbol{v}}{T} \tag{1.60}$$

Thus, if the temperature T is positive then $\dot{S}$ is also positive, which means that entropy increases over time.

1.10 Worked Solutions

1.10.1 Velocity and Angular Velocity

1. Establish relation (1.18) in which the velocity $\boldsymbol{v}$ is the intensive quantity conjugated to the momentum $\boldsymbol{P}$.
2. Establish relation (1.21) in which that the angular velocity $\boldsymbol{\omega}$ is the intensive quantity conjugated to the angular momentum $\boldsymbol{L}$.

Solution:

1. *Using the momentum definition (1.16), the expression (1.22) for the energy E of the system is written as,*

$$E\left(\boldsymbol{P}, \boldsymbol{L}, X_0, X_1, \ldots, X_n\right) = \frac{\boldsymbol{P}^2}{2\,M} + \frac{1}{2}\,\boldsymbol{\omega} \cdot \boldsymbol{L} + U(X_0, X_1, \ldots, X_n)$$

From it, we derive relation (1.18), i.e.

$$v = \frac{P}{M} = \frac{\partial E\left(P,L,X_0,X_1,\ldots,X_n\right)}{\partial P}$$

2. Using the angular momentum definition (1.19), the expression (1.22) of the energy E of the system is written as,

$$E\left(P,L,X_0,X_1,\ldots,X_n\right) = \frac{1}{2}\,v\cdot P + \frac{L^2}{2I} + U(X_0,X_1,\ldots,X_n)$$

From it, we derive relation (1.21), i.e.

$$\omega = \frac{L}{I} = \frac{\partial E\left(P,L,X_0,X_1,\ldots,X_n\right)}{\partial L}$$

1.10.2 Man in a Boat

A man moves across the horizontal deck of a boat. The man and the boat are initially at rest with respect to the water. Then, the man moves across the deck and eventually stops.

1. In the absence of friction between the boat and the water, describe the motion of the boat when the man stops.
2. In the presence of friction between the boat and the water, describe the motion of the boat when the man stops.

Solution:

1. *The total momentum P is the sum of the momenta of the boat P_B and of the man with respect to the water P_M. In the absence of friction, the net force applied to the system vanishes and the momentum is conserved. Indeed, the centre-of-mass theorem (1.13) implies that,*

$$\dot{P} = 0 \quad \Rightarrow \quad P = \text{const}$$

   Initially, all the momenta vanish. By conservation, the total momentum is constant. Thus, when the man stops, his momentum vanishes, which implies that the momentum of the boat vanishes as well. Therefore, the boat is at rest with respect to the water.
2. *While the man is moving on the boat, the boat is moving in the opposite direction. The boat is subjected to a viscous friction force F^{fr} opposed to the motion and thus directed along with the displacement of the man (Fig. 1.6). According to the centre-of-mass theorem (1.13), the action of the friction force modifies the total momentum of the system, i.e.*

$$\dot{P} = F^{\mathrm{fr}} \neq 0$$

   Thus, when the man stops, the momentum of the total system is in the direction of the friction force and thus in the direction of the displacement of the man. But the man is at rest with respect to the boat, and the boat moves in the direction of the friction force. However, the norm of the momentum diminishes due to the viscous friction force.

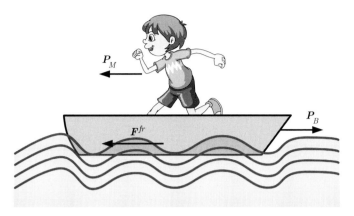

Figure 1.6 A man moves with a momentum $\boldsymbol{P}_M$ across the horizontal deck of a boat that moves with a momentum $\boldsymbol{P}_B$. The water exerts a friction force $\boldsymbol{F}^{\text{fr}}$, which is opposed to the motion of the boat.

1.10.3 Perfectly Inelastic Collision

We consider two colliding objects that remain attached after impact. It can be shown that this type of collision has the maximum kinetic energy change. We treat the colliding objects as two point masses M_1 and M_2 that constitute an isolated system. The point mass M_1 has an initial momentum $\boldsymbol{P}_1$ and the point mass M_2 is initially at rest. The state variables are the momentum and the extensive variable X_0 associated to an internal property of the system (i.e. the entropy S as we will see in the following chapter).

Let $E_i\left(\boldsymbol{P}, X_0\right)$ be the energy and $U_i\left(X_0\right)$ the internal energy just before the impact. Let $E_f\left(\boldsymbol{P}, X_0\right)$ be the energy and $U_f\left(X_0\right)$ be the internal energy just after the impact. Using the conservation laws of energy (1.9) and momentum (1.12), determine the variation of internal energy of the system $\Delta U \equiv U_f\left(X_0\right) - U_i\left(X_0\right)$.

Solution:
According to the first law (1.12) *for an isolated system, the momentum $\boldsymbol{P}$ is a constant, i.e. $\boldsymbol{P} = \boldsymbol{P}_1$. The energy of the system before and after the impact reads,*

$$E_i\left(\boldsymbol{P}, X_0\right) = \frac{\boldsymbol{P}_1^2}{2\,M_1} + U_i\left(X_0\right)$$

$$E_f\left(\boldsymbol{P}, X_0\right) = \frac{\boldsymbol{P}_1^2}{2\left(M_1 + M_2\right)} + U_f\left(X_0\right)$$

According to the first law (1.9) *for an isolated system, the energy $E\left(\boldsymbol{P}, X_0\right)$ is also a constant, i.e. $E_i\left(\boldsymbol{P}, X_0\right) = E_f\left(\boldsymbol{P}, X_0\right)$. This is expressed as,*

$$\frac{\boldsymbol{P}_1^2}{2\,M_1} + U_i\left(X_0\right) = \frac{\boldsymbol{P}_1^2}{2\left(M_1 + M_2\right)} + U_f\left(X_0\right)$$

Thus,

$$\Delta U = U_f\left(X_0\right) - U_i\left(X_0\right) = \frac{\boldsymbol{P}_1^2}{2}\left(\frac{1}{M_1} - \frac{1}{M_1 + M_2}\right) > 0$$

In conclusion, the internal energy of this isolated system increases during the collision in order to compensate for the loss of kinetic energy.

1.10.4 Gas and Piston

Let us consider an adiabatic closed system consisting of a homogeneous gas and a piston that closes the cylinder containing the gas (Fig. 1.7). The piston has a mass M and it is set in motion through the action of an external force $\boldsymbol{F}^{\text{ext}}$. The gas exerts a viscous friction force $\boldsymbol{F}^{\text{fr}}$ on the piston. It is assumed that the mass of the gas is negligible.

The following quantities are chosen as state variables: the momentum of the piston $\boldsymbol{P}$ and a variable X_0 associated with an internal property of the gas (i.e. the entropy S, as we will see in the next chapter). The energy of the system is given by,

$$E\left(\boldsymbol{P}, X_0\right) = \frac{\boldsymbol{P}^2}{2\,M} + U\left(X_0\right)$$

Applying the first law to the system and the centre-of-mass theorem to the piston, show that,

$$\dot{X}_0 = -\frac{\boldsymbol{F}^{\text{fr}} \cdot \boldsymbol{v}}{Y_0}$$

which is identical to equation (1.59) obtained for a damped harmonic oscillator.

Solution:
According to the first law (1.11), the energy variation of a rigid, adiabatic closed system is due to the power of the external forces, i.e. $\dot{E} = P^{ext}$. Furthermore, as the energy $E\left(\boldsymbol{P}, X_0\right)$ is a function of $\boldsymbol{P}$ and X_0, we can write,

$$\dot{E}\left(\boldsymbol{P}, X_0\right) = \frac{\partial E\left(\boldsymbol{P}, X_0\right)}{\partial \boldsymbol{P}} \cdot \dot{\boldsymbol{P}} + \frac{\partial U\left(X_0\right)}{\partial X_0}\, \dot{X}_0 = P^{\text{ext}}$$

Using the definitions (1.18), (1.26) and (1.46), we obtain,

$$\boldsymbol{v} \cdot \dot{\boldsymbol{P}} + Y_0 \dot{X}_0 = \boldsymbol{F}^{\text{ext}} \cdot \boldsymbol{v}$$

The centre of mass theorem (1.13) applied to the piston is expressed as,

$$\dot{\boldsymbol{P}} = \boldsymbol{F}^{\text{ext}} + \boldsymbol{F}^{\text{fr}}$$

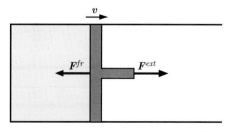

Figure 1.7   A gas is contained in a cylinder closed by a piston of mass M. An external force $\boldsymbol{F}^{\text{ext}}$ is exerted on the piston. A viscous friction force $\boldsymbol{F}^{\text{fr}}$ opposes the motion of the piston which moves at velocity $\boldsymbol{v}$.

Thus, the evolution equation is reduced to,

$$v \cdot F^{\text{fr}} + Y_0 \dot{X}_0 = 0$$

The pressure of the gas does not intervene in the application of the centre-of-mass theorem, because pressure generates an internal force to the system consisting of the gas and the piston. In the following chapter, we will examine an analogous problem in which the gas will be the system and its pressure, a state variable (see § 2.9).

1.10.5 Air and Fan

Let us consider a system consisting of a fan and the surrounding air in an adiabatic closed room (Fig. 1.8). The propeller has a moment of inertia I with respect to the rotational axis. It is set in rotation using an external electric engine that exerts an external torque M^{ext}. A viscous friction torque M^{fr} is opposed to the rotation of the propeller. We assume that the air is homogeneous and its mass is negligible.

The following quantities are chosen as state variables: the angular momentum of the propeller of the fan L and a variable X_0 associated with an internal property of air (i.e. the entropy S, as we will see in the next chapter). The energy of the system is given by,

$$E(L, X_0) = \frac{L^2}{2\,I} + U(X_0)$$

Applying the first law to this system and the angular momentum theorem to the fan, show that,

$$\dot{X}_0 = -\frac{M^{\text{fr}} \cdot \omega}{Y_0}$$

which is the rotational analog to the equation obtained for the system consisting of gas contained by a piston in a cylinder.

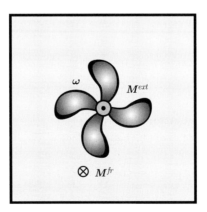

Figure 1.8 A room filled with air contains a fan where the propeller has a moment of inertia I with respect to its axis of rotation. An external torque M^{ext} is exerted on the propeller. A viscous friction torque M^{fr} is opposed to the rotational motion of the fan propeller that turns with an angular velocity ω.

Solution:

According to the first law (1.11), the energy variation of a rigid, adiabatic, closed system is due to the power of the external torques, $\dot{E} = P^{\text{ext}}$. Furthermore, since the energy E is a function of $\boldsymbol{L}$ and X_0, we can write,

$$\dot{E}(\boldsymbol{L}, X_0) = \frac{\partial E(\boldsymbol{L}, X_0)}{\partial \boldsymbol{L}} \cdot \dot{\boldsymbol{L}} + \frac{\partial U(X_0)}{\partial X_0} \dot{X}_0 = P^{\text{ext}}$$

Using the definitions (1.21), (1.26) and (1.46), we obtain

$$\boldsymbol{\omega} \cdot \dot{\boldsymbol{L}} + Y_0 \dot{X}_0 = \boldsymbol{M}^{\text{ext}} \cdot \boldsymbol{\omega}$$

The angular momentum theorem (1.15) applied to the fan implies that

$$\dot{\boldsymbol{L}} = \boldsymbol{M}^{\text{ext}} + \boldsymbol{M}^{\text{fr}}$$

Thus, the evolution equation is reduced to

$$\boldsymbol{\omega} \cdot \boldsymbol{M}^{\text{fr}} + Y_0 \dot{X}_0 = 0$$

Exercises

1.1 State function: Mathematics

Consider the function $f(x, y) = y \exp(ax) + xy + bx \ln y$ where a and b are constants.

a) Calculate $\dfrac{\partial f(x, y)}{\partial x}$, $\dfrac{\partial f(x, y)}{\partial y}$ and $df(x, y)$

b) Calculate $\dfrac{\partial^2 f(x, y)}{\partial x \, \partial y}$

1.2 State Function: Ideal Gas

An ideal gas is characterised by the relation $pV = NRT$ where p is the pressure of the gas, V is the volume, T is the temperature, N is the number of moles of gas and R is a constant.

a) Calculate the differential $dp(T, V)$

b) Calculate $\dfrac{\partial}{\partial T}\left(\dfrac{\partial p(T, V)}{\partial V}\right)$ and $\dfrac{\partial}{\partial V}\left(\dfrac{\partial p(T, V)}{\partial T}\right)$

1.3 State Function: Rubber Cord

A rubber cord of length L, which is a known state function $L(T, F)$ of the temperature T of the cord and of the forces of magnitude F applied at each end to stretch it. Two physical properties of the cord are :

a) the Young modulus, defined as $E = \dfrac{L}{A}\left(\dfrac{\partial L}{\partial F}\right)^{-1}$, where A is the cord cross section area.

b) the thermal expansion coefficient $\alpha = \dfrac{1}{L}\dfrac{\partial L}{\partial T}$.

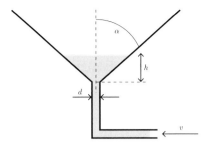

Figure 1.9 A liquid enters into a funnel with a laminar flow of velocity v in a tube of diameter d. The funnel is a cone of opening angle α. The cone axis is vertical.

Determine how much the length of the cord varies if its temperature changes by ΔT and at the same time the force F changes by ΔF. Assume that $\Delta T \ll T$ and $\Delta F \ll F$. Express ΔL in terms of E and α.

1.4   State Function: Volume

A liquid is filling a container that has the form of a cone of angle α around a vertical axis (Fig. 1.9). The liquid enters the cone from the apex through a hole of diameter d at a velocity $v(t) = kt$ where k is a constant. When the surface of the liquid is at height $h(t)$, the volume is $V(t) = \dfrac{1}{3}\pi \tan^2 \alpha \, h^3(t)$. Initially, at time $t = 0$, the height $h(0) = 0$. Find an expression for the rate of change of volume $\dot{V}(t)$ and determine $h(t)$.

1.5   Cyclic Rule for the Ideal Gas

An ideal gas is characterised by the relation $pV = NRT$ as in § 1.2 where the pressure $p(T, V)$ is a function of T and V, the temperature $T(p, V)$ is a function of p and V and the volume (T, p) is a function of T and p. Calculate,

$$\frac{\partial p(T, V)}{\partial T} \frac{\partial T(p, V)}{\partial V} \frac{\partial V(T, p)}{\partial p}$$

1.6   Evolution of Salt Concentration

A basin contains $N_s(t)$ moles of salt dissolved in $N_w(t)$ moles of water. The basin receives fresh water at a constant rate Ω_w^{in}. This water is assumed to be thoroughly mixed in the basin so that the salt concentration can be considered homogeneous. The salty water comes out of the basin at a constant rate $\Omega_{sw}^{\text{out}} = \Omega_s^{\text{out}}(t) + \Omega_w^{\text{out}}(t)$, where $\Omega_s^{\text{out}}(t)$ and $\Omega_w^{\text{out}}(t)$ are the salt and water outflow rates. Determine the salt concentration,

$$c_s(t) = \frac{N_s(t)}{N_s(t) + N_w(t)}$$

as a function of time t for the given initial conditions $N_s(0)$ and $N_w(0)$.

1.7   Capilarity: Contact Angle

Capilarity effects are taken into account by considering that the energy of the system contains contributions that are proportional to the surface area of the interfaces between the different parts of the system. For a drop of wetting liquid on a horizontal surface (Fig. 1.10), where the drop is assumed to have a spherical shape, the internal

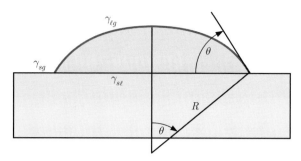

Figure 1.10 A drop of liquid on a horizontal substrate is assumed to have a spherical shape. The angle θ is called the contact angle. Surface tension is defined for each of the three interfaces: solid–liquid ($\gamma_{s\ell}$), solid–gas (γ_{sg}) and liquid–gas ($\gamma_{\ell g}$).

energy is expressed as $U(h,R) = \left(\gamma_{s\ell} - \gamma_{sg}\right)\pi a^2 + \gamma_{\ell g}A$ where $a = R\sin\theta = \sqrt{2Rh - h^2}$ is the radius and $A = 2\pi Rh$ is the surface area of the spherical cap of height h at the intersection of the sphere of radius R and the solid substrate. The parameters $\gamma_{s\ell}$, γ_{sg}, $\gamma_{\ell g}$ characterise the substances and are independent of the drop shape. Show that the contact angle θ is given by,

$$\left(\gamma_{s\ell} - \gamma_{sg}\right) + \gamma_{\ell g}\cos\theta = 0$$

by minimising the internal energy $U(h,R)$ under the condition that the volume $V(h,R) = \frac{\pi}{3}h^2(3R - h) = V_0$ is constant.

1.8 Energy: Thermodynamics versus Mechanics

A weight of mass M is hanging from a rope. The force $\boldsymbol{F}$ applied to the rope is such that the weight is lowered vertically at a velocity $\boldsymbol{v}$, which may vary with time.

a) Determine the expression for the time evolution of the mechanical energy E', which is the sum of the kinetic and potential energies.

b) Determine the time evolution of the energy E of the system according to the first law (1.11).

1.9 Damped Harmonic Oscillator

A one-dimensional harmonic oscillator of mass M and spring constant k is subjected to a friction force $\boldsymbol{F}_f(t) = -\lambda\boldsymbol{v}(t)$ where $\boldsymbol{v}(t)$ is the velocity of the point mass and $\lambda > 0$. Using a coordinate axis Ox where the origin O corresponds to the position of the point mass when the harmonic oscillator is at rest, the equation of motion reads,

$$\ddot{x} + 2\gamma\dot{x} + \omega_0^2 x = 0$$

where $\omega_0^2 = k/M$ and $\gamma = \lambda/(2M)$. In the weak damping regime, where $\gamma \ll \omega_0$, the position can be expressed as

$$x(t) = Ce^{-\gamma t}\cos(\omega_0 t + \phi)$$

where C and ϕ are integration constants.

a) Express the mechanical energy $E(t)$ in terms of the coefficients k, C and γ.

b) Calculate the power $P(t)$ dissipated due to the friction force $\boldsymbol{F}_f(t)$ during one oscillation period.

2 Entropy and Second Law

Rudolf Julius Emmanuel Clausius, 1822–1888

R. J. E. Clausius taught at the universities of Zurich, Würzburg and Bonn. During his career, he developed the heat engine theory that is now part of most thermodynamics courses. In 1854, he extended the calculation of the thermodynamic efficiency that Carnot had established for an infinitesimal cycle and derived the well-known formula $1 - T^-/T^+$. In 1865, he suggested the introduction of a new quantity, entropy, a word he created after the Greek word meaning '*transformation*'.

2.1 Historical Introduction

Temperature is a central thermodynamic concept that was developed over centuries. Around 250 B.C., thermoscopes appeared, which allowed the detection of a temperature difference. These devices did not provide a measurement on a pre-established scale. More than 18 centuries later, Galileo built a thermometer that consisted of a water column containing balls of slightly different masses with only one that was at equilibrium in the middle for a given temperature, the others having sunk to the bottom or floated to the top.

In the middle of the seventeenth century, a thermometer was created that measured alcohol thermal expansion. Thereafter, different models of thermometers with various scales appeared until Daniel Gabriel **Fahrenheit** invented the scale that bears his name. The value of $0°F$ corresponded to the temperature of a mixture of ice, water and ammonium chloride in equal amounts. Fahrenheit imposed 64 divisions between the melting point of pure ice and the temperature of the human body. Anders **Celsius**, 24 years later, set a scale that was defined by the fusion point of pure ice at $0°C$ and the melting point of water at $100°C$. Willliam Thomson, who became Lord "**Kelvin**", invented the scale that eventually became widely used and is now the norm in science and technology.

The concept of temperature was further refined by the introduction of the **zeroth law** of thermodynamics. It is called this because it was introduced about 100 years after the first and second laws. This law needs to be introduced in the framework of thermostatics. It establishes the notion of equilibrium states and formalises the notion of temperature. According to this law, if two systems are at thermal equilibrium with a third, then these two systems are at thermal equilibrium with each other [16]. Furthermore, if one system is considered as a reference, then all systems at thermal equilibrium with it have a property in common, called temperature. In other words, systems at thermal equilibrium with one another have the same temperature.

Pressure, like temperature, is a fundamental concept of thermodynamics that has been explored since antiquity. In ancient Greece, Aristotle sought to demonstrate that a vacuum could not exist. The idea of a universe filled with matter everywhere remained quite widespread during the sixteenth century. Galileo postulated the effect of a 'force of vacuum', a totally new idea at that time. He introduced this concept in order to solve the problem of Florentine gardeners who could not raise water above a height of about 10.5 m. In contrast, Giovanni Battista Baliani suggested that this effect was due to the weight of air. To illustrate his point of view, he took a glass tube, open at one end and closed at the other, filled with water and then immersed vertically in a pond. He observed that the water remained at a high level within the tube. Later on, Evangelista Torricelli, a student of Gailieo, performed the same experiment with mercury. Torricelli postulated that the space left free at the top of the mercury column was a vacuum and that the ascension of mercury in the tube was due to the weight of the atmosphere acting on the free surface of mercury.

Blaise Pascal repeated this experiment in France, performing it with different kinds of liquids. He noticed a relation between the height of the liquid in the tube and its specific weight. To show that it is the weight of the atmosphere that raises the liquid in the tube, he varied the atmospheric pressure. His brother-in-law, Florian Périer, performed the experiment at the top of a mountain, and then at a lower elevation. Blaise Pascal collected his findings in a treatise [17] in which he defined **pressure**. It is in memory of his work that the unit of pressure, the **pascal** [Pa], was named after him.

At the end of the eighteenth century, there was no agreement as to what heat really was. The dominant theory invoked the the notion of 'caloric'. For Lavoisier, the caloric had to be considered as an indestructible substance, conserved in every thermal process. Thermodynamicists such as Joule, Rumford, Mayer and Helmholtz realised that something was wrong with the idea that heat flowed simply from the furnace to the condenser. One had to concede that energy changed in the process, because part of the heat was 'converted' into work.

Figure 2.1   The final states of a thermodynamic system that can be reached from a given initial state are characterised by an increase in entropy through work only. The process can be violent. Here, the cube (state X) gets deformed and smashed (state Y). Adapted from [21].

Clausius introduced the notion of internal energy as a state function U. Then he realised that heat was not a state function and established the formula : $dU = \delta Q + \delta W$. At the time, it was known that the infinitesimal mechanical work performed on a gas was expressed as $\delta W = -p\,dV$. Clausius sought to define the infinitesimal heat δQ in terms of the differential of a state function. He realised that the quantity $dU + p\,dV$ was not equal to the differential of a state function, because the integral from the initial state to final state depended on the process. In mathematical terms, one would say that $dU + p\,dV$ is not an 'exact differential'. To the contrary, Clausius found that the integral of $T^{-1}\,(dU + p\,dV)$ was independent of the process between the initial and final states. Thus, he was led to postulate the existence of a state function S, the differential of which was $dS = T^{-1}\,(dU + p\,dV)$.

In this chapter, we will establish these results based on a formal statement of the second law that postulates the existence of a state function S named entropy. It is remarkable that the notion of entropy is still currently giving rise to fundamental works in the field of thermodynamics [18]. In a hundred-page long theorem published in 1999, Lieb and Yngvason deduced the existence of the state function entropy, starting from the axiomatic properties of an order relation between thermodynamic states (Fig. 2.1) [19].

Gibbs, one of the founding fathers of thermodynamics, stated that in order to make thermodynamics clear, the notion of entropy should be introduced from the beginning. In a recent textbook, Fuchs noticed that the developments of continuous media physics during the twentieth century justified such an approach [20]. He recalled that Gibbs fought already against the practices of his day, 'A method involving the notion of entropy, the very existence of which depends upon the second law of thermodynamics, will doubtless seem to many far-fetched, and may repel beginners as obscure and difficult of comprehension'. The inconvenience of introducing a new physical property is, however, counter-balanced by the advantage of 'a method which makes the second law of thermodynamics so prominent, and gives it so clear and elementary an expression' [21].

2.2 Temperature

In everyday life, sensations are labelled as more or less 'warm', the full range going from 'cold' to 'hot'. The temperature is the coordinate that defines the position on this ordered

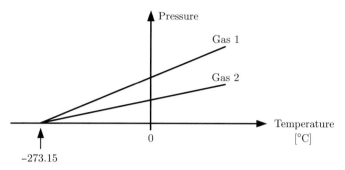

In first approximation, the pressure of all gases is a linear function of temperature. The zero pressure, common to all gases, defines the zero of the kelvin temperature scale.

sequence. Thermometers detect volume changes of liquids according to their temperature variation. Liquids, such as mercury or alcohol found in thermometers, change volume continuously with temperature over a broad temperature range. It is worth noting the empirical nature of this definition of the temperature scale. Notably, not all substances have a specific volume that depends monotonously on temperature. For instance, the specific volume of water reaches a minimum value at about 4°C.

The **centigrade** scale is defined from 0 degree for melting ice and 100 degrees for boiling water and an atmospheric pressure of one atmosphere (1.015 kPa). The celsius scale has the same definition of zero degree, but the **celsius** is a unit derived from the international system of units, introduced in 1948, according to which 1°C is the 1/273.15 fraction of the temperature of the triple point of water, a notion that we will see in Chapter 6. With this definition, water boils at 99.975°C when the pressure is one atmosphere.

The **kelvin** is the temperature unit of the international system (SI). It is defined so that a variation of 1 K is equal to 1°C. For diluted gas, experiments show that pressure is a linear function of temperature (Fig. 2.2). The straight line extends until the pressure reaches zero at the limit temperature of 0 K. The zero of the Kelvin scale is at -273.15°C.

2.3 Heat and Entropy

There are many ways to change the energy of a system. If a stone is thrown horizontally, its kinetic energy increases. If a sphere is electrically charged, its electrostatic energy increases. In the first case, the state of the system changed because the momentum changed, in the second case, because the electric charge changed. These two state variables are extensive quantities. We need to establish what state variable we need to describe 'heat'.

2.3.1 Heat

Let us begin with our experience of everyday life to illustrates various forms of heat transfers. When a stone warms up because it is exposed to the sun, it receives 'heat' by a thermal process that occurs in the absence of any macroscopic external force. It is due to heat transfer by *radiation*. When two objects at different temperatures are connected to each other, heat transfer takes place by *conduction* until both objects reach the same temperature. A third type of heat transfer takes place through *convection*. It is due to matter transfer from one region of space to another, which is at a different temperature. It is clearly distinct from heat transfer by radiation or conduction, since in both of these no matter transfer occurs.

2.3.2 Entropy

We realise now that temperature cannot be the only thermodynamic variable that characterises thermal processes. The melting process of an ice cube suggests the need to introduce a thermodynamic quantity that is different from temperature. It is indeed clear that an ice cube needs heat for it to melt. However, the melting occurs at a fixed temperature. Thus, a variable different than temperature needs to be used to characterise this process.

From this basic experiment, from our analysis of the damped harmonic oscillator (c.f. §1.9) and from the exercises of the previous chapter, we see that it is necessary to introduce a state variable different from temperature. This quantity is *entropy*, denoted S. This notion will be formalised by the formulation of the 2nd law (c.f. § 2.4).

Let us consider some numerical values of the entropy S. The unit of entropy is the ratio of the units of energy and temperature, as we will see later on in this chapter. In SI units, the entropy is given in joule per kelvin ($\mathrm{JK^{-1}}$). Here are some values of entropy changes for some thermal processes:

1. When $1\ \mathrm{cm^3}$ of ice melts, its entropy changes by $\Delta S = 1\ \mathrm{J\,K^{-1}}$
2. When $1\ \mathrm{cm^3}$ of water is brought from room temperature to boiling, its entropy increases by $\Delta S = 1\ \mathrm{J\,K^{-1}}$
3. When $1\ \mathrm{cm^3}$ of water at $100°C$ is evaporated, its entropy changes by $\Delta S = 6\ \mathrm{J\,K^{-1}}$
4. When $1\ \mathrm{kg}$ of coal is burnt, its entropy increases by $\Delta S = 105\ \mathrm{J\,K^{-1}}$

The entropy of a system is a state function (notion formalised in §2.4). For instance, we can say the following:

1. The entropy of a litre of gas in normal conditions of temperature and pressure (i.e. 10^5 Pa and $0°$ C) is about 10 $\mathrm{J\,K^{-1}}$
2. The entropy of a litre of a liquid or a solid at room temperature (i.e. $20°$ C) is about 100 $\mathrm{J\,K^{-1}}$

2.3.3 Internal Entropy Production

During heat transfer to a system by conduction or during a phase transition between ice and water, entropy increases or decreases depending on the direction of the process. However, this reversibility is not always possible.

Simple experiments points to the existence of entropy production, no matter in which way the process is carried out. For instance, entropy is produced in a fire or by rubbing hands together. It is also produced by an electric current flowing through a resistance. Many historical experiments, such as that of Earl Rumford (hollowing out of canons) or that of Humphry Davy (two ice blocks rubbed against each other), led to the conclusion that there are processes which cause an **entropy production**.

In an isolated system, entropy can increase, but it cannot decrease. All experiments lead to this conclusion. For instance, a drill produces heat while it is drilling. Conversely, angular momentum can be provided to the drill to generate electricity. In both cases, there is entropy production and heat transfer towards the environment. A car left to itself ends up stopping, and heat transfer occurs spontaneously from hot to cold. These processes cannot be reversed spontaneously; they are irreversible.

In light of the above considerations, we can now state the second law of thermodynamics.

2.4 Second Law of Thermodynamics

The **second law of thermodynamics** states that:

For every system, there is a scalar and extensive state function called *entropy* (S) **that satisfies the following two conditions:**

1. **Evolution condition: The entropy of an adiabatic system is a monotonously non-decreasing function of time.**

$$\dot{S} = \Pi_S \geq 0 \quad \text{(adiabatic system)} \tag{2.1}$$

2. **Equilibrium condition: In the distant future, the entropy of an isolated system tends to a finite local maximum that is compatible with the constraints imposed on the system, such as its internal walls or characteristics of the enclosure.**

The quantity Π_S is called the **entropy production rate** of the system. A process is called **reversible** if $\Pi_S = 0$ and **irreversible** if $\Pi_S > 0$. When a process is irreversible, i.e. $\Pi_S > 0$, we say that there is **dissipation**. When a system interacts with its environment, the evolution of the entropy is due to two distinct causes:

- irreversible processes taking place inside the system, described by the **entropy production rate** Π_S,
- reversible processes taking place at the system enclosure, described by an **entropy exchange rate** I_S.

Thus, the evolution of the entropy is described by,

$$\dot{S} = \Pi_S + I_S \tag{2.2}$$

Whenever dissipation takes place at the enclosure itself, then the system under study should be redefined so that it encompasses this region. This way we avoid any possible confusion.

An equation such as (2.2), where the evolution of an extensive variable is expressed in terms of an exchange with the environment and an internal rate of production, is called a **balance equation**.

2.4.1 Time Reversal

The second law helps us to distinguish two types of physical processes: reversible and irreversible. The reversibility of processes is defined in terms of a fundamental physical symmetry: **time reversal**. The time reversal transformation T is a bijective function that maps time to its opposite,

$$T : t \longrightarrow -t \tag{2.3}$$

For example, position is invariant under time reversal. This implies that velocity $\boldsymbol{v} = \dot{\boldsymbol{r}}$ changes sign and acceleration $\boldsymbol{a} = \ddot{\boldsymbol{r}}$ is invariant under time reversal,

$$\begin{aligned} \boldsymbol{r} &\longrightarrow \boldsymbol{r} \\ \boldsymbol{v} &\longrightarrow -\boldsymbol{v} \\ \boldsymbol{a} &\longrightarrow \boldsymbol{a} \end{aligned} \tag{2.4}$$

The evolution of a thermodynamic system undergoing physical processes is reversible if the evolution equation that characterises this evolution is invariant under time reversal. In the opposite case, it is irreversible. Thus, the evolution condition of the 2nd law implies that the evolution of a closed adiabatic system is **invariant** under **time reversal** if and only if $\dot{S} = \Pi_S = 0 \ \forall t$.

Given that physical processes happening in the universe are in general irreversible, the evolution of the universe is not invariant under time reversal. Thus, there is an order relation between different events happening at different times. If a 1st event happens at time t_1, a 2nd at time t_2 and an nth at time t_n then

$$t_1 < t_2 < \ldots < t_n \tag{2.5}$$

This order relation is called the **time arrow**. Time is a scalar parameter that only increases from an initial event, called the 'Big Bang', which defines the origin of time $t_0 = 0$ [22].

2.5 Simple System

A **homogeneous system** where the state is determined by a single state variable entropy S is called **simple system**. For such a system, entropy is defined globally. We recall that a simple system is called homogeneous because it has the same intensive properties (i.e. temperature, pressure, etc ...) at every point. Inhomogeneous systems will be treated in Chapters 3 and 10.

We will now describe the state of a simple system consisting of r different substances. We consider mechanical actions and matter transfers that are sufficiently slow so that kinetic energy variations are negligible. Then the system can be described with respect to a frame of reference where it is at rest. As entropy S is an extensive state function,

$$S \equiv X_0 \tag{2.6}$$

It can also be taken as one of the extensive state variable since a state variable is a trivial state function. The second extensive state variable is taken as the volume of the system,

$$V \equiv X_1 \tag{2.7}$$

The other extensive state variables describe the amount of substances,

$$N_A \equiv X_{A+1} \qquad \text{where} \qquad A = 1, \ldots, r \tag{2.8}$$

Thus the internal energy $U(S, V, N_1, \ldots, N_r)$ becomes a function of entropy S, volume V and number of moles $N_1, \ldots, N_r$ of the r substances. The **mole** [mol] is a fundamental unit of the international system representing the amount of matter of a system containing as many elementary constituents as there are atoms in 12 g of ^{12}C. Thus a mole of substance contains $6.022 \cdot 10^{23}$ elementary constituents, atoms or molecules. This number is called the Avogadro number $\mathcal{N}_A$.

Thermodynamicists compared the predictions of the two laws of thermodynamics pertaining to gases. They concluded that the intensive quantity Y_0, which is the conjugate of the entropy S, is the **temperature**, commonly denoted T and defined as,

$$T(S, V, N_1, \ldots, N_r) \equiv \frac{\partial U(S, V, N_1, \ldots, N_r)}{\partial S} \tag{2.9}$$

In general, the temperature T of a system is not negative. However, there are magnetic systems (e.g. spins) or optical systems (e.g. a laser) that may have a negative temperature [14]. In this textbook, we will only consider systems that have a positive temperature, i.e. $T \geq 0$.

As we shall see (2.38), the predictions of the two laws of thermodynamics pertaining to gases lead to the conclusion that the intensive quantity Y_1, which is the conjugate of the volume V, is the **pressure**, denoted p and defined as,

$$p(S, V, N_1, \ldots, N_r) \equiv -\frac{\partial U(S, V, N_1, \ldots, N_r)}{\partial V} \tag{2.10}$$

Finally, the intensive quantity Y_{A+1}, which is the conjugate of the number of moles of substance A, is the **chemical potential**, denoted μ_A and defined as,

$$\mu_A(S, V, N_1, \ldots, N_r) \equiv \frac{\partial U(S, V, N_1, \ldots, N_r)}{\partial N_A} \qquad \text{where} \qquad A = 1, \ldots, r \tag{2.11}$$

The notion of chemical potential will be explained in Chapters 3 and 8. In Chapter 1, the time derivative of the internal energy (1.47) was written in terms of the non-specific extensive state variables $X_0, X_1, X_2 \ldots X_n$ and their conjugates $Y_0, Y_1, Y_2, \ldots, Y_n$. Now that we have identified the extensive state variables $S, V, N_1, \ldots, N_r$ in equations (2.6)–(2.8) and their conjugates $T, -p, \mu_1 \ldots \mu_r$ in equations (2.9)–(2.11), equation (1.47) becomes,

$$\dot{U} = T\dot{S} - p\dot{V} + \sum_{A=1}^{r} \mu_A \dot{N}_A \tag{2.12}$$

The time evolution equations (1.28) and (2.12) for the internal energy require the powers of the physical processes and the evolution of the state variables to satisfy equation,

$$T\dot{S} - p\dot{V} + \sum_{A=1}^{r} \mu_A \dot{N}_A = P_W + P_Q + P_C \quad \text{(open system)} \tag{2.13}$$

For a closed system, i.e. $P_C = 0$, equation (2.13) reduces to,

$$T\dot{S} - p\dot{V} + \sum_{A=1}^{r} \mu_A \dot{N}_A = P_W + P_Q \quad \text{(closed system)} \tag{2.14}$$

For an adiabatic system, the thermal power vanishes, i.e. $P_Q = 0$. Combining the evolution equation (2.1) of the second law with equation (2.13), the entropy production rate is expressed as,

$$\Pi_S = \frac{1}{T} \left(P_W + p\dot{V} + P_C - \sum_{A=1}^{r} \mu_A \dot{N}_A \right) \geq 0 \, \text{(adiabatic system)} \tag{2.15}$$

For a reversible process, i.e. $\Pi_S = 0$, equation (2.15) imposes a condition on the mechanical action of the environment on the system,

$$P_W = -p\dot{V} \quad \text{(reversible process)} \tag{2.16}$$

and another condition on the transformations of substances inside the system,

$$P_C = \sum_{A=1}^{r} \mu_A \dot{N}_A \quad \text{(reversible process)} \tag{2.17}$$

We will see in Chapter 3 that the chemical potentials have to be equal on both sides of the enclosure in order for the chemical process to be reversible. If dissipation occurs at the enclosure because of matter transfer, the hypothesis that the system is simple becomes invalid. Therefore, in analysing simple systems, only reversible chemical work can be taken into account.

In light of the entropy production rate (2.15), equation (2.13) can be written as,

$$\dot{S} = \Pi_S + \frac{P_Q}{T} \tag{2.18}$$

The time evolution equation (2.18) is applicable to a homogeneous system that is not simple; however the system must have the same temperature T everywhere.

Comparing the balance equations (2.2) and (2.18), we conclude that the thermal power P_Q is proportional to the entropy exchange rate I_S with,

$$P_Q = T I_S \tag{2.19}$$

The state of the system is called **stationary** if its state variables are time independent. Thus, in a stationary state, entropy is constant,

$$\dot{S} = 0 \quad \text{(stationary state)} \tag{2.20}$$

In this case, the entropy balance equation (2.18) is reduced to,

$$P_Q = -T\Pi_S \leq 0 \quad \text{(stationary state)} \tag{2.21}$$

This means that the thermal power P_Q accounts for the heat evacuated out of the system in order to compensate for the internal entropy production Π_S.

The thermal power P_Q is sometimes called **heat current**. This denomination can be misleading for the following reason. For some systems consisting of distinct subsystems, the thermal powers that they exert on one another do not compensate each other whereas by definition heat currents would (e.g. § 3.6.4).

After setting the stage for the thermodynamic analysis of a simple system, we will now give several examples of simple systems consisting of a homogeneous gas made of one substance.

2.6 Closed and Rigid Simple System

Let us consider a simple system that is **closed and rigid**. This system consists of a homogeneous gas, made of a single substance, contained in an enclosure allowing heat transfer and subjected to the given thermal power P_Q (Fig. 2.3). Since it is rigid, $P_W = 0$.

In order to describe the heat transfer between the system and its environment, we use the entropy S as the unique extensive state variable. Thus, the evolution equations (2.12) and (2.14) reduce to,

$$\dot{U}(S) = T(S)\,\dot{S} = P_Q \tag{2.22}$$

Taking into account the entropy balance equation (2.18), we find that the entropy production rate is zero in this system,

$$\Pi_S = 0 \tag{2.23}$$

This also means that the heat transfer between the system and its environment is a reversible process. In Chapter 3, we will see that if a rigid system is inhomogeneous, for instance if its temperature depends on position, then heat transfer leads to irreversibility.

P_Q

Figure 2.3   A closed, and rigid simple system consisting of a homogeneous gas subjected to a thermal power P_Q.

According to the definition (1.35) of δQ and taking into account (2.22), the infinitesimal heat provided by the environment during an infinitesimal time interval dt is defined as,

$$\delta Q = P_Q\, dt = T(S)\, dS \qquad \text{(reversible process)} \qquad (2.24)$$

When integrating equation (2.24) from an initial state i to a final state f, we obtain the heat provided by a reversible process,

$$Q_{if} = \int_i^f \delta Q = \int_{S_i}^{S_f} T(S)\, dS \qquad \text{(reversible process)} \qquad (2.25)$$

2.7 Adiabatic and Closed Mechanical System

Let us now examine a simple system that is **closed, adiabatic** and subjected to a **mechanical action** (Fig. 2.4). The system consists of a homogeneous gas contained in a cylinder enclosed between two pistons that together define the enclosure, which is assumed to be impermeable and adiabatic, i.e. $P_Q = 0$. We will consider only **reversible** displacements of the pistons where the forces $\pm F^{\text{cont}}$ applied to the pistons do not change the centre of mass motion of the system.

Whenever a process occurs without heat transfer, i.e. $P_Q = 0$, it is an **adiabatic** process. If this adiabatic process is reversible, i.e. $\Pi_S = 0$, then the evolution equation (2.18) implies that the system entropy is constant, i.e. $\dot{S} = 0$.

In order to describe the mechanical action of the environment on the system, the volume V is chosen as the unique extensive state variable. Thus, the evolution equations (2.12) and (2.14) reduce to,

$$\dot{U}(V) = -p\,(V)\,\dot{V} = P_W \qquad \text{(reversible adiabatic process)} \qquad (2.26)$$

According to equation (1.34) and taking into account (2.26), the infinitesimal work δW performed on the system is defined as,

$$\delta W = P_W\, dt = -p\,(V)\, dV \qquad \text{(reversible adiabatic process)} \qquad (2.27)$$

When integrating equation (2.27) from an initial state i to a final state f, we obtain the work performed on the system during a reversible adiabatic process,

$$W_{if} = \int_i^f \delta W = -\int_{V_i}^{V_f} p\,(V)\, dV \qquad \text{(reversible adiabatic process)} \qquad (2.28)$$

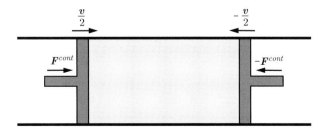

Figure 2.4 An adiabatically closed mechanical system consisting of a homogeneous gas enclosed between two pistons.

Figure 2.5   A rigid open system consisting of a homogeneous gas subjected to a chemical power P_C.

Here, we are examining processes on gases; however it is also possible to describe mechanical work done in other cases. For instance, a solid body subjected to a longitudinal or shear stress. In that case, the state of the system cannot simply be described by its volume, because the local deformations also must be specified [23].

2.8 Open, Rigid and Adiabatic System

Let us now consider an ***open*** simple system consisting of a homogeneous gas. The gas enters and leaves the system symmetrically such that the centre of mass remains at rest. The enclosure is rigid, i.e. $P_W = 0$, and heat transfer by conduction is negligible, i.e. $P_Q = 0$ (Fig. 2.5).

A heat transfer occurs through convection, i.e. it is due to the matter transfer. It is described by the chemical power P_C. Since in a simple system, we must restrict our analysis to reversible matter transfers, relation (2.18) implies that entropy is constant. Therefore, we only need the number of moles of gas N as a state variable to describe this process. Thus, the evolution equations (2.12) and (2.14) reduce to,

$$\dot{U}(N) = \mu(N)\,\dot{N} = P_C \tag{2.29}$$

According to the definition (1.36) of δC and (2.29), the infinitesimal chemical work performed on the system is,

$$\delta C = P_C\,dt = \mu\,dN \tag{2.30}$$

Integrating equation (2.30) from an initial state i to a final state f, we obtain the chemical work performed on the system during a reversible process,

$$C_{if} = \int_i^f \delta C = \int_{N_i}^{N_f} \mu(N)\,dN \tag{2.31}$$

2.9 Closed Simple System

Let us consider a ***closed*** simple system consisting of a homogeneous gas contained in a cylinder closed by two pistons. The cylinder is diathermal, allowing heat transfer between the system and its environment (Fig. 2.6). The forces $\pm F^{\text{cont}}$ applied on the pistons are equal and opposite to each other, and hence, do not change the centre of mass motion.

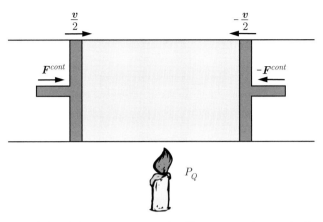

Figure 2.6 A closed simple system consisting of a homogeneous gas enclosed between two pistons and subjected to a thermal power P_Q.

In order to describe the heat transfer and the mechanical action, the extensive state variables entropy S and volume V are chosen as state variables. The processes are characterised by the thermal power P_Q and the mechanical power P_W. Thus, the evolution equations (2.12) and (2.14) reduce to,

$$\dot{U}(S, V) = T(S, V)\,\dot{S} - p\,(S, V)\,\dot{V} = P_W + P_Q \tag{2.32}$$

and the entropy production (2.15) is given by,

$$\Pi_S = \frac{1}{T(S, V)}\left(P_W + p\,(S, V)\,\dot{V}\right) \geq 0 \tag{2.33}$$

The mechanical power P_W is due to the forces exerted by the pistons on the gas,

$$P_W = \boldsymbol{F}^{\mathrm{cont}} \cdot \boldsymbol{v} \tag{2.34}$$

where $\boldsymbol{v}$ is the relative velocity of the left piston relative to the right piston. Actually, the left piston has a speed $\boldsymbol{v}/2$ and the right piston has a speed $-\boldsymbol{v}/2$ with respect to the centre of mass frame of reference. The total external pressure p^{ext} exerted by the pistons on the gas and the rate of change of the volume $\dot{V}$ satisfy the relations,

$$\begin{aligned} \boldsymbol{F}^{\mathrm{cont}} &= p^{\mathrm{ext}}\,\boldsymbol{A} \\ \dot{V} &= -\boldsymbol{A} \cdot \boldsymbol{v} \end{aligned} \tag{2.35}$$

where $\boldsymbol{A}$ is collinear to the vector $\boldsymbol{F}^{\mathrm{cont}}$ and its norm A is the surface area of a piston, i.e. $A = \|\boldsymbol{A}\|$. Using equations (2.34) and (2.35), the mechanical power is written as,

$$P_W = -p^{\mathrm{ext}}\,\dot{V} \tag{2.36}$$

Combining the mechanical power (2.36) with the expression (2.33) for the entropy production rate, we deduce that,

$$\Pi_S = \frac{1}{T(S, V)}\left(p\,(S, V) - p^{\mathrm{ext}}\right)\dot{V} \geq 0 \tag{2.37}$$

Therefore, if the process is reversible, i.e. $\Pi_S = 0$, the pressure $p(S, V)$ of the gas is equal to the external pressure p^{ext} exerted by the pistons on the gas,

$$p(S, V) = p^{\text{ext}} \quad \text{(reversible process)} \tag{2.38}$$

It follows that in this case the mechanical power (2.36) can be expressed in terms of the pressure p, a state function of the system,

$$P_W = -p(S, V)\,\dot{V} \quad \text{(reversible process)} \tag{2.39}$$

Furthermore, in the absence of entropy production, the balance equation (2.18) implies:

$$P_Q = T(S, V)\,\dot{S} \quad \text{(reversible process)} \tag{2.40}$$

We conclude that the entropy rate of change $\dot{S}$ is due to the thermal power P_Q only when the process is reversible.

We are now able to find expressions for the work and the heat transfer in reversible processes. According to the definition (1.34) of the infinitesimal work δW performed on the system equation (2.39) implies,

$$\delta W = P_W\,dt = -p(S, V)\,dV \quad \text{(reversible process)} \tag{2.41}$$

When integrating equation (2.41) from an initial state i to a final state f, we obtain the work performed on the system during a reversible process,

$$W_{if} = \int_i^f \delta W = -\int_{V_i}^{V_f} p(S, V)\,dV \quad \text{(reversible process)} \tag{2.42}$$

According to the definition (1.35) of the infinitesimal heat δQ provided to the system and taking into account expression (2.40) we have,

$$\delta Q = P_Q\,dt = T(S, V)\,dS \quad \text{(reversible process)} \tag{2.43}$$

When integrating equation (2.43) from an initial state i to a final state f, we obtain the heat provided to the system during a reversible process,

$$Q_{if} = \int_i^f \delta Q = \int_{S_i}^{S_f} T(S, V)\,dS \quad \text{(reversible process)} \tag{2.44}$$

2.10 Open, Rigid and Diathermal System

Finally, let us consider an **open, rigid and diathermal** simple system consisting of a homogeneous gas. The gas enters and leaves the system symmetrically such that the centre of mass remains at rest. The enclosure is rigid, i.e. $P_W = 0$, and allows heat transfer and matter transfer (Fig. 2.7).

In order to describe the heat transfer and the matter transfer, the extensive state variables entropy S and number of moles of gas N are chosen as state variables. The processes are

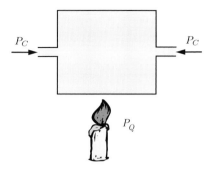

An open simple system consisting of a gas subjected to a thermal power P_Q and to a chemical power P_C.

characterised by the thermal power P_Q and the chemical power P_C. Thus, the evolution equations (2.12) and (2.14) reduce to,

$$\dot{U}(S, N) = T(S, N)\,\dot{S} + \mu\,(S, N)\,\dot{N} = P_Q + P_C \tag{2.45}$$

and the entropy production (2.15) is given by,

$$\Pi_S = \frac{1}{T(S, N)}\left(P_C - \mu\,(S, N)\,\dot{N}\right) \geq 0 \tag{2.46}$$

The chemical power P_C due to the matter transfer from the environment is written as,

$$P_C = \mu^{\text{ext}}\,\dot{N} \tag{2.47}$$

where μ^{ext} is the chemical power of the gas in the environment. Combining the chemical power (2.47) with the expression (2.46) for the entropy production rate, we deduce that,

$$\Pi_S = \frac{1}{T(S, N)}\left(\mu^{\text{ext}} - \mu\,(S, N)\right)\dot{N} \geq 0 \tag{2.48}$$

Therefore, if the process is reversible, i.e. $\Pi_S = 0$, the chemical potential $\mu\,(S, N)$ of the gas in the system is equal to the chemical potential μ^{ext} of the gas outside the system,

$$\mu\,(S, N) = \mu^{\text{ext}} \quad \text{(reversible process)} \tag{2.49}$$

It follows that in this case the chemical power (2.47) can be expressed in terms of the chemical potential μ, a state function of the system,

$$P_C = \mu\,(S, N)\,\dot{N} \quad \text{(reversible process)} \tag{2.50}$$

Furthermore, in the absence of entropy production, the balance equation (2.18) implies:

$$P_Q = T(S, N)\,\dot{S} \quad \text{(reversible process)} \tag{2.51}$$

We conclude that the entropy rate of change $\dot{S}$ is due to the thermal power P_Q only when the process is reversible.

We are now able to find expressions for the matter transfer and the heat transfer in reversible processes. According to the definition (1.36) of the infinitesimal chemical work δC performed on the system equation (2.50) implies,

$$\delta C = P_C\,dt = \mu\,(S, N)\,dN \quad \text{(reversible process)} \tag{2.52}$$

When integrating equation (2.52) from an initial state i to a final state f, we obtain the chemical work performed on the system during a reversible process,

$$C_{if} = \int_i^f \delta C = \int_{N_i}^{N_f} \mu(S, N)\, dN \qquad \text{(reversible process)} \qquad (2.53)$$

According to the definition (1.35) of the infinitesimal heat δQ provided to the system and taking into account expression (2.51) we have,

$$\delta Q = P_Q\, dt = T(S, N)\, dS \qquad \text{(reversible process)} \qquad (2.54)$$

When integrating equation (2.54) from an initial state i to a final state f, we obtain the heat provided to the system during a reversible process,

$$Q_{if} = \int_i^f \delta Q = \int_{S_i}^{S_f} T(S, N)\, dS \qquad \text{(reversible process)} \qquad (2.55)$$

2.11 Worked Solutions

2.11.1 Time Symmetry of the Harmonic Oscillator

1. Show that the evolution equation of a harmonic oscillator of mass M, subjected to an elastic force $\boldsymbol{F}^{\text{el}} = -k\boldsymbol{r}$ and free of friction, is invariant under time reversal.
2. Demonstrate that the evolution equation of a damped harmonic oscillator of mass M, subjected to an elastic force $\boldsymbol{F}^{\text{el}} = -k\boldsymbol{r}$ and a viscous friction force $\boldsymbol{F}^{\text{fr}} = -b\boldsymbol{v}$, is not invariant under time reversal.

Solution:

1. *According to the definition (1.16) of momentum, the equation of motion (1.13) applied to the harmonic oscillator is written as,*

$$-k\boldsymbol{r} = M\boldsymbol{a}$$

   According to the transformation laws (2.4) for the position $\boldsymbol{r}$ and acceleration $\boldsymbol{a}$, the equation of motion is invariant under time reversal. Thus, this evolution is reversible.
2. *The equation of motion of the damped harmonic oscillator is written,*

$$-k\boldsymbol{r} - b\boldsymbol{v} = M\boldsymbol{a}$$

   According to the transformation laws (2.4) for the position $\boldsymbol{r}$, the velocity $\boldsymbol{v}$ and the acceleration $\boldsymbol{a}$, the equation of motion becomes,

$$-k\boldsymbol{r} + b\boldsymbol{v} = M\boldsymbol{a}$$

   under time reversal. This equation is different from the previous one, because the sign of the second term on the left-hand side changed, which implies that the equation of motion is not invariant under time reversal. Thus, this evolution is not reversible. The irreversibility is due to the viscous friction force.

2.11.2 Entropy Variation in Water

Water is heated up with a small electric heater and the temperature of the water is monitored. The heater provides heat to the system with a thermal power P_Q. The container is a calorimeter that is assumed to absorb no heat. Before turning on the heater, the temperature of the water is T_0 and its entropy is S_0. A linear temperature evolution given by $T(t) = T_0 + A t$ is observed. Deduce the entropy variation ΔS during that process.

Solution:

Since the system is simple, its evolution is reversible (2.23). The first law (2.22) implies that,

$$\dot{S} = \frac{P_Q}{T(t)} = \frac{P_Q}{T_0 + A t}$$

Thus, the entropy differential is expressed as,

$$dS = \frac{P_Q}{A} \left(\frac{\dfrac{A}{T_0} \, dt}{1 + \dfrac{A}{T_0} t} \right)$$

When integrating this equation over time, we obtain,

$$S(t) = \int_{S_0}^{S(t)} dS = \int_0^t \frac{P_Q}{A} \left(\frac{\dfrac{A}{T_0} \, dt'}{1 + \dfrac{A}{T_0} t'} \right) = S_0 + \frac{P_Q}{A} \ln \left(1 + \frac{A}{T_0} t \right)$$

Thus, the entropy variation in water is given by,

$$\Delta S = S(\Delta t) - S_0 = \frac{P_Q}{A} \ln \left(1 + \frac{A}{T_0} \Delta t \right)$$

2.11.3 Reversible Adiabatic Process on a Gas

An ideal gas is such that its internal energy is given by $U = c p V$, where c is a dimensionless constant, p is its pressure and V is its volume. Determine the pressure $p(V)$ for a reversible adiabatic compression or expansion.

Solution:

Since the process is reversible and adiabatic, there is no entropy variation. We can use the volume V as the unique state variable. Thus, the internal energy reads $U(V) = c p(V) V$. The derivative of the internal energy with respect to the volume is given by,

$$\frac{dU}{dV} = c \frac{dp}{dV} V + c p = -p$$

which can be recast as,

$$\frac{dp}{p} + \gamma \frac{dV}{V} = 0$$

where $\gamma = (c+1)/c$. The integration of this expression from an initial state (p_i, V_i) to a final state (p_f, V_f) is written as,

$$\int_{p_i}^{p_f} \frac{dp}{p} + \gamma \int_{V_i}^{V_f} \frac{dV}{V} = 0$$

which yields,

$$\ln\left(\frac{p_f}{p_i}\right) + \gamma \ln\left(\frac{V_f}{V_i}\right) = 0 \qquad \text{thus} \qquad \ln\left(\frac{p_f V_f^{\gamma}}{p_i V_i^{\gamma}}\right) = 0$$

Therefore, the initial and final variables are related by,

$$p_i V_i^{\gamma} = p_f V_f^{\gamma}$$

which yields the identity,

$$p V^{\gamma} = \text{const}$$

The gas constant γ will be formally introduced in § 5.7.

2.11.4 Work and Internal Energy

In the experiment illustrated in Fig. 2.8, a mass M is chosen such that the end of the string attached to the spring is subjected to almost no tension, when the operator turns the crank. This means that force resulting from the dry friction F^{fr} exerted on the whole surface of the drum is equal to the weight Mg.

The temperature T is the same everywhere in the system consisting of the copper wire, of the copper drum and of the water at all times. Furthermore, the system is adiabatic.

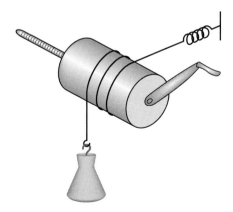

Figure 2.8   A given mass is held by a copper wire making several turns around a copper cylinder containing water. A thermometer plunged into the water provides a measurement of its temperature.

The variation of internal energy is given by $\Delta U = C_V \Delta T$ where C_V, called the specific heat, is a constant.

1. Determine the work W performed in a single turn of the drum of radius R_0.
2. Compute the temperature variation ΔT of the system per drum turn.

Solution:

1. *The work performed per drum turn is the opposite of the work performed by the friction force $\boldsymbol{F}^{fr}$ that is opposed to the motion,*

$$W = - \int \boldsymbol{F}^{\text{fr}} \cdot d\boldsymbol{s} = \int_0^{2\pi} \|\boldsymbol{F}^{\text{fr}}\| R_0 \, d\theta = 2\pi R_0 \|\boldsymbol{F}^{\text{fr}}\| > 0$$

Since the norm of the friction force is equal to the norm of the weight of the hanging mass, i.e. $\|\boldsymbol{F}^{fr}\| = Mg$, the work provided per drum turn is,

$$W = 2\pi R_0 Mg > 0$$

2. *Since the system is adiabatic, the heat provided to the system per drum turn vanishes, i.e. $Q = 0$. The first law (1.39) for each drum turn is written,*

$$\Delta U = W + Q = 2\pi R_0 Mg > 0$$

Hence,

$$\Delta T = \frac{\Delta U}{C_V} = \frac{2\pi R_0 Mg}{C_V}$$

2.11.5 Thermal Compression of a Spring

Consider a piston sliding without friction in a cylinder of surface A, which is attached to a spring of elastic constant k (Fig. 2.9). When the cylinder is empty, the piston is at position x_0. The cylinder is filled with a gas that satisfies the law $pV = NRT$. The internal energy of the gas is given by $U = cNRT$ where $c > 0$ and $R > 0$ are constants. After filling the

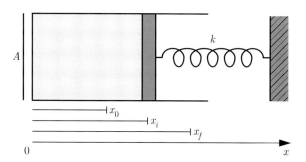

Figure 2.9 A piston closing a cylinder is displaced from position x_i to position x_f when the gas contained inside the cylinder is heated up. The piston is held by a compressed spring of elastic constant k. The rest position of the spring would be at x_0.

cylinder with gas, the piston is at equilibrium at the initial position x_i. Then, the cylinder heats up and reaches the final equilibrium position at x_f. The process is assumed to be reversible and the system is in a vacuum chamber, i.e. the pressure vanishes outside the system. The mass of the piston is not taken into consideration here.

1. Determine the volume V_a, pressure p_a and temperature T_a of the gas in any equilibrium state a in terms of the parameters k, A, x_0 and x_a.
2. Show that the derivative of the pressure p with respect to the volume V is given by,

$$\frac{dp}{dV} = \frac{k}{A^2}$$

3. Determine the work $-W_{if}$ performed by the gas on the spring when the piston moves from x_i to x_f in terms of the parameters k, x_i and x_f.
4. Determine the internal energy variation ΔU_{if} of the gas when the piston moves from x_i to x_f in terms of the parameters k, c, x_0, x_i and x_f.

Solution:

1. *In the equilibrium state a, where $a \in \{i,f\}$, the volume of the ideal gas is given by,*

$$V_a = Ax_a$$

and the pressure of the ideal gas is given by

$$p_a = -\frac{F_a}{A} = \frac{k}{A}(x_a - x_0)$$

where F_a is the projection on the x-axis of the elastic force acting on the spring. Given that the process is reversible, relation (2.35) implies that the pressure of the ideal gas is equal to the pressure exerted by the spring. The temperature of the ideal gas is given by,

$$T_a = \frac{p_a V_a}{NR} = \frac{k}{NR}(x_a - x_0)x_a$$

2. *The equations for the volume and the pressure imply,*

$$dp = -\frac{1}{A}dF = \frac{k}{A}dx = \frac{k}{A^2}dV$$

Thus,

$$\frac{dp}{dV} = \frac{k}{A^2} = \text{const}$$

3. *Hence, the pressure p is expressed in terms of the volume V as,*

$$p = \frac{k}{A^2}(V - V_0)$$

The work $-W_{if}$ performed by the ideal gas on the spring is given in terms of V_i and V_f by,

$$-W_{if} = \int_{V_i}^{V_f} p\,dV = \frac{k}{A^2}\int_{V_i}^{V_f}(V - V_0)\,dV$$

$$= \frac{k}{2A^2}(V_f^2 - V_i^2) - \frac{k}{A^2}V_0(V_f - V_i)$$

This result can be recast in terms of x as,

$$-W_{if} = \frac{k}{2}\left(x_f^2 - x_i^2\right) - kx_0\left(x_f - x_i\right) = \frac{k}{2}\left(\left(x_f - x_0\right)^2 - \left(x_i - x_0\right)^2\right)$$

The work performed by the ideal gas on the spring is equal to the variation of the elastic energy of the spring during the compression. In other words, the work of the spring is entirely used to compress the gas. This is a consequence of the fact that the thermal expansion of the gas is a reversible process.

4. *The internal energy variation ΔU_{if} is given by,*

$$\Delta U_{if} = cNR\left(T_f - T_i\right) = ck\left(\left(x_f - x_0\right)x_f - \left(x_i - x_0\right)x_i\right)$$

Exercises

2.1 Entropy as a State Function

Determine which of the following functions may represent the entropy of a system of positive temperature. In these expressions, E_0 and V_0 are constants representing an energy and a volume, respectively.

a)

$$S\left(U, V, N\right) = NR\ln\left(1 + \frac{U}{NE_0}\right) + \frac{RU}{E_0}\ln\left(1 + \frac{NE_0}{U}\right)$$

b)

$$S\left(U, V, N\right) = \frac{RU}{E_0}\exp\left(-\frac{UV}{NE_0}\right)$$

c)

$$S\left(U, V, N\right) = \frac{NRU}{\left(V^3/V_0^3\right)E_0}$$

d)

$$S\left(U, V, N\right) = R\frac{U^{3/5}V^{2/5}}{E_0^{3/5}V_0^{2/5}}$$

2.2 Work as a Process Dependent Quantity

Three processes are performed on a gas from a state given by (p_1, V_1) to a state given by (p_2, V_2) :

a) an isochoric process followed by an isobaric process,
b) an isobaric process followed by an isochoric process,
c) a process where $p\,V$ remains constant.

Compute for the three processes the work performed on the gas from the initial to the final state. These processes are assumed to be reversible. Determine the analytical results first, then give numerical values in joules.

Numerical Application:

$p_1 = p_0 = 1$ bar, $V_1 = 3 V_0, p_2 = 3 p_0, V_2 = V_0 = 1$ l.

2.3 Bicycle Pump

Air is compressed inside the inner tube of a bike using a manual bicycle pump. The handle of the pump is brought down from an initial position x_2 to a final position x_1 where $x_1 < x_2$ and the norm of the force is assumed to be given by,

$$F(x) = F_{max} \frac{x_2 - x}{x_2 - x_1}$$

The process is assumed to be reversible and the cylinder of the pump has a cross section A. Determine in terms of the atmospheric pressure p_0,

a) the work W_h performed by the hand on the handle of the pump,
b) the pressure $p(x)$,
c) the work W_{12} performed on the system according to relation (2.42).

Numerical Application:

$F_{max} = 10$ N, $x_1 = 20$ cm, $x_2 = 40$ cm, $A = 20$ cm^2 and $p_0 = 10^5$ Pa.

2.4 Rubbing Hands

Rubbing hands together is a dissipative process that we would like to model and quantify.

a) Determine the mechanical power P_W dissipated by friction during this process, in terms of the friction force $\boldsymbol{F}^{fr}$ and the the mean relative velocity $\boldsymbol{v}$ of a hand with respect to the other.
b) At room temperature T, determine the entropy production rate Π_S of this process.

Numerical Application:

$\|\boldsymbol{F}^{fr}\| = 1$ N, $\|\boldsymbol{v}\| = 0.1$ m/s and $T = 25°$C

2.5 Heating by Stirring

In an experiment similar to the Joule experiment (Fig. 1.1), an electric motor is used instead of a weight to stir the liquid. The thermal power P_Q, assumed to result from the friction, is known. The coefficient c_M, which represents the heat per unit mass and temperature, is known and it is assumed to be independent of temperature.

a) Deduce the temperature rise ΔT after stirring for a time Δt.
b) Find an expression for the entropy variation ΔS during this process, which started at temperature T_0.

Numerical Application:

$M = 200$ g, $P_Q = 19$ W, $c_m = 3$ J g^{-1}K^{-1}, $\Delta t = 120$ s, $T_0 = 300$ K.

2.6 Swiss Clock

A Swiss watchmaker states in a flyer the mechanical power P_W dissipated by a specific clock (Fig. 2.10). The work provided to the clock is due to the temperature

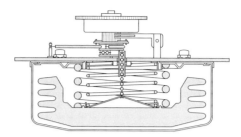

Figure 2.10 A clock receives its energy from the gas capsule (shaded area). The gas expands and contracts under the effects of the temperature fluctuations in the room.

fluctuations ΔT around room temperature T that let the clock run during a time t. We consider that the atmospheric pressure p_{ext} is equal to the pressure of the gas p, i.e. $p_{ext} = p$. The pressure p and the volume V of the gas are related by the ideal gas law $pV = NRT$ where R is the ideal gas constant. Consider that the gas in the capsule is always at equilibrium with the air outside of the capsule (pressure and temperature are the same inside and outside). From the watchmaker's information, estimate the volume V of the gas capsule used to run this clock.

Numerical Application:

$P_W = 0.25 \cdot 10^{-6}$ W, $T = 25°C$, $\Delta T = 1°C$, $p_{ext} = 10^5$ Pa and $t = 1$ day.

2.7 Reversible and Irreversible Gas Expansion

A mole of gas undergoes an expansion through two different processes. The gas satisfies the equation of state $pV = NRT$ where R is a constant, N the number of moles, p the pressure, T the temperature and V the volume of the gas. The initial and final temperatures are T_0. The walls of the gas container are diathermal. However, if a process takes place extremely fast, the walls can be considered as adiabatic. The initial pressure of the gas is p_1, the final pressure is p_2. Express the work performed on the gas in terms of p_1, p_2 and T_0 for the following processes:

a) a reversible isothermal process,
b) an extremely fast pressure variation, such that the external pressure on the gas is p_2 during the expansion, then an isochoric process during which the temperature again reaches the equilibrium temperature T_0,

3

Thermodynamics of Subsystems

Ernst Carl Gerlach Stückelberg, 1905–1984

After completing a thesis in experimental physics in Basel, E. C. G. Stückelberg went to Princeton before returning to Switzerland as a professor at the universities of Zurich, Geneva and Lausanne. He made three fundamental discoveries that later on led to the Nobel Prize awarded to others: the potential for nuclear physics (Yukawa), quantum electrodynamics (Tomonaga, Schwinger, Feynman) and the renormalisation group (Wilson). Feyman, during a talk he gave at CERN shortly after being awarded the Nobel Prize, acknowledged the precedence of Stückelberg by openly stating: 'Stückelberg did the work and walks alone toward the sunset; and, here I am, covered in all the glory, which rightfully should be his!'

3.1 Historical Introduction

This chapter shows how to apply the first and second laws of thermodynamics to systems consisting of two simple subsystems. We will see that the foundations laid out in the first two chapters are sufficient to derive evolution equations for these systems. We will determine entropy production rates and thus illustrate the notion that heat and matter transfers are fundamentally dissipative processes. Before engaging getting into the calculations, let us reflect on the standing of thermodynamics as a physical theory.

For Einstein, *'Physics constitutes a logical system of thought which is in a state of evolution, whose basis (principles) cannot be distilled, as it were, from experience by an inductive method, but can only be arrived at by free invention'* [24]. The validity of the approach is provided by the experimental verification of the propositions derived from it. The scientific framework evolves in the direction of an ever greater simplicity of its logical foundations. A theory is all the more impressive when its premises are simple, that the variety of things it connects is ever greater and the range of its applicability is broader. *'Hence, the deep impression that classical thermodynamics made upon me'*, said Einstein. *'It is the only physical theory of universal content. I am convinced, that, within the framework of applicability of its basic concepts, it will never be overthrown'* [25].

Over time, the exposition of thermodynamics has been improved in order to bring out the features that Einstein so admired. The pioneers of thermodynamics were concerned with a description of heat engines (a subject we will address in Chapter 7). They did not feel the need to specify that they were considering simple systems, nor that δQ and δW described actions that were taking place at the enclosure of the system. This kind of approach, based on analyses of cycles and on many implicit assumptions, had become common practice and perpetuated the reputation that thermodynamics was vague or lacked rigor.

According to Ingo Müller [26], it was Duhem, Jaumann [27] and Lohr [28] who formalised the foundations of thermodynamics, In particular, they presented the two laws of thermodynamics as balance equations for energy and entropy, similarly to the balance equations commonly written for mass or momentum (as we will see in Chapter 10). E. C. G. Stückelberg followed a similar approach when he was teaching thermodynamics in Geneva and Lausanne [14].

In this chapter, we will analyse the evolution of a system composed of two homogeneous subsystems separated by different types of walls. This series of examples illustrates the notion that irreversibility takes place whenever transport of heat, matter or work occurs between regions of a system that have different values of intensive state variables. Our analysis allows us to introduce empirical laws of transport in discrete systems, i.e. systems composed of homogeneous subsystems. In Chapter 11, we will see these transport laws for systems in which the intensive variables change continuously in space.

3.2 Rigid and Impermeable Diathermal Wall

Let us consider an isolated system consisting of two simple subsystems, labelled 1 and 2. These subsystems consist of a homogeneous gas separated by a fixed and impermeable diathermal wall (Fig. 3.1).

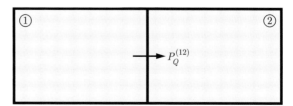

Figure 3.1   Two simple subsystems separated by a diathermal wall. $P_Q^{(12)}$ denotes the thermal power that subsystem 1 exerts on subsystem 2.

The state variable of subsystem 1 is its entropy S_1, and that of subsystem 2, its entropy S_2. Since entropy is an extensive quantity, the entropy S of the system is the sum of the subsystems entropies,

$$S = S_1 + S_2 \tag{3.1}$$

The time derivative of the entropy (3.1) yields,

$$\dot{S} = \dot{S}_1 + \dot{S}_2 \tag{3.2}$$

Since the internal energy $U(S_1, S_2)$ is an extensive state function, it is the sum of the subsystems internal energies $U_1(S_1)$ and $U_2(S_2)$,

$$U(S_1, S_2) = U_1(S_1) + U_2(S_2) \tag{3.3}$$

The internal energies $U_1(S_1)$ and $U_2(S_2)$ are state functions that have the same dependence on their state variable, since both subsystems contain the same gas. The time derivative of the internal energy (3.3) is given by,

$$\dot{U}(S_1, S_2) = \dot{U}_1(S_1) + \dot{U}_2(S_2) \tag{3.4}$$

Since the system is isolated, the first law (1.30) implies,

$$\dot{U}_2(S_2) = - \dot{U}_1(S_1) \tag{3.5}$$

The expression (2.22) of the first law can be applied to subsystems 1 and 2, yielding,

$$\dot{U}_1(S_1) = P_Q^{(21)} \quad \text{and} \quad \dot{U}_2(S_2) = P_Q^{(12)} \tag{3.6}$$

where $P_Q^{(12)}$ is defined as positive if the heat transfer occurs from subsystem 1 to subsystem 2. Comparing relations (3.5) and (3.6), we see that the thermal powers satisfy,

$$P_Q^{(12)} = - P_Q^{(21)} \tag{3.7}$$

because the system is isolated. As each subsystem is a rigid and simple system, we can use equation (2.22) to derive entropy evolution equations for each subsystem,

$$\dot{S}_1 = \frac{\dot{U}_1(S_1)}{T_1(S_1)} \quad \text{and} \quad \dot{S}_2 = \frac{\dot{U}_2(S_2)}{T_2(S_2)} \tag{3.8}$$

Combining relations (3.2) and (3.5), the sum of the equations (3.8) can be written as,

$$\dot{S} = \left(\frac{1}{T_1(S_1)} - \frac{1}{T_2(S_2)} \right) \dot{U}_1(S_1) \tag{3.9}$$

Since we have the differentials $dS = \dot{S}\,dt$ and $dU_1 = \dot{U}_1\,dt$, equation (3.9) implies the partial derivative,

$$\frac{dS}{dU_1} = \frac{1}{T_1(S_1)} - \frac{1}{T_2(S_2)} \tag{3.10}$$

The equilibrium condition of the second law requires that the entropy of the isolated system is maximum at equilibrium, i.e.

$$\frac{dS}{dU_1} = 0 \quad \text{(equilibrium)} \tag{3.11}$$

The equilibrium condition (3.11) and equation (3.10) imply the thermal equilibrium condition

$$T_1(S_1) = T_2(S_2) \qquad \text{(thermal equilibrium)} \tag{3.12}$$

The first and second laws imply that the subsystems temperatures become equal at thermal equilibrium.

Since the system is isolated, the entropy production rate is $\Pi_S = \dot{S} \geq 0$ (eq. (2.1)). The entropy production rate of the whole system can be found by combining (3.6) and (3.9). Thus we have,

$$\Pi_S = \left(\frac{1}{T_1(S_1)} - \frac{1}{T_2(S_2)} \right) P_Q^{(21)} \geq 0 \tag{3.13}$$

Notice that $\Pi_S \neq 0$, although the entropy production rates of the subsystems vanish, according to (2.23), i.e. $\Pi_{S_1} = \Pi_{S_2} = 0$. Thus, the system entropy production rate is not the sum of the subsystems entropy production rates,

$$\Pi_S \neq \Pi_{S_1} + \Pi_{S_2} \tag{3.14}$$

We see clearly in this example that **the entropy production rate is not an extensive quantity**.

Result (3.13) allows us to infer two important features of heat transfer. First, (3.13) implies that if $T_2 > T_1$, the thermal power $P_Q^{(21)} > 0$, which means that the heat transfer occurs from subsystem 2 to subsystem 1. Thus, **the second law implies that the heat transfer takes place from hot to cold in an isolated system.**

Second, we can show that heat transfer is proportional to the temperature difference between the subsystems. The reasoning expands on the argument we have already laid out in Chapter 1 regarding the friction force (1.60). The entropy production rate (3.13) can be recast as,

$$\Pi_S = \frac{T_2(S_2) - T_1(S_1)}{T_1(S_1) \, T_2(S_2)} P_Q^{(21)} \geq 0 \tag{3.15}$$

where the temperatures $T_1(S_1) > 0$ and $T_2(S_2) > 0$.

In view of equation (3.15), we see that a heat transfer between subsystems 1 and 2 is reversible if and only if the temperatures are equal, i.e. $T_1(S_1) = T_2(S_2)$. When choosing subsystem 1 as the simple system (at temperature T) and subsystem 2 as its environment (at temperature T^{ext}), we obtain the condition $T(S) = T^{\text{ext}}$ under which a matter transfer to a simple system is reversible.

In order to satisfy inequality (3.15) in the neighbourhood of the equilibrium state, the entropy production rate Π_S has to be a positive-definite quadratic form. Thus, we can write,

$$P_Q^{(21)} = \kappa \frac{A}{\ell} \left(T_2(S_2) - T_1(S_1) \right) \tag{3.16}$$

where the coefficient $\kappa > 0$ is called the **thermal conductivity**, A is the contact area between the subsystems and ℓ is a specific length of the system. The coefficient κ is defined by the **Fourier law**, which will be discussed in detail in Chapter 11 devoted to the continuum formulation of thermodynamics.

Equation (3.16) is a "discrete" formulation of the Fourier law, which states that the thermal power is proportional to the temperature difference. We write (3.16) in terms of κ, because it is a property intrinsic in all material. Then, dimensional analysis implies that the coefficient ℓ is a length.

3.3 Moving, Impermeable and Diathermal Wall

Let us now consider the same system as in § 3.2, except that the wall can move (Fig. 3.1). The wall mass and the variations of the gas kinetic energy are supposed to be negligible and the subsystems are in thermal equilibrium at a common temperature T at all times.

The state variables of subsystem 1 are its entropy S_1 and volume V_1. Likewise, the state variables of subsystem 2 are its entropy S_2 and volume V_2. Since the volume is an extensive quantity, the volume of the system V is the sum of the subsystems volumes,

$$V = V_1 + V_2 \tag{3.17}$$

Since the system volume V is fixed, i.e. $\dot{V} = 0$, the time derivative of the volume (3.17) implies that,

$$\dot{V}_2 = -\dot{V}_1 \tag{3.18}$$

Since the internal energy $U(S_1, S_2, V_1, V_2)$ is an extensive state function, it is the sum of the subsystems internal energies,

$$U(S_1, S_2, V_1, V_2) = U_1(S_1, V_1) + U_2(S_2, V_2) \tag{3.19}$$

The internal energies $U_1(S_1, V_1)$ and $U_2(S_2, V_2)$ are state functions that have the same dependence on their state variables since the two subsystems contain the same gas. The time derivative of the internal energy (3.19) is expressed as

$$\dot{U}(S_1, S_2, V_1, V_2) = \dot{U}_1(S_1, V_1) + \dot{U}_2(S_2, V_2) \tag{3.20}$$

Since the system is isolated, the first law (1.30) implies that

$$\dot{U}_2(S_2, V_2) = -\dot{U}_1(S_1, V_1) \tag{3.21}$$

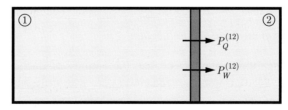

Figure 3.2   Two simple subsystems containing a homogeneous gas are separated by an impermeable, moving and diathermal wall. $P_Q^{(12)}$ denotes the thermal power and $P_W^{(12)}$ the mechanical power exerted by subsystem 1 on subsystem 2.

The first law expressed in the form (2.32) can be applied to subsystems 1 and 2, as they are closed simple systems. This gives,

$$\dot{U}_1(S_1, V_1) = P_W^{(21)} + P_Q^{(21)} \qquad \text{and} \qquad \dot{U}_2(S_2, V_2) = P_W^{(12)} + P_Q^{(12)} \tag{3.22}$$

Here $P_W^{(12)}$ is defined as positive if the mechanical action is exerted by subsystem 1 on subsystem 2. Since the system is isolated (3.21), the mechanical and thermal powers are equal and opposite to each other, i.e.

$$P_W^{(12)} + P_Q^{(12)} = -P_W^{(21)} - P_Q^{(21)} \tag{3.23}$$

The fact that the two subsystems are at thermal equilibrium implies,

$$T(S_1, V_1) = T(S_2, V_2) \qquad \text{(thermal equilibrium)} \tag{3.24}$$

At thermal equilibrium (3.24), the internal energy evolution equation (2.32) for subsystems 1 and 2 can be recast as,

$$\dot{S}_1 = \frac{1}{T(S_1, V_1)} \left(\dot{U}_1(S_1, V_1) + p_1(S_1, V_1)\, \dot{V}_1 \right)$$
$$\dot{S}_2 = \frac{1}{T(S_2, V_2)} \left(\dot{U}_2(S_2, V_2) + p_2(S_2, V_2)\, \dot{V}_2 \right) \tag{3.25}$$

By using relations (3.2), (3.18), (3.21) and (3.24), the sum of the two equations in (3.25) reduces to,

$$\dot{S} = \frac{1}{T(S_1, V_1)} \left(p_1(S_1, V_1) - p_2(S_2, V_2) \right) \dot{V}_1 \tag{3.26}$$

Since the differentials dS and dV_1 can be obtained from $dS = \dot{S}\,dt$ and $dV_1 = \dot{V}_1\,dt$, equation (3.26) yields the partial derivative,

$$\frac{\partial S}{\partial V_1} = \frac{1}{T(S_1, V_1)} \left(p_1(S_1, V_1) - p_2(S_2, V_2) \right) \tag{3.27}$$

The equilibrium condition of the second law requires the entropy of the isolated system to be maximum at equilibrium, i.e.

$$\frac{\partial S}{\partial V_1} = 0 \qquad \text{(equilibrium)} \tag{3.28}$$

We obtain the mechanical equilibrium condition by combining the condition (3.28) with our result (3.27),

$$p_1(S_1, V_1) = p_2(S_2, V_2) \qquad \text{(mechanical equilibrium)} \tag{3.29}$$

The first and second laws require that the pressures of the subsystems reach the same value at mechanical equilibrium.

Since the system is isolated, the entropy production rate of the system is given by $\Pi_S = \dot{S} \geq 0$. Combining equation (3.26) with the evolution condition (2.1) of the second law, we obtain the entropy production rate:

$$\Pi_S = \frac{1}{T(S_1, V_1)} \left(p_1(S_1, V_1) - p_2(S_2, V_2) \right) \dot{V}_1 \geq 0 \tag{3.30}$$

Therefore, if $p_1 > p_2$, then $\dot{V}_1 > 0$, which means that the volume of subsystem 1 increases and that of subsystem 2 decreases.

In view of equation (3.30), we see that a mechanical action between subsystems 1 and 2 is reversible if and only if the pressures are equal, i.e. $p_1(S_1, V_1) = p_2(S_2, V_2)$. When choosing subsystem 1 as the simple system and subsystem 2 as its environment, we obtain the condition (2.38) under which a mechanical action on a simple system is reversible.

In order to satisfy the inequality (3.30) in the neighbourhood of the equilibrium state, the entropy production rate Π_S has to be a positive definite quadratic form. Thus,

$$\dot{V}_1 = \frac{1}{R_{\text{th}}} \Big(p_1(S_1, V_1) - p_2(S_2, V_2) \Big) \tag{3.31}$$

where the coefficient $R_{\text{th}} > 0$ is a ***thermo-hydraulic resistance*** (see exercise 3.4). Its form is analogous to ***Poiseuille's law***, which describes the laminar flow of a liquid in a cylinder. The thermo-hydraulic resistance R_{th} in that case is a function of the liquid viscosity and the cylinder radius [29].

3.4 Rigid and Permeable Diathermal Wall

Let us continue our analysis of systems composed of two subsystems by considering a system as in § 3.2, except that the wall is diathermal, fixed and permeable. The subsystems are supposed to remain in thermal equilibrium at the same temperature T at all times (Fig. 3.3).

The state variables of subsystem 1 are its entropy S_1 and number of moles N_1. Likewise for subsystem 2, the state variables are the entropy S_2 and number of moles N_2. Since the number of moles of gas is an extensive quantity, the number of moles N of the system is the sum of the subsystems number of moles N_1 and N_2,

$$N = N_1 + N_2 \tag{3.32}$$

As the number of moles of gas in the whole system N is fixed, i.e. $\dot{N} = 0$, the time derivative of the number of moles (3.32) implies,

$$\dot{N}_2 = -\dot{N}_1 \tag{3.33}$$

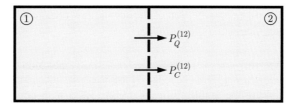

Figure 3.3 Two simple subsystems containing a homogeneous gas are separated by a fixed, permeable and diathermal wall. $P_Q^{(12)}$ and $P_C^{(12)}$ denote the thermal and chemical powers exerted by subsystem 1 on subsystem 2.

Since the internal energy $U(S_1, S_2, N_1, N_2)$ is an extensive state function, it is the sum of the internal energies $U_1(S_1, N_1)$ and $U_2(S_2, N_2)$ of the two subsystems,

$$U(S_1, S_2, N_1, N_2) = U_1(S_1, N_1) + U_2(S_2, N_2) \tag{3.34}$$

The internal energies $U_1(S_1, N_1)$ and $U_2(S_2, N_2)$ are state functions that have the same dependence on their state variables since the two subsystems contain the same gas. The time derivative of the internal energy (3.34) is written,

$$\dot{U}(S_1, S_2, N_1, N_2) = \dot{U}_1(S_1, N_1) + \dot{U}_2(S_2, N_2) \tag{3.35}$$

As the system is isolated, the first law (1.30) implies,

$$\dot{U}_2(S_2, N_2) = -\dot{U}_1(S_1, N_1) \tag{3.36}$$

The first law (1.28) applied to subsystems 1 and 2 yields,

$$\dot{U}_1(S_1, N_1) = P_Q^{(21)} + P_C^{(21)} \qquad \text{and} \qquad \dot{U}_2(S_2, N_2) = P_Q^{(12)} + P_C^{(12)} \tag{3.37}$$

Each subsystem is rigid and open. Thermal equilibrium between the two subsystems implies,

$$T(S_1, N_1) = T(S_2, N_2) \qquad \text{(thermal equilibrium)} \tag{3.38}$$

The time derivatives of the subsystems entropies S_1 and S_2 are deduced from the general evolution equation (2.12) for internal energy,

$$
\begin{aligned}
\dot{S}_1 &= \frac{1}{T(S_1, N_1)} \left(\dot{U}_1(S_1, N_1) - \mu_1(S_1, N_1)\,\dot{N}_1 \right) \\
\dot{S}_2 &= \frac{1}{T(S_2, N_2)} \left(\dot{U}_2(S_2, N_2) - \mu_2(S_2, N_2)\,\dot{N}_2 \right)
\end{aligned}
\tag{3.39}
$$

The evolution of the total entropy is obtained by summing the equations in (3.39). According to relations (3.33), (3.36) and (3.38), we find,

$$\dot{S} = \frac{1}{T(S_1, N_1)} \left(\mu_2(S_2, N_2) - \mu_1(S_1, N_1) \right) \dot{N}_1 \tag{3.40}$$

Since the differentials dS and dN_1 can be written as $dS = \dot{S}\,dt$ and $dN_1 = \dot{N}_1\,dt$, equation (3.40) yields the partial derivative,

$$\frac{\partial S}{\partial N_1} = \frac{1}{T(S_1, N_1)} \left(\mu_2(S_2, N_2) - \mu_1(S_1, N_1) \right) \tag{3.41}$$

The equilibrium condition of the second law requires the entropy of the isolated system to be maximum at equilibrium, i.e.

$$\frac{\partial S}{\partial N_1} = 0 \qquad \text{(equilibrium)} \tag{3.42}$$

Comparing the equilibrium condition (3.42) with equation (3.41), we find the chemical equilibrium condition, i.e.

$$\mu_1(S_1, N_1) = \mu_2(S_2, N_2) \qquad \text{(chemical equilibrium)} \tag{3.43}$$

The first and the second law require the chemical potentials to have the same value at equilibrium.

We use here the term chemical equilibrium to mean that the amount of substance that could go through the wall is constant in both subsystems. This equilibrium should not be confused with the equilibrium between different chemical substances undergoing chemical reactions. This latter case shall be treated in Chapter 8.

Since the system is isolated, its entropy production rate is positive, i.e. $\Pi_S = \dot{S} \geq 0$. Combining equation (3.40) with the evolution condition (2.1) of the second law, we find the entropy production rate to be,

$$\Pi_S = \frac{1}{T(S_1, N_1)} \Big(\mu_2(S_2, N_2) - \mu_1(S_1, N_1) \Big) \dot{N}_1 \geq 0 \qquad (3.44)$$

Thus, if $\mu_2 > \mu_1$, then $\dot{N}_1 > 0$, which means that the number of moles of subsystem 1 increases while that of subsystem 2 decreases. Thus, **the second law implies that matter transfer occurs from a higher chemical potential to a lower chemical potential**.

In view of equation (3.44), we see that a matter transfer between subsystems 1 and 2 is reversible if and only if the chemical potentials are equal, i.e. $\mu_1(S_1, N_1) = \mu_2(S_2, N_2)$. When choosing subsystem 1 as the simple system and subsystem 2 as its environment, we obtain the condition (2.49) under which a matter transfer to a simple system is reversible.

This result for chemical potentials is analogous to what is familiar to us regarding the gravitational potential. Let us consider two pools filled with water that have molar potential energies mgh_1 and mgh_2, where m is the molar mass of water, h_1 and $h_2 > h_1$ are the levels of the pools above a position that defines the potential reference. When these pools are connected by a canal, the gravitational potential difference between these pools causes a water flow from the higher pool at level h_2 to the lower one at level h_1. Similarly, the chemical potential difference generates the gas transfer from the subsystem with the higher chemical potential μ_2 to the one with the lower chemical potential μ_1 (Fig. 3.4).

In order to satisfy the inequality (3.44) in the neighbourhood of the equilibrium state, the entropy production rate Π_S has to be a positive definite quadratic form. Thus, we write,

$$\dot{N}_1 = F\frac{A}{\ell} \Big(\mu_2(S_2, N_2) - \mu_1(S_1, N_1) \Big) \qquad (3.45)$$

where the coefficient $F > 0$ is Fick's **diffusion coefficient**, A is the contact area between the subsystems and ℓ is a specific length.

The reason for writing the prefactor as FA/ℓ is the following: equation (3.45) is a 'discrete' formulation of **Fick's law**. It is "discrete" in the sense that there are just two

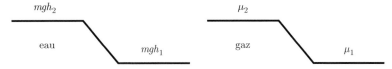

Figure 3.4 Comparison between water that goes "down" a gravitational potential mgh and gas that goes "down" a chemical potential μ.

values of the chemical potential associated with two areas of space. We will examine in Chapter 11 what happens when the chemical potential varies continuously over space. Fick's law in that case involves only the diffusion coefficient F.

3.5 Movable and Permeable Diathermal Wall

Let us consider as a final example a system composed of two subsystems as in § 3.2, separated by a wall that lets heat and matter pass through, and that can also move, i.e. the wall is movable, diathermal and permeable (Fig. 3.5). Each subsystem is an open simple system. We neglect the kinetic energy variation of the system due to the displacement of the wall or the gas. The state variables of subsystem 1 are the entropy S_1, the volume V_1 and the number of moles N_1. Likewise, the state variables of subsystem 2 are the entropy S_2, the volume V_2 and the number of moles N_2.

We can deduce the evolution equation for entropy from that of internal energy (2.12),

$$\dot{S}_1 = \frac{1}{T_1(S_1, V_1, N_1)} \left(\dot{U}_1(S_1, V_1, N_1) + p_1(S_1, V_1, N_1)\, \dot{V}_1 - \mu_1(S_1, V_1, N_1)\, \dot{N}_1 \right)$$
$$\dot{S}_2 = \frac{1}{T_2(S_2, V_2, N_2)} \left(\dot{U}_2(S_2, V_2, N_2) + p_2(S_2, V_2, N_2)\, \dot{V}_2 - \mu_2(S_2, V_2, N_2)\, \dot{N}_2 \right) \tag{3.46}$$

Since the internal energy $U(S_1, S_2, V_1, V_2, N_1, N_2)$ is an extensive state function, it is the sum of the internal energies $U_1(S_1, V_1, N_1)$ and $U_2(S_2, V_2, N_2)$ of the two subsystems,

$$U(S_1, S_2, V_1, V_2, N_1, N_2) = U_1(S_1, V_1, N_1) + U_2(S_2, V_2, N_2) \tag{3.47}$$

The internal energies $U_1(S_1, V_1, N_1)$ and $U_2(S_2, V_2, N_2)$ are state functions that have the same dependence on their state variables since the two subsystems contain the same gas. The time derivative of the internal energy (3.47) is written,

$$\dot{U}(S_1, S_2, V_1, V_2, N_1, N_2) = \dot{U}_1(S_1, V_1, N_1) + \dot{U}_2(S_2, V_2, N_2) \tag{3.48}$$

Since the system is isolated, the first law (1.30) implies that

$$\dot{U}_2(S_2, V_2, N_2) = -\dot{U}_1(S_1, V_1, N_1) \tag{3.49}$$

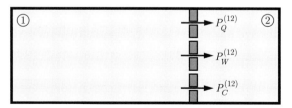

Figure 3.5 Two simple subsystems containing a homogeneous gas are separated by a moving and permeable diathermal wall. $P_Q^{(12)}$ denotes the thermal power, $P_W^{(12)}$ the mechanical power and $P_C^{(12)}$ the chemical power exerted by the subsystem 1 on the subsystem 2.

$\dot{S}$ is obtained in (3.2) by summing the equations in (3.46). According to relations (3.18), (3.33) and (3.49), we find,

$$\dot{S} = \left(\frac{1}{T_1(S_1, V_1, N_1)} - \frac{1}{T_2(S_2, V_2, N_2)} \right) \dot{U}_1(S_1, V_1, N_1)$$
$$+ \left(\frac{p_1(S_1, V_1, N_1)}{T_1(S_1, V_1, N_1)} - \frac{p_2(S_2, V_2, N_2)}{T_2(S_2, V_2, N_2)} \right) \dot{V}_1 \tag{3.50}$$
$$- \left(\frac{\mu_1(S_1, V_1, N_1)}{T_1(S_1, V_1, N_1)} - \frac{\mu_2(S_2, V_2, N_2)}{T_2(S_2, V_2, N_2)} \right) \dot{N}_1$$

Multiplying equation (3.50) by dt, we can identify in the resulting equation the differentials $dS = \dot{S}\,dt$, $dU = \dot{U}_1\,dt$, $dV_1 = \dot{V}_1\,dt$ and $dN_1 = \dot{N}_1\,dt$, and hence, deduce from equation (3.50) the partial derivatives,

$$\frac{\partial S}{\partial U_1} = \frac{1}{T_1(S_1, V_1, N_1)} - \frac{1}{T_2(S_2, V_2, N_2)}$$
$$\frac{\partial S}{\partial V_1} = \frac{p_1(S_1, V_1, N_1)}{T_1(S_1, V_1, N_1)} - \frac{p_2(S_2, V_2, N_2)}{T_2(S_2, V_2, N_2)} \tag{3.51}$$
$$\frac{\partial S}{\partial N_1} = \frac{\mu_2(S_2, V_2, N_2)}{T_2(S_2, V_2, N_2)} - \frac{\mu_1(S_1, V_1, N_1)}{T_1(S_1, V_1, N_1)}$$

According to the equilibrium condition of the second law, the entropy of the isolated system is maximum at equilibrium, i.e.

$$\frac{\partial S}{\partial U_1} = 0, \qquad \frac{\partial S}{\partial V_1} = 0 \quad \text{and} \quad \frac{\partial S}{\partial N_1} = 0 \qquad \text{(equilibrium)} \tag{3.52}$$

Comparing the equilibrium conditions (3.52) with the equations of (3.51), we find the thermal, mechanical and chemical equilibrium conditions,

$$T_1(S_1, V_1, N_1) = T_2(S_2, V_2, N_2) \qquad \text{(thermal equilibrium)}$$
$$p_1(S_1, V_1, N_1) = p_2(S_2, V_2, N_2) \qquad \text{(mechanical equilibrium)} \tag{3.53}$$
$$\mu_1(S_1, V_1, N_1) = \mu_2(S_2, V_2, N_2) \qquad \text{(chemical equilibrium)}$$

When the system is at equilibrium, it is at thermal, mechanical and chemical equilibrium. This equilibrium is characterised by the fact that the intensive quantities conjugated to the extensive state variables, i.e. temperature, pressure and chemical potential, have the same value in each subsystem.

3.6 Worked Solutions

3.6.1 Stationary Heat Transfer between Two Blocks

Consider a system consisting of two simple systems that are rigid blocks in thermal contact (Fig. 3.6). Assume that block 1 is kept at temperature T_1, block 2 at temperature $T_2 < T_1$ and that the heat transfer between both blocks occurs in a stationary regime.

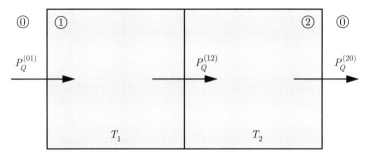

Figure 3.6 A heat transfer occurs in a stationary regime between block 1 at temperature T_1 and block 2 at temperature T_2.

Use the following notation: $P_Q^{(01)}$ is the heat transfer from the environment (labelled 0) to block 1, $P_Q^{(12)}$ is that from block 1 to block 2 and $P_Q^{(20)}$ is that from block 2 to the environment on the other side of the system. Show that in the stationary regime, the thermal power each block exerts on the other and the thermal power exerted by the blocks on the environment are equal and written as,

$$P_Q \equiv P_Q^{(01)} = P_Q^{(12)} = P_Q^{(20)}$$

Solution:

In a stationary state, the entropy of each block is constant,

$$\dot{S}_1 = 0 \qquad and \qquad \dot{S}_2 = 0$$

The blocks are simple systems. Thus, their entropy production rates (2.23) vanish. In this case, the entropy evolution (2.18) of each subsystem is given by,

$$\dot{S}_1 = \frac{P_Q^{(01)} - P_Q^{(12)}}{T_1(S_1)} = 0 \qquad \Rightarrow \qquad P_Q^{(01)} = P_Q^{(12)}$$

$$\dot{S}_2 = \frac{P_Q^{(12)} - P_Q^{(20)}}{T_2(S_2)} = 0 \qquad \Rightarrow \qquad P_Q^{(12)} = P_Q^{(20)}$$

Thus,

$$P_Q \equiv P_Q^{(01)} = P_Q^{(12)} = P_Q^{(20)}$$

3.6.2 Thermalisation of Two Blocks

An isolated system consists of two homogeneous metallic blocks, labelled 1 and 2, that can be considered as rigid simple systems. These blocks contain N_1 and N_2 moles of a metal. The blocks, initially separated, are at temperatures T_1 and T_2. When they are brought into contact, they evolve asymptotically towards a thermal equilibrium at final temperature T_f. The internal energy U_i of each block ($i = 1, 2$) is a function of its temperature T_i and the number of moles N_i of metal in each block,

$$U_i = 3N_i R T_i$$

where R is a positive constant.

1. Determine the final temperature T_f of the system.
2. Compute the entropy variation ΔS of the system during the process that leads to its thermal equilibrium.

Solution:

1. *Given that the system is isolated, the internal energy is constant (1.30). Thus, the internal energy variation ΔU of the system is,*

$$\Delta U = \Delta U_1 + \Delta U_2 = 0$$

The variation of the internal energy of each block is given by,

$$\Delta U_1 = 3N_1 R \int_{T_1}^{T_f} dT = 3N_1 R \left(T_f - T_1\right)$$

$$\Delta U_2 = 3N_2 R \int_{T_2}^{T_f} dT = 3N_2 R \left(T_f - T_2\right)$$

Thus, the final temperature T_f of the system is,

$$T_f = \frac{N_1 T_1 + N_2 T_2}{N_1 + N_2}$$

2. *The entropy variation ΔS of the system is,*

$$\Delta S = \Delta S_1 + \Delta S_2$$

The infinitesimal entropy variation of each block is written as,

$$dS_1 = \frac{dU_1}{T_1} = 3N_1 R \frac{dT_1}{T_1}$$

$$dS_2 = \frac{dU_2}{T_2} = 3N_1 R \frac{dT_2}{T_2}$$

which implies that the entropy variation of each block is given by,

$$\Delta S_1 = 3N_1 R \int_{T_1}^{T_f} \frac{dT}{T} = 3N_1 R \ln\left(\frac{T_f}{T_1}\right)$$

$$\Delta S_2 = 3N_2 R \int_{T_2}^{T_f} \frac{dT}{T} = 3N_2 R \ln\left(\frac{T_f}{T_2}\right)$$

Thus, the entropy variation during the process yields,

$$\Delta S = 3N_1 R \ln\left(\frac{T_f}{T_1}\right) + 3N_2 R \ln\left(\frac{T_f}{T_2}\right)$$

3.6.3 Heating by Kinetic Friction in Translation

Consider a system consisting of a block of length L_1 that moves with a constant velocity $\boldsymbol{v}_1$ on top of a fixed block of length $L_2 > L_1$. Subsystem 1 is the moving block and subsystem 2 is the fixed block. Both blocks are made of the same metal and can be treated as simple systems. The moving block is subjected to the action of an external force $\boldsymbol{F}^{\text{ext}}$ (Fig. 3.7) and the friction force $\boldsymbol{F}^{\text{fr}}$ due to its contact with the fixed block. The whole system is adiabatic and closed, but the interface between the blocks is diathermal. Moreover, the heat transfer between the blocks is such that they have the same temperature at all times. The internal energy U of the whole system is a function of its temperature T and number of moles of substance N,

$$U = 3NRT$$

where R is a positive constant. Note that the whole system is not a simple system since there is no frame of reference where the translational kinetic energy vanishes.

1. Show that the time derivative of the temperature is a positive constant

$$\dot{T} = \frac{\boldsymbol{F}^{\text{ext}} \cdot \boldsymbol{v}_1}{3NR}$$

    Thus, the system temperature increases linearly with time.
2. Show that the heat transfer $P_Q^{(12)}$ from block 1 to block 2 and the heat transfer $P_Q^{(21)}$ from block 2 to block 1 are related by,

$$P_Q^{(12)} + P_Q^{(21)} = \boldsymbol{F}^{\text{ext}} \cdot \boldsymbol{v}_1$$

    Thus, the heat transfers do not compensate for each other.
3. Show that the entropy production rate Π_S is given by,

$$\Pi_S = -\frac{\boldsymbol{F}^{\text{fr}} \cdot \boldsymbol{v}_1}{T} > 0$$

It is entirely due to the friction between the blocks. They do not have intrinsic contributions to the dissipation, since they are simple systems. The expression thus found is analogous to the one obtained in § 1.10.4 for a closed system consisting of a gas contained in a cylinder closed by a piston subjected to friction; note that X_0 is identified as entropy ($X_0 \equiv S$) and Y_0 as temperature ($Y_0 \equiv T$).

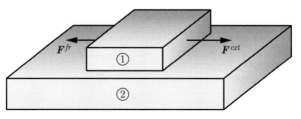

Figure 3.7   A block moves with a velocity $\boldsymbol{v}_1$ on top of a fixed block because of an external force $\boldsymbol{F}^{\text{ext}}$. Friction between the block gives rise to the force $\boldsymbol{F}^{\text{fr}}$.

Solution:

1. *The state variables of the system are the momentum P_1 of block 1 and the entropies S_1 and S_2 of the blocks. According to (1.22), the energy of the system $E(P_1, S_1, S_2)$ is the sum of its translational kinetic energy and its internal energy,*

$$E(P_1, S_1, S_2) = \frac{1}{2} P_1 \cdot v_1 + U(S_1, S_2)$$

According to the first law (1.11) and to the definition (1.26), we have,

$$\dot{E}(P_1, S_1, S_2) = P^{\text{ext}} = F^{\text{ext}} \cdot v_1$$

because the mechanical and thermal powers exerted on the system vanish, i.e. $P_W = P_Q = 0$. The momentum P_1 is given by $P_1 = M_1 v_1$, where M_1 is the mass of the moving block. The velocity is constant, i.e. $\dot{v}_1 = 0$. The momentum is constant, i.e $\dot{P}_1 = 0$. Thus,

$$\dot{E}(P_1, S_1, S_2) = \dot{U}(S_1, S_2)$$

which implies that,

$$\dot{U}(S_1, S_2) = T\dot{S}_1 + T\dot{S}_2 = T\dot{S} = F^{\text{ext}} \cdot v_1$$

where $S = S_1 + S_2$ is the entropy of the system. Moreover, the derivative of the internal energy of the system is given by,

$$\dot{U} = 3NR\dot{T}$$

where $U \equiv U(S_1, S_2)$ and $T \equiv T(S_1, S_2)$. Combining these two equations, we conclude that,

$$\dot{T} = \frac{F^{\text{ext}} \cdot v_1}{3NR}$$

2. *The two blocks are simple systems. Thus, the evolution equations of their internal energies are expressed as in (3.6):*

$$\dot{U}_1(S_1) = P_Q^{(21)} \qquad \text{and} \qquad \dot{U}_2(S_2) = P_Q^{(12)}$$

The extensivity of the internal energy (3.4) implies that,

$$P_Q^{(12)} + P_Q^{(21)} = F^{\text{ext}} \cdot v_1$$

3. *Since the system is adiabatic and closed, $\Pi_S = \dot{S}$. Thus,*

$$\Pi_S = \frac{F^{\text{ext}} \cdot v_1}{T} > 0$$

The forces acting on the moving block are the driving force F^{ext} and the friction force F^{fr}. Since the momentum of the moving block P_1 is constant, the centre of mass theorem (1.13) implies,

$$\dot{P}_1 = F^{\text{ext}} + F^{\text{fr}} = 0$$

Hence, we have,

$$F^{\text{fr}} = -F^{\text{ext}}$$

and therefore,

$$\Pi_S = -\frac{F^{\text{fr}} \cdot v_1}{T} > 0$$

3.6.4 Heating by Kinetic Friction in Rotation

Consider a system consisting of a cylinder of radius R_1 that rotates around its symmetry axis, on top of a fixed cylinder of radius $R_2 > R_1$, at a constant angular velocity ω_1. The two cylinders are made of the same metal. An external force $\boldsymbol{F}^{\text{ext}}$ is applied to a string of negligible mass, wrapped around the rotating cylinder (Fig. 3.8) and causes its rotation. The rotating cylinder is also subjected to the action of a friction torque $\boldsymbol{M}^{\text{fr}}$ due to the contact with the fixed cylinder [30]. The internal energy U of the system is a function of its temperature T and number of moles N:

$$U = 3NRT$$

where R is a positive constant. The system consisting of the two cylinders is adiabatic and closed, but the interface between these cylinders is diathermal. The heat transfer between the cylinders is such that they have the same temperature at all times. Subsystem 1 is the rotating cylinder and subsystem 2 the fixed cylinder. Both can be treated as simple systems. Note that the whole system is not a simple system because only one of the cylinders moves and there is no frame of reference where the rotational kinetic energy vanishes.

1. Show that the time derivative of the temperature is a positive constant

$$\dot{T} = \frac{\left(\boldsymbol{R}_1 \times \boldsymbol{F}^{\text{ext}}\right) \cdot \boldsymbol{\omega}_1}{3NR}$$

where $\boldsymbol{R}_1$ it the position vector of the application point of the external force $\boldsymbol{F}^{\text{ext}}$. Thus, the temperature of the system increases linearly with time.

2. Show that the heat transfers $P_Q^{(12)}$ from cylinder 1 to cylinder 2 and $P_Q^{(21)}$ from cylinder 2 to cylinder 1 are related by,

$$P_Q^{(12)} + P_Q^{(21)} = \left(\boldsymbol{R}_1 \times \boldsymbol{F}^{\text{ext}}\right) \cdot \boldsymbol{\omega}_1$$

Thus, the heat transfers do not compensate for each other.

3. Show that the entropy production rate Π_S is expressed as,

$$\Pi_S = -\frac{\boldsymbol{M}^{\text{fr}} \cdot \boldsymbol{\omega}_1}{T} > 0$$

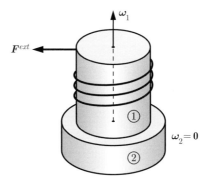

Figure 3.8   Friction of a cylinder rotating on top of a fixed cylinder with an angular velocity ω_1 due to an external force $\boldsymbol{F}^{\text{ext}}$.

It is due to the friction between the two cylinders that are simple systems and therefore, do not contribute separately to the dissipation. This expression for the dissipation is analogous to the one obtained in § 1.10.5 for a closed system consisting of a rotating fan; once X_0 and Y_0 are identified as entropy and temperature.

Solution:

1. *The state variables of the system are the angular momentum L_1 of the cylinder and the entropies S_1 and S_2 of the two cylinders. According to (1.22), the energy of the system $E(L_1, S_1, S_2)$ is the sum of the rotational kinetic energy and the internal energy,*

$$E(L_1, S_1, S_2) = \frac{1}{2} L_1 \cdot \omega_1 + U(S_1, S_2)$$

According to the first law (1.11) and the definition (1.26), we have,

$$\dot{E}(L_1, S_1, S_2) = P^{\text{ext}} = M^{\text{ext}} \cdot \omega_1$$

because there are no mechanical and thermal powers exerted on the system, i.e. $P_W = P_Q = 0$. The external torque is,

$$M^{\text{ext}} = R_1 \times F^{\text{ext}}$$

The angular momentum is $L_1 = I_1 \omega_1$ where I_1 is the moment of inertia of the moving cylinder. The angular velocity is constant, i.e. $\dot{\omega}_1 = 0$, and the angular momentum also, i.e $\dot{L}_1 = 0$. Thus, we have,

$$\dot{E}(L_1, S_1, S_2) = \dot{U}(S_1, S_2)$$

which implies that,

$$\dot{U}(S_1, S_2) = T\dot{S}_1 + T\dot{S}_2 = T\dot{S} = \left(R_1 \times F^{\text{ext}} \right) \cdot \omega_1$$

Moreover, the derivative of the internal energy of the system is given by,

$$\dot{U} = 3NR\dot{T}$$

where $U \equiv U(S_1, S_2)$ and $T \equiv T(S_1, S_2)$. Combining the two equations, we conclude that,

$$\dot{T} = \frac{\left(R_1 \times F^{\text{ext}} \right) \cdot \omega_1}{3NR}$$

2. *The two cylinders are simple systems. Thus the evolution equations (3.6) of their internal energies are,*

$$\dot{U}_1(S_1) = P_Q^{(21)} \qquad and \qquad \dot{U}_2(S_2) = P_Q^{(12)}$$

The extensivity of the internal energy (3.4) implies that,

$$P_Q^{(12)} + P_Q^{(21)} = \left(R_1 \times F^{\text{ext}} \right) \cdot \omega_1$$

3. *Since the system is adiabatic and closed, $\Pi_S = \dot{S}$. Thus,*

$$\Pi_S = \frac{\left(R_1 \times F^{\text{ext}} \right) \cdot \omega_1}{T} > 0$$

The torques acting on the rotating cylinder are the driving torque $\boldsymbol{M}^{ext}$ and the friction torque $\boldsymbol{M}^{fr}$. Since the angular momentum of the rotating cylinder $\boldsymbol{L}_1$ is constant, the angular momentum theorem (1.15) implies that,

$$\dot{\boldsymbol{L}}_1 = \boldsymbol{M}^{\text{ext}} + \boldsymbol{M}^{\text{fr}} = \boldsymbol{0}$$

so that,

$$\boldsymbol{M}^{\text{fr}} = -\boldsymbol{R}_1 \times \boldsymbol{F}^{\text{ext}}$$

and finally,

$$\Pi_S = -\frac{\boldsymbol{M}^{\text{fr}} \cdot \boldsymbol{\omega}_1}{T} > 0$$

3.6.5 Entropy Production Rate between Several Blocks

Let us consider an isolated system, consisting of several rigid subsystems (or blocks) labelled A, B, etc. We assume that each subsystem A can be characterised by its entropy S_A and the number of moles of substance N_A. Thus, each subsystem A has an internal energy $U_A(S_A, N_A)$. According to relations (3.37) and (3.39), the state variables of every subsystem satisfy the relation,

$$\dot{S}_A = \sum_B \frac{1}{T_A} \left(P_Q^{(BA)} + P_C^{(BA)} \right) - \frac{\mu_A}{T_A} \dot{N}_A$$

where $P_Q^{(BA)}$ and $P_C^{(BA)}$ are the thermal and chemical powers exerted by subsystem B on subsystem A. We express the time derivative $\dot{N}_A$ of the number of moles in subsystem A as the sum over the subsystems B of the matter inflow rates $\Omega^{(BA)}$ from subsystem B to subsystem A,

$$\dot{N}_A = \sum_B \Omega^{(BA)}$$

Show that the entropy production rate can be written as,

$$\Pi_S = \frac{1}{2} \sum_{A,B} \left(\frac{1}{T_A} - \frac{1}{T_B} \right) \left(P_Q^{(BA)} + P_C^{(BA)} \right) - \frac{1}{2} \sum_{A,B} \left(\frac{\mu_A}{T_A} - \frac{\mu_B}{T_B} \right) \Omega^{(BA)}$$

This result can be used to describe heat and mass transfer inside an isolated system consisting of simple subsystems (e.g. § 8.8.5).

Solution:

Since the whole system is isolated, the time derivative of the total entropy reduces to the entropy production rate,

$$\Pi_S = \dot{S} = \sum_A \dot{S}_A = \sum_{A,B} \frac{1}{T_A} \left(P_Q^{(BA)} + P_C^{(BA)} \right) - \sum_{A,B} \frac{\mu_A}{T_A} \Omega^{(BA)}$$

which can be recast as,

$$\Pi_S = \frac{1}{2} \sum_{A,B} \left(\frac{1}{T_A} \left(P_Q^{(BA)} + P_C^{(BA)} \right) + \frac{1}{T_B} \left(P_Q^{(AB)} + P_C^{(AB)} \right) \right)$$
$$- \frac{1}{2} \sum_{A,B} \left(\frac{\mu_A}{T_A} \Omega^{(BA)} + \frac{\mu_B}{T_B} \Omega^{(AB)} \right)$$

Furthermore, the time derivatives of the internal energy and number of moles of substance vanish,

$$\dot{U} = \sum_A \dot{U}_A = \sum_{A,B} \left(P_Q^{(BA)} + P_C^{(BA)} \right) = 0$$
$$\dot{N} = \sum_A \dot{N}_A = \sum_{A,B} \Omega^{(BA)} = 0$$

In order for these identites to always be verified, we require that,

$$P_Q^{(AB)} = -P_Q^{(BA)} \quad \text{and} \quad P_C^{(AB)} = -P_C^{(BA)} \quad \text{and} \quad \Omega^{(AB)} = -\Omega^{(BA)}$$

Using these relations, the entropy production rate reduces to the correct result.

Exercises

3.1 Thermalisation of Two Separate Gases

An isolated system consisting of two closed subsystems A and B is separated by a diathermal wall. Initially, they are held at temperatures T_A^i and T_B^i. Subsystem A contains N_A moles of gas. The internal energy of the gas is given by $U_A = c_A N_A R \, T_A$, where T_A is the temperature of the gas, R is a positive constant and c_A is a dimensionless coefficient. Likewise, there are N_B moles of gas in subsystem B and the internal energy of the gas is given by $U_B = c_B N_B R \, T_B$.

a) Determine the change of the internal energy U_A due to the thermalisation process.
b) Compare the initial temperature T_B^i and the final temperature T_f of the system if the size of subsystem B is much larger than that of subsystem A.

3.2 Thermalisation of Two Separate Substances

The entropy of a particular substance is given in terms of its internal energy U and number of moles N as [31],

$$S(U, V, N) = NR \ln \left(1 + \frac{U}{NE_0} \right) + \frac{RU}{E_0} \ln \left(1 + \frac{NE_0}{U} \right)$$

where R and E_0 are positive constants. A system consists of two subsystems containing such a substance, with N_A moles in subsystem A and N_B moles of it in subsystem B. When the subsystems are set in thermal contact, their initial temperatures are T_A^i and T_B^i. Determine the final temperature T_f of the system.

3.3   Diffusion of a Gas through a Permeable Wall

Analyse the time evolution of a gas consisting of one substance that diffuses through a permeable wall. Thus, consider an isolated system containing N moles of gas, consisting of two subsystems of equal volumes separated by a fixed and permeable wall, with $N_1(t)$ moles of gas in subsystem 1 and $N_2(t)$ in subsystem 2. In order to be able to find the time evolution, model the chemical potentials in each subsystem by assuming that they are proportional to the amount of matter. In order to simplify the expressions in the solution, write,

$$\mu_1(N_1) = \frac{\ell}{FA}\frac{N_1}{2\tau}$$
$$\mu_2(N_2) = \frac{\ell}{FA}\frac{N_2}{2\tau}$$

where $\tau > 0$ will be identified as a specific diffusion time, $F > 0$ is the Fick diffusion coefficient and $\ell > 0$ is a specific length. Initially, there are N_0 moles in the subsystem 1, i.e. $N_1(0) = N_0$, and $N - N_0$ moles in the subsystem 2, i.e. $N_2(0) = N - N_0$. Determine the evolution of the number of moles $N_1(t)$ and $N_2(t)$. Find the number of moles in each subsystem when equilibrium is reached.

3.4   Mechanical Damping by Heat Flow

An isolated system of volume V_0 consists of two subsystems, labelled 1 and 2, separated by an impermeable and moving diathermal wall of mass M and of negligible volume. Both subsystems contain a gas. The presure p, the volume V, the number of moles N and the temperature T of the gas are related by the equation $pV = NRT$ where R is a positive constant (see § 5.6). The internal energy of this gas is given by $U = cNRT$ where c is a dimensionless coefficient (see § 5.7). Initially, both subsystems are at the temperature T_i. Subsystem 1 is in a state characterised by a volume V_{1i} and a pressure p_{1i}. Likewise, subsystem 2 is characterised by a pressure p_{2i} and a volume V_{2i}. Determine,

a) The number of moles N_1 and N_2 in subsystems 1 and 2.
b) The final temperature T_f when the system has reached equilibrium.
c) The final volumes V_{1f} and V_{2f} of the subsystems when the system has reached equilibrium.
d) The final pressure p_f when the system has reached equilibrium.
e) Determine the entropy variation between the initial state and the final equilibrium state and show that, for the particular case where $N_1 = N_2 = N$, the result implies an increase in entropy.
f) As in § 3.3, assume that the wall is able to transfer heat fast enough that the temperature T stays the same on both sides of the wall, which means that the heat transfer is reversible. Take into account the kinetic energy of the wall, neglect any heat stored inside the wall. Show that the wall comes to its equilibrium position with a velocity v that decays exponentially with a time constant τ inversely proportional to the thermo-hydraulic resistance R_{th}.

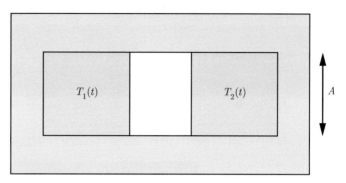

Two blocks of the same material face each other, separated by air. Thermal conduction through the air and convection are neglected. The blocks come to a thermal equilibrium due to heat exchange by radiation.

3.5 Entropy Production by Thermalisation

In exercise 3.6.2 devoted to the thermalisation of two blocks, show for the particular case where $N_1 = N_2 = N$ that the entropy variation,

$$\Delta S = 3N_1 R \ln \left(\frac{T_f}{T_1} \right) + 3N_2 R \ln \left(\frac{T_f}{T_2} \right)$$

is strictly positive.

3.6 Entropy Production by Heat Transfer

An isolated system consists of two subsystems labelled 1 and 2 analysed in exercise 3.6.1. Using the second law (2.2), show that in a stationary state when $T_1 > T_2$ the entropy production rate Π_S of the whole system is positive when heat flows across these two subsystems despite the fact that, according to equation (2.23), $\Pi_{S_1} = \Pi_{S_2} = 0$.

3.7 Thermalisation by Radiation

An isolated system consists of two blocks made of the same substance (Fig. 3.9). The internal energies of blocks 1 and 2 are $U_1 = C_1 T_1$ and $U_2 = C_2 T_2$ where C_1 and C_2 are two positive constants. Two sides of the blocks face each other exactly. The area of each side is A and they are separated by a fixed air gap. We neglect the heat conductivity of the air in the gap. The radiative thermal power that block i exerts on block j, where $i, j = 1, 2$, is given by,

$$P_Q^{(ij)} = \sigma A \left(T_i (t)^4 - T_j (t)^4 \right)$$

where σ is a constant coefficient.

a) Determine the final temperature T_f of the system when it reaches equilibrium.
b) Derive the time evolution equation for $T_1 (t)$ and $T_2 (t)$.
c) Consider the particular case where $C_1 = C_2 = C$ and the limit of small temperature variations, i.e. $T_1 (t) = T_f + \Delta T_1 (t)$ and $T_2 (t) = T_f + \Delta T_2 (t)$ with $\Delta T_1 (t) \ll T_f$ and $\Delta T_2 (t) \ll T_f$ at all times. Show that the temperature difference $\Delta T (t) = \Delta T_1 (t) - \Delta T_2 (t)$ is exponentially decreasing.

4

Thermodynamic Potentials

Hermann Ludwig Ferdinand von Helmholtz, 1821–1894

H. L. F. von Helmoltz was professor of physiology, first in Königsberg, then in Bonn,
Heidelberg and Berlin. He saw friction and inelastic collisions as processes of energy redistribution,
which he explained in his first book on thermodynamics : 'Über die Erhaltung der Kraft'.
In addition to the vast research that made him famous, he was interested in the sensory
perception of colours and music.

4.1 Historical Introduction

The nineteenth century witnessed many developments in thermodynamics, most notably
the introduction of the *thermodynamic potentials* that will be presented in this chapter.

The study of chemical equilibria raised the question of knowing whether it was possible
to define a potential that would play in thermodynamics an analogous role to that of
the potential in mechanics. Then, equilibrium would be obtained by minimising the
potential [32].

In 1869, François-Jacques-Dominique Massieu, general inspector of "Ecole des Mines" and professor at the Science Faculty of the University of Rennes, presented to the "Académie des sciences" his work on what he called the *characteristic functions of fluids*. He considered entropy as a function of internal energy, volume and number of moles of each constituent. He then defined transformations of this function according to a method developed by the French mathematician Adrien-Marie Legendre [33].

In the 1870s, the American physico-chemist and mathematician Josiah Williard Gibbs published a series of articles on the equilibrium of heterogeneous fluids. While his analysis was based on the condition of entropy maximisation, he used the functions denoted nowadays by F, G and H in order to clarify the equilibrium conditions he had found. He acknowledged the contribution of Massieu in his work. Gibbs considered how much energy changed when the mass of one of the constituents changed. Thus, he introduced the notion of chemical potential, but did not use this term. He insisted on avoiding the adjective 'chemical', as he wished to emphasise that his equilibrium conditions pertained not only to chemical substances, but also to phase transitions, electrical effects or elastic media [34].

In 1882, Hermann von Helmholtz determined the potential that defined the equilibrium state of a system when its temperature is kept constant [35]. Thus, he defined the ***free energy***, denoted by the letter F, that is still used nowadays. Another potential, designated today by the letter H, was named ***enthalpy*** by Kammerling Onnes [36].

In 1884, the Frenchman Pierre Duhem presented a thesis on the application of thermodynamic potentials in chemistry. Purportedly, he had to withdraw it under the pressure of Marcelin Berthelot. Later, he published his works in the form of three books, in which he described what he called *chemical mechanics*, which in effect was a description of chemical reactions based on thermodynamics.

In this chapter, we introduce the free energy also called the "Helmholtz free energy", denoted F, and the "Gibbs free energy" denoted G. The international organisation IUPAC recommends using the expressions "***Helmholtz free energy***" A and "***Gibbs free energy***" G. We will discuss the meaning of the adjective "free" in these designations. Although the IUPAC convention denotes Helmholtz free energy by the letter A, the physics community perseveres in the usage of F and we will follow this convention.

4.2 Fundamental Relations

In this chapter, we will consider a simple thermodynamic system described by the following state variables: entropy S, volume V and number of moles $N_1, \ldots, N_r$ of the r constituents. From now on, we shall denote by $\{N_A\}$ the set of numbers of moles of the r chemical constituents ($A = 1, \ldots, r$).

4.2.1 Gibbs Relation

According to the definitions (2.9), (2.10) and (2.11) of temperature T, pressure p and chemical potentials μ_A, the differential of $U(S, V, \{N_A\})$ is obtained by applying relation (1.7),

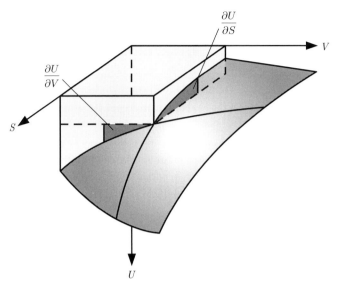

Figure 4.1 Convex surface corresponding to the internal energy $U(S, V)$ as a function of entropy S and volume V in state space (S, U, V). The vertical planes at fixed S or fixed V define curves on the surface $U(S, V)$. The slopes of these curves correspond to the partial derivatives.

$$dU = T\,dS - p\,dV + \sum_{A=1}^{r} \mu_A\,dN_A \qquad (4.1)$$

Equation (4.1) is called the **Gibbs relation**. This equation can also be obtained by multiplying (2.12) by the infinitesimal time interval dt. The Gibbs relation (4.1) expresses the fact that the internal energy U is a state function, i.e. $U = U(S, V, \{N_A\})$, and that the intensive quantities T, p and μ_A are also state functions, i.e.

$$\begin{aligned} T &= T(S, V, \{N_A\}) \\ p &= p(S, V, \{N_A\}) \\ \mu_A &= \mu_A(S, V, \{N_A\}) \end{aligned} \qquad (4.2)$$

The relations that appear in (4.2) are called **equations of state**, because they constitute relationships between the state variables S, V and $\{N_A\}$.

As an ilustration, the internal energy $U(S, V)$ is graphically represented as a state function of the variables entropy S and volume V (Fig. 4.1). This allows a geometric interpretation of the temperature $T(S, V)$ and pressure $p(S, V)$. In general, the internal energy $U(S, V)$ is represented as a surface in the abstract 3-dimensional space (S, U, V), called **state space**. The temperature $T(S, V)$ and pressure $p(S, V)$ are state functions evaluated at a point on the surface $U(S, V)$. The temperature, defined by $T = \partial U(S, V)/\partial S$, is the derivative with respect to S of the curve located at the intersection of the surface $U(S, V)$ and the plane $V = \text{const}$. The pressure $p = -\partial U(S, V)/\partial V$ is the opposite of the derivative with respect to V of the curve located at the intersection of the surface $U(S, V)$ and the plane $S = \text{const}$. The curvature of the surface $U(S, V)$ is such that $T > 0$ and

$p > 0$. Notice that the equations of state $U(S, V)$ and $S(U, V)$ are represented by the same surface in state space.

4.2.2 Euler Relation

Let us consider now a system consisting of λ identical subsystems, each characterised by its entropy S, volume V and number of moles N_A for substances $A = 1, ..., r$. Since the state variables are extensive, the system has a total entropy λS, a total volume λV and λN_A moles of substance A. Since the internal energy of the system is an extensive state function, it is the sum of the internal energies of the subsystems, i.e.

$$U(\lambda S, \lambda V, \{\lambda N_A\}) = \lambda U(S, V, \{N_A\}) \tag{4.3}$$

The derivative of the relation (4.3) with respect to λ can be written as,

$$\frac{\partial U}{\partial (\lambda S)} \frac{d(\lambda S)}{d\lambda} + \frac{\partial U}{\partial (\lambda V)} \frac{d(\lambda V)}{d\lambda} + \sum_{A=1}^{r} \frac{\partial U}{\partial (\lambda N_A)} \frac{d(\lambda N_A)}{d\lambda} = U \tag{4.4}$$

Since S, V and N_A are independent of λ, equation (4.4) reduces to,

$$\frac{\partial U}{\partial (\lambda S)} S + \frac{\partial U}{\partial (\lambda V)} V + \sum_{A=1}^{r} \frac{\partial U}{\partial (\lambda N_A)} N_A = U \tag{4.5}$$

Equation (4.5) has to be satisfied for every value of λ. Thus, choosing $\lambda = 1$,

$$U = \frac{\partial U}{\partial S} S + \frac{\partial U}{\partial V} V + \sum_{A=1}^{r} \frac{\partial U}{\partial N_A} N_A \tag{4.6}$$

Using the definitions (2.9)–(2.11), relation (4.6) yields,

$$U = TS - pV + \sum_{A=1}^{r} \mu_A N_A \tag{4.7}$$

which is called the ***Euler relation***.

4.2.3 Gibbs–Duhem Relation

The infinitesimal variation of internal energy dU is obtained by differentiating the Euler equation (4.7),

$$dU = T dS + S dT - p dV - V dp + \sum_{A=1}^{r} \left(\mu_A dN_A + N_A d\mu_A \right) \tag{4.8}$$

Then, according to the Gibbs relation (4.1), the expression (4.8) requires that,

$$S dT - V dp + \sum_{A=1}^{r} N_A d\mu_A = 0 \tag{4.9}$$

which is called the ***Gibbs–Duhem relation***.

4.3 Legendre Transformation

So far, we have considered the state function internal energy, which is a function of extensive state variables. The conjugated quantities obtained from internal energy by partial derivation are intensive quantities. Often in practical situations, it is these intensive quantities that are measured or controlled. The most obvious case is entropy and its conjugated variable temperature. In a measurement, temperature is more likely to be controlled than entropy.

This raises the question whether it is possible to reformulate our mathematical description in order to use intensive quantities as state variables. The extensive variables will appear then as conjugated quantities. We will now see that such a mathematical operation is possible [37]. This approach will lead us to introduce the thermodynamic potentials called free energy, enthalpy and Gibbs free energy. The simplest case is a state function $F(X_0, X_1, \ldots, X_n)$. Our goal is to use one or more partial derivatives $Y_i(X_0, X_1, \ldots, X_n) = \partial F / \partial X_i$ as independent state variables without loosing the information contained in the function $F(X_0, X_1, \ldots, X_n)$.

Let us consider first the state function $F(X)$ of a single extensive variable X. The slope of the tangent to the state function $F(X)$ at X is given by,

$$Y(X) = \frac{dF(X)}{dX} \tag{4.10}$$

The state function $F(X)$ is a curve determined by a set of points (X, F) corresponding to values of the variable X and the associated function F. The tangent of slope $Y(X)$ to this curve at a fixed point (X, F) intersects the vertical axis at point $(0, G)$ (Fig. 4.2).

Thus, the tangent is determined by a couple of values (Y, G) that correspond to the slope Y and to the coordinate of the intersect on the vertical axis G. Graphically (Fig. 4.2), the slope Y of the tangent is given by

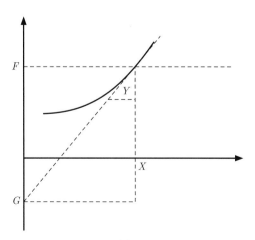

Figure 4.2 $Y(X)$ is the slope of the tangent to the state function $F(X)$ at point X. The tangent intersects the vertical axis at point G.

$$Y = \frac{F - G}{X - 0} \tag{4.11}$$

When inverting the equation of state $Y = Y(X)$, we obtain $X = X(Y)$. Thus, the state function F can be expressed as $F(X(Y))$ and equation (4.11) becomes

$$G(Y) = F(X(Y)) - YX(Y) \tag{4.12}$$

The expression (4.12) is called the **Legendre transform** of the state function F with respect to the state variable X. The Legendre transform $G(Y)$ is a state function of Y. The Legendre transformation of $G(Y)$ with respect Y yields again the function $F(X)$. Thus, the state functions $F(X)$ and $G(Y)$ contain the same information.

The Legendre transformation can be generalised from one to several variables. The partial derivative of the state function $F(X_0, X_1, \ldots, X_n)$ with respect to X_i, where $i = 0, 1, \ldots, n$, evaluated in the state $(X_0, X_1, \ldots, X_n)$ is given by,

$$Y_i(X_0, X_1, \ldots, X_n) = \frac{\partial F(X_0, X_1, \ldots, X_n)}{\partial X_i} \tag{4.13}$$

By analogy with the Legendre transformation with respect to a single state variable, we can write,

$$Y_i = \frac{F - G}{X_i - 0} \tag{4.14}$$

When inverting the equation of state $Y_i = Y_i(X_0, X_1, \ldots, X_n)$, we obtain $X_i = X_i(X_0, X_1, \ldots, Y_i, \ldots, X_n)$. Thus, the state function F can be expressed as $F\left(X_0, X_1, \ldots, X_i\right.$ $(X_0, X_1, \ldots, Y_i, \ldots, X_n), \ldots, X_n\left.\right)$ and equation (4.11) becomes,

$$G(X_0, X_1, \ldots, Y_i, \ldots, X_n) = F\left(X_0, X_1, \ldots, X_i(X_0, X_1, \ldots, Y_i, \ldots, X_n), \ldots, X_n\right)$$
$$- Y_i X_i(X_0, X_1, \ldots, Y_i, \ldots, X_n) \tag{4.15}$$

According to the definition (4.13), the partial derivative of equation (4.15) with respect to Y_i yields,

$$\frac{\partial G}{\partial Y_i} = \frac{\partial F}{\partial X_i}\frac{\partial X_i}{\partial Y_i} - X_i - Y_i\frac{\partial X_i}{\partial Y_i} = -X_i \tag{4.16}$$

There is a simple relationship between the curvatures of the functions $G(Y)$ and $F(X)$ that we can demonstrate as follows. We obtain the second order partial derivative of F with respect to X_i and of G with respect to Y_i, by differentiating relations (4.13) and (4.16),

$$\frac{\partial^2 F}{\partial X_i^2} = \frac{\partial Y_i}{\partial X_i} \quad \text{and} \quad \frac{\partial^2 G}{\partial Y_i^2} = -\frac{\partial X_i}{\partial Y_i} \tag{4.17}$$

This implies that,

$$\frac{\partial^2 G}{\partial Y_i^2} = -\left(\frac{\partial^2 F}{\partial X_i^2}\right)^{-1} \tag{4.18}$$

Hence, we find that the sign of the curvature of the state function G with respect to the variable Y_i is the opposite of the sign of the curvature of the state function F with respect to the variable X_i.

Equation (4.12) defines the Legendre transform of the state function F with respect to the state variable X_i. To simplify the notation, we simply write,

$$G = F - Y_i X_i \qquad (4.19)$$

We obtain the state function $H(Y_0, Y_1, \ldots, Y_n)$ by performing successive Legendre transformations on the state function $F(X_0, X_1, \ldots, X_n)$, i.e.

$$H = F - \sum_{i=0}^{n} Y_i X_i \qquad (4.20)$$

It is also possible to perform Legendre transformations on a few variables only, which is often the case in thermodynamics. In the following section, we will obtain the thermodynamic potentials by performing several Legendre transformations of the internal energy with respect to various extensive state variables.

4.3.1 Link between Thermodynamics and Mechanics

In analytical mechanics, a Legendre transformation defined with the opposite sign relates the Lagrangian $L(r_1, \ldots, r_n, v_1, \ldots, v_n)$ of a system containing n material points to the Hamiltonian $H(r_1, \ldots, r_n, P_1, \ldots, P_n)$ of this system [9], i.e.

$$H = \sum_{i=1}^{n} P_i \cdot v_i - L \qquad (4.21)$$

Since the Legendre transform of a state function is a state function, its opposite is also a state function and contains the same information.

4.4 Thermodynamic Potentials

The extensive state function that plays a central role in thermodynamics is the internal energy $U(S, V, \{N_A\})$. We call ***thermodynamic potential*** every state function obtained by a Legendre transformation of $U(S, V, \{N_A\})$. By extension, the internal energy itself can also be called a thermodynamic potential since it can be obtained by two successive Legendre transformations starting from itself. The three other thermodynamic potentials that we will define in this section are the free energy, the enthalpy and the Gibbs free energy.

4.4.1 Free Energy

The ***free energy*** $F(T, V, \{N_A\})$, also called ***Helmholtz free energy***, is defined as the Legendre transform (4.19) of the internal energy $U(S, V, \{N_A\})$ with respect to the entropy S, i.e.

$$F = U - TS \qquad (4.22)$$

According to the Euler relation (4.7), the free energy can be written as,

$$F = -pV + \sum_{A=1}^{r} \mu_A N_A \tag{4.23}$$

The differentiation of expression (4.22) is written,

$$dF = dU - T\,dS - S\,dT \tag{4.24}$$

Substituting for dU its value given by the Gibbs relation (4.1), expression (4.24) becomes,

$$dF = -S\,dT - p\,dV + \sum_{A=1}^{r} \mu_A\,dN_A \tag{4.25}$$

In conformity with the mathematical description of Legendre transformations, the differential (4.25) shows that the free energy is a state function of the variables T, V and N_A. Moreover, the variables that are conjugates of these state variables are defined as,

$$S\,(T, V, \{N_A\}) = -\frac{\partial F\,(T, V, \{N_A\})}{\partial T} \tag{4.26}$$

$$p\,(T, V, \{N_A\}) = -\frac{\partial F\,(T, V, \{N_A\})}{\partial V} \tag{4.27}$$

$$\mu_A\,(T, V, \{N_A\}) = \frac{\partial F\,(T, V, \{N_A\})}{\partial N_A} \tag{4.28}$$

4.4.2 Enthalpy

The **enthalpy** $H\,(S, p, \{N_A\})$ is defined as the Legendre transform (4.19) of the internal energy $U\,(S, V, \{N_A\})$ with respect to the volume V, i.e.

$$H = U + pV \tag{4.29}$$

According to the Euler relation (4.7), the enthalpy H can be written as,

$$H = TS + \sum_{A=1}^{r} \mu_A N_A \tag{4.30}$$

The differentiation of expression (4.29) yields,

$$dH = dU + p\,dV + V\,dp \tag{4.31}$$

Substituting for dU its value given by the relation (4.1), expression (4.31) becomes,

$$dH = T\,dS + V\,dp + \sum_{A=1}^{r} \mu_A\,dN_A \tag{4.32}$$

The differential (4.32) shows that the enthalpy is a state function of the variables S, p and N_A. Moreover, the conjugates of these state variables are,

$$T(S,p,\{N_A\}) = \frac{\partial H(S,p,\{N_A\})}{\partial S} \tag{4.33}$$

$$V(S,p,\{N_A\}) = \frac{\partial H(S,p,\{N_A\})}{\partial p} \tag{4.34}$$

$$\mu_A(S,p,\{N_A\}) = \frac{\partial H(S,p,\{N_A\})}{\partial N_A} \tag{4.35}$$

4.4.3 Gibbs Free Energy

The **Gibbs free energy** $G(T,p,\{N_A\})$ also called **free enthalpy** is defined as the Legendre transform (4.19) of the internal energy $U(S,V,\{N_A\})$ with respect to entropy S and volume V or, equivalently, as the Legendre transform of the enthalpy $H(S,p,\{N_A\})$ with respect to entropy S, i.e.

$$G = U + pV - TS = H - TS \tag{4.36}$$

Taking into account the Euler relation (4.7), the Gibbs free energy reads,

$$G = \sum_{A=1}^{r} \mu_A N_A \tag{4.37}$$

The differentiation of (4.36) yields,

$$dG = dU - T\,dS - S\,dT + p\,dV + V\,dp \tag{4.38}$$

Substituting for dU its value given by the Gibbs relation (4.1), the expression (4.38) for dG becomes,

$$dG = -S\,dT + V\,dp + \sum_{A=1}^{r} \mu_A\,dN_A \tag{4.39}$$

The differential (4.39) shows that the Gibbs free energy is a state function of the variables T, p and N_A. Moreover, the conjugates of these state variables are defined as,

$$S(T,p,\{N_A\}) = -\frac{\partial G(T,p,\{N_A\})}{\partial T} \tag{4.40}$$

$$V(T,p,\{N_A\}) = \frac{\partial G(T,p,\{N_A\})}{\partial p} \tag{4.41}$$

$$\mu_A(T,p,\{N_A\}) = \frac{\partial G(T,p,\{N_A\})}{\partial N_A} \tag{4.42}$$

In the following sections of this chapter, we shall see that these thermodynamic potentials enable the characterisation of equilibrium states of systems subjected to constraints imposed by their interaction with an environment that we shall define as reservoirs or baths.

4.5 Equilibrium of Subsystems Coupled to a Reservoir

In this section, we seek to characterise the approach to equilibrium of a system consisting of two simple subsystems separated by a wall and coupled with a very large simple system called a **reservoir** or a **bath**. A reservoir is characterised by one or several constant intensive variables.

Each subsystem is assumed to be at equilibrium with the reservoir at all times. The whole system consisting of both subsystems and the reservoir is considered isolated. The extensive state variables of the system consisting of the two subsystems 1 and 2 are its entropy S and volume V, i.e.

$$S = S_1 + S_2 \quad \text{and} \quad V = V_1 + V_2 \tag{4.43}$$

The thermodynamic potentials of the system are,

$$U = U_1 + U_2, \quad F = F_1 + F_2, \quad H = H_1 + H_2, \quad G = G_1 + G_2 \tag{4.44}$$

since they also are extensive quantities.

4.5.1 Minimum of Free Energy

Let us consider the approach to equilibrium when both subsystems are kept at all times in thermal equilibrium with a reservoir at constant temperature T_{ext}. Such a reservoir is called a **heat reservoir** or a **thermal bath**. Furthermore, we assume that the walls separating the reservoir and the subsystems are fixed, so that the volume of the system $V = V_1 + V_2$ is constant (Fig. 4.3).

Since each subsystem is at thermal equilibrium with the heat reservoir, each subsystem has the temperature T_{ext} of the reservoir, i.e.

$$T_{\text{ext}} = T_1 = T_2 \equiv T \tag{4.45}$$

From now on, we will speak of the temperature T of both the system and the bath. Taking into account the free energy definition (4.22) and the thermal equilibrium condition (4.45), the evolution of the free energy F of the system at constant temperature T is given by,

$$\dot{F} = \dot{U} - T\dot{S} \tag{4.46}$$

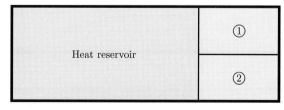

Figure 4.3　A system and a heat reservoir form a whole that is isolated. The subsystems 1 and 2 are separated from the reservoir by fixed walls that ensure the thermal equilibrium of each subsystem with the reservoir.

According to relation (2.36), the mechanical power of the system vanishes since its volume is constant, i.e. $P_W = 0$. Taking into account the first law (1.29), the second law (2.1) and the evolution equation for the entropy (2.18), relation (4.46) reduces to [38],

$$\dot{F} = P_Q - T\dot{S} = -T\Pi_S \leq 0 \qquad (4.47)$$

Multiplying the evolution equation (4.47) by the infinitesimal time interval dt, we obtain the condition,

$$dF \leq 0 \qquad \text{(constant temperature and volume)} \qquad (4.48)$$

The strict inequalities of the relations (4.47) and (4.48) describe the decrease of free energy F during irreversible processes in which the system is kept at constant temperature T and volume V, while the constraint on the inner wall is relaxed. The final equilibrium state, predicted by the second law, is described by the equalities of relations (4.47) and (4.48). Thus, we draw the following conclusion:

If a rigid system is kept at constant temperature as a consequence of its interaction with a heat reservoir, the equilibrium state minimises the free energy of the system.

4.5.2 Minimum of Enthalpy

Let us consider now the approach to equilibrium of a closed system composed of two subsystems which are kept in mechanical equilibrium with a **work reservoir**, also called **mechanical bath**, which is at constant pressure p_{ext}. The walls separating the reservoir and the subsystems are moveable and the action of the reservoir on the system is assumed to be reversible in such a way that its entropy $S = S_1 + S_2$ is constant (Fig. 4.4).

Since each subsystem is at mechanical equilibrium with the work reservoir, each subsystem has the pressure p_{ext} of the reservoir, i.e.

$$p_{\text{ext}} = p_1 = p_2 \equiv p \qquad (4.49)$$

Taking into account the enthalpy definition (4.29) and the mechanical equilibrium condition (4.49), the evolution of the system enthalpy H at constant pressure p is given by,

$$\dot{H} = \dot{U} + p\dot{V} \qquad (4.50)$$

According to relations (2.36) and (4.49), the mechanical power on the system is,

$$P_W = -p\dot{V} \qquad (4.51)$$

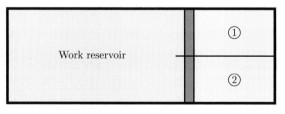

A system and a work reservoir form a whole that is isolated. The two subsystems 1 and 2 are separated from the reservoir by moving walls that ensure the mechanical equilibrium of each subsystem with the reservoir.

When a system has a constant entropy, i.e. $\dot{S} = 0$, the entropy evolution equation (2.2) implies,

$$I_S = -\Pi_S \leq 0 \tag{4.52}$$

where the inequality is an expression of the second law. By analogy with relation (2.21) where we characterised a simple system in a stationary regime, we assume that the thermal power P_Q has the same sign as the entropy exchange rate I_S. Then, relation (4.52) implies,

$$P_Q \leq 0 \tag{4.53}$$

Taking into account the mechanical power (4.51), the first law (1.29) and the thermal power (4.53), the evolution equation for enthalpy (4.50) reads [39],

$$\dot{H} = P_Q + P_W + p\,\dot{V} = P_Q \leq 0 \tag{4.54}$$

Multiplying the evolution equation (4.54) by an infinitesimal time interval dt, we obtain the condition,

$$dH \leq 0 \qquad \text{(constant entropy and pressure)} \tag{4.55}$$

The strict inequalities in relations (4.54) and (4.55) describe the decrease of enthalpy H during irreversible processes in which the system is kept at constant entropy S and pressure p, while the constraints on the inner wall are relaxed. The final equilibrium state implied by the second law is described by the equalities of relations (4.54) and (4.55). Thus, we draw the following conclusion:

If a system is kept at constant pressure with a work reservoir through reversible processes, the equilibrium state minimises the enthalpy of the system.

4.5.3 Minimum of Gibbs Free Energy

Let us consider finally the approach to equilibrium of a closed system composed of two subsystems that are kept at constant temperature T and pressure p because they are each in equilibrium with a **heat and work reservoir**. The walls separating the reservoir and the subsystems are moveable and diathermal (Fig. 4.5).

Taking into account the Gibbs free energy definition (4.36), the thermal equilibrium condition (4.45) and the mechanical equilibrium condition (4.49), the evolution of the Gibbs free energy G at constant pressure p and temperature T is given by,

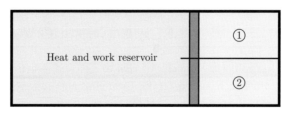

Figure 4.5   A system is composed of two subsystems which are coupled to a heat and work reservoir. The reservoir and the system form a whole that is isolated. Each subsystem is separated from the reservoir by a moveable, diathermal wall that ensures its thermal and mechanical equilibrium with the reservoir.

$$\dot{G} = \dot{U} - T\dot{S} + p\dot{V} \tag{4.56}$$

According to the first law (1.29), the second law (2.1), the entropy evolution equation (2.18) and taking into account relation (4.51), equation (4.56) reduces to [39],

$$\dot{G} = P_Q + P_W - T\dot{S} + p\dot{V} = P_Q - T\dot{S} = -T\Pi_S \leq 0 \tag{4.57}$$

Multiplying the evolution equation (4.57) by an infinitesimal time interval dt, we obtain the condition,

$$dG \leq 0 \qquad \text{(constant temperature and pressure)} \tag{4.58}$$

The strict inequalities in the relations (4.57) and (4.58) describe the decrease of Gibbs free energy G during irreversible processes in which the system is kept at constant T and pressure p while the constraint on the wall separating the subsystems is relaxed. The final equilibrium state prescribed by the second law is given by the equalities of the relations (4.57) and (4.58). Thus, we draw the following conclusion:

If a system is kept at a constant temperature and pressure with a heat and work reservoir, the equilibrium state between its subsystems minimises the Gibbs free energy of the system.

The foregoing examples show that the equilibrium of a system coupled to a bath can be characterised in the following way: when a system is coupled to a reservoir, then one or several intensive state variables are kept constant. The equilibrium state corresponds to the minimum of the thermodynamic potential that is a state function of these intensive state variables.

4.6 Heat and Work of Systems Coupled to Reservoirs

In the previous section, we saw how the thermodynamic potentials F, G and H took on a physical meaning as the potentials that were minimal at equilibrium under certain constraints. In this section, we illustrate another aspect of free energy and enthalpy by seeking to characterise two processes: the heat provided to a system in mechanical equilibrium with a work reservoir, and the work performed on a system which is in thermal equilibrium with a heat reservoir.

4.6.1 System Coupled to a Work Reservoir

Let us consider the transfer of heat to a closed system that is in equilibrium with a work reservoir. This heat transfer occurs at constant pressure, since the pressure p of the system is equal to the pressure p_{ext} of the reservoir,

$$p = p_{\text{ext}} \tag{4.59}$$

The system and the reservoir considered as a whole constitute a rigid system (Fig. 4.6).

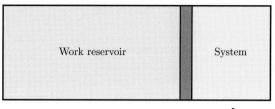

Figure 4.6 A closed system is in mechanical equilibrium with a work reservoir. The wall between them is movable and adiabatic. A heat transfer occurs between the environment and the system. The system and the reservoir together constitute a rigid system.

According to relations (1.38), (2.41), (4.31) and (4.1), the infinitesimal heat provided to the system at constant pressure p is given by,

$$\delta Q = dU - \delta W = dU + p\,dV = d\,(U + p\,V) = dH \qquad (4.60)$$

The heat transfer that brought the system from an initial state i to a final state f is found by integrating (4.60),

$$Q_{if} = \Delta H_{if} \qquad (4.61)$$

where

$$\Delta H_{if} = \int_{H_i}^{H_f} dH = H_f - H_i \qquad (4.62)$$

and Q_{if} was defined in (1.41). Thus, we draw the following conclusion:

> **The heat provided to a system that is kept at constant pressure is equal to the difference in the enthalpy of the initial and final states.**

The result $\delta Q = dH$ obtained for a system at a fixed pressure, sheds more light on the equilibrium of a system coupled to a work reservoir (§ 4.5.2). When the system reaches an equilibrium state, the reservoir ceases to exchange heat with the system, i.e. $\delta Q = 0$, therefore $dH = 0$, which means that enthalpy $H(S,p)$ at equilibrium is an extremum. We will show in § 6.5 that the enthalpy H is a convex function of the entropy S and thus, the enthalpy H is minimal at the equilibrium reached at constant pressure p.

4.6.2 System Coupled to a Heat Reservoir

Let us now consider the mechanical action exerted on a closed system in equilibrium with a heat reservoir. This mechanical action occurs at constant temperature, since the temperature T of the system is equal to the temperature T_{ext} of the reservoir,

$$T = T_{\text{ext}} \qquad (4.63)$$

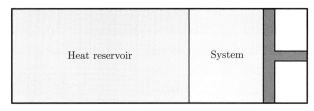

Figure 4.7 A system at thermal equilibrium with a heat reservoir is separated from it by a fixed diathermal wall. A mechanical action is exerted by the environment on the system. The system and the reservoir taken as a whole constitute an adiabatic system.

The system and the reservoir taken as a whole constitute an adiabatic closed system (Fig. 4.7).

Taking into account the elementary relations (1.38), (2.43), (4.24) and (4.1), the infinitesimal work performed on the system at constant temperature T is given by,

$$\delta W = dU - \delta Q = dU - T\,dS = d\,(U - TS) = dF \tag{4.64}$$

When integrating the differential equation (4.64) from an initial state i to a final state f, we obtain,

$$W_{if} = \Delta F_{if} \tag{4.65}$$

where

$$\Delta F_{if} = \int_{F_i}^{F_f} dF = F_f - F_i \tag{4.66}$$

and W_{if} is defined according to (1.40). Thus, we come to the following conclusion:

The work performed on a system that is kept at constant temperature is equal to the difference in the free energy of the initial and final states.

The result (4.64) allows us to characterise the approach to equilibrium of a system which is coupled to a heat reservoir and separated from it by a movable diathermal wall. When the system reaches an equilibrium state, the reservoir ceases to perform work on the system, i.e. $\delta W = 0$, therefore $dF = 0$, which implies in this case that the free energy $F(T, V)$ is an extremum. We will show in § 6.5 that the free energy F is a convex function of the volume V. Thus, at constant temperature T the free energy F is minimal at equilibrium.

4.7 Maxwell Relations

The Schwarz theorem states that the second order partial derivatives of a function of several variables are related to each other. The application of this theorem to thermodynamic potentials reveals important relations between physical quantities. These relations are called the Maxwell relations. They allow us to relate a partial derivative that has an unclear physical meaning to another partial derivative that has a much more intuitive meaning, as we will see with a few examples.

First, let us consider a continuous function $f(x, y)$ that has continuous partial derivatives with respect to the variables x and y. The Schwarz theorem states that the order of the partial derivatives of $f(x, y)$ can be interchanged, i.e.

$$\frac{\partial}{\partial x}\left(\frac{\partial f(x, y)}{\partial y}\right) = \frac{\partial}{\partial y}\left(\frac{\partial f(x, y)}{\partial x}\right) \tag{4.67}$$

Applying the Schwarz theorem to the thermodynamic potentials $U(S, V)$, $F(T, V)$, $H(S, p)$ and $G(T, p)$, we obtain the Maxwell relations.

Our first example pertains to a system that is described by the state variables entropy S and volume V. Applying the Schwarz theorem (4.67) to the internal energy $U(S, V)$ we obtain,

$$\frac{\partial}{\partial S}\left(\frac{\partial U}{\partial V}\right) = \frac{\partial}{\partial V}\left(\frac{\partial U}{\partial S}\right) \tag{4.68}$$

Using definition (2.10) of the pressure $p(S, V)$ and definition (2.9) of the temperature $T(S, V)$, the differential equation (4.68) leads to the Maxwell relation,

$$-\frac{\partial p}{\partial S} = \frac{\partial T}{\partial V} \tag{4.69}$$

Now, let us consider a system that is described by the state variables temperature T and volume V. Applying the Schwarz theorem (4.67) to the free energy $F(T, V)$ we obtain,

$$\frac{\partial}{\partial T}\left(\frac{\partial F}{\partial V}\right) = \frac{\partial}{\partial V}\left(\frac{\partial F}{\partial T}\right) \tag{4.70}$$

Using definition (4.26) of the entropy $S(T, V)$ and definition (4.27) of the pressure $p(T, V)$, the differential equation (4.70) leads to the Maxwell relation,

$$\frac{\partial p}{\partial T} = \frac{\partial S}{\partial V} \tag{4.71}$$

The Maxwell relation (4.71) enables us to express the variation of the entropy S with respect to the volume V at constant temperature T – not a very intuitive physical quantity – in terms of the variation of the pressure p with respect to the temperature T at constant volume V – a much more intuitive physical quantity.

When we describe the state of a system by the state variables entropy S and pressure p, the Schwarz theorem (4.67) applied to enthalpy $H(S, p)$ yields,

$$\frac{\partial}{\partial S}\left(\frac{\partial H}{\partial p}\right) = \frac{\partial}{\partial p}\left(\frac{\partial H}{\partial S}\right) \tag{4.72}$$

Using definition (4.34) of the volume $V(S, p)$ and definition (4.33) of the temperature $T(S, p)$, the differential equation (4.72) leads to the Maxwell relation,

$$\frac{\partial V}{\partial S} = \frac{\partial T}{\partial p} \tag{4.73}$$

As a final example, let us consider a system that is described by the state variables temperature T and pressure p. Applying the Schwarz theorem (4.67) to the Gibbs free energy $G(T, p)$ we obtain,

$$\frac{\partial}{\partial T}\left(\frac{\partial G}{\partial p}\right) = \frac{\partial}{\partial p}\left(\frac{\partial G}{\partial T}\right) \tag{4.74}$$

Using definition (4.41) of the volume $V(T,p)$ and definition (4.40) of the entropy $S(T,p)$, the differential equation (4.74) leads to the Maxwell relation,

$$\frac{\partial V}{\partial T} = -\frac{\partial S}{\partial p} \tag{4.75}$$

Notice that the Maxwell relations are not restricted to state functions of two variables. They can also be used with state functions of several variables. However, according to the Schwarz theorem, they are always the result of partial derivatives of a couple of state variables [40].

4.7.1 Partial Derivatives of a Function of Functions

Let us consider the function of functions $f\left(x(y,z),y\right)$ that is a function of the function $x(y,z)$ and of the variable y. The differential $df\left(x(y,z),y\right)$ is written in terms of the variables y and z as,

$$df\left(x(y,z),y\right) = \left(\frac{\partial f\left(x(y,z),y\right)}{\partial x(y,z)}\frac{\partial x(y,z)}{\partial y} + \frac{\partial f\left(x(y,z),y\right)}{\partial y}\right) dy$$

$$+ \left(\frac{\partial f\left(x(y,z),y\right)}{\partial x(y,z)}\frac{\partial x(y,z)}{\partial z}\right) dz \tag{4.76}$$

Taking into account the differential (4.76), the partial derivatives of the function of functions $f\left(x(y,z),y\right)$ with respect to the variables y and z are defined respectively as,

$$\left.\frac{df}{dy}\right|_z \equiv \frac{\partial f\left(x(y,z),y\right)}{\partial x(y,z)}\frac{\partial x(y,z)}{\partial y} + \frac{\partial f\left(x(y,z),y\right)}{\partial y}$$

$$\left.\frac{\partial f}{\partial z}\right|_y \equiv \frac{\partial f\left(x(y,z),y\right)}{\partial x(y,z)}\frac{\partial x(y,z)}{\partial z} \tag{4.77}$$

In this book, the notation introduced by (4.77) is used to describe partial derivatives of a thermodynamic potential f only when the state variable z is not one of the variables that stems from the Legendre transformation that defines f.

4.7.2 Cyclic Chain Rule

The differentials of the invertible state functions $x(y,z)$, $y(z,x)$ and $z(x,y)$ are given by,

$$dx = \frac{\partial x}{\partial y} dy + \frac{\partial x}{\partial z} dz$$

$$dy = \frac{\partial y}{\partial z} dz + \frac{\partial y}{\partial x} dx \tag{4.78}$$

$$dz = \frac{\partial z}{\partial x} dx + \frac{\partial z}{\partial y} dy$$

Here, the explicit dependence in terms of the state variables x, y and z is not mentioned in order to simplify the notation. Substituting the second and third equation of (4.78) for dy and dz in the first equation, we obtain,

$$dx \left(1 - \frac{\partial x}{\partial y} \frac{\partial y}{\partial x} \right) = dz \left(\frac{\partial x}{\partial z} + \frac{\partial x}{\partial y} \frac{\partial y}{\partial z} \right)$$

$$dx \left(1 - \frac{\partial x}{\partial z} \frac{\partial z}{\partial x} \right) = dy \left(\frac{\partial x}{\partial y} + \frac{\partial x}{\partial z} \frac{\partial z}{\partial y} \right) \tag{4.79}$$

In order to satisfy (4.79) for any value of dx, dy and dz, the terms in brackets have to vanish. The terms on the left-hand side of the identities (4.79) thus provide,

$$\frac{\partial x\,(y,z)}{\partial y} = \left(\frac{\partial y\,(z,x)}{\partial x} \right)^{-1}$$

$$\frac{\partial x\,(y,z)}{\partial z} = \left(\frac{\partial z\,(x,y)}{\partial x} \right)^{-1} \tag{4.80}$$

Applying relations (4.80) to the terms in brackets on the right-hand side of the identities (4.79) yields,

$$\frac{\partial x\,(y,z)}{\partial y} \frac{\partial y\,(z,x)}{\partial z} \frac{\partial z\,(x,y)}{\partial x} = -1 \tag{4.81}$$

This is often referred to as the triple product rule or the **cyclic chain rule**.

4.8 Worked Solutions

4.8.1 Black Body Radiation

A black body is an object at equilibrium with the radiation it emits. This radiation is characterised by the fact that the internal energy density depends only on the temperature at thermal equilibrium. The internal energy of this radiation is given by,

$$U\,(S, V) = \frac{3}{4} \left(\frac{3c}{16\sigma} \right)^{1/3} S^{4/3} V^{-1/3}$$

where σ is called the Stefan-Boltzmann constant.

1. Determine the free energy $F\,(T, V)$ of this radiation.
2. Show that the internal energy $U\,(S, V)$ of the radiation can be obtained by performing an inverse Legendre transformation on the free energy $F\,(T, V)$.
3. Find the expressions $p\,(T, V)$ and $p\,(S, V)$ of the radiation pressure.

Solution:

1. *The black body temperature (2.9) is defined as, i.e.*

$$T\,(S, V) = \frac{\partial U\,(S, V)}{\partial S} = \left(\frac{3c}{16\sigma} \right)^{1/3} S^{1/3} V^{-1/3}$$

When inverting this relation, we obtain the radiation entropy $S(T, V)$ as a function of the temperature T and of the volume V, i.e.

$$S(T, V) = \left(\frac{16\sigma}{3c}\right) T^3 V$$

When substituting this result into the expression for the internal energy of the radiation $U(S, V)$, we find,

$$U = \frac{4\sigma}{c} T^4 V$$

The free energy $F(T, V)$ is obtained by Legendre transformation (4.22) of the internal energy $U(S, V)$ with respect to the entropy S. Using the two previous equations this transformation is written explicitly as, i.e.

$$F(T, V) = U - TS = -\frac{4\sigma}{3c} T^4 V$$

2. *For the black body radiation, the entropy defined by (4.26) is found to be,*

$$S(T, V) = -\frac{\partial F(T, V)}{\partial T} = \frac{16\sigma}{3c} T^3 V$$

When inverting this relation, we obtain the radiation temperature $T(S, V)$ as a function of the entropy S and of the volume V, i.e.

$$T(S, V) = \left(\frac{3c}{16\sigma}\right)^{1/3} S^{1/3} V^{-1/3}$$

When substituting for T in the radiation free energy $F(T, V)$, we find,

$$F = -\frac{1}{4}\left(\frac{3c}{16\sigma}\right)^{1/3} S^{4/3} V^{-1/3}$$

The internal energy $U(S, V)$ is obtained by Legendre transformation (4.22) of the free energy $F(T, V)$ with respect to the temperature T. Using the two previous equations, this transformation is written explicitly as,

$$U(S, V) = F + ST = \frac{3}{4}\left(\frac{3c}{16\sigma}\right)^{1/3} S^{4/3} V^{-1/3}$$

3. *According to definition (2.10), the black body radiation pressure $p(S, V)$ is expressed in terms of S and V as,*

$$p(S, V) = -\frac{\partial U(S, V)}{\partial V} = \frac{1}{4}\left(\frac{3c}{16\sigma}\right)^{1/3} S^{4/3} V^{-4/3}$$

According to definition (4.27), the black body radiation pressure $p(T, V)$ is expressed in terms of T and V as,

$$p(T, V) = -\frac{\partial F(T, V)}{\partial V} = \frac{4\sigma}{3c} T^4$$

4.8.2 Joule Expansion

Consider a rigid, adiabatic closed system consisting of two compartments initially separated by a fixed and impermeable horizontal wall (Fig. 4.8). Initially, the top compartment is filled with N moles of a gas at temperature T and the bottom compartment is empty. After removing a magnet that holds a ball on top of the container, the ball falls and breaks the separation wall. The gas is then distributed in the whole system, the pressure quickly reaches an equilibrium and we observe that the temperature decreases. The experiment is configured in such a way that the system reaches a new equilibrium state before a thermal equilibrium with the walls can occur. The relative change in temperature is of the order of 10^{-4} [41].

In this experiment, called the **Joule expansion**, the internal energy is constant, i.e. $dU = 0$, because the work performed on the system and the heat provided to the system are both zero. The observed temperature drop is surprising because if we assume that the gas is ideal (§ 5.6), then we would predict $dU = cNR\,dT = 0$. In order to account for the temperature change, we need a more elaborate model like that of van der Waals (§ 6.10) that takes into account the interaction energy between the gas molecules.

Show that for a gas that undergoes a Joule expansion, the **Joule coefficient**, defined as the partial derivative of the temperature with respect to the volume, is given by,

$$\frac{\partial T}{\partial V} = \frac{1}{C_V}\left(p - T\frac{\partial p}{\partial T}\right)$$

where C_V is the specific heat at constant volume (see Chapter 5), defined as,

$$C_V = \left.\frac{\partial U}{\partial T}\right|_V = T\frac{\partial S}{\partial T}$$

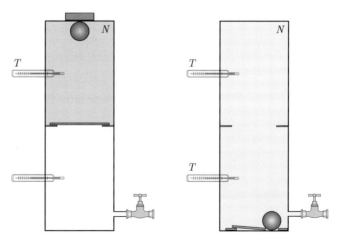

Figure 4.8 Initially, a gas is at temperature T in the top compartment. Just after the fall of the ball that breaks the wall, the gas is colder on the top than on the bottom, but the system evolves quickly towards an equilibrium state at a temperature very slightly inferior to T.

The last equality is a consequence of the definitions (2.9) and (4.77). In these equalities, the pressure p and the entropy S are functions of the state variables temperature T and volume V.

Solution:

The differential of entropy $S(T, V)$ is written,

$$dS = \frac{\partial S}{\partial T} dT + \frac{\partial S}{\partial V} dV$$

Substituting this differential for dS in the Gibbs relation (4.1), we obtain,

$$dU = T\frac{\partial S}{\partial T} dT + \left(T\frac{\partial S}{\partial V} - p\right) dV = 0$$

The Maxwell relation (4.71) implies that,

$$\frac{\partial S}{\partial V} = \frac{\partial p}{\partial T}$$

According to the definition of the specific heat at constant volume C_V, the differential of the internal energy yields,

$$C_V dT + \left(T\frac{\partial p}{\partial T} - p\right) dV = 0$$

Thus, we obtain the expression for the Joule coefficient,

$$\frac{\partial T}{\partial V} = \frac{1}{C_V} \left(p - T\frac{\partial p}{\partial T}\right)$$

In the particular case of an ideal gas that satisfies the equation of state $pV = NRT$, the Joule coefficient vanishes.

4.8.3 Joule–Thomson Expansion

Consider a cylinder closed by two sliding pistons separated by a permeable fixed wall. The cylinder contains N moles of an ideal gas passing through the wall under the effect of pistons 1 and 2 that keep the pressures p_1 and p_2 constant in the subsystems 1 and 2 on both sides of the wall (Fig. 4.9). The device is an adiabatic closed system.

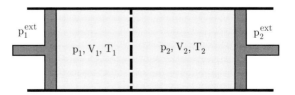

Figure 4.9 A gas is contained in a cylinder closed by two pistons. A permeable wall allows a pressure difference between the left and right compartments. The pressures p_1 and p_2 are kept constant through the action of external forces exerted on the pistons.

1. Show that the enthalpy H is conserved if the external pressures exerted by the pistons are equal to the pressures in the corresponding subsystems at all times, i.e. $p_1^{\text{ext}} = p_1$ and $p_2^{\text{ext}} = p_2$. This is called the Joule–Thomson expansion.

2. For an arbitrary gas and an infinitesimal pressure difference dp, show that the **Joule–Thomson coefficient**, defined as the partial derivative of temperature T with respect to the pressure p, is given by,

$$\frac{\partial T}{\partial p} = \frac{(T\alpha - 1)\, V}{C_p}$$

where α is the **thermal expansion** coefficient, defined by,

$$\alpha = \frac{1}{V} \frac{\partial V}{\partial T}$$

and C_p, the specific heat at constant pressure (see Chapter 5), defined as,

$$C_p = \left.\frac{\partial H}{\partial T}\right|_p = T \frac{\partial S}{\partial T}$$

The last equality is a consequence of the definitions (4.33) and (4.77). In these equations, the volume V and the entropy S are functions of the state variables temperature T and pressure p.

Solution:

1. *According to the first law (1.38) applied to the case of an adiabatic closed system and in view of definition (2.27) for an infinitesimal work, we have,*

$$dU = dU_1 + dU_2 = \delta W_1 + \delta W_2 = -p_1\, dV_1 - p_2\, dV_2$$

Since the pressures p_1 and p_2 are kept constant,

$$d(U_1 + p_1\, V_1) + d(U_2 + p_2\, V_2) = 0$$

The definition of the Legendre transform (4.29) implies that,

$$dH_1 = d(U_1 + p_1\, V_1) \qquad \text{and} \qquad dH_2 = d(U_2 + p_2\, V_2)$$

Thus, the total enthalpy H is constant, i.e.

$$dH = dH_1 + dH_2 = 0$$

2. *The differential of the entropy $S(T,p)$ is written,*

$$dS = \frac{\partial S}{\partial T}\, dT + \frac{\partial S}{\partial p}\, dp$$

Substituting for dS this expression in the enthalpy differential (4.32), we obtain,

$$dH = T \frac{\partial S}{\partial T}\, dT + \left(T \frac{\partial S}{\partial p} + V \right) dp = 0$$

The Maxwell relation (4.75) *and the definition of the thermal expansion coefficient* α *imply that,*

$$\frac{\partial S}{\partial p} = -\frac{\partial V}{\partial T} = -\alpha V$$

According to the definition of the specific heat at constant pressure C_p, *the enthalpy differential yields,*

$$C_p \, dT + (1 - \alpha T) \, V \, dp = 0$$

Thus, we obtain the expression of the Joule–Thomson coefficient,

$$\frac{\partial T}{\partial p} = \frac{(\alpha T - 1) \, V}{C_p}$$

In the particular case of an ideal gas that satisfies the equation of state $pV = NRT$, *the Joule–Thomson coefficient vanishes.*

4.8.4 Fundamental Relations for Specific Quantities

It is common practice, especially in chemistry, to use state functions that are specific quantities, i.e. extensive quantities per unit of volume or mass. We consider a simple system that has an internal energy $U(S, V, \{N_A\})$ where $A = 1, \ldots, r$. The volume densities u, s and n_A are defined as,

$$u = \frac{U}{V}, \qquad s = \frac{S}{V} \qquad \text{and} \qquad n_A = \frac{N_A}{V}$$

The mass densities u^*, s^* and v^* are defined as,

$$u^* = \frac{U}{M}, \qquad s^* = \frac{S}{M} \qquad \text{and} \qquad v^* = \frac{V}{M}$$

The mass chemical potential μ_A^* and mass concentration c_A^* of substance A are defined as,

$$\mu_A^* = \frac{\mu_A}{M_A^*} \qquad \text{and} \qquad c_A^* = \frac{N_A M_A^*}{M}$$

where M_A^* is the molar mass of substance A.

1. Determine the Gibbs, Euler and Gibbs–Duhem relations expressed in terms of u, s and n_A.
2. Determine the Gibbs, Euler and Gibbs–Duhem relations expressed in terms of u^*, s^*, v^* and c_A^*.

Solution:

1. *The Gibbs relation* (4.1) *divided by the volume* V *is written as,*

$$\frac{dU}{V} = T \frac{dS}{V} - p \frac{dV}{V} + \sum_{A=1}^{r} \mu_A \frac{dN_A}{V}$$

Moreover, the differentials can be expanded as,

$$du = d\left(\frac{U}{V}\right) = \frac{dU}{V} - U\frac{dV}{V^2}$$

$$ds = d\left(\frac{S}{V}\right) = \frac{dS}{V} - S\frac{dV}{V^2}$$

$$dn_A = d\left(\frac{N_A}{V}\right) = \frac{dN_A}{V} - N_A\frac{dV}{V^2}$$

Thus,

$$du - T\,ds - \sum_{A=1}^{r} \mu_A\,dn_A + \left(U - TS + pV - \sum_{A=1}^{r} \mu_A N_A\right)\frac{dV}{V^2} = 0$$

Using the Euler relation (4.7), the term in brackets vanishes and we find the volume-specific Gibbs relation, i.e.

$$du = T\,ds + \sum_{A=1}^{r} \mu_A\,dn_A$$

The Euler relation (4.1) divided by the volume V is written as,

$$\frac{U}{V} = T\frac{S}{V} - p + \sum_{A=1}^{r} \mu_A\frac{N_A}{V}$$

which yields the volume specific Euler relation, i.e.

$$u = Ts - p + \sum_{A=1}^{r} \mu_A n_A$$

The differentiation of the volume specific Euler relation is written as,

$$du = T\,ds + s\,dT - dp + \sum_{A=1}^{r} (\mu_A\,dn_A + n_A\,d\mu_A)$$

Taking into account the volume-specific Gibbs relation, we obtain the volume-specific Gibbs–Duhem relation, i.e.

$$s\,dT - dp + \sum_{A=1}^{r} n_A\,d\mu_A = 0$$

2. The Gibbs relation (4.1) divided by the mass M is written as

$$\frac{dU}{M} = T\frac{dS}{M} - p\frac{dV}{M} + \sum_{A=1}^{r} \mu_A\frac{dN_A}{M}$$

Moreover, the differentials can be written as,

$$du^* = d\left(\frac{U}{M}\right) = \frac{dU}{M} - U\frac{dM}{M^2}$$

$$ds^* = d\left(\frac{S}{M}\right) = \frac{dS}{M} - S\frac{dM}{M^2}$$

$$dv^* = d\left(\frac{V}{M}\right) = \frac{dV}{M} - V\frac{dM}{M^2}$$

$$\frac{dc_A^*}{M_A} = d\left(\frac{N_A}{M}\right) = \frac{dN_A}{M} - N_A\frac{dM}{M^2}$$

Thus,

$$du^* - T\,ds^* + p\,dv^* - \sum_{A=1}^{r}\mu_A^*\,dc_A^* + \left(U - TS + pV - \sum_{A=1}^{r}\mu_A N_A\right)\frac{dM}{M^2} = 0$$

Using the Euler relation (4.7), the term in brackets vanishes and we find the mass-specific Gibbs relation, i.e.

$$du^* = T\,ds^* - p\,dv^* + \sum_{A=1}^{r}\mu_A^*\,dc_A^*$$

The Euler relation (4.1) divided by the mass M is written as,

$$\frac{U}{M} = T\frac{S}{M} - p\frac{V}{M} + \sum_{A=1}^{r}\frac{\mu_A}{M_A}\frac{N_A M_A}{M}$$

which yields the mass-specific Euler relation, i.e.

$$u^* = T s^* - p\,v^* + \sum_{A=1}^{r}\mu_A^*\,c_A^*$$

The differentiation of the mass-specific Euler relation is written as,

$$du^* = T\,ds^* + s^*\,dT - p\,dv^* - v^*\,dp + \sum_{A=1}^{r}(\mu_A^*\,dc_A^* + c_A^*\,d\mu_A^*)$$

Taking into account the mass specific Gibbs relation, we obtain the mass specific Gibbs–Duhem relation, i.e.

$$s^*\,dT - v^*\,dp + \sum_{A=1}^{r}c_A^*\,d\mu_A^* = 0$$

4.8.5 Simple Subsystems in a Thermal Bath

We consider a rigid closed system containing a homogeneous gas. The system is divided in two simple subsystems that are separated by a moveable, impermeable diathermal wall. The system is at thermal equilibrium with a thermal bath at temperature $T = \text{cste}$ (Fig. 4.10). The kinetic and internal energies of the wall are negligible.

1. Express the differential of the free energy dF as a function of the infinitesimal heat δQ provided to the system.
2. Express the differential of the free energy dF as a function of the pressures p_1 and p_2 of the gas in subsystems 1 and 2. Deduce that $dF \leq 0$.

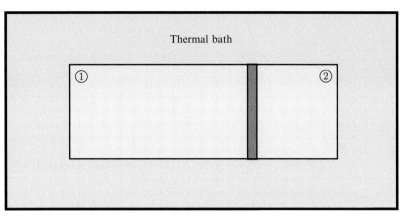

Figure 4.10 A rigid closed system containing a homogeneous gas is divided in two subsystems separated by a moveable, impermeable diathermal wall. The system is coupled to a thermal bath at constant temperature T.

Solution:

1. *Taking into account definition* (4.22) *and the fact that the temperature T is constant, the differential of the free energy dF can be written as,*

$$dF = dU - T\,dS$$

The mechanical power exerted by the thermal bath on the rigid subsystem vanishes, i.e. $P_W = 0$. Thus, in this case, the first law (1.38) *is reduced to,*

$$dU = \delta Q$$

which implies that the differential of the free energy can be recast as,

$$dF = \delta Q - T\,dS$$

2. *The free energy of the whole system $F(T_1, T_2, V_1, V_2)$ is the sum of the free energies $F_1(T_1, V_1)$ and $F_2(T_2, V_2)$ of the two subsystems,*

$$F(T_1, T_2, V_1, V_2) = F_1(T_1, V_1) + F_2(T_2, V_2)$$

At thermal equilibrium the temperatures of the two subsystems T_1 and T_2 are equal to the temperature T of the bath, i.e.

$$T_1 = T_2 = T$$

The differential of the free energy dF at temperature T is written,

$$dF = \frac{\partial F_1}{\partial V_1}\,dV_1 + \frac{\partial F_2}{\partial V_2}\,dV_2$$

Given that the system is rigid, the differential of the total volume dV vanishes, i.e.

$$dV = dV_1 + dV_2 = 0 \qquad thus \qquad dV_2 = -\,dV_1$$

Moreover, applying definition (4.27) *to both subsystems yields,*

$$p_1 = -\frac{\partial F_1}{\partial V_1} \quad and \quad p_2 = -\frac{\partial F_2}{\partial V_2}$$

Thus, the free energy differential dF at temperature T reduces to,

$$dF = -\left(p_1 - p_2\right) dV_1$$

This equation yields three types of solutions. In the case where $p_1 > p_2$, *the wall moves from the left to the right, i.e.*

$$p_1 > p_2 \quad and \quad dV_1 > 0 \quad thus \quad dF < 0$$

In the case where $p_1 < p_2$, *the wall moves from the right to the left, i.e.*

$$p_1 < p_2 \quad and \quad dV_1 < 0 \quad thus \quad dF < 0$$

In the case where $p_1 = p_2$, *i.e. at the mechanical equilibrium* (3.29), *the wall does not move, i.e.*

$$p_1 = p_2 \quad and \quad dV_1 = 0 \quad thus \quad dF = 0$$

Exercises

4.1   Adiabatic Compression
A gas is characterised by the enthalpy $H(S,p) = C_p T$, where C_p is a constant (called heat capacity and defined in § 5.2), and by $pV = NRT$, where p is its pressure, V its volume, T its temperature and N the number of moles of gas. An adiabatic reversible compression brings the pressure from p_1 to p_2 where $p_2 > p_1$. The initial temperature is T_1. Determine the temperature T_2 at the end of the compression.

4.2   Irreversible Heat Transfer
A cylinder closed by a piston contains N moles of a diatomic gas characterised by $U = (5/2) NRT$ and by $pV = NRT$, as in exercise 4.1. The gas has a temperature T when it is brought in contact with a heat reservoir at tempertaure T_{ext}, causing an irreversible process to occur. Determine the amount of heat exchanged.

Numerical application:

$N = 0.5\,mol$, $T = 450\,K$ and $T_{ext} = 300\,K$.

4.3   Internal Energy as Function of T and V
Establish the expression of the differential of the internal energy $dU\left(S(T, V), V\right)$ as a function of the temperature T and the volume V. In the particular case of a gas that satisfies the relation $pV = NRT$, show that $dU\left(S(T, V), V\right)$ is proportional to dT.

4.4   Grand Potential
The **grand potential** $\Phi(T, V, \{\mu_A\})$, also known as the **Landau free energy**, is a thermodynamical potential obtained by performing Legendre transformations of

the internal energy $U(S, V, \{N_A\})$. Use Legendre transformations to express the thermodynamical potential $\Phi(T, V, \{\mu_A\})$ in terms of the thermodynamical potential F. Also determine the differential $d\Phi(T, V, \{\mu_A\})$.

4.5 Massieu Functions

Two of the *Massieu functions* are functions of the following state variables:

1. $J\left(\dfrac{1}{T}, V\right)$

2. $Y\left(\dfrac{1}{T}, \dfrac{p}{T}\right)$

The Massieu functions are obtained by performing Legendre transformations of the state function entropy $S(U, V)$ with respect to the state variables U and V. Use Legendre transformations to express the Massieu functions $J\left(\dfrac{1}{T}, V\right)$ and $Y\left(\dfrac{1}{T}, \dfrac{p}{T}\right)$ in terms of the thermodynamical potentials F and G. Determine also the differentials $dJ\left(\dfrac{1}{T}, V\right)$ and $dY\left(\dfrac{1}{T}, \dfrac{p}{T}\right)$.

4.6 Gibbs–Helmoltz Equations

a) Show that

$$U(S, V) = -T^2 \frac{\partial}{\partial T}\left(\frac{F(T, V)}{T}\right)$$

where $T \equiv T(S, V)$ is to be understood as a function of S and V.

b) Show that

$$H(S, p) = -T^2 \frac{\partial}{\partial T}\left(\frac{G(T, p)}{T}\right)$$

where $T \equiv T(S, p)$ is to be understood as a function of S and p.

4.7 Pressure in a Soap Bubble

A soap bubble is a system consisting of two subsystems. Subsystem (f) is the thin film and subsystem (g) is the gas enclosed inside the film. The surrounding air is a thermal bath. The equilibrium is characterised by the minimum of the free energy F of the system. The differential of the free energy dF reads,

$$dF = -(S_g + S_f)\, dT + 2\gamma\, dA - (p - p_0)\, dV$$

where A is the surface area of the soap film and V the volume of the bubble. The parameter γ is called the **surface tension**. It characterises the interactions at the interface between the liquid and the air. Since the soap film has two such interfaces, there is a factor 2 in front of the parameter γ. The surface tension γ is an intensive variable that plays an analogous role for a surfacic system as the pressure p for a volumic system. However, the force due to pressure of a gas is exerted outwards whereas the force due to the surface tension is exerted inwards. This is the reason why the signs of the corresponding two terms in dF differ. The term $p - p_0$ is the

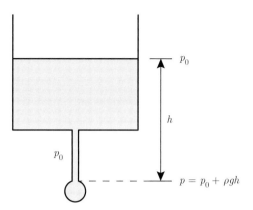

Figure 4.11 Principle of a setup that could be used to estimate the influence of surface tension on the pressure inside a liquid drop. The container is wide enough so that when a drop is forming the change in height of the liquid is negligible. The system is in a thermal bath at constant temperature T.

pressure difference between the pressure p inside the bubble and the atmospheric pressure p_0. Consider the bubble to be a sphere of radius r and show that,

$$p - p_0 = \frac{4\gamma}{r}$$

4.8   Pressure in a Droplet

Determine the hydrostatic pressure p inside a droplet, as a function of its radius r (Fig. 4.11). Assume that the drop (d) forms at the end of a short thin tube mounted at the end of vertical cylindrer containing the liquid (l). When a drop forms at the end of the tube, the change in the container height is negligible. If the height of the liquid above the tip of the tube is h, then the hydrostatic pressure is $p = p_0 + \rho g h$, where ρ is the volumetric mass density of the liquid, and g characterises the gravitation at the surface of the Earth. For this liquid, the differential of the free energy reads,

$$dF = -(S_l + S_d)\, dT + \gamma\, dA - (p - p_0)\, dV$$

Show that

$$p - p_0 = \frac{2\gamma}{r} = \rho g h$$

4.9   Isothermal Heat of Surface Expansion

A system consists of a thin film of surface area A, of internal energy $U(S, A)$, where

$$dU = T\, dS + \gamma\, dA$$

Hence, the surface tension is given by

$$\gamma(S, A) = \frac{\partial U(S, A)}{\partial A}$$

Express the heat Q_{if} to provide to the film for a variation $\Delta A_{if} = A_f - A_i$ of the surface of the film through an isothermal process at temperature T, that brings the film from an initial state i to a final state f, in terms of $\gamma(T, A)$ and its partial derivatives.

4.10 Thermomechanical Properties of an Elastic Rod

A state of an elastic rod is described by the state variables entropy S and length L. The differential of the internal energy $U(S,L)$ of the rod is written as,

$$dU = \frac{\partial U(S,L)}{\partial S} dS + \frac{\partial U(S,L)}{\partial L} dL = T(S,L)\,dS + f(S,L)\,dL$$

Note that $f(S,L)$ has the units of a force. The longitudinal stress τ on the rod is $\tau = \dfrac{f}{A}$, where A is the cross-section of the rod. We neglect any change of A due to f. The physical properties of the rod material are given by the linear thermal expansion coefficient at constant stress,

$$\alpha = \frac{1}{L}\frac{\partial L(T,f)}{\partial T},$$

and the isothermal Young modulus,

$$E = \frac{L}{A}\frac{\partial f(T,L)}{\partial L}.$$

Make use of these two physical properties of the material to answer the following questions:

a) Compute the partial derivative of the rod's stress τ in the rod changes with respect to its temperature when its length is fixed. Consider that the cross-section A is independent of the temperature.

b) Determine the heat transfer during an isothermal variation of the length of the rod ΔL_{if} from an initial state i to a final state f in terms of α and E.

c) Compute the partial derivative of the rod's length L with respect to its temperature T.

4.11 Chemical Power

An open system consists of a fluid of a single substance kept between two pistons sliding inside a cylinder with adiabatic walls. Matter enters and exits the cylinder in two specific locations. These two matter flows generate a chemical power P_C. The pressure at the entrance is p^+ and the pressure at the exit is p^-. The pistons exert a mechanical power P_W on the fluid. Since the walls are adiabatic there is a heat transfer through convection but not through conduction, i.e. $P_Q = 0$ (Fig. 4.12). For this open system, show that the chemical power P_C generated by the matter flow can be written as,

$$P_C = h^+ \dot{N}^+ - h^- \dot{N}^-$$

where $\dot{N}^+$ and $\dot{N}^-$ are the rates of substance entering and exiting the system and h^+ and h^- are the molar enthalpies entering and exiting the system.

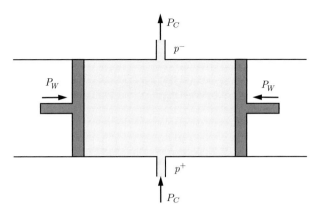

Figure 4.12 Two pistons slide in a cylinder that contains a fluid that enters and exits the system. The pressure at the entrance is p^+ and the pressure at the exit is p^-. The mechanical power generated by the pistons on the system is P_W and the chemical power generated by the matter flows is P_C.

Part II

PHENOMENOLOGY

5 Calorimetry

Benoît Paul Emile Clapeyron, 1799–1864

B. P. E. Clapeyron was a Frenchman who today is known for his representation
of thermal cycles by (p, V) diagrams. He found a relation between the slope of a $p\,(T)$ line
in a phase diagram and the corresponding latent heat.

5.1 Historical Introduction

In this chapter, we will introduce the notion of specific heat, also known as heat capacity.
Antoine Laurent de Lavoisier and Pierre-Simon de Laplace designed an experiment that
allowed them to quantify heat phenomena [42]. They built an ice *calorimeter* (Fig. 5.1)

Figure 5.1 Calorimeter of Lavoisier and Laplace. When a thermodynamic process takes place inside the inner chamber, heat is released which makes the ice melt. The heat transfer is quantified by the amount of water collected.

which they used to determine the heat released in a chemical reaction. The reaction took place inside a sphere of ice kept at $0°C$. From the amount of ice that melted in that process, they measured the heat of the reaction, also called ***reaction enthalpy***.

The heat transfer that occurred when a set amount of gas passed through the ice was measured by the amount of ice that melted. They also recorded the temperature of the gas at the inlet and the outlet. The heat value thus obtained, divided by the temperature drop, yielded the heat capacity or the ***specific heat*** of the gas.

These famous scientists also used their calorimeter to analyse the breathing of animals and concluded from their observations that breathing must be a combustion process.

Experimentally, it is observed that the specific heat of any substance decreases to zero as the temperature tends to zero. As we will see in this chapter, this general property can be derived from what is known as the ***third law of thermodynamics***. It was formulated by Walther Nernst in 1904. After his studies in Zürich, Berlin and Graz, he founded the Institute of physical chemistry and electrochemistry in Göttingen and received the Nobel Prize in 1920 in recognition for his work on thermochemistry. The third law was important, because at the time it was formulated, physicists faced a challenge concerning specific heat. Pierre Louis Dulong (1785–1838) and Alexis Thérèse Petit (1791–1820) established an empirical law according to which the specific heat of a metal depends on its molar mass, but not on its temperature. This temperature independence could be derived from statistical physics, using the physical concepts known at the turn of the twentieth century. This failure to predict the correct temperature dependence was one of the fundamental problems physics faced at the beginning of the twentieth century. Einstein resolved the issue by a derivation of the specific heat that was based on quantum mechanical arguments.

In this chapter, we will illustrate the notion of specific heat and of ***latent heat*** with the ***ideal gas*** model. The ideal gas equation of state can be derived from empirical laws established by Robert Boyle (1621–1697), Edme Mariotte (1620–1684), Jacques Charles (1746–1823), Louis Gay-Lussac (1778–1850) and Amedeo Avogadro (1776–1856). The Boyle–Mariotte law states that at a constant temperature T, the pressure p of a dilute gas is inversely proportional to its volume V [43, 44]. Charles' law states that at constant pressure

p, the volume V of a dilute gas is proportional to its temperature T [45]. The law of Gay-Lussac states that at fixed volume V, the pressure p of a dilute gas is proportional to its temperature T [46]. Avogadro's law states that at constant temperature T and pressure p, the volume V of N moles is the same for all dilute gases [47]. These laws imply that for a dilute gas, consisting of only one substance, the product of the pressure p and the volume V is proportional to the product of the number of moles N and the temperature T of the gas. This can be expressed by the **ideal gas equation of state**.

5.2 Thermal Response Coefficients

Let us consider a simple closed thermodynamic system consisting of N moles of a single substance. The system is subjected to a heat transfer that can be considered reversible, in other words, characterised by zero entropy production rate, i.e. $\Pi_S = 0$. The **thermal response coefficients** are empirical coefficients that characterise the response of the system to a heat transfer.

5.2.1 Temperature T and Volume V as State Variables

In order to characterise the thermal response of a system that is described by the state variables temperature T and volume V, we treat entropy $S(T, V)$ and pressure $p(T, V)$ as functions of T and V. In terms of the state variables T and V, the first law (2.32) reads,

$$P_Q + P_W = T\dot{S}(T, V) - p(T, V)\,\dot{V} \tag{5.1}$$

and expression (2.40) for a reversible heat transfer becomes,

$$P_W = -p(T, V)\,\dot{V} \tag{5.2}$$

After substitution of this expression of P_W in (5.1), the only power that remains is P_Q. We introduce the coefficients $C_V(T, V)$ and $L_V(T, V)$ to characterise the thermal response of the system that reads,

$$P_Q = T\dot{S}(T, V) \equiv C_V(T, V)\,\dot{T} + L_V(T, V)\,\dot{V} \tag{5.3}$$

$C_V(T, V)$ is called the **specific heat** at constant volume and $L_V(T, V)$ the **latent heat of expansion** [48].

The term 'specific' qualifies a heat transfer that changes the temperature of the system. Thus, the specific heat of a system is equal to the heat provided to the system in order to increase its temperature by one unit while the other state variables are constant.

It is possible to carry out a heat transfer without modifying the temperature of a system. This is called 'latent' heat. The latent heat of expansion (or of compression, see § 5.2.2) are equal to the heat that has to be provided to the system at constant temperature in order to generate an increase in volume by one unit (or an increase in pressure by one unit).

Given the definition (1.35) of the infinitesimal heat δQ, we can get an expression for δQ in terms of the thermal response coefficients when multiplying equation (5.3) by an infinitesimal time interval dt:

$$\delta Q = T\,dS\,(T,V) \equiv C_V\,(T,V)\,dT + L_V\,(T,V)\,dV \tag{5.4}$$

In order to express $C_V\,(T,V)$ and $L_V\,(T,V)$ in terms of partial derivatives of the fundamental state functions entropy $S\,(T,V)$ and internal energy $U\,(S,V)$, we write δQ as,

$$\delta Q = T\,dS\,(T,V) = T\left(\frac{\partial S\,(T,V)}{\partial T}\,dT + \frac{\partial S\,(T,V)}{\partial V}\,dV\right) \tag{5.5}$$

Furthermore, we write the partial derivative of $U\left(S\,(T,V)\,,V\right)$ with respect to temperature at constant volume V by using definition (4.77) for the partial derivatives of functions of functions. Thus, we can write,

$$\left.\frac{\partial U}{\partial T}\right|_V = \frac{\partial U\left(S\,(T,V)\,,V\right)}{\partial S\,(T,V)}\frac{\partial S\,(T,V)}{\partial T} = T\frac{\partial S\,(T,V)}{\partial T} \tag{5.6}$$

where in the last step we took into account the definition (2.9) of T. In view of (5.6), we can write (5.5) as,

$$\delta Q = T\,dS\,(T,V) = \left.\frac{\partial U}{\partial T}\right|_V\,dT + T\frac{\partial S\,(T,V)}{\partial V}\,dV \tag{5.7}$$

By equating in (5.4) and (5.7) the terms proportional to dT and dV, we obtain,

$$C_V\,(T,V) = \left.\frac{\partial U}{\partial T}\right|_V = T\frac{\partial S\,(T,V)}{\partial T} \tag{5.8}$$

$$L_V\,(T,V) = T\frac{\partial S\,(T,V)}{\partial V} \tag{5.9}$$

According to the Maxwell relation (4.71), the latent heat of expansion $L_V\,(T,V)$ can be written as,

$$L_V\,(T,V) = T\frac{\partial p\,(T,V)}{\partial T} \tag{5.10}$$

In view of the importance of the partial derivatives of $S\,(T,V)$ in the description of the thermal response of a system, we define,

$$K_V\,(T,V) = \frac{C_V\,(T,V)}{T} = \frac{\partial S\,(T,V)}{\partial T} \tag{5.11}$$

$$\Lambda_V\,(T,V) = \frac{L_V\,(T,V)}{T} = \frac{\partial S\,(T,V)}{\partial V} \tag{5.12}$$

Then, the differential of the entropy $dS\,(T,V)$ reads,

$$dS\,(T,V) = K_V\,(T,V)\,dT + \Lambda_V\,(T,V)\,dV \tag{5.13}$$

$K_V\,(T,V)$ is called the **entropy capacity** at constant volume and $\Lambda_V\,(T,V)$, the **latent entropy of expansion**.

In physics, a **capacity** is always defined as the derivative of an extensive quantity with respect to the intensive variable that is the conjugate of this extensive quantity. For example, we will see in Chapter 9, the capacity of a capacitor is the ratio of its charge (an extensive quantity) to the electric potential (the intensive quantity conjugated to the charge).

5.2.2 Temperature T and Pressure p as State Variables

In order to characterise the thermal response coefficients of a system that is described by the state variables temperature T and pressure p, we treat entropy $S(T,p)$ and volume $V(T,p)$ as functions of T and p. With respect to the state variables T and p, the first law (2.32) reads,

$$P_Q + P_W = T\dot{S}(T,p) - p\dot{V}(T,p) \tag{5.14}$$

and the expression (2.40) for a reversible heat transfer becomes,

$$P_W = -p\dot{V}(T,p) \tag{5.15}$$

As we did in relation (5.3), we introduce coefficients describing the response of the system to the thermal power P_Q. Thus, by substituting (5.15) for P_W in relation (5.14), we write,

$$P_Q = T\dot{S}(T,p) \equiv C_p(T,p)\,\dot{T} + L_p(T,p)\,\dot{p} \tag{5.16}$$

where $C_p(T,p)$ is the **specific heat** of the system at constant pressure and $L_p(T,p)$, the **latent heat of compression** of the system [48]. Multiplying equation (5.16) by the infinitesimal time interval dt and taking into account definition (1.35) of the infinitesimal heat δQ, we obtain an expression for δQ in terms of the thermal response coefficients,

$$\delta Q = T\,dS(T,p) \equiv C_p(T,p)\,dT + L_p(T,p)\,dp \tag{5.17}$$

Furthermore, expressing the differential $dS(T,p)$ explicitly in terms of partial derivatives, the infinitesimal heat δQ reads,

$$\delta Q = T\,dS(T,p) = T\left(\frac{\partial S(T,p)}{\partial T}\,dT + \frac{\partial S(T,p)}{\partial p}\,dp\right) \tag{5.18}$$

The partial derivative of the enthalpy $H\big(S(T,p),p\big)$ with respect to temperature T at constant pressure p can be obtained by means of definition (4.77) for the partial derivative of a function of functions. Taking into account definition (2.9) for T, we thus find,

$$\frac{\partial H}{\partial T}\bigg|_p = \frac{\partial H\big(S(T,p),p\big)}{\partial S(T,p)}\frac{\partial S(T,p)}{\partial T} = T\frac{\partial S(T,p)}{\partial T} \tag{5.19}$$

Using (5.19), the expression (5.18) of the infinitesimal heat δQ becomes,

$$\delta Q = T\,dS(T,p) = \frac{\partial H}{\partial T}\bigg|_p dT + T\frac{\partial S(T,p)}{\partial p}\,dp \tag{5.20}$$

By comparing relations (5.17) and (5.20) and taking into account relation (5.19), we obtain explicit expressions for the thermal response coefficients, i.e.

$$C_p(T,p) = \frac{\partial H}{\partial T}\bigg|_p = T\frac{\partial S(T,p)}{\partial T} \tag{5.21}$$

$$L_p(T,p) = T\frac{\partial S(T,p)}{\partial p} \tag{5.22}$$

Using the Maxwell relation (4.75), the latent heat of compression $L_p(T,p)$ can be written as,

$$L_p(T,p) = -T \frac{\partial V(T,p)}{\partial T} \tag{5.23}$$

We define the **entropy capacity** $K_p(T,p)$ at constant pressure and the **latent entropy of compression** $\Lambda_p(T,p)$ by,

$$K_p(T,p) = \frac{C_p(T,p)}{T} = \frac{\partial S(T,p)}{\partial T} \tag{5.24}$$

$$\Lambda_p(T,p) = \frac{L_p(T,p)}{T} = \frac{\partial S(T,p)}{\partial p} \tag{5.25}$$

In terms of (5.24) and (5.25), the entropy differential $dS(T,p)$ reads,

$$dS(T,p) = K_p(T,p)\, dT + \Lambda_p(T,p)\, dp \tag{5.26}$$

5.2.3 Molar Specific Heat and Specific Heat per Unit Mass

The specific heat $C_V(T,V)$ and its counterpart $C_p(T,p)$ are extensive quantities, as can readily be seen by their expressions (5.8) and (5.21) as derivatives of thermodynamic potentials. A **molar specific heat** is the specific heat of a mole of substance. Thus,

$$c_V(T,V) = \frac{C_V(T,V)}{N} = \frac{1}{N} \left. \frac{\partial U}{\partial T} \right|_V \tag{5.27}$$

$$c_p(T,p) = \frac{C_p(T,p)}{N} = \frac{1}{N} \left. \frac{\partial H}{\partial T} \right|_p \tag{5.28}$$

where N is the number of moles in the system. A specific heat per unit mass is the specific heat of a unit mass of substance, i.e.

$$c_V^*(T,V) = \frac{C_V(T,V)}{M} = \frac{1}{M} \left. \frac{\partial U}{\partial T} \right|_V \tag{5.29}$$

$$c_p^*(T,p) = \frac{C_p(T,p)}{M} = \frac{1}{M} \left. \frac{\partial H}{\partial T} \right|_p \tag{5.30}$$

where M is the mass of the system.

5.3 Third Law of Thermodynamics

The **third law of thermodynamics** states that:

At the limit of absolute zero temperature, which one could never reach, the equilibrium entropy of a system tends towards a constant which is independent of all other intensive parameters. This constant ought to be taken as zero.

The third law requires that,

$$\lim_{T \to 0} S(T, V) = 0 \qquad \text{and} \qquad \lim_{T \to 0} S(T, p) = 0 \qquad (5.31)$$

Therefore, the integral of equations (5.13) and (5.26) with respect to the temperature yields,

$$S(T, V) = \int_0^T K_V(T', V)\, dT' = \int_0^T \frac{C_V(T', V)}{T'}\, dT'$$
$$S(T, p) = \int_0^T K_p(T', p)\, dT' = \int_0^T \frac{C_p(T', p)}{T'}\, dT' \qquad (5.32)$$

Thus, the third law (5.31) requires that,

$$\lim_{T \to 0} C_V(T, V) = 0 \qquad \text{and} \qquad \lim_{T \to 0} C_p(T, p) = 0 \qquad (5.33)$$

5.4 Mayer Relations

It is possible to find relationships among thermal response coefficients. These are often referred to as **Mayer relations**. Here we show a method of deriving them that might appear cumbersome, but that has the advantage of being systematic. We illustrate it by finding the Mayer relation between $C_V(T, V)$ and $C_p(T, p)$.

Let us consider a simple system described by the state variables temperature T and pressure p. Taking into account definition (4.29) of enthalpy H, we express it as a function of the state functions S and V, which are functions of the state variables T and p,

$$H\Big(S(T, p), p\Big) = U\Big(S\big(T, V(T, p)\big), V(T, p)\Big) + p\, V(T, p) \qquad (5.34)$$

According to definition (5.21), the specific heat at constant pressure is written,

$$C_p(T, p) = \left. \frac{\partial H\Big(S(T, p), p\Big)}{\partial T} \right|_p \qquad (5.35)$$

The derivative of equation (5.34) with respect to the temperature T while keeping the pressure p constant yields,

$$C_p(T, p) = \frac{\partial U\Big(S\big(T, V(T, p)\big), V(T, p)\Big)}{\partial S\big(T, V(T, p)\big)} \frac{dS\big(T, V(T, p)\big)}{dT}$$
$$+ \frac{\partial U\Big(S\big(T, V(T, p)\big), V(T, p)\Big)}{\partial V(T, p)} \frac{\partial V(T, p)}{\partial T} + p\, \frac{\partial V(T, p)}{\partial T} \qquad (5.36)$$

In view of the definitions (2.9) and (2.10) for temperature T and pressure p, we can write,

$$T = \frac{\partial U\Big(S\big(T, V(T,p)\big), V(T,p)\Big)}{\partial S\big(T, V(T,p)\big)}$$

$$p = -\frac{\partial U\Big(S\big(T, V(T,p)\big), V(T,p)\Big)}{\partial V(T,p)} \tag{5.37}$$

Making use of the definitions (5.37) in (5.36), we see that the last two terms on the right hand side of relation (5.36) cancel out and that equation (5.36) is reduced to,

$$C_p(T,p) = T\frac{dS\big(T, V(T,p)\big)}{dT} \tag{5.38}$$

The total derivative of entropy in relation (5.38) with respect to temperature can be expressed as a sum of partial derivatives, i.e.

$$\frac{dS\big(T, V(T,p)\big)}{dT} = \frac{\partial S\big(T, V(T,p)\big)}{\partial T} + \frac{\partial S\big(T, V(T,p)\big)}{\partial V(T,p)}\frac{\partial V(T,p)}{\partial T}$$

$$= \frac{\partial S(T, V)}{\partial T} + \frac{\partial S(T, V)}{\partial V}\frac{\partial V(T,p)}{\partial T} \tag{5.39}$$

Thus, equation (5.38) can be rewritten as,

$$C_p(T,p) = T\frac{\partial S(T, V)}{\partial T} + T\frac{\partial S(T, V)}{\partial V}\frac{\partial V(T,p)}{\partial T} \tag{5.40}$$

The specific heat at constant pressure is given by,

$$C_V(T, V) = T\frac{\partial S(T, V)}{\partial T} \tag{5.41}$$

Making use of the Maxwell relation (4.71) and of the specific heat (5.41) in relation (5.40), we obtain the Mayer relation,

$$C_p(T,p) = C_V(T, V) + T\frac{\partial p(T, V)}{\partial T}\frac{\partial V(T,p)}{\partial T} \tag{5.42}$$

5.5 Specific Heat of Solids

For many solids at high enough temperature, the molar specific heat at constant volume c_V is roughly given by,

$$c_V = 3R \tag{5.43}$$

where R is a constant defined in § 5.6. This expression for c_V is called the **Dulong–Petit law**. According to this law and definition (5.27), the specific heat C_V of a solid containing N moles is given by,

$$C_V = 3NR \tag{5.44}$$

Often, a variation in a solid internal energy U due to thermal expansion can be neglected in comparison to the energy variation due to a change of the solid temperature. Hence, combining relations (5.8) and (5.44), the differential of the internal energy can be written,

$$dU = 3NR\,dT \qquad (5.45)$$

When integrating (5.45), we obtain the internal energy of a solid in the Dulong–Petit approximation as,

$$U = 3NR\,T \qquad (5.46)$$

It is worth emphasising that the internal energy U is a state function of entropy S and volume V. The temperature in equation (5.46) should be understood as the function $T(S, V)$.

5.6 Ideal Gas

In this section, we define the ideal gas model, and in the next section we will determine the thermal response coefficients. The **ideal gas** is defined by the equation of state,

$$p\,V = NR\,T \qquad (5.47)$$

where N is the number of moles of the gas contained in the volume V, p the pressure, T the temperature and R is the **ideal gas constant** given by,

$$R = 8.31 \text{ J mol}^{-1}\text{ K}^{-1} \qquad (5.48)$$

The constant R is closely connected to the **Boltzmann constant** k_B via the Avogadro number $\mathcal{N}_A = 6.022 \cdot 10^{23}$, i.e.

$$R = \mathcal{N}_A\,k_B \qquad (5.49)$$

In order to derive an expression for the internal energy U of the ideal gas, we express the differential of the entropy S as a function of temperature T and volume V. That is, we write

$$dS = \frac{\partial S}{\partial T}\,dT + \frac{\partial S}{\partial V}\,dV \qquad (5.50)$$

Thus, the Gibbs relation (4.1) reads,

$$dU = T\frac{\partial S}{\partial T}\,dT + \left(T\frac{\partial S}{\partial V} - p\right)dV \qquad (5.51)$$

Taking into account definition (5.8) for the specific heat at constant volume C_V and the Maxwell relation (4.71), the internal energy differential (5.51) becomes,

$$dU = C_V\,dT + \left(T\frac{\partial p}{\partial T} - p\right)dV \qquad (5.52)$$

The ideal gas equation of state (5.47) implies that in equation (5.52) the terms in brackets cancel each other out. Hence, the internal energy differential (5.52) for the ideal gas is independent of the volume, i.e.

$$dU = C_V \, dT \tag{5.53}$$

In order to determine an expression for the enthalpy H of the ideal gas, we substitute in the enthalpy differential (4.32) the entropy differential expressed as a function of temperature T and pressure p,

$$dS = \frac{\partial S}{\partial T} \, dT + \frac{\partial S}{\partial p} \, dp \tag{5.54}$$

Hence, the enthalpy differential (4.32) reads,

$$dH = T\frac{\partial S}{\partial T} \, dT + \left(T\frac{\partial S}{\partial p} + V \right) dp \tag{5.55}$$

Taking into account definition (5.21) for the specific heat at constant pressure C_p and the Maxwell relation (4.75), the enthalpy differential (5.55) becomes,

$$dH = C_p \, dT + \left(-T\frac{\partial V}{\partial T} + V \right) dp \tag{5.56}$$

The ideal gas equation of state (5.47) implies that the terms inside the brackets of equation (5.56) cancel each other out. Hence, the enthalpy differential (5.56) of an ideal gas is independent of pressure, i.e.

$$dH = C_p \, dT \tag{5.57}$$

The linearity of the ideal gas equation of state (5.47) implies,

$$\frac{\partial p\,(T, V)}{\partial T} = \frac{p}{T} \quad \text{and} \quad \frac{\partial V\,(T, p)}{\partial T} = \frac{V}{T} \tag{5.58}$$

Thus, by using relations (5.58) and the equation of state (5.47) in the Mayer relation (5.42), we find that for an ideal gas,

$$C_p = C_V + NR \tag{5.59}$$

Statistical physics provides further insight about the ideal gas. It can be shown that for an ideal gas, the interactions between the elementary constituents of the gas (atoms or molecules) can be modelled by elastic collisions (kinetic energy conservation). Furthermore, for an ideal gas, the volume occupied by the constituents is negligible with respect to the volume of gas. Dilute gases follow the ideal gas model when their pressure is sufficiently low. In general, real gases can be described by the van der Waals model that will be presented in § 6.10.

5.7 Thermal Response Coefficients of the Ideal Gas

The specific heat at constant volume C_V of the ideal gas is the amount of heat needed to increase its temperature by dT while its volume V is kept constant. Such a process is called **isochoric**. Experimentally, it is found that the specific heat at constant volume C_V of a gas approaching the ideal gas model is independent of temperature T. C_V is also proportional to the number N of moles of gas. Therefore, the specific heat at constant volume C_V of the ideal gas is given by,

$$C_V = c\,NR \tag{5.60}$$

where $c > 0$, a dimensionless parameter, depends on the nature of the elementary constituents of the gas. For a gas of atoms, we can take $c = 3/2$, for a gas of rigid diatomic molecules, $c = 5/2$. An argument of statistical physics can account for these values. Substituting the expression (5.60) for C_V in the internal energy differential (5.53), we obtain,

$$dU = c\,NR\,dT \tag{5.61}$$

When integrating the differential (5.61), we obtain the internal energy of the ideal gas,

$$U = c\,NR\,T \tag{5.62}$$

The specific heat at constant pressure C_p of the ideal gas is the amount of heat needed to increase its temperature by dT while its pressure p is kept constant. Such a process is called **isobaric**. Taking into account the Mayer relation (5.59) and the expression (5.60) for the specific heat at constant volume C_V, the specific heat at constant pressure C_p reads,

$$C_p = (c + 1)\,NR \tag{5.63}$$

It is to be expected that the value of the specific heat depends on the constraints under which the heat transfer is carried out. When the temperature increases by dT while the pressure p of the gas is kept constant, the volume of the gas increases by an amount dV, as implied by the equation of state $pV = NRT$. Thus, there is a negative work $-p\,dV$ associated with this isobaric heat transfer. In view of (5.62), dT determines the value of $dU = \delta Q + \delta W$, and therefore, the heat provided to the ideal gas must be greater for this isobaric process ($\delta W \le 0$) than for the isochoric process ($\delta W = 0$).

Substituting the expression (5.63) for C_p in the enthalpy differential of the ideal gas (5.57), we obtain,

$$dH = (c + 1)\,NR\,dT \tag{5.64}$$

When integrating the differential (5.64), we obtain the enthalpy of the ideal gas,

$$H = (c + 1)\,NR\,T \tag{5.65}$$

From definitions (5.11) and (5.24) of the entropy capacities and from relations (5.60) and (5.63), we can derive expressions for the entropy capacities of the ideal gas,

$$K_V(T) = \frac{C_V}{T} = \frac{c\,NR}{T} \tag{5.66}$$

$$K_p(T) = \frac{C_p}{T} = \frac{(c + 1)\,NR}{T} \tag{5.67}$$

The ratio of the specific heat at constant pressure and the specific heat at constant volume depends only on the dimensionless parameter c. This ratio is called the **gamma coefficient** of the ideal gas,

$$\frac{C_p}{C_V} = \frac{K_p}{K_V} = \frac{c+1}{c} = \gamma > 1 \tag{5.68}$$

We now turn to the latent heat and latent entropy coefficients. Using definition (5.10) for the latent heat of expansion and the ideal gas equation of state (5.47), we find,

$$L_V(T, V) = T \frac{\partial p(T, V)}{\partial T} = \frac{NRT}{V} = p(T, V) \tag{5.69}$$

The latent heat of expansion $L_V(T, V) > 0$ is the amount of heat provided to the ideal gas in order to increase its volume by dV while its temperature T is kept constant. Such a process is called **isothermal**.

Using definition (5.23) of the latent heat of compression and the ideal gas equation of state (5.47), we find,

$$L_p(T, p) = -T \frac{\partial V(T, p)}{\partial T} = -\frac{NRT}{p} = -V(T, p) \tag{5.70}$$

The latent heat of compression $L_p(T, p) < 0$ is the amount of heat provided to the ideal gas in order to increase its pressure by dp when its temperature T is kept constant.

From definitions (5.12) and (5.25) of latent heat and from relations (5.69) and (5.70), we find expressions for the latent entropies of expansion and compression,

$$\Lambda_V(V) = \frac{L_V(T, V)}{T} = \frac{p(T, V)}{T} = \frac{NR}{V} \tag{5.71}$$

$$\Lambda_p(p) = \frac{L_p(T, p)}{T} = -\frac{V(T, p)}{T} = -\frac{NR}{p} \tag{5.72}$$

5.8 Entropy of the Ideal Gas

Using the thermal response coefficients that we established in the previous section, we will now determine expressions for the entropy of the ideal gas as a function of either T and V, T and p, or V and p.

5.8.1 Entropy S as a Function of T, p and V

Let us begin with temperature T and volume V as state variables. Using the entropy capacity (5.66) and the latent entropy of expansion (5.71), the entropy differential (5.13) of the ideal gas can be written as,

$$dS = c\,NR\,\frac{dT}{T} + NR\,\frac{dV}{V} \tag{5.73}$$

For a process that goes from an initial state i to a final state f, the entropy variation ΔS_{if} is defined as,

$$\Delta S_{if} \equiv \int_{S_i}^{S_f} dS \tag{5.74}$$

Thus, the integral of equation (5.73) from the initial state i to the final state f yields,

$$\Delta S_{if} = c\,NR \int_{T_i}^{T_f} \frac{dT}{T} + NR \int_{V_i}^{V_f} \frac{dV}{V} \tag{5.75}$$

Computing the integrals of the left-hand side of this equation, we obtain

$$\Delta S_{if} = c\,NR \ln\left(\frac{T_f}{T_i}\right) + NR \ln\left(\frac{V_f}{V_i}\right) \tag{5.76}$$

In terms of the γ coefficient, expression (5.76) can be reformulated as,

$$\Delta S_{if} = c\,NR \left(\ln\left(\frac{T_f}{T_i}\right) + \ln\left(\frac{V_f}{V_i}\right)^{\frac{1}{c}} \right) = c\,NR \ln\left(\frac{T_f V_f^{\gamma-1}}{T_i V_i^{\gamma-1}}\right) \tag{5.77}$$

Thus, for the case of an **isentropic** process, which is defined as a process taking place at constant entropy, i.e. $\Delta S_{if} = 0$, relation (5.77) imposes the following condition,

$$T V^{\gamma-1} = \text{const} \tag{5.78}$$

Now we turn to temperature T and pressure p as state variables. Substituting for the entropy capacity (5.67) and for the latent entropy of compression (5.72) in (5.26), the entropy differential of the ideal gas becomes,

$$dS = (c+1)\,NR\,\frac{dT}{T} - NR\,\frac{dp}{p} \tag{5.79}$$

In a process going from the initial state i to the final state f, the entropy variation ΔS_{if} can be written as,

$$\Delta S_{if} \equiv \int_{S_i}^{S_f} dS = (c+1)\,NR \int_{T_i}^{T_f} \frac{dT}{T} - NR \int_{p_i}^{p_f} \frac{dp}{p} \tag{5.80}$$

which after computing the integrals yields,

$$\Delta S_{if} = (c+1)\,NR \ln\left(\frac{T_f}{T_i}\right) - NR \ln\left(\frac{p_f}{p_i}\right) \tag{5.81}$$

Using once again the definition of the γ coefficient, expression (5.81) can be written as,

$$\Delta S_{if} = c\,NR \left(\ln\left(\frac{p_f}{p_i}\right)^{-\frac{1}{c}} + \ln\left(\frac{T_f}{T_i}\right)^{\frac{c+1}{c}} \right) = c\,NR \ln\left(\frac{p_f^{1-\gamma} T_f^{\gamma}}{p_i^{1-\gamma} T_i^{\gamma}}\right) \tag{5.82}$$

Thus, during an isentropic process, for which $\Delta S_{if} = 0$, relation (5.82) requires that,

$$p^{1-\gamma} T^{\gamma} = \text{const} \tag{5.83}$$

The expressions (5.76) and (5.81) for entropy are equivalent. Indeed, the ideal gas equation of state (5.47) implies that,

$$\ln\left(\frac{V_f}{V_i}\right) = \ln\left(\frac{T_f\,p_i}{p_f\,T_i}\right) = \ln\left(\frac{T_f}{T_i}\right) - \ln\left(\frac{p_f}{p_i}\right) \qquad (5.84)$$

which establishes this equivalence.

Finally, we use volume V and pressure p as state variables. Differentiating the ideal gas equation of state (5.47) yields the following identity,

$$\frac{dT}{T} = \frac{p\,dV + V\,dp}{NRT} = \frac{dV}{V} + \frac{dp}{p} \qquad (5.85)$$

Taking into account relation (5.85), the entropy differentials (5.81) or (5.76) can be turned into the differential of a function of V and p,

$$dS = (c+1)\,NR\,\frac{dV}{V} + c\,NR\,\frac{dp}{p} \qquad (5.86)$$

For a process that goes from the initial state i to the final state f, the entropy variation ΔS_{if} thus becomes,

$$\Delta S_{if} \equiv \int_{S_i}^{S_f} dS = (c+1)\,NR\,\int_{V_i}^{V_f} \frac{dV}{V} + c\,NR\,\int_{p_i}^{p_f} \frac{dp}{p} \qquad (5.87)$$

which after integration yields,

$$\Delta S_{if} = (c+1)\,NR\,\ln\left(\frac{V_f}{V_i}\right) + c\,NR\,\ln\left(\frac{p_f}{p_i}\right) \qquad (5.88)$$

In terms of the γ coefficient, expression (5.88) can be rewritten as,

$$\Delta S_{if} = c\,NR\left(\ln\left(\frac{p_f}{p_i}\right) + \ln\left(\frac{V_f}{V_i}\right)^{\frac{c+1}{c}}\right) = c\,NR\,\ln\left(\frac{p_f\,V_f^{\gamma}}{p_i\,V_i^{\gamma}}\right) \qquad (5.89)$$

Thus, during an isentropic process, $\Delta S_{if} = 0$ and relation (5.89) requires that,

$$p\,V^{\gamma} = \text{const} \qquad (5.90)$$

The numerous expressions obtained in this section will be frequently referred to in Chapter 7. In the next section, we illustrate the significance of these results with three examples. In the first one, we consider a simple experiment in which the entropy of a gas is changed at constant temperature. In the second example, we derive an expression for pressure $p(S, V)$ as a function of entropy S and volume V. In the third example, we point out that expressions such as (5.76) and (5.81), although they include the number of moles N, are not expressions of S as a function of N.

5.8.2 Reversible Isothermal Processes

We consider an ideal gas contained within a cylinder-piston device such as the syringe illustrated in Fig. 5.2. While pulling on the piston, the user can monitor the gas temperature T measured by a thermocouple. Let us assume that the user pulls on the piston so slowly that, during this expansion, the gas is in thermal equilibrium at all times with the

Figure 5.2   A large syringe made of glass contains a fixed quantity of gas. A thermocouple is used to measure the temperature of the gas.

environment. Thus, this process is isothermal. At constant temperature, equations (5.76) and (5.81) imply that the entropy variation ΔS_{if} is given by,

$$\Delta S_{if} = NR \ln \left(\frac{V_f}{V_i} \right) = - NR \ln \left(\frac{p_f}{p_i} \right) \tag{5.91}$$

Therefore, during an isothermal process, the ideal gas satisfies the condition,

$$p V = \text{const} \tag{5.92}$$

which can be derived more readily from the ideal gas equation of state (5.47).

5.8.3 Entropy and Pressure

In order to express the pressure $p\,(S, V)$ of the ideal gas as a function of the state variables entropy S and volume V, we consider an initial state i of pressure $p_i = p_0\,(S_0, V_0)$ assumed to be known and a final state f of pressure $p_f = p\,(S, V)$ to be determined. Equation (5.89) can be written as,

$$S - S_0 = c\,NR \ln \left(\frac{p\,V^\gamma}{p_0\,V_0^\gamma} \right) \tag{5.93}$$

When inverting relation (5.93), we obtain the expression we are looking for,

$$p\,(S, V) = p_0\,(S_0, V_0) \left(\frac{V_0}{V} \right)^\gamma \exp \left(\frac{S - S_0}{c\,NR} \right) \tag{5.94}$$

5.8.4 Entropy and Amount of Substance

Let us consider an initial state i of known entropy $S_i = S_0\,(T_0, V_0)$ and a final state f of entropy $S_f = S\,(T, V)$ to be determined. Equation (5.89) then takes the following form,

$$S\,(T, V) - S_0\,(T_0, V_0) = c\,NR \ln \left(\frac{T\,V^{\gamma - 1}}{T_0\,V_0^{\gamma - 1}} \right) \tag{5.95}$$

Relation (5.95) provides the entropy variation as a function of temperature and volume for a fixed number of moles of substance N. Taking into account definition (5.68) of the γ coefficient, equation (5.95) can be rewritten as,

$$S\,(T, V) = NR \ln \left(\frac{T^c\,V}{\alpha} \right) \tag{5.96}$$

where the coefficient α is given by,

$$\alpha = T_0^c \, V_0 \, \exp\left(-\frac{S_0}{NR}\right) \tag{5.97}$$

Expression (5.96) for $S(T, V)$ contains the number N of moles of substance, but this result cannot be interpreted as an analytical expression of the state function $S(T, V, N)$. The notation $S(T, V)$ is meant to convey that we are not writing the state function $S(T, V, N)$. In order to obtain an explicit expression of $S(T, V, N)$, a process would have to be carried out on an open system that would increase its number of moles N. This process could not be used to determine $S(T, V, N)$, because it would be irreversible since the chemical potentials inside and outside have to be different (see Chapter 3). Nonetheless, from fundamental principles, we can say that when N is a state variable, the parameter $\alpha(N)$ must be a function of the number of moles N. The extensive characters of entropy S, volume V and number of moles N implies,

$$S(T, \lambda V, \lambda N) = \lambda S(T, V, N) \tag{5.98}$$

Condition (5.98) and equation (5.96) require that the coefficient $\alpha(N)$ must be of the form,

$$\alpha(N) = \Phi N \tag{5.99}$$

where Φ is a coefficient that is independent of T, V and N. Substituting (5.99) for $\alpha(N)$ in expression (5.97) for entropy yields,

$$S(T, V, N) = NR \, \ln\left(\frac{T^c V}{\Phi N}\right) \tag{5.100}$$

This result can be used to find an expression for temperature $T(S, V, N)$ by solving relation (5.100) for T. Then, the internal energy equation of state (5.62) becomes a function of S, V and N.

5.9 Worked Solutions

5.9.1 Temperature Rise Due to an Impact

A solid of mass M falls from a height h, hits the ground and stays there. We assume that at the point of impact there is no macroscopic deformation of the solid and that no heat transfer takes place between the solid and the ground. We denote by i the initial state of the solid, that is, before impact and by f the state just after impact. Determine the temperature variation ΔT_{if} caused by the impact.

Solution:

Owing to the conservation of mechanical energy before the impact, the kinetic energy just before the impact is equal to the gravitational potential energy, characterised by the constant g. After the impact, the kinetic energy is entirely converted into internal energy. Therefore, the internal energy variation is given by,

$$\Delta U_{if} = Mgh$$

Owing to (5.46), the internal energy variation is,

$$\Delta U_{if} = 3NR\,\Delta T_{if}$$

Thus, the temperature variation of the solid upon impact is estimated as,

$$\Delta T_{if} = \frac{Mgh}{3NR}$$

5.9.2 Measuring the Specific Heat of Water

A few students heat up water with an electric immersion water heater, presumed to be made of N moles of iron. Using a thermometer, they record the temperature $T(t)$ of the water as a function of time. They notice a linear increase in time, i.e.

$$T(t) = T_0 + \alpha t$$

where T_0 is the ambient temperature and $\alpha > 0$ is a positive constant. The electric power of the heater is presumed to be entirely converted into heating power P_Q due to Joule heating (see § 11.4.11). The increase in the volume of water is negligible and its specific heat C_V is independent of temperature.

1. Determine an expression for the specific heat of water C_V at constant volume in terms of the thermal power P_Q of the heater and the experimental coefficient α, taking into account the fact that the heater must also be heated.
2. Determine an expression for the entropy variation ΔS of the water during the time interval Δt in terms of C_V and α.

Solution:

1. *The specific heat at constant volume is the sum of that of the heater, given by equation (5.44), and that of the water. Thus, according to relation (5.3) at constant volume, i.e. $\dot{V} = 0$, the thermal power becomes,*

$$P_Q = (C_V + 3NR)\,\dot{T}$$

From these two equations and from the time dependence of the temperature $T(t)$, we deduce an expression for the specific heat of water,

$$C_V = \frac{P_Q}{\alpha} - 3NR$$

2. *Using equation (5.32), the entropy variation ΔS writes,*

$$\Delta S = C_V \int_{T_0}^{T_0 + \alpha\,\Delta t} \frac{dT}{T} = C_V \ln\left(1 + \alpha\,\frac{\Delta t}{T_0}\right)$$

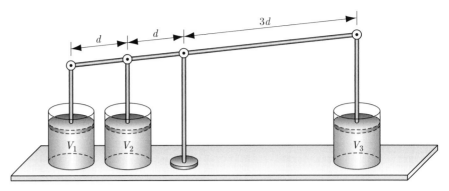

Figure 5.3   Three cylinders contain N moles of gas each. The table ensures a uniform and constant temperature T in the three cylinders.

5.9.3 Three Cylinders

Three cylinders labelled i (where $i = 1, 2, 3$), each of section A and each containing N moles of ideal gas (Fig. 5.3), are linked to a table that ensures a thermal contact between the three cylinders. The system is thus maintained at constant temperature T. The pistons, which keep the gas inside each cylinder, are mounted on a lever. The mass of the lever and the thermal transfers with this mechanical system are negligible.

1. Determine the norm F_i of the forces exerted by the ith piston on the lever.
2. Applying a principle of classical mechanics, establish the conditions on the pressure values p_i when the lever is horizontal at equilibrium.
3. Determine the relations between the infinitesimal variations of volume dV_i ($i = 1, 2, 3$) imposed by the lever.
4. Determine the infinitesimal variation of internal energy dU of the system during an infinitesimal motion of the lever.
5. Determine the infinitesimal variation of entropy dS of the system during an infinitesimal motion of the lever, using the equilibrium condition for the pressure in each cylinder.

Solution:

1. *The N moles of gas contained in the ith cylinder satisfy the ideal gas equation of state (5.47), i.e.*

$$p_i V_i = NR T$$

   The norm F_i exerted on the lever by the N moles of gas contained in the ith cylinder is,

$$F_i = \|\boldsymbol{F}_i\| = p_i A = \frac{NR TA}{V_i}$$

2. *The mechanical condition implied by the equilibrium is that the net torque $\boldsymbol{M}_i^{ext}$ with respect to the fulcrum of the lever must vanish. When the lever is horizontal, this condition implies,*

$$\sum_{i=1}^{3} M_i^{\text{ext}} = 0 \quad \Rightarrow \quad (2F_1 + F_2 - 3F_3)\, d = 0$$

Dividing this condition by Ad, we obtain a relation between the pressures in each cylinder, i.e.

$$2p_1 + p_2 - 3p_3 = 0$$

3. *The infinitesimal variations dV_i of the volume of gas in the ith cylinder is given by,*

$$dV_i = A\, dh_i \tag{5.101}$$

where dh_i is the height variation and A is the cross-section of the piston. The data on the figure imply that,

$$dh_1 = 2dh_2 \quad \text{and} \quad dh_3 = -3dh_2$$

which in turn implies that after multiplication by the cross-section A,

$$dV_1 = 2dV_2 \quad \text{and} \quad dV_3 = -3dV_2$$

Therefore, the total volume of gas V is constant, i.e.

$$dV_1 + dV_2 + dV_3 = 0 \quad \text{and thus} \quad V = V_1 + V_2 + V_3 = \text{const}$$

4. *Taking into account the extensive character of the internal energy, during an infinitesimal isothermal process (i.e. $dT = 0$), the internal energy variation dU_i of the gas contained in the ith cylinder vanishes. This implies that the infinitesimal internal energy variation (5.62) of the system vanishes, i.e.*

$$dU_i = c\, NR\, dT = 0 \quad \text{and thus} \quad dU = \sum_{i=1}^{3} dU_i = 0$$

5. *Taking into account the Gibbs relation (4.1) and the extensive characters of entropy and volume during an infinitesimal isothermal process associated with the motion of the lever, the infinitesimal entropy variation dS of the system is given by,*

$$dS = \sum_{i=1}^{3} dS_i = \sum_{i=1}^{3} \frac{p_i\, dV_i}{T} = NR \sum_{i=1}^{3} \frac{dV_i}{V_i} = NR \left(\frac{dV_1}{V_1} + \frac{dV_2}{V_2} + \frac{dV_3}{V_3} \right)$$

Thus, we find that the energy variation of the system during an infinitesimal change in position of the lever vanishes, i.e.

$$dS = NR\, dV_2 \left(\frac{2}{V_1} + \frac{1}{V_2} - \frac{3}{V_3} \right) = \frac{dV_2}{T} \left(2p_1 + p_2 - 3p_3 \right) = 0$$

5.9.4 Reversible and Irreversible Adiabatic Compression

A vertical cylinder contains N moles of ideal gas. It is closed by a piston of cross-section A and mass M. The gas and the piston constitute an adiabatic closed system. It is initially at equilibrium. An additional mass M' is placed on the piston (Fig. 5.4). The initial state when

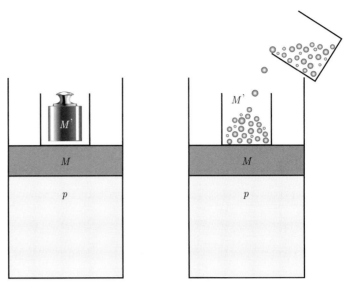

Figure 5.4 Left: a mass M' is placed suddenly on the piston of mass M that was initially in equilibrium with a gas at pressure p. Right: the mass M' is slowly poured on the piston.

the additional mass M' is just placed on the piston is labelled i and the final state when the piston does not move anymore is labelled f. The atmospheric pressure is considered negligible in comparison to the pressure due to the mass M.

1. Show that the process by which the mass M' is added all at once is irreversible.
2. For this irreversible process, determine the ratio of the heights of the piston, i.e. h_f/h_i.
3. For this irreversible process, determine the entropy variation ΔS_{if} and show that $\Delta S_{if} > 0$.
4. Show that the adiabatic compression of the gas is reversible and isentropic if the mass is poured slowly enough so that the pressure of the gas inside the cylinder is equal to the pressure outside at all times.
5. For this reversible process, determine the ratio of the heights of the pistons, i.e. h_f/h_i.
6. For this reversible process, show that the entropy variation vanishes, $\Delta S_{if} = 0$.

Notice that for an irreversible process, i.e. $\Pi_S > 0$, the evolution of an adiabatic closed system ($P_Q = 0$), is not isentropic, since $\dot{S} = \Pi_S$. For a reversible process, i.e. $\Pi_S = 0$, the evolution of an adiabatic closed system is isentropic since $\dot{S} = 0$.

Solution:

1. *The entropy production rate is given by* (2.37)*:*

$$\Pi_S = \frac{\dot{V}}{T}\left(p - p^{\text{ext}}\right) \geq 0$$

Initially, before the additional mass M' is added, the system consisting of the piston and the gas is at equilibrium. This implies that the initial pressure p_i of the ideal gas (5.47) *is equal to the pressure p_i^{ext} exerted by the piston (since the atmospheric pressure is negligible),*

$$p_i = p_i^{ext} \quad \text{where} \quad p_i = \frac{NR\, T_i}{V_i} \quad \text{and} \quad p_i^{ext} = \frac{Mg}{A}$$

This implies that,

$$NR\, T_i = \frac{Mg}{A}\, V_i$$

Finally, after the compression, when the piston is at rest, the system consisting of the piston and the gas is at equilibrium. This means that the final pressure p_f of the ideal gas is then equal to the final pressure p_f^{ext} exerted by the piston and the additional mass,

$$p_f = p_f^{ext} \quad \text{where} \quad p_f = \frac{NR\, T_f}{V_f} \quad \text{and} \quad p_f^{ext} = \frac{(M + M')\, g}{A}$$

This implies,

$$NR\, T_f = \frac{(M + M')\, g}{A}\, V_f$$

Thus, before and after the adiabatic compression, the system is at equilibrium and the entropy production rate vanishes, i.e. $\Pi_S = 0$. However, during the adiabatic compression, when $p_i < p < p_f$ and $p^{ext} = p_f^{ext}$, the entropy production rate is positive, i.e. $\Pi_S > 0$ as $p < p^{ext}$ for $\dot{V} < 0$ and $T > 0$. This implies that the adiabatic compression is irreversible.

2. *As the compression is adiabatic, the heat provided to the system vanishes, i.e. $Q_{if} = 0$. Thus, according to the first law (1.44), the internal energy variation of the gas is only due to the work W_{if} performed on the gas by the weight of the piston and the additional mass,*

$$\Delta U_{if} = W_{if}$$

The internal energy variation is given by,

$$\Delta U_{if} = c\, NR\, (T_f - T_i)$$

Taking into account the fact that the pressure is constant during the compression process, the work W_{if} performed on the gas by the weight of the piston and of the additional mass is expressed as,

$$W_{if} = -p_f^{ext}\, (V_f - V_i)$$

Thus, according to the first law,

$$c\, NR\, (T_f - T_i) = -\frac{(M + M')\, g}{A}\, (V_f - V_i)$$

This can be expressed in terms of the initial and final equilibrium states, respectively,

$$c\left((M + M')\, V_f - M V_i \right) = -(M + M')\, (V_f - V_i)$$

Finally, as $V_i = h_i A$ and $V_f = h_f A$, the piston height ratio yields,

$$\frac{h_f}{h_i} = \frac{V_f}{V_i} = \frac{1 + \frac{1}{c+1}\frac{M'}{M}}{1 + \frac{M'}{M}} < 1$$

3. *According to expression (5.88) for the ideal gas, the entropy variation ΔS_{if} reads,*

$$\Delta S_{if} = (c + 1)\, NR \ln\left(\frac{V_f}{V_i}\right) + c\, NR \ln\left(\frac{p_f}{p_i}\right)$$

Taking into account that,

$$\frac{p_f}{p_i} = \frac{p_f^{\text{ext}}}{p_i^{\text{ext}}} = 1 + \frac{M'}{M}$$

the entropy variation ΔS_{if} for this irreversible adiabatic process is,

$$\Delta S_{if} = (c + 1)\, NR \ln\left(\frac{1 + \frac{1}{c+1}\frac{M'}{M}}{1 + \frac{M'}{M}}\right) + c\, NR \ln\left(1 + \frac{M'}{M}\right)$$

$$= NR \ln\left(\frac{\left(1 + \frac{1}{c+1}\frac{M'}{M}\right)^{c+1}}{1 + \frac{M'}{M}}\right) > 0$$

The last step is obtained by performing a series expansion of the numerator and denominator.

4. *When the pressure p of the gas is equal to the outside pressure p^{ext}, the entropy production rate vanishes, i.e.*

$$\Pi_S = 0$$

Thus, the process is reversible. As the process is also adiabatic, i.e. $P_Q = 0$, the relation (2.40) implies that it is isentropic,

$$\dot{S} = 0$$

5. *For an isentropic process (5.90), pressure and volume satisfy the following condition,*

$$p_i V_i^\gamma = p_f V_f^\gamma$$

Thus, the volume ratio is,

$$\frac{V_f}{V_i} = \left(\frac{p_i}{p_f}\right)^{1/\gamma}$$

For a reversible compression, the outside pressures p_i^{ext} and p_f^{ext} are equal to the pressures p_i and p_f of the gas, respectively. Thus,

$$\frac{p_i}{p_f} = \frac{p_i^{\text{ext}}}{p_f^{\text{ext}}} = \frac{1}{1 + \frac{M'}{M}}$$

Therefore, the height ratio, which is equal to the volume ratio, can be written as,

$$\frac{h_f}{h_i} = \frac{V_f}{V_i} = \left(\frac{p_i}{p_f}\right)^{1/\gamma} = \left(\frac{1}{1 + \frac{M'}{M}}\right)^{1/\gamma} < 1$$

6. *According to equation (5.88) for the ideal gas, the entropy variation ΔS_{if} for a reversible adiabatic process vanishes,*

$$\Delta S_{if} = (c+1)\, NR \ln\left(\left(\frac{p_i}{p_f}\right)^{1/\gamma}\right) + c\, NR \ln\left(\frac{p_f}{p_i}\right) = 0$$

This exercise illustrates the point that the irreversibility of a process can be related to the difference in pressure between a system and its environment. Likewise, in § 3.2, we found that irreversibility was related to the difference in temperature between two subsystems, one of which could be considered as the environment of the other.

5.9.5 Measure of the γ Coefficient

To a large spherical container filled with gas, we add a fine tube of section A in which a steel ball of mass M can slide (Fig. 5.5). We wish to determine the γ coefficient of the gas. The ball is dropped in the tube and oscillates at a frequency f, which can easily be measured. This process is assumed reversible. However, the measurement of the motion of the ball is sufficiently fast that the process can be considered adiabatic. We denote by V_0 the volume and p_0 the pressure at equilibrium and by p^{ext} the external pressure, considered constant.

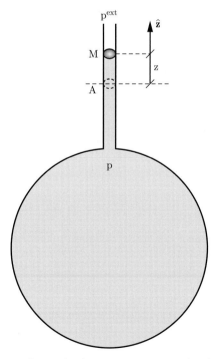

Figure 5.5   A steel ball of mass M oscillates vertically in a tube of cross-section A connected to a large container filled with gas.

1. Determine to first-order the volume $V(z)$ and the pressure $p(z)$ as a function of the vertical displacement z of the ball in the limit of small displacements, i.e. $Az \ll V_0$.
2. The elastic force exerted by the gas on the ball is given by $\boldsymbol{F} = (p(z) - p^{\text{ext}}) A \hat{\boldsymbol{z}}$. Determine the ball equation of motion.
3. Deduce from it the expression for the γ coefficient in terms of the frequency f of the oscillations around the equilibrium position, the pressure p_0 and the volume V_0 of the gas, and the cross-section A of the tube.

Solution:

1. *The volume of the gas $V(z)$ is the sum of the volume V_0 at equilibrium and of the change in volume Az due to the displacement z of the ball, i.e.*

$$V(z) = V_0 + Az$$

As the process is isentropic (5.90), the pressure and the volume satisfy the following relation,

$$p(z) V(z)^\gamma = p_0 V_0^\gamma$$

Performing a first-order series expansion in terms of the small dimensionless term Az/V_0, the pressure is given by,

$$p(z) = p_0 \left(\frac{1}{1 + \frac{A}{V_0} z} \right)^\gamma \simeq p_0 \left(1 - \gamma \frac{A}{V_0} z \right)$$

2. *Newton's second law applied to the ball yields,*

$$M\boldsymbol{g} + \boldsymbol{F} = M\boldsymbol{a}$$

Projecting this equation on a vertical axis of unit vector $\hat{\boldsymbol{z}}$ yields,

$$-Mg + (p(z) - p^{\text{ext}}) A = M\ddot{z}$$

This equation can be rewritten in the usual manner for a harmonic oscillator,

$$\ddot{z} + \frac{\gamma A^2 p_0}{MV_0} z - \frac{1}{M} \left((p_0 - p^{\text{ext}}) A - Mg \right) = 0$$

3. *The frequency f of the oscillations around the equilibrium position is given by,*

$$f = \frac{\omega}{2\pi} = \frac{1}{2\pi} \sqrt{\frac{\gamma A^2 p_0}{MV_0}}$$

When solving this equation for γ, we obtain,

$$\gamma = \frac{4\pi^2 f^2 MV_0}{A^2 p_0}$$

Exercises

5.1   **Heat Transfer as a Function of V and p**

The infinitesimal heat transfer δQ is expressed as a function of the state variables T and V in equation (5.4). It was done as a function of the state variables T and p in equation (5.17). Express the infinitesimal heat transfer δQ as a function of V and p.

5.2   **Bicycle Pump**

A bicycle pump takes a volume ΔV of air at atmospheric pressure p_0 and constant temperature T_0 and compresses it so that it enters a tyre that has a volume V_0. The air inside the tyre is initially at atmospheric pressure p_0 and can be considered as an ideal gas. Determine the number of times n the user has to pump air into the tyre to reach a pressure p_f. Assume that the pump is designed so that the air in the tyre is always at temperature T_0.

Numerical Application:

$V_0 = 50 \, \text{l}$, $\Delta V = 1.2 \, \text{l}$ and $p_f = 2.5 \, p_0$.

5.3   **Heat Transfer at Constant Pressure**

A gas container is thermally isolated except for a small hole that insures that the pressure inside the container is equal to the atmospheric pressure p_0. Initially, the container holds N_i moles of gas at a temperature T_i. The molar specific heat of the gas at constant pressure is c_p. The gas is heated up to a temperature T_f by a resistive coil in the cylinder. As the gas temperature rises, some of the gas is released through the small hole. Assume that for the gas remaining in the cylinder the process is reversible and neglect the specific heat of the heater. Determine:

a) the volume V_0 of the container.
b) the number of moles ΔN leaving the container in this process.
c) the heat transfer Q_{if} to accomplish this process.

Numerical Application:

$p_0 = 10^5 \, \text{Pa}$, $N_0 = 10 \, \text{moles}$, $T_0 = 273 \, \text{K}$, $c_p = 29.1 \, \text{J K}^{-1} \, \text{mol}^{-1}$, $T_f = 293 \, \text{K}$.

5.4   **Specific Heat of a Metal**

A metallic block of mass M is brought to a temperature T_0 and plunged into a calorimeter filled with a mass M' of water. The system consisting of the metallic block and the water container is considered as isolated. In this process, the water temperature rises from T_i to T_f, the equilibrium temperature. The specific heat of the water per unit mass is $c_{M'}^*$. Determine the specific heat per unit mass of the metal c_M^* in the temperature range used in this experiment. Consider that the water container is made of a material with a negligible specific heat.

Numerical Application:

$M = 0.5 \, \text{kg}$, $M' = 1 \, \text{kg}$, $T_0 = 120°\text{C}$, $T_i = 16°\text{C}$, $T_f = 20°\text{C}$ and $c_{M'}^* = 4187 \, \text{J kg}^{-1} \, \text{K}^{-1}$.

5.5 Work in Adiabatic Compression

An ideal gas undergoes a reversible adiabatic compression from an initial volume V_i and initial pressure p_i to a final pressure p_f. Determine the work W_{if} performed on the gas during this process.

Numerical Application:

$V_i = 1\,1, p_i = 5 \cdot 10^5$ Pa, $p_f = 2p_i$, $c = 5/2$ (see definition (5.62)).

5.6 Slopes of Isothermal and Adiabatic Processes

For an ideal gas, show that at any point on a Clapeyron (p, V) diagram, the absolute value of the slope is greater for an adiabatic process (A) than an isothermal process (I).

5.7 Adsorption Heating of Nanoparticles

The process whereby molecules bind to a metallic surface is called adsorption. Here, molecules are adsorbed on Pt nanoparticles. The specific heat of an average Pt nanoparticle is C_V. The heat transferred to an average Pt nanoparticle during the adsorption of molecules is Q_{if}. Determine the temperature increase $\Delta T_{if} = T_f - T_i$ of an average Pt nanoparticle, assuming that the system consisting of the nanoparticles and the gas is isolated.

Numerical Application:

$C_V = 1.4 \cdot 10^{-18}$ J K^{-1}, $Q_{if} = 6.5 \cdot 10^{-16}$ J.

5.8 Thermal Response Coefficients

The thermal response of a homogeneous system subjected to an infinitesimal heat transfer δQ is characterised by coefficients defined in equations (5.4) and (5.17) when either the state variables (T, V) or (T, p) are used.

a) Find a relation between the latent heat of expansion $L_V(T, V)$ and the latent heat of compression $L_p(T, p)$.

b) Express the latent heat of compression $L_p(T, p)$ in terms of the specific heat at constant volume $C_V(T, V)$ and the specific heat at constant pressure $C_p(T, p)$.

Phase Transitions

Johannes Diderik van der Waals, 1837–1923

While teaching in secondary school, J.D. van der Waals submitted a
doctoral thesis on a model for a gas of interacting molecules. In 1910,
he was awarded the Nobel prize in physics for this insight.

6.1 Historical Introduction

In this chapter, we will describe phase transitions such as melting ice. Heat needs to be
provided for this phase transition to occur. As the temperature remains constant in the
process, the transferred heat is a form of latent heat.

The notion of latent heat became clear during the eighteenth century. Jean André Deluc, born in Geneva in 1727, was a trader who explored the Alps and conducted experiments about ice melting. He observed that the temperature of ice remained constant when it was melting, despite the thermal energy input. He noticed that evaporation was also a process that required thermal energy at constant temperature [49]. Joseph Black made similar observations. This Scottish chemistry professor, born in Bordeaux in 1728, noticed the great amount of heat that comes into play in the process of melting ice, while the temperature appeared not to change at all. So, he proposed to call it *latent heat*. Thus, by the beginning of the nineteenth century, people discerned two aspects of heat: one which is manifest, as it provokes a raise in temperature, the other which is latent and strongly dependent on the chemical nature of the substance [50].

While some scientists studied how liquids evaporate, others attempted the *liquefaction* of gases. In 1823, Michael Faraday managed to liquify chlorine at $0°$C, thanks to a setup that allowed him to reach a very high pressure. Once the chlorine liquified, he would lower the pressure, the liquid would evaporate in part, which in turn cooled it down. He thus obtained chlorine at $-34.5°$C. Applying the same principle, Jean-Charles Thilorier liquified CO_2; lowering the pressure caused the solid we call nowadays dry ice. At ambient pressure, this solid transforms directly into a gas at $-78.5°$C, a phase transition called *sublimation*. He also mixed dry ice with ether and reached a temperature of $-110°$C.

Thomas Andrews studied the transition of CO_2 at high pressure from liquid to gas. He noticed that the greater the pressure was, the harder it was for him to distinguish the liquid from the gas. The distinction became impossible at a *critical point* of 73 atm and $31°$C.

Louis Cailletet liquified oxygen at a temperature of 90 K by first compressing a gas maintained at ambient temperature, then cooling it by an adiabatic expansion. In 1898, James Dewar liquified hydrogen, which has a boiling temperature of 20.3 K at ambient pressure. He designed a container with a double wall forming a vacuum jacket. Nowadays, these cryogenic vessels bear his name. Up to this point, liquefaction was achieved through a discontinuous process. A continuous process for the cooling of gases was then invented by Carl von Linde. He subjected the gases to a Joule–Thomson expansion after pre-cooling them by adiabatic expansion. The company founded by von Linde in 1879 continues to use this process today.

In this chapter, we introduce the *van der Waals equation of state* because it predicts a phase transition between the liquid and gaseous states of a substance. Van der Waals realised that two molecules attract each other when they are sufficiently close, even if they are neutral. Of course, they repel each other at very short distances. He obtained the equation of state that now bears his name by a statistical derivation. Boltzmann was full of admiration over his result and Maxwell, another founder of statistical physics, established a graphical method based on the van der Waals equation of state to determine the pressure at which the phase transition occurs for a given temperature.

Cryogenics refers to the techniques used to obtain very low temperatures. Reaching low temperatures can reveal all kinds of fascinating phase transitions. For example, Heike Kammerlingh-Onnes succeeded in liquifying helium at a temperature of 4.2 K in 1908. For this, he received the Nobel prize in physics in 1913. Thanks to the low temperature reached, he discovered that the electrical resistance of mercury, tin and lead dropped suddenly as the temperature was lowered sufficiently, a phenomenon called *superconductivity*. By decreasing the pression over the liquid helium, it is possible to reach temperatures lower than 4.2 K. At 2.19 K, liquid helium becomes *superfluid*, which means that its viscosity is zero. Peter Debye and William Francis Giauque independently invented cooling by *adiabatic demagnetisation* (see § 9.5). Nowadays, using this method temperatures as low as 3 mK can be reached. A temperature of $10 \, \mu$K can be maintained for days by means of a dilution refrigerator containing a mixture of two isotopes of helium, ^{3}He and ^{4}He. Heinz London demonstrated this principle in 1962. If in addition to using a dilution refrigerator, adiabatic demagnetisation is applied to the copper nuclei located in the coldest region of the cryostat, a temperature of $1.5 \, \mu$K can be reached [51].

6.2 The Concavity of Entropy

In order to engage in a thermodynamic description of phase transitions, we need to understand what thermodynamics says about the stability of a system. In order to do so, we first need to realise that thermodynamics imposes conditions on the curvatures of the state functions $U(S, V)$ and $S(U, V)$.

In this section, we show that the entropy S of a given amount of substance is a concave function of internal energy U and volume V. We consider an isolated system consisting of two subsystems labelled 1 and 2, separated by a wall. The state space of this system is the set of points characterised by entropy, internal energy and volume. When the wall is removed, the first law (1.30) implies that the internal energy of the system does not vary, nor does its volume. Therefore, when the system reaches a new equilibrium after the wall is removed, the internal energy U_0 and the volume V_0 are equal to the sum of the initial internal energies U_1 and U_2 and volumes V_1 and V_2 of the subsystems:

$$U_0 = U_1 + U_2 \qquad V_0 = V_1 + V_2 \tag{6.1}$$

According to the equilibrium condition of the second law, the entropy of the system grows towards a maximum, which is the entropy value that the system has at equilibrium. Therefore, the final value of entropy of the system, $S(U_0, V_0)$, is greater or equal to the sum of the initial values $S(U_1, V_1)$ and $S(U_2, V_2)$ of the subsystems,

$$S(U_0, V_0) \geq S(U_1, V_1) + S(U_2, V_2) \tag{6.2}$$

The equality in relation (6.2) refers to the situation when the subsystems in their initial states are already at equilibrium. If they are not, then relation (6.2) is a strict inequality [52].

When the subsystems have different values of internal energy, but the same volume, we write,

$$U_1 = U - \Delta U \qquad U_2 = U + \Delta U \qquad V_1 = V_2 = V \tag{6.3}$$

Using conditions (6.1) and (6.3), the extensivity of entropy implies,

$$S(U_0, V_0) = S(2U, 2V) = 2S(U, V) \tag{6.4}$$

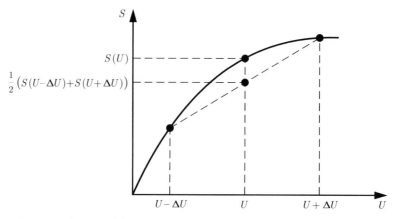

Figure 6.1 At constant volume V, the function $S(U)$ which has a negative or zero curvature, satisfies the concavity condition (6.5).

Taking into account conditions (6.3) and (6.4), relation (6.2) becomes,

$$S(U - \Delta U, V) + S(U + \Delta U, V) \leq 2S(U, V) \qquad (6.5)$$

This means that entropy S is a concave function of internal energy U (Fig. 6.1).

When the subsystems have the same value of internal energy but different volumes, we write,

$$U_1 = U_2 = U \qquad V_1 = V - \Delta V \qquad V_2 = V + \Delta V \qquad (6.6)$$

Taking into account conditions (6.6) and (6.4), relation (6.2) becomes,

$$S(U, V - \Delta V) + S(U, V + \Delta V) \leq 2S(U, V) \qquad (6.7)$$

This means that entropy S is a concave function of volume V.

When internal energy and volume differ in both subsystems, we write,

$$U_1 = U - \Delta U \qquad U_2 = U + \Delta U \qquad V_1 = V - \Delta V \qquad V_2 = V + \Delta V \qquad (6.8)$$

Taking into account conditions (6.8) and (6.4), relation (6.2) becomes,

$$S(U - \Delta U, V - \Delta V) + S(U + \Delta U, V + \Delta V) \leq 2S(U, V) \qquad (6.9)$$

This means that entropy S is a concave function of both internal energy U and volume V.

The conditions (6.5), (6.7) and (6.9) of the concavity of S are global conditions. Below, we will analyse phase transitions that are characterised by discontinuities in the partial derivatives of U, S and V. Then, it will be necessary to determine local conditions as well. A local condition is defined in the neighbourhood of a point in state space, whereas a global condition is defined over the entire state space. In order to establish the local conditions for the concavity of entropy, we now consider the limits $\Delta U \to 0$ and $\Delta V \to 0$ in the global conditions for the concavity of entropy. Passing the right-hand side term of relation (6.5) to the left-hand side, then dividing the equation resulting from it by ΔU^2 and finally, taking the limit $\Delta U \to 0$ yields,

$$\lim_{\Delta U \to 0} \frac{1}{\Delta U} \left(\frac{S(U+\Delta U, V) - S(U,V)}{\Delta U} - \frac{S(U,V) - S(U-\Delta U, V)}{\Delta U} \right) \leq 0 \quad (6.10)$$

In a similar way, passing the right-hand side term of relation (6.7) to the left-hand side, then dividing the result by ΔV^2 and taking the limit $\Delta V \to 0$ yields,

$$\lim_{\Delta V \to 0} \frac{1}{\Delta V} \left(\frac{S(U, V+\Delta V) - S(U,V)}{\Delta V} - \frac{S(U,V) - S(U, V-\Delta V)}{\Delta V} \right) \leq 0 \quad (6.11)$$

According to the definition (1.1) of partial derivatives of a function, relations (6.10) and (6.11) provide two local conditions for the concavity of entropy in the neighbourhood of equilibrium,

$$\frac{\partial^2 S}{\partial U^2} \leq 0 \quad \text{and} \quad \frac{\partial^2 S}{\partial V^2} \leq 0 \quad (6.12)$$

We leave it as an exercise (see § 6.11.1) to show that a second-order series expansion of condition (6.9) with respect to ΔU and ΔV, together with the local condition (6.12) provide a third local condition for the concavity of entropy in the neighbourhood of equilibrium [53],

$$\frac{\partial^2 S}{\partial U^2} \frac{\partial^2 S}{\partial V^2} - \left(\frac{\partial^2 S}{\partial U \partial V} \right)^2 \geq 0 \quad (6.13)$$

The concavity condition (6.13) ensures that the equation of state $S(U,V)$ describes locally a concave surface in the state space (S, U, V). The left-hand side of the inequality (6.13) corresponds to the **Gauss curvature** of the surface $S(U,V)$, up to a constant [54].

It is worth noting that the local concavity conditions (6.12) and (6.13), which have to be satisfied for $\Delta U \to 0$ and $\Delta V \to 0$, are less stringent that the global conditions (6.5), (6.7) and (6.9) that have to be satisfied for any value of ΔU and ΔV.

6.3 The Convexity of Internal Energy

Now we show that the internal energy U of a given amount of substance is a convex function of entropy S and volume V. We base our argument on the concavity of the entropy S as a function of internal energy U and volume V. At constant volume, the equation of state $U(S, V)$ is the inverse of the equation of state $S(U, V)$. Thus, these functions have a symmetry with respect to the bisector of the first quadrant in the (S, U) plane (Fig. 6.2). According to the first condition in (6.12), the curvature of the entropy function $S(U, V)$ with respect to internal energy U is negative. Therefore, in view of the symmetry with respect to the bisector, the curvature of the internal energy $U(S, V)$ as a function of entropy S is positive,

$$\frac{\partial^2 U}{\partial S^2} \geq 0 \quad (6.14)$$

In order to determine the curvature of U as a function of volume, we use Gibbs relation (4.1), which we rewrite as,

$$dS = \frac{1}{T} dU + \frac{p}{T} dV \quad (6.15)$$

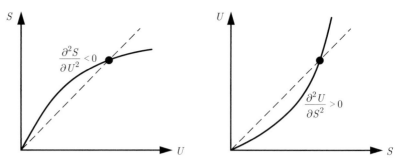

At constant volume, entropy S as a function of internal energy U has a negative curvature (left image). The internal energy U as a function of the entropy S can be obtained by a symmetry operation with respect to the dotted line (bisector). This implies that this function has a positive curvature (right image).

Since the entropy of the system $S(U, V)$ is a function of internal energy U and volume V, its differential is written as,

$$dS = \frac{\partial S}{\partial U} dU + \frac{\partial S}{\partial V} dV \tag{6.16}$$

Equating the terms proportional to dU and dV in expressions (6.15) and (6.16) of dS yields,

$$\frac{\partial S}{\partial U} = \frac{1}{T} \quad \text{and} \quad \frac{\partial S}{\partial V} = \frac{p}{T} \tag{6.17}$$

In view of definition (2.10) for pressure, the second identity (6.17) can be recast as,

$$\frac{\partial U}{\partial V} = -T \frac{\partial S}{\partial V} \tag{6.18}$$

The partial derivative of identity (6.18) with respect to volume V yields,

$$\frac{\partial^2 U}{\partial V^2} = -\frac{\partial T}{\partial V} \frac{\partial S}{\partial V} - T \frac{\partial^2 S}{\partial V^2} \tag{6.19}$$

The cyclic chain rule (4.81) applied to S, V and T reads,

$$\frac{\partial S}{\partial V} \frac{\partial V}{\partial T} \frac{\partial T}{\partial S} = -1 \quad \text{thus} \quad \frac{\partial S}{\partial V} = -\frac{\partial T}{\partial V} \frac{\partial S}{\partial T}$$

Using the chain rule (6.3) and the specific heat at constant volume (5.8), the condition (6.19) is rewritten as,

$$\frac{\partial^2 U}{\partial V^2} = \frac{C_V}{T} \left(\frac{\partial T}{\partial V} \right)^2 - T \frac{\partial^2 S}{\partial V^2} \tag{6.20}$$

where the ratio C_V/T is expressed as,

$$\frac{C_V}{T} = \frac{\partial S}{\partial T} = \left(\frac{\partial T}{\partial S} \right)^{-1} = \left(\frac{\partial^2 U}{\partial S^2} \right)^{-1} \geq 0$$

according to condition (6.14). Thus, the first term on the right-hand side of condition (6.20) is positive. Furthermore, since the temperature T is positive, inequality (6.12) implies that

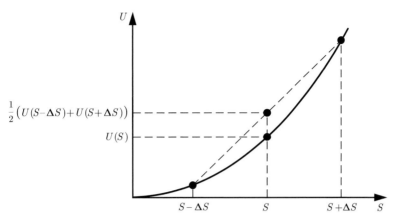

Figure 6.3   At constant volume V, a function $U(S)$ of positive or zero curvature satisfies the convexity condition (6.23).

the second term is also positive. Thus, we found that the curvature of the internal energy $U(S, V)$ as a function of volume V is positive in the neighbourhood of equilibrium, i.e.

$$\frac{\partial^2 U}{\partial V^2} \geq 0 \tag{6.21}$$

In order to ensure that U is a convex function of entropy S and volume V with a positive Gauss curvature in the state space (U, S, V), we need to add to the local convexity conditions (6.14) and (6.21) the condition,

$$\frac{\partial^2 U}{\partial S^2} \frac{\partial^2 U}{\partial V^2} - \left(\frac{\partial^2 U}{\partial S \partial V} \right)^2 \geq 0 \tag{6.22}$$

As U is a monotonous convex function of entropy S and volume V, its positive curvature is also defined globally. Therefore, to the local conditions of convexity (6.14), (6.21) and (6.22) we can add the global conditions of convexity,

$$\begin{aligned}
U(S - \Delta S, V) + U(S + \Delta S, V) &\geq 2U(S, V) \\
U(S, V - \Delta V) + U(S, V + \Delta V) &\geq 2U(S, V) \\
U(S - \Delta S, V - \Delta V) + U(S + \Delta S, V + \Delta V) &\geq 2U(S, V)
\end{aligned} \tag{6.23}$$

These relations mean that $U(S, V)$ is a convex function of entropy S and volume V (Fig. 6.3).

6.4 Stability and Entropy

Based on the curvature properties of entropy and internal energy derived in the previous section, we are now in a position to discuss the stability of a given amount of matter in a certain state. In this section, we focus out attention on the concavity condition for entropy. In the next section, we will consider stability in terms of the thermodynamic potentials.

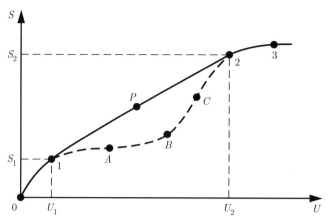

Figure 6.4   A line going through the points 0, 1, A, B, C, 2 and 3 is defined by an equation of state $S\,(U,V)$ at constant volume V. It describes a set of states that are globally unstable. The full line going through the points 0, 1, P, 2 and 3 corresponds to globally stable states.

We begin by assuming an equation of state $S(U,V)$ that has a second partial derivative with respect to internal energy U that changes sign as U varies (Fig. 6.4). On a plane of constant volume V in state space, equation of state $S(U,V)$ corresponds to a curve passing through the points 0, 1, A, B, C, 2 and 3 (Fig. 6.4). This curve has a negative curvature between points 0 and A, a positive curvature between points A and C, and a negative curvature between points C and 3. The points A and C are inflection points, i.e. points where the curvature vanishes. The states corresponding to points on the line between A and C are locally unstable, since the local condition (6.12) of concavity is not satisfied. The states corresponding to the points of the curve passing through 0, 1, A, B, C, 2 and 3 are globally unstable, as the global condition (6.5) of concavity is not satisfied [55].

According to the global concavity condition, in order for the points on the curve to correspond to stable states, these points have to saturate the inequality (6.5) from $U - \Delta U$ to $U + \Delta U$,

$$S(U - \Delta U, V) + S(U + \Delta U, V) = 2S(U, V) \tag{6.24}$$

where the values $U - \Delta U$ and $U + \Delta U$ have yet to be determined. Taking into account relations (6.5), (6.10) and (6.12), the local form of (6.24) reads,

$$\frac{\partial^2 S}{\partial U^2} = 0 \tag{6.25}$$

This means that the curve has to contain a segment of straight line between the points of abscissa $U - \Delta U$ and $U + \Delta U$. Now, we find the ends of the segment by the following argument. In order to satisfy the local concavity condition (6.12), the second partial derivative of entropy with respect to internal energy U has to be defined. This implies that the partial derivative has to be differentiable. Therefore, the segment has to be tangent to the curve at its ends, which are the points 1 and 2 of abscissa $U_1 = U - \Delta U$ and $U_2 = U + \Delta U$ (Fig. 6.4). The equation of state for the stable states located on this segment is given by,

$$S(U, V) = \lambda S(U_1, V) + (1 - \lambda) S(U_2, V) \quad \text{where} \quad \lambda \in [0, 1] \qquad (6.26)$$

Thus, the curve corresponding to the states that are globally stable passes through the points 0, 1, P, 2 and 3. This curve is characterised by the property of having no positive curvature at any point.

The stability criteria for entropy S as a function of volume V follows the same derivation as above. It can be obtained by simply replacing the internal energy U by the volume V and vice versa.

6.5 Stability and Thermodynamic Potentials

We establish now the stability criteria for systems characterised by a minimum of one of the thermodynamic potentials. We assume an equation of state $U(S, V)$ that has a second partial derivative with respect to entropy S that changes sign when S varies (Fig. 6.5). On a state space plane of constant volume, i.e. $V = $ const, the equation of state $U(S, V)$ is described by a curve passing through the points 0, 1, A, B, C, 2 and 3. This curve has a positive curvature between points 0 and A, a negative curvature between points A and C, and a positive curvature between points C and 3. The points A and C are inflexion points, i.e. points where the curvature vanishes. The curve between points A and C corresponds to locally unstable states, as the local convexity condition (6.14) is not fulfilled. The curve passing through the points 0, 1, A, B, C, 2 and 3 corresponds to globally unstable states, as the global convexity condition (6.5) is not satisfied. By analogy with the previous section, it can be shown that the curve of globally stable states, characterised by not having a negative curvature at any point, is the one passing through the points 0, 1, P, 2 and 3. In particular, the equation of state that lies on the segment spanning from point 1 to point 2 (Fig. 6.5) is,

$$U(S, V) = \lambda U(S_1, V) + (1 - \lambda) U(S_2, V) \qquad \text{where} \quad \lambda \in [0, 1] \qquad (6.27)$$

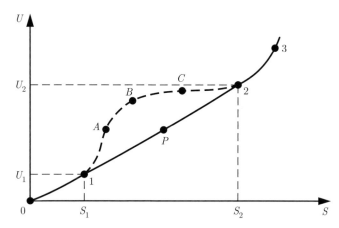

Figure 6.5 The curve (full and dotted line) passing through the points 0, 1, A, B, C, 2 and 3 is defined by the equation of state $U(S, V)$ at constant volume V. It corresponds to a set of states that are globally unstable. The curve (full line) passing through the points 0, 1, P, 2 and 3 corresponds to globally stable states.

The stability criterium for the internal energy U as a function of the volume V derives from a discussion similar to the one just made, in which the entropy S is replaced by the volume V and vice versa.

We now turn our attention to the thermodynamic potentials free energy F, enthalpy H, and Gibbs free energy G. In order to establish stability criteria for them, we define the **compressibility coefficients** at constant entropy and temperature,

$$\kappa_S = -\frac{1}{V}\frac{\partial V(S,p)}{\partial p} \quad \text{and} \quad \kappa_T = -\frac{1}{V}\frac{\partial V(T,p)}{\partial p} \tag{6.28}$$

Taking into account the definitions of temperature (2.9), pressure (2.10), specific heat at constant volume (5.8) and compressibility coefficient at constant entropy (6.28), the conditions (6.14) and (6.21) can be expressed as,

$$\frac{\partial^2 U}{\partial S^2} = \frac{\partial T}{\partial S} = \frac{T}{C_V} \geq 0 \quad \text{and} \quad \frac{\partial^2 U}{\partial V^2} = -\frac{\partial p}{\partial V} = \frac{1}{\kappa_S V} \geq 0 \tag{6.29}$$

For positive temperatures, the conditions (6.29) imply that the specific heat at constant volume C_V and the compressibility coefficient at constant entropy κ_S cannot be negative in a stable system,

$$C_V \geq 0 \quad \text{and} \quad \kappa_S \geq 0 \tag{6.30}$$

For positive temperatures, the specific heat at constant pressure C_p and the compressibility coefficient at constant temperature κ_T cannot be negative in a stable system either (see exercise 6.6),

$$C_p \geq 0 \quad \text{and} \quad \kappa_T \geq 0 \tag{6.31}$$

We are now ready to express the conditions of stability expressed in terms of the free energy F, the enthalpy H and the Gibbs free energy G. Using the differential (4.25) of the free energy $F(T,V)$, the definitions of entropy (4.26), pressure (4.27), specific heat at constant volume (5.8), compressibility coefficient at constant temperature (6.28) and in view of the conditions (6.30) and (6.31), we derive the local stability conditions for free energy F,

$$\frac{\partial^2 F}{\partial T^2} = -\frac{\partial S}{\partial T} = -\frac{C_V}{T} \leq 0 \quad \text{and} \quad \frac{\partial^2 F}{\partial V^2} = -\frac{\partial p}{\partial V} = \frac{1}{\kappa_T V} \geq 0 \tag{6.32}$$

Using the differential (4.32) of enthalpy $H(S,p)$, the definitions of temperature (4.33), volume (4.34), specific heat at constant pressure (5.21) and compressibility coefficient at constant entropy (6.28), we derive from the conditions (6.30) and (6.31) the local stability conditions for enthalpy H,

$$\frac{\partial^2 H}{\partial S^2} = \frac{\partial T}{\partial S} = \frac{T}{C_p} \geq 0 \quad \text{and} \quad \frac{\partial^2 H}{\partial p^2} = \frac{\partial V}{\partial p} = -\kappa_S V \leq 0 \tag{6.33}$$

Finally, using the differential (4.39) of the Gibbs free energy $G(T,p)$, the definitions of entropy (4.26), volume (4.41), specific heat at constant pressure (5.21), compressibility coefficient at constant temperature (6.28), we derive from the conditions (6.31) the local stability conditions for the Gibbs free energy G,

$$\frac{\partial^2 G}{\partial T^2} = -\frac{\partial S}{\partial T} = -\frac{C_p}{T} \leq 0 \quad \text{and} \quad \frac{\partial^2 G}{\partial p^2} = \frac{\partial V}{\partial p} = -\kappa_T V \leq 0 \tag{6.34}$$

In order to ensure that the Gibbs free energy G is a concave function of temperature T and pressure p with a positive Gauss curvature in state space (G, T, p), we have to add to the local concavity conditions (6.34) the condition,

$$\frac{\partial^2 G}{\partial T^2} \frac{\partial^2 G}{\partial p^2} - \left(\frac{\partial^2 G}{\partial T \partial p}\right)^2 \geq 0 \tag{6.35}$$

In summary, the local stability conditions (6.14), (6.21), (6.32), (6.33), (6.34) and the local concavity condition (6.35) show that **the thermodynamic potentials U, F, H and G are convex functions of their extensive variables S and V and concave functions of their intensive variables T and p.**

The structure of these curvature conditions stems from relation (4.18) which implies that when a thermodynamic potential is a convex function of an extensive variable, its Legendre transform is a thermodynamic potential that has to be a concave function of the corresponding intensive variable. The converse is of course also true: when a thermodynamic potential is a concave function of an intensive variable, its Legendre transform has to be, according to relation (4.18), a convex function of the corresponding extensive variable.

6.6 Phase Transitions

In a thermodynamic system, a **phase** is a state of matter which occupies a subspace of the state space characterised by physical properties of the system.

As an illustration, let us consider a system consisting of a mixture of water and ice in a closed glass container. The ice corresponds to the solid phase, the water in the container to the liquid phase and the vapour above the water-ice mixture, to the gaseous phase. A simple system is necessarily in a single phase, since by definition it is homogeneous.

The three main phases of matter are the solid state, the liquid state and the gaseous state. If we place a **solid** in a container, it keeps its volume and its geometrical shape. If we place a **liquid** in a container, it preserves its volume but it adapts to the shape of the container. If we place a **gas** in a container, it fills the entire volume of the container and thus, takes the geometrical shape of the container. We call **fluid** the state of a substance that is liquid, gaseous or a mixture of both phases.

Matter can also be in other phases, such as a plasma, a ferromagnetic state, a superconducting state or a superfluid state. In a **plasma**, matter is ionised; this means it consists of electrically charged particles. In a **ferromagnetic** state, the magnetisation is aligned in a certain direction, even when no external magnetic field is applied. In a **superconducting** state, the electrical resistance is zero, implying that electrical currents can circulate indefinitely. In a **superfluid** state, the viscosity is zero, which implies that matter currents can circulate indefinitely.

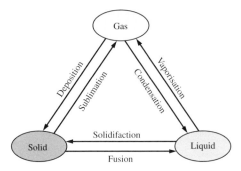

Figure 6.6   Phase transitions between the solid, liquid and gaseous states.

A **phase transition** is a change of one or several phases of the substance in response to a physical process. According to the Ehrenfest classification, there are two kinds of phase transitions, referred to as first and second-order phase transitions [56].

First-order phase transitions are characterised by discontinuities in some of the first partial derivatives of the Gibbs free energy G, namely, volume V and entropy S. In general, phase transitions between solid, liquid and gaseous states are of this kind.

Second-order phase transitions are characterised by discontinuities in the second derivatives of the Gibbs free energy G, that is, the compressibility at constant temperature κ_T and the specific heat at constant pressure C_p. For instance, at the critical point defined in this chapter, a second-order phase transition takes place.

The phase transition from a solid to a liquid state is called **melting** and the reverse transition is known as **solidification** (Fig. 6.6). The phase transition from a liquid to a gaseous state is called **vaporisation** and the reverse transition is known as **condensation**. The phase transition from a solid to a gaseous state is called **sublimation** and the reverse transition is known as **deposition**.

Two or more phases can coexist at equilibrium. On the curve defined by U as a function of the entropy S (Fig. 6.5), the segment connecting points 1 and 2 describes the coexistence of two phases denoted α and β. At point 1, the system is in the pure phase α and at point 2, it is in the pure phase β. Between these two points, the system is a mixture of the phases α and β. The internal energy of this state is a linear combination of the internal energies of the system in phases α and β, as shown in equation (6.27). The points on the segment between 1 and 2 are **points of phase coexistence**. At these points, the proportion of each phase is given by the parameter λ of (6.27). On this segment, the derivative of internal energy U with respect to entropy S is a constant equal to the temperature of the phase transition.

In the following, we consider the Gibbs free energy G as a function of temperature T for a substance that can be in a solid, liquid or gaseous state. The three curves (Fig. 6.7) represent the Gibbs free energies of the solid, liquid and gaseous phases. Phase coexistence occurs at points M and V located at the intersections of these curves. According to relation (4.58), the state of equilibrium is the one that minimises the Gibbs free energy G. Since the Gibbs free energy of the solid phase is the smallest of the Gibbs free energies below T_m, the equilibrium state of the system is a solid below T_m. Likewise, the liquid phase Gibbs free energy is the smallest between temperatures T_m and T_v. Therefore, the equilibrium state of

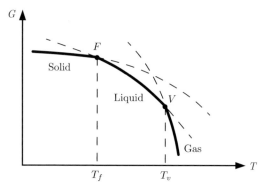

Figure 6.7 Curve (full line) representing the minimum of Gibbs free energy G as a function of temperature, at a fixed pressure p. In the temperature range $T < T_m$, the system is in the solid phase. In the temperature range $T_m < T < T_v$, it is in the liquid phase and in the temperature range temperature $T > T_v$, it is in the gaseous phase.

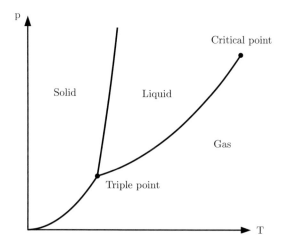

Figure 6.8 Phase diagram (p, T) of one substance: the phase coexistence curves denote the limit of the phases, the triple point is where the three phases merge, and the critical point is at the other end of the liquid–gas coexistence line.

the system is a liquid in this temperature range. Finally, since the Gibbs free energy of the gaseous phase is the smallest above temperature T_v, the equilibrium state above T_v is a gas.

The phase transition between solid and liquid takes place at the melting temperature T_m and between liquid and gas at the vaporisation temperature T_v. According to definition (4.40), the entropy of the system is equal to the opposite of the slope on the (G, T) diagram. Since entropy varies suddenly at the melting temperature T_m and at the vaporisation temperature T_v, these phase transitions are of first-order.

On the (G, T) diagram (Fig. 6.7), pressure is constant. The melting point, M, and vaporisation point, V, of the (G, T) diagram become curves when varying p, that is, when considering the three-dimensional state space (G, T, p). We can represent the phases on a diagram (p, T) known as a ***phase diagram*** (Fig. 6.8). The phase diagram can be divided into three regions corresponding to the three phases (Fig. 6.8).

In each phase, the chemical potential $\mu_\alpha(T,p)$ is defined as the partial derivative of the Gibbs free energy with respect to the amount of matter, in accordance with (4.42). We label the solid, liquid and gaseous phases by $\alpha = s, \ell, g$. The solid phase $\alpha = s$ corresponds to the region of the phase diagram where,

$$\mu_s \leq \mu_\ell \quad \text{and} \quad \mu_s \leq \mu_g \tag{6.36}$$

The liquid phase $\alpha = \ell$ corresponds to the region of the phase diagram where,

$$\mu_\ell \leq \mu_g \quad \text{and} \quad \mu_\ell \leq \mu_s \tag{6.37}$$

Similarly, the gas phase $\alpha = g$ corresponds to the region of the phase diagram where,

$$\mu_g \leq \mu_s \quad \text{and} \quad \mu_g \leq \mu_\ell \tag{6.38}$$

These regions are separated by curves where the two phases coexist. These lines are called **phase coexistence lines**. Both phases are at equilibrium (3.43) along the phase coexistence lines. The solid–liquid coexistence line is defined by,

$$\mu_s = \mu_\ell < \mu_g \tag{6.39}$$

The liquid–gas coexistence line is defined by,

$$\mu_\ell = \mu_g < \mu_s \tag{6.40}$$

Similarly, the solid–gas coexistence line is defined by,

$$\mu_s = \mu_g < \mu_\ell \tag{6.41}$$

In the (p, T) phase diagram, there is a particular point where the three phases coexist. This point is called the **triple point** and is defined by,

$$\mu_s = \mu_\ell = \mu_g \tag{6.42}$$

There is also a point at which the liquid–gas coexistence curve ends. This point is called the **critical point** and the corresponding state, the **critical state**. The temperature and pressure at which it occurs depends on the substance. When the system pressure and temperature are above that of the critical point, the substance is said to be in a **supercritical state**.

For a given substance, there can be several solid or liquid phases, hence several critical points. But there is only one gaseous phase [14]. It is generally accepted that there is no critical point on the solid–liquid coexistence line. However, recent studies on nanostructures seem to indicate that this might not always be the case [57, 58].

When a system evolves through a phase coexistence line, it undergoes a first-order phase transition as its volume and entropy change abruptly. When the state of the system goes through the critical point, it undergoes a second-order phase transition and its volume and entropy vary in a continuous manner. When the state of the system passes around the critical point, it goes continuously from a liquid to a gaseous state, passing through supercritical states. There is no phase transition in this case. Therefore, it is not possible to define strictly liquid and gaseous phases, since it is possible to go continuously from one to the other.

6.7 Latent Heat

In Chapter 5, we defined the latent heat of expansion L_V and of compression L_p in order to characterise the thermal response of a simple system to a heat transfer at constant temperature. During a phase transition between a solid and a liquid or between a liquid and a gas, the temperature remains constant (Fig. 6.7). These phase transitions can be thought of as the response of the system to a heat transfer. *Latent heat* is defined as the heat received from the environment during a process at constant temperature. The heat provided to the system undergoing a phase transition from solid to liquid phases is called the ***latent heat of melting***. The heat provided to the system during a phase transition from liquid to gaseous states is called ***latent heat of vaporisation***.

Thus, in this chapter, we have heat transferred to a system in which two phases coexist. This causes some of the substance to pass from one phase to another while the system remains at a constant temperature, equal to the phase transition temperature. We can express the heat transfer in terms of the entropy of each phase. We note in passing that since different parts of the system have different entropies, the system is not simple, even though the temperature is the same in every phase. Nonetheless, definition (2.25) applies and the heat provided to the system causing melting or vaporisation is given by,

$$Q_{s\ell} = T(S_\ell - S_s) \qquad \text{and} \qquad Q_{\ell g} = T(S_g - S_\ell) \qquad (6.43)$$

where T is the phase transition temperature, i.e. the melting temperature T_m or the vaporisation temperature T_v, S_s is the entropy of the solid phase, S_ℓ the entropy of the liquid phase and S_g the entropy of the gaseous phase.

The molar latent heat of melting $\ell_{s\ell}$ and the molar latent heat of vaporisation $\ell_{\ell g}$ are defined as the heat required for one mole of the substance to undergo these phase transitions,

$$\ell_{s\ell} = T(s_\ell - s_s) \qquad \text{and} \qquad \ell_{\ell g} = T(s_g - s_\ell) \qquad (6.44)$$

where s_s is the molar entropy of the solid phase, s_ℓ is the molar entropy of the liquid phase and s_g is the molar entropy of the gaseous phase. We have written molar quantities using small letters in order to distinguish them from the corresponding extensive quantities.

We can illustrate the notion of latent heat by considering a (T, s) diagram representing the temperature of one mole of water as a function of its entropy (Fig. 6.9). Between $0\,K$ and $273\,K$, the entropy of ice increases with temperature. At $273\,K$, ice begins to melt. This means that it undergoes a phase transition and that it is transforming into liquid water as the molar entropy s is increasing at fixed temperature. This phase transition is represented by the first step on the (T, s) diagram. According to the first definition in (6.44), the area under the first step corresponds to the latent heat of melting $\ell_{s\ell} = 6 \cdot 10^3\,J$, that is, the heat that has to be provided to the mole of ice so that it becomes a mole of water. Between $273\,K$ and $373\,K$, the entropy of water increases with increasing temperature. At $373\,K$, water begins to boil. This means that a phase transition occurs and that some of the water becomes vapour. The area under the second step corresponds to the latent heat of

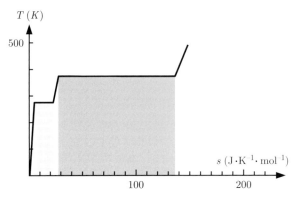

Figure 6.9 (T, s) diagram of H_2O. The first step corresponds to the melting of ice, and the second step to the boiling of water. The shaded areas correspond to the latent heat of melting and of vaporisation.

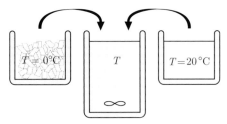

Figure 6.10 Ice at a temperature $T = 0°C$ is mixed with water at room temperature, $T = 20°C$. The final temperature T of the water once thermal equilibrium is reached is noticeably lower than the weighted average of the absolute initial temperatures, because of the latent heat of melting.

vaporisation $\ell_{\ell g} = 4 \cdot 10^4$ J. It is the heat that has to be provided to the water so that one mole of liquid turns into vapour.

The latent heat of melting can be made evident by considering the simple experiment illustrated in Fig. 6.10. If ice at $0°C$ is added to water at room temperature, the resulting mixture will be colder than if an equal amount of water at $0°C$ is added instead. In order to melt ice, a heat transfer from the water to the ice is required. Thus the final temperature is lower when starting with ice at $0°C$ than with water at $0°C$.

6.8 The Clausius–Clapeyron Equation

In this section and the following one, we present two particular features of the phenomenon of phase coexistence. First, we derive a relationship between latent heat and the slope of the phase coexistence curves in a (p, T) diagram (Fig. 6.8). In the next section, we will see that the coexistence of several phases in a system imposes restrictions on the number of independent state variables.

Let us consider a substance where two phases (labelled α and β) coexist. This means that the phases are at chemical equilibrium at all points on the coexistence curve. Hence, their chemical potentials are equal,

$$\mu_\alpha = \mu_\beta \qquad (6.45)$$

This equality was expressed by equation (6.39) for the solid–liquid transition, and by (6.40) and (6.41) for the liquid–gas and solid–gas transitions, respectively. While the chemical potentials are equal, they are not constant along the coexistence curve. The differential of the identity (6.45) implies that along the coexistence curve, we have,

$$d\mu_\alpha = d\mu_\beta \qquad (6.46)$$

We apply the Gibbs–Duhem equation (4.9) to one mole of substance in each phase to write,

$$d\mu_\alpha = -s_\alpha \, dT + v_\alpha \, dp \qquad \text{and} \qquad d\mu_\beta = -s_\beta \, dT + v_\beta \, dp \qquad (6.47)$$

where v_α and v_β are the molar volumes of the substance in the α and β phases, respectively, and $\alpha, \beta \in \{s, \ell, g\}$. Combining equations (6.46) and (6.47), we obtain,

$$-(s_\alpha - s_\beta) \, dT + (v_\alpha - v_\beta) \, dp = 0 \qquad (6.48)$$

This equation can be recast as,

$$\frac{dp}{dT} = \frac{s_\alpha - s_\beta}{v_\alpha - v_\beta} \qquad (6.49)$$

Thus, we find an expression for the slope of the phase coexistence curve $p\,(T)$ in terms of the molar entropies and the molar volumes of both phases. Inserting definitions (6.44) in equation (6.49), we obtain expressions for this slope which are known as the **Clausius–Clapeyron equations**,

$$\frac{dp}{dT} = \frac{\ell_{s\ell}}{T(v_\ell - v_s)} \qquad \text{and} \qquad \frac{dp}{dT} = \frac{\ell_{\ell g}}{T(v_g - v_\ell)} \qquad (6.50)$$

The slope of a coexistence curve between solid and liquid phases is positive if the molar volume v_ℓ of the liquid phase is greater than the molar volume v_s of the solid phase. This is the case for most substances. Water is an exception, as the molar volume of ice is greater than that of water, implying that the denominator of the right-hand side of equation (6.48) is negative. Therefore, by increasing pressure at constant temperature, it is possible to melt ice.

6.9 Gibbs Phase Rule

Let us show now that the equilibrium condition (6.45) between phases imposes a restriction on the number of state variables that describe the state of a system. Thus, we consider a system consisting of r substances that can be present in m phases at thermal, chemical and mechanical equilibrium. Each substance can be in any of the phases. Each phase can be thought of as a simple subsystem and described by a set of state variables. The system is characterised by a temperature T and a pressure p. As the Gibbs free energy is an extensive quantity, definition (4.37) can be generalised in order to account for the presence of m phases,

$$G\,(T, p, \{N_A^\alpha\}) = \sum_{\alpha=1}^{m} \left(\sum_{A=1}^{r} \mu_A^\alpha \, N_A^\alpha \right) \qquad (6.51)$$

where N_A^α is the number of moles of substance A in phase α and μ_A^α is the chemical potential of substance A in phase α. The number of moles N^α of all substances in phase α is given by,

$$N^\alpha = \sum_{A=1}^{r} N_A^\alpha \tag{6.52}$$

Taking into account definitions (6.52), the Gibbs free energy (6.51) is written as,

$$G\left(T, p, \{N_A^\alpha\}\right) = \sum_{\alpha=1}^{m} N^\alpha g^\alpha\left(T, p, \{N_A^\alpha\}\right) \tag{6.53}$$

where the molar Gibbs free energy $g^\alpha\left(T, p, \{N_A^\alpha\}\right)$ of phase α is given by,

$$g^\alpha\left(T, p, \{N_A^\alpha\}\right) = \sum_{A=1}^{r} \mu_A^\alpha \frac{N_A^\alpha}{N^\alpha} \tag{6.54}$$

The molar Gibbs free energy $g^\alpha\left(T, p, \{N_A^\alpha\}\right)$ of phase α is an intensive quantity that must satisfy the identity,

$$g^\alpha\left(T, p, \{N_A^\alpha\}\right) = g^\alpha\left(T, p, \{\lambda N_A^\alpha\}\right) \tag{6.55}$$

which can be recast in terms of the concentration c_A^α of substance A in phase α defined as,

$$c_A^\alpha = \frac{N_A^\alpha}{N^\alpha} \tag{6.56}$$

Taking into account relation (6.56) and evaluating identity (6.55) for $\lambda = 1/N^\alpha$, we find,

$$g^\alpha\left(T, p, \{N_A^\alpha\}\right) = g^\alpha\left(T, p, \{c_A^\alpha\}\right) \tag{6.57}$$

According to the definition (6.56) of the concentration c_A^α, the molar Gibbs free energy (6.54) becomes,

$$g^\alpha\left(T, p, \{c_A^\alpha\}\right) = \sum_{A=1}^{r} \mu_A^\alpha c_A^\alpha \tag{6.58}$$

In view of relation (6.57), the Gibbs free energy (6.53) can be written as,

$$G\left(T, p, \{N_A^\alpha\}\right) = \sum_{\alpha=1}^{m} N^\alpha g^\alpha\left(T, p, \{c_A^\alpha\}\right) \tag{6.59}$$

Relations (6.52) and (6.56) require,

$$\sum_{A=1}^{r} c_A^\alpha = 1 \qquad \forall \quad \alpha = 1, \ldots, m \tag{6.60}$$

Equation (6.60) imposes a condition on the concentrations c_A^α of each substance in each phase. Therefore, there are $r - 1$ independent variables c_A^α for each phase α. Thus, the state of the system is defined by $m\left(r - 1\right)$ independent variables c_A^α for $A = 1, \ldots, r$ and $\alpha = 1, \ldots, m$. The temperature T and the pressure p of the system have to be included as state variables also.

As the system is at chemical equilibrium, the chemical potentials of each substance A have to be equal in each phase α. Thus, for each substance, this equilibrium condition imposes $m - 1$ constraints, i.e.

$$\mu_A^1 = \mu_A^2 = \ldots = \mu_A^m \qquad \forall \quad A = 1, \ldots, r \tag{6.61}$$

Therefore, the condition of chemical equilibrium imposes $r\,(m - 1)$ constraints on the r substances. The number of degrees of freedom f, that is, the number of state variables that can be independently specified, given the number of phases in coexistence, is defined as the difference between the number of state variables and the number of constraints [59],

$$f = 2 + m\,(r - 1) - r\,(m - 1) = r - m + 2 \tag{6.62}$$

This equation is called the **Gibbs phase rule**. If the system consists of two substances that coexist in two phases, then $r = m = 2$ and the rule implies $f = 2$. Therefore, if the pressure p and the temperature T are given, the composition of each phase is uniquely determined. If the system consists of a single substance that is present in three phases, then $r = 1$, $m = 3$ and the rule implies $f = 0$. Therefore, the temperature and the pressure are determined. This is the case of the triple point (Fig. 6.8).

6.10 Van der Waals Gas

In Chapter 5, we defined the ideal gas model. That model did not take into account the interaction forces between the elementary constituents of the gas, that is, its atoms or molecules. The ideal gas model describes correctly many features of dilute gases, but when a gas is dense, this model is often insufficient to account for actual observations. For instance, it cannot describe a phase transition between gaseous and a liquid state. In this section, we introduce a more realistic model established by van der Waals because it also predicts two phases and their coexistence.

This model is defined by the **van der Waals equation of state** [60]. For a gas consisting of N moles of substance, it reads,

$$\left(p + \frac{N^2\,a}{V^2}\right)(V - N\,b) = NR\,T \tag{6.63}$$

where a and b are constants that depend on the substance. This equation of state can be derived in the framework of statistical physics, assuming that the elementary constituents are hard spheres of a given radius, subject to mutual attractive forces. Considering hard spheres amounts to taking into account the repulsive forces that dominate when the elementary constituents of the gas are very close to one another. The order of magnitude of this distance at which repulsion dominates determines the size of the hard spheres. The term proportional to a expresses the effect of the attractive forces and the constant b in (6.63) represents the volume occupied by one mole of hard spheres.

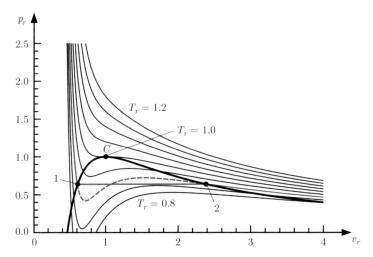

Figure 6.11 Phase coexistence as predicted by the van der Waals isothermal. Points 1 and 2 are the molar volumes of both phases. The black line, defined by the locus of points such as 1 and 2, delimits a "forbidden zone" in the $(p_r\, v_r)$ diagram. Point C is the critical point. The reduced quantities are defined as the ratio of the quantities p and v and their corresponding critical values, i.e. $v_r = v/v_c, p_r = p/p_c, T_r = T/T_c$.

When the attractive and repulsive forces are negligible, i.e. $N^2 a/V^2 \ll p$ and $Nb \ll V$, the equation of state (6.63) of the van der Waals gas reduces to the equation of state (5.47) of the ideal gas.

The van der Waals equation (6.63) can be expressed in terms of the molar volume $v = V/N$ as,

$$p = \frac{RT}{v - b} - \frac{a}{v^2} \tag{6.64}$$

In order to proceed further without reference to any specific value of the constants a and b, we proceed to recast the van der Waals equation (6.64) in a reduced form that does not depend explicitly on these two material parameters. As we shall see, the van der Waals model predicts a phase transition between liquid and gaseous states. Furthermore, it accounts for the existence of a critical point (p_c, v_c, T_c) in the state space (p, v, T). This point is defined as an inflection point with a vanishing first derivative in the (p, v) plane of which the temperature is equal to the critical temperature T_c (Fig. 6.11). Thus, this point is characterised by,

$$\left. \frac{\partial p}{\partial v} \right|_{p_c, v_c, T_c} = 0 \qquad \left. \frac{\partial^2 p}{\partial v^2} \right|_{p_c, v_c, T_c} = 0 \tag{6.65}$$

Using the van der Waals equation (6.64), these conditions yield,

$$-\frac{RT_c}{(v_c - b)^2} + \frac{2a}{v_c^3} = 0 \quad \text{and} \quad \frac{2RT_c}{(v_c - b)^3} - \frac{6a}{v_c^4} = 0 \tag{6.66}$$

We can solve (6.66) for the molar volume v_c,

$$v_c = 3b \tag{6.67}$$

Substituting expression (6.67) for v_c in the first condition in (6.66), we find the critical temperature T_c,

$$T_c = \frac{8a}{27Rb} \tag{6.68}$$

Finally, substituting expression (6.67) and (6.68) for v_c and T_c into the van der Waals equation (6.64) evaluated at the critical point (p_c, v_c, T_c), we obtain an expression for the critical pressure p_c,

$$p_c = \frac{a}{27b^2} \tag{6.69}$$

We introduce now dimensionless parameters, namely, a reduced pressure p_r, a reduced molar volume v_r and a reduced temperature T_r defined as,

$$p_r = \frac{p}{p_c} \quad \text{and} \quad v_r = \frac{v}{v_c} \quad \text{and} \quad T_r = \frac{T}{T_c} \tag{6.70}$$

Expressing the van der Waals equation (6.64) in terms of these reduced quantities (6.70), and taking into account expressions (6.67), (6.69) and (6.68) for the critical quantities, we obtain,

$$\frac{a\,p_r}{27\,b^2} = \frac{8\,a\,T_r}{27\,b^2\,(3\,v_r - 1)} - \frac{a}{9\,b^2\,v_r^2} \tag{6.71}$$

Multiplying relation (6.71) by $27b^2/a$ yields the reduced form of the van der Waals equation (Fig. 6.11),

$$p_r = \frac{8\,T_r}{3\,v_r - 1} - \frac{3}{v_r^2} \tag{6.72}$$

This reduced form of the van der Waals equation (6.72) is independent of the composition of the van der Waals gas, in the sense that it does not depend explicitly on a and b.

On the (p_r, v_r) diagram (Fig. 6.11), the curves drawn with a fine black line describe isothermal processes at reduced temperatures T_r with $0.8 < T_r < 1.2$. The critical point C lies on the isothermal curve $T_r = 1.0$.

During a supercritical isothermal process, i.e. $T_r > 1.0$, when the reduced molar volume v_r increases, the system evolves continuously from the liquid phase to the gaseous phase, and vice versa when the reduced molar volume v_r decreases. No phase transition takes place in this case. One can show that during an isothermal process which goes through the critical point C, i.e. $T_r = 1.0$, the system undergoes a phase transition of second-order.

During an isothermal process below the critical point, i.e. $T_r < 1.0$, there is a portion of the curve over which its derivative is positive, i.e. $\partial p/\partial v > 0$. This implies that part of the isothermal process described by the van der Waals equation (6.72) is unstable, because the stability criterium (6.29) requires that $\partial p/\partial v \leq 0$. Therefore, this part of the curve has to be replaced by a segment of zero slope that saturates the inequality, i.e. $\partial p/\partial v = 0$. According to relations (2.42) and (4.65), the change in free energy variation ΔF_{12} between 1 and 2 during an isothermal process is equal to the work W_{12} performed on the system. For one mole of substance, this equality is written,

$$\Delta F_{12} = W_{12} = - \int_1^2 p\,dv \tag{6.73}$$

Since the free energy F is a state function, the free energy variation ΔF_{12} of one mole of substance going from state 1 to state 2 is independent of the path followed in the (p_r, v_r) diagram. Taking into account definitions (6.70) for the reduced quantities, dividing equation (6.73) by the critical pressure p_c and the molar volume v_c yields,

$$\int_1^2 p_r \, dv_r = p_{r1} \, (v_{r2} - v_{r1}) \tag{6.74}$$

The left-hand side of equation (6.74) corresponds to the area under the state function $p_r \, (v_r)$ while the right-hand side corresponds to the area under the segment of zero slope between states 1 and 2 (Fig. 6.11). In order for these two areas to be equal, the area delimited by the part of the isothermal curve above the segment has to be equal to the area delimited by the part of the isothermal curve below the segment. This geometrical consideration is called the ***Maxwell construction*** [61]. The equal areas are shaded in Fig. 6.11 for the subcritical isothermal curve $T_r = 0.9$.

The locus of points such as 1 and 2 for all the subcritical isotherms is called the ***saturation curve***, marked by a thick black line (Fig. 6.11). Even if the construction of this curve seems rather intuitive from a geometrical standpoint, finding its mathematical expression leads to a transcendental equation. A good approximation can be obtained with a function that is the ratio of two cubic polynomials of the reduced molar volume v_r [62].

For a subcritical reduced temperature, i.e. for $T_r < 1.0$, the part of the phase diagram (p_r, v_r) that lies to the left of the saturation curve corresponds to the liquid phase and the part that lies to the right of it corresponds to the gaseous phase (Fig. 6.11). Inside the saturation curve of the (p_r, v_r) phase diagram, a unique reduced molar volume v_r cannot be attributed to the system since it is consisting of a liquid phase and of a gaseous phase. When the system undergoes a vaporisation, the substance passes from point 1 of the phase coexistence to point 2 without passing through the states of the segment. We could say that this part of the (p_r, v_r) phase diagram corresponds to a 'forbidden zone'. During a subcritical isothermal process, when the system reaches the phase coexistence point 1, the gaseous phase starts appearing. The proportion of that phase increases progressively until the liquid phase has completely disappeared, and at that point, the system is fully at point 2 of the phase coexistence segment. During the reverse subcritical isothermal process, i.e. during condensation, what happens at points 1 and 2 is reversed.

Thus, the van der Waals gas model describes transitions between liquid and gas phases. In order to describe phase transitions between solid, liquid and gas phases, a three-dimensional phase diagram (p, T, v) can be used, as shown (Fig. 6.12). It contains surfaces of phase coexistence between solid, liquid and gas phases. These surfaces can be projected into two-dimensional phase diagrams in the (p, T) or (p, v) planes (Fig. 6.12). When the phase coexistence surfaces shown on the three-dimensional (p, T, v) phase diagram are projected on a plane of constant v, we obtain a two-dimensional (p, T) phase diagram (Fig. 6.8). Likewise, the projection of the phase coexistence curves of the (p, T, v) phase diagram on a plane of constant T (Fig. 6.12) provides a generalisation of the (p, v) phase diagram of the van der Waals gas (Fig. 6.11) which includes a solid phase.

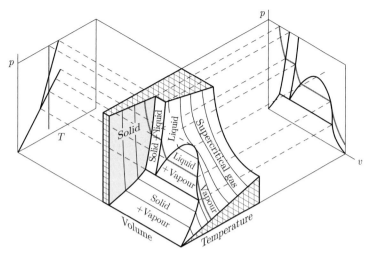

Figure 6.12   Three-dimensional (p, T, v) phase diagram. Solid, liquid and gas phases can be distinguished. They are separated by phase coexistence surfaces, of which the projections onto two-dimensional (p, T) and (p, v) planes give phase coexistence curves. An isothermal process is represented on all three phase diagrams.

6.11 Worked Solutions

6.11.1 Entropy Concavity, Internal Energy Convexity

1. Perform a second-order series expansion with respect to ΔU and ΔV and show that the global condition for the concavity of entropy (6.9) as a function of internal energy and volume can be expressed as,

$$\frac{\partial^2 S}{\partial U^2} \Delta U^2 + 2 \frac{\partial^2 S}{\partial U \partial V} \Delta U \Delta V + \frac{\partial^2 S}{\partial V^2} \Delta V^2 \leq 0$$

2. Take into account the local concavity condition for the entropy (6.12) as a function of internal energy and deduce from the above result that,

$$\left(\frac{\partial^2 S}{\partial U^2} \Delta U + \frac{\partial^2 S}{\partial U \partial V} \Delta V \right)^2 + \left(\frac{\partial^2 S}{\partial U^2} \frac{\partial^2 S}{\partial V^2} - \left(\frac{\partial^2 S}{\partial U \partial V} \right)^2 \right) \Delta V^2 \geq 0$$

3. Deduce from this the local concavity condition for entropy (6.13) as a function of internal energy and volume.

4. Henceforth, consider entropy $S(U)$ as a function of internal energy U only. Then, the global condition of concavity for the entropy can be written as,

$$\lambda S(U_1) + (1 - \lambda) S(U_2) \leq S(\lambda U_1 + (1 - \lambda) U_2) \qquad \text{where} \quad \lambda \in [0, 1]$$

Use the property that the inverse of a monotonously increasing function is also a monotonously increasing function to show graphically that the global concavity condition for entropy implies the global convexity condition for internal energy,

$$\lambda U(S_1) + (1 - \lambda) U(S_2) \geq U(\lambda S_1 + (1 - \lambda) S_2) \qquad \text{where} \quad \lambda \in [0, 1]$$

5. Show that the global conditions of entropy concavity and internal energy convexity can be written as,

$$S(U - \Delta U) + S(U + \Delta U) \leq 2S(U)$$
$$U(S - \Delta S) + U(S + \Delta S) \geq 2U(S)$$

Solution:

1. *The global condition for the concavity of entropy (6.9) as a function of internal energy and volume is written,*

$$S(U - \Delta U, V - \Delta V) + S(U + \Delta U, V + \Delta V) \leq 2S(U, V)$$

The series expansion of entropy $S(U + \Delta U, V + \Delta V)$ to second-order in ΔU and ΔV is given by,

$$S(U + \Delta U, V + \Delta V) \simeq S(U, V) + \frac{\partial S(U, V)}{\partial U} \Delta U + \frac{\partial S(U, V)}{\partial V} \Delta V$$
$$+ \frac{1}{2!} \left(\frac{\partial^2 S(U, V)}{\partial U^2} \Delta U^2 + 2 \frac{\partial^2 S(U, V)}{\partial U \partial V} \Delta U \Delta V + \frac{\partial^2 S(U, V)}{\partial V^2} \Delta V^2 \right)$$

and the series expansion of entropy $S(U - \Delta U, V - \Delta V)$ to second-order in ΔU and ΔV by,

$$S(U - \Delta U, V - \Delta V) \simeq S(U, V) - \frac{\partial S(U, V)}{\partial U} \Delta U - \frac{\partial S(U, V)}{\partial V} \Delta V$$
$$+ \frac{1}{2!} \left(\frac{\partial^2 S(U, V)}{\partial U^2} \Delta U^2 + 2 \frac{\partial^2 S(U, V)}{\partial U \partial V} \Delta U \Delta V + \frac{\partial^2 S(U, V)}{\partial V^2} \Delta V^2 \right)$$

Substituting these two series expansions into the global condition for the entropy concavity, the zeroth-order terms and the first-order terms cancel each other out and the inequality, to second-order, becomes,

$$\frac{\partial^2 S(U, V)}{\partial U^2} \Delta U^2 + 2 \frac{\partial^2 S(U, V)}{\partial U \partial V} \Delta U \Delta V + \frac{\partial^2 S(U, V)}{\partial V^2} \Delta V^2 \leq 0$$

In order to simplify this expression, the dependence of the partial derivatives on the state variables U and V will no longer be indicated explicitly. Thus, the previous result is written,

$$\frac{\partial^2 S}{\partial U^2} \Delta U^2 + 2 \frac{\partial^2 S}{\partial U \partial V} \Delta U \Delta V + \frac{\partial^2 S}{\partial V^2} \Delta V^2 \leq 0$$

2. *Multiplying this inequality by the local condition for entropy concavity (6.12) as a function of internal energy,*

$$\frac{\partial^2 S}{\partial U^2} \leq 0$$

yields the following inequality,

$$\left(\frac{\partial^2 S}{\partial U^2} \right)^2 \Delta U^2 + 2 \frac{\partial^2 S}{\partial U^2} \frac{\partial^2 S}{\partial U \partial V} \Delta U \Delta V + \frac{\partial^2 S}{\partial U^2} \frac{\partial^2 S}{\partial V^2} \Delta V^2 \geq 0$$

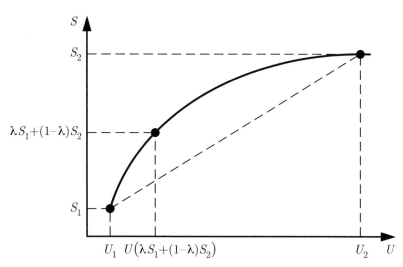

S_2

$\lambda S_1 + (1-\lambda)S_2$

S_1

$U_1 \quad U\!\left(\lambda S_1 + (1-\lambda)S_2\right)$ ⠀⠀⠀⠀ $U_2 \quad U$

Figure 6.13 The entropy $S(U)$ is a concave function of the internal energy U.

It can be recast as,

$$\left(\frac{\partial^2 S}{\partial U^2}\,\Delta U + \frac{\partial^2 S}{\partial U\,\partial V}\,\Delta V\right)^2 + \left(\frac{\partial^2 S}{\partial U^2}\,\frac{\partial^2 S}{\partial V^2} - \left(\frac{\partial^2 S}{\partial U\,\partial V}\right)^2\right)\Delta V^2 \geq 0$$

3. *The internal energy variation ΔU and volume variation ΔV can always be chosen such that the terms in the first brackets on the left-hand side vanish. Thus, in order for this inequality to be alway satisfied, the difference of terms in the brackets above has to be non-negative, i.e.*

$$\frac{\partial^2 S}{\partial U^2}\,\frac{\partial^2 S}{\partial V^2} - \left(\frac{\partial^2 S}{\partial U\,\partial V}\right)^2 \geq 0$$

This corresponds to the local condition for entropy concavity (6.13) as a function of internal energy and volume.

4. *Since the inverse of the entropy function $S(U)$ is monotonous and increasing, the inverses on both sides of the global condition for the entropy concavity (Fig. 6.13) have to satisfy the inequality,*

$$S^{-1}\left(\lambda S_1 + (1-\lambda)\,S_2\right) \leq \lambda\,U_1 + (1-\lambda)\,U_2 \qquad \text{and} \quad \lambda \in [0,1]$$

where $S_1 = S(U_1)$, $S_2 = S(U_2)$, $U_1 = U(S_1)$ and $U_2 = U(S_2)$. Moreover,

$$U\left(\lambda S_1 + (1-\lambda)\,S_2\right) = S^{-1}\left(\lambda S_1 + (1-\lambda)\,S_2\right) \qquad \text{where} \quad \lambda \in [0,1]$$

These two equations imply the global condition for the internal energy convexity,

$$\lambda\,U(S_1) + (1-\lambda)\,U(S_2) \geq U(\lambda S_1 + (1-\lambda)\,S_2) \qquad \text{where} \quad \lambda \in [0,1]$$

5. *In the particular case where $\lambda = 1/2$, $U_1 = U - \Delta U$, $U_2 = U + \Delta U$, $S_1 = S - \Delta S$ and $S_2 = S + \Delta S$, the global conditions for entropy concavity and internal energy convexity become,*

$$\frac{1}{2} S(U - \Delta U) + \frac{1}{2} S(U + \Delta U) \le S(U)$$

$$\frac{1}{2} U(S - \Delta S) + \frac{1}{2} U(S + \Delta S) \ge U(S)$$

and correspond to the inequalities stated in the problem, up to a factor of 2.

6.11.2 Model of Phase Coexistence

Let us consider a concrete model of phase coexistence, defined here for a liquid solution containing two substances at a given pressure. Let N_A be the number of moles of substance A and N_B the number of moles of substance B. The concentration of substance A is defined as $c = N_A / (N_A + N_B)$ where $0 \le c \le 1$. The Gibbs free energy of this model is given by,

$$G(T, N_A, N_B) = N_A R T \ln \left(\frac{N_A}{N_A + N_B} \right) + N_B R T \ln \left(\frac{N_B}{N_A + N_B} \right)$$
$$+ \frac{N_A N_B}{N_A + N_B} \Delta U$$

where $\Delta U > 0$ is an energy term representing the interaction between the substances. The first two terms will be derived in the framework of Chapter 8 [63]. The global condition of stability requires that the Gibbs free energy of the system $G(T, N_A, N_B)$ is a convex function of the extensive variables N_A and N_B.

1. Analyse the behaviour of the dimensionless function $g(\beta, c) = G(T, N_A, N_B) / (R T (N_A + N_B))$ with respect to the dimensionless parameters c and $\beta = \Delta U / R T > 0$.
2. Draw a graph of the function $g(\beta, c)$ where $2 < \beta \le 4 \ln(2)$ is constant.
3. Show that if $2 < \beta \le 4 \ln(2)$, there is a domain of concentrations c in which the system splits into two phases. Determine the proportions r_1 and r_2 of phases 1 and 2 as a function of the concentration c, in terms of the concentrations c_0 and $1 - c_0$ at the minima of the function $g(\beta, c)$.

Solution:

1. *The dimensionless function $g(\beta, c)$ writes,*

$$g(\beta, c) = c \ln(c) + (1 - c) \ln(1 - c) + \beta c (1 - c)$$

where the concentration $0 \le c \le 1$. The function $g(\beta, c)$ is symmetric with respect to $c = 1/2 \; \forall \; \beta$. This can be verified analytically by showing that the function $g(\beta, c)$ is invariant under the exchange of c and $1 - c$. The function $g(\beta, c)$ vanishes when $c \to 0$ and $c \to 1$,

$$\lim_{c \to 0} g(\beta, c) = \lim_{c \to 0} \ln \left(c^c (1 - c)^{1 - c} \right) = 0$$

$$\lim_{c \to 1} g(\beta, c) = \lim_{c \to 1} \ln \left(c^c (1 - c)^{1 - c} \right) = 0$$

The first derivative of function $g(\beta, c)$ with respect to c is given by,

$$\frac{dg}{dc}(\beta, c) = \ln(c) - \ln(1 - c) + \beta(1 - 2c)$$

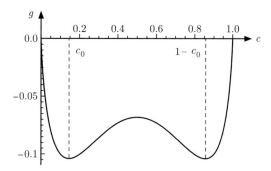

Figure 6.14 The function $g\,(\beta,c)$ for $0 \leq c \leq 1$ and $\beta = 2.5$ has two minima symmetrically located at c_0 and $1 - c_0$.

This implies that,

$$\lim_{c \to 1/2} \frac{dg}{dc}\,(\beta,c) = 0$$

Therefore, function $g\,(\beta,c)$ has an extremum if $c = 1/2\ \forall\ \beta$. The second derivative of function $g\,(\beta,c)$ with respect to c is given by,

$$\frac{d^2 g}{dc^2}\,(\beta,c) = \frac{1}{c} + \frac{1}{1-c} - 2\,\beta$$

which implies that,

$$\lim_{c \to 1/2} \frac{d^2 g}{dc^2}\,(\beta,c) = 2\,(2 - \beta)$$

Therefore, if $0 < \beta < 2$, the point $g\,(\beta,1/2)$ is a minimum and if $\beta > 2$, the point $g\,(\beta,1/2)$ is a maximum. Furthermore,

$$\lim_{c \to 1/2} g\,(\beta,c) = \ln\left(\frac{1}{2}\right) + \frac{\beta}{4}$$

Thus, if $\beta = 4\ln(2)$, then $g\,(\beta,1/2) = 0$. Therefore, if $2 < \beta \leq 4\ln(2)$, then function $g\,(\beta,c) \leq 0\ \forall\ 0 \leq c \leq 1$. Furthermore, this function has a maximum at $c = 1/2$. So, there has to be two minima, located symmetrically with respect to $c = 1/2$ (Fig 6.14). We will label the concentrations of these minima c_0 and $1 - c_0$ where $c_0 < 1/2$. These minima are function of β. The locus of all the minima, obtained by varying β, constitutes the saturation curve.

2. *The graph of $g\,(\beta,c)$ for $0 \leq c \leq 1$ is drawn for $\beta = 2.5$ in Fig. 6.14.*
3. *The global condition of stability for the Gibbs free energy requires that $g\,(\beta,c)$ is a convex function of c. At a concentration $c \leq c_0$ where $c \geq 1 - c_0$, the model is stable. In this case, the solution is made up of one phase which contains both substances A and B. At a concentration $c_0 < c < 1 - c_0$, the model is unstable. This instability manifests itself by the fact that the actual minimum is obtained for a solution consisting of two phases, each containing both substances A and B. We speak of phase segregation to express the fact that they are located in different spatial regions. At concentrations $c \leq c_0$ only phase 1 exists, containing solute A in solvent B, and at concentrations $c \geq 1 - c_0$ only phase 2 exists, containing solute B in*

solvent A. At concentrations $c_0 < c < 1 - c_0$, the proportion r_1 of phase 1 and the proportion $r_2 = 1 - r_1$ of phase 2 vary linearly between 0 and 1 as c varies between the concentrations at both minima, i.e.

$$r_1 = \frac{(1 - c_0) - c}{1 - 2\,c_0} \qquad and \qquad r_2 = \frac{c - c_0}{1 - 2\,c_0}$$

*This linear variation is known as the **lever rule**.*

6.11.3 Melting Point of Salt Water

Let us consider a bloc of ice in equilibrium with salt water. The chemical potential $\mu_s\,(T)$ of the ice depends on temperature. The chemical potential of water $\mu_\ell\,(T, 1 - c)$ depends on temperature T and on the concentration of salt c according to the following model,

$$\mu_\ell\,(T, 1 - c) = \mu_\ell\,(T) + RT \ln\,(1 - c)$$

This expression will be justified in Chapter 8. Determine the change ΔT in the melting temperature as a function of the salt concentration c. The latent heat of melting is $\ell_{s\ell}$ is known. Assume that $c \ll 1$ and $\Delta T \ll T_m$.

Solution:

For pure water, i.e. $c = 0$, the chemical equilibrium between the water and the ice at the melting temperature T_m is written,

$$\mu_s\,(T_m) = \mu_\ell\,(T_m)$$

For salt water, i.e. $c > 0$, the chemical equilibrium between salt water and ice at the melting temperature $T_m + \Delta T$ is written,

$$\mu_s\,(T_m + \Delta T) = \mu_\ell\,(T_m + \Delta T) + R\,(T_m + \Delta T) \ln\,(1 - c)$$

In the limit $c \ll 1$, this expression expanded to 1st order in c amounts to,

$$\mu_s\,(T_m + \Delta T) = \mu_\ell\,(T_m + \Delta T) - R\,(T_m + \Delta T)\,c$$

as $\ln\,(1 - c) = -c + \mathcal{O}\,(c^2)$. Taking into account expression (6.47) for the chemical potential differential of a phase, in the limit $\Delta T / T_m \ll 1$ the series expansion of the chemical potentials $\mu_\ell\,(T_m + \Delta T)$ and $\mu_s\,(T_m + \Delta T)$ as well as that of $R\,(T_m + \Delta T)\,c$ to 1st order in $\Delta T / T_m$ yield,

$$\mu_\ell\,(T_m + \Delta T) = \mu_\ell\,(T_m) + \frac{\partial \mu_\ell}{\partial T}\,\Delta T = \mu_\ell\,(T_m) - s_\ell\,\Delta T$$

$$\mu_s\,(T_m + \Delta T) = \mu_s\,(T_m) + \frac{\partial \mu_s}{\partial T}\,\Delta T = \mu_s\,(T_m) - s_s\,\Delta T$$

Therefore, taking into account the chemical equilibrium between ice and pure water,

$$-s_s\,\Delta T = -s_\ell\,\Delta T - R\,(T_m + \Delta T)\,c$$

which implies that, in the limit $\Delta T / T_m \ll 1$,

$$\Delta T = -\frac{R\, T_m\, c}{s_\ell - s_s}$$

Using definition (6.44) for the latent heat of melting $\ell_{s\ell}$, this temperature change can be written as,

$$\Delta T = -\frac{R\, T_m^2\, c}{\ell_{s\ell}} < 0$$

Therefore, the melting point decreases with the concentration of salt in water. This effect is the reason for adding salt on icy roads.

6.11.4   Liquid–Gas Equilibrium

Let us consider a mole of a substance at temperature T_0 and pressure p_0 such that both liquid and gas phases coexist. The gaseous phase can be considered as an ideal gas. The molar volume v_ℓ of the liquid phase is negligible compared to the molar volume v_g of the gas phase, i.e. $v_\ell \ll v_g$. Furthermore, we assume that the latent heat of vaporisation $\ell_{\ell g}$ is independent of the temperature and pressure. Determine an expression for the pressure $p\,(T)$ along the phase coexistence curve.

Solution:

According to equation of state (5.47) of the ideal gas, the molar volume of the ideal gas can be written as,

$$v_g = \frac{V}{N} = \frac{R\,T}{p}$$

Since the molar volume of the liquid phase is negligible compared to that of the gas phase, i.e. $v_\ell \ll v_g$, the Clausius–Clapeyron equation (6.50) reduces to,

$$\frac{dp}{dT} = \frac{\ell_{\ell g}}{T\, v_g}$$

and can be recast in the form,

$$\frac{dp}{p} = \frac{\ell_{\ell g}}{R}\,\frac{dT}{T^2}$$

Integrating between state (p_0, T_0) and state (p, T) yields,

$$\ln\left(\frac{p}{p_0}\right) = -\frac{\ell_{\ell g}}{R}\left(\frac{1}{T} - \frac{1}{T_0}\right)$$

Therefore, the pressure $p\,(T)$ along the phase coexistence curve is written,

$$p\,(T) = p_0 \exp\left(\frac{\ell_{\ell g}}{R\, T_0}\left(1 - \frac{T_0}{T}\right)\right)$$

which implies that the pressure increases exponentially with increasing temperature if the latent heat of vaporisation is independent of temperature.

6.11.5 van der Waals Gas

1. Determine the expression of the internal energy $U(T, V)$ of the van der Waals gas, assuming that the specific heat at constant volume C_V is independent of temperature.
2. Show that during a Joule expansion, where the volume of gas increases while the system is isolated, the temperature T of the gas decreases.

Solution:

1. *Using Maxwell relation (4.71) and the van der Waals equation of state (6.63), we obtain the latent entropy of expansion,*

$$\frac{\partial S}{\partial V} = \frac{\partial p}{\partial T} = \frac{NR}{V - Nb}$$

Furthermore, taking into account the expression (5.11) for the entropy capacity, the entropy differential can be written,

$$dS = \frac{\partial S}{\partial T} dT + \frac{\partial S}{\partial V} dV = \frac{C_V}{T} dT + \frac{NR}{V - Nb} dV$$

Using the internal energy differential,

$$dU = \left.\frac{\partial U}{\partial T}\right|_V dT + \left.\frac{dU}{dV}\right|_T dV$$

the entropy differential (6.15) is written,

$$dS = \frac{dU}{T} + \frac{p}{T} dV = \frac{1}{T}\left.\frac{\partial U}{\partial T}\right|_V dT + \left(\frac{1}{T}\left.\frac{dU}{dV}\right|_T + \frac{p}{T}\right) dV$$

Equating the equivalent terms in both expressions of the differential dS, and taking into account the van der Waals equation of state (6.63), we find,

$$\left.\frac{\partial U}{\partial T}\right|_V = C_V$$

$$\left.\frac{dU}{dV}\right|_T = \frac{NRT}{V - Nb} - p = \frac{N^2 a}{V^2}$$

Substituting these expressions in the internal energy differential, dU becomes,

$$dU = C_V dT + \frac{N^2 a}{V^2} dV$$

Now, we show that the specific heat at constant volume C_V of the van der Waals gas is independent of the volume. Taking into account the first principle (1.38), the infinitesimal heat (5.4) provided to the system and the infinitesimal work (2.41) performed on the gas during a reversible process, we can rewrite the differential of the internal energy as,

$$dU = \delta Q + \delta W = C_V dT + (L_V - p) \, dV$$

The Schwarz theorem applied to the internal energy $U\left(S\left(T,V\right),V\right)$ reads,

$$\frac{\partial}{\partial V}\left(\left.\frac{\partial U}{\partial T}\right|_{V}\right)=\frac{\partial}{\partial T}\left(\left.\frac{dU}{dV}\right|_{T}\right)$$

and gives the following Maxwell relation,

$$\frac{\partial C_V}{\partial V}=\frac{\partial}{\partial T}\left(L_V-p\right)$$

Taking into account expression (5.10) for the coefficient L_V and the van der Waals equation of state (6.63), we obtain,

$$L_V-p=T\frac{\partial p\left(T,V\right)}{\partial T}-p=\frac{NRT}{V-Nb}-p=\frac{N^2a}{V^2}$$

Thus, the Maxwell relation reduces to,

$$\frac{\partial C_V}{\partial V}=\frac{\partial}{\partial T}\left(\frac{N^2a}{V^2}\right)=0$$

This shows that the specific heat at constant volume C_V of the van der Waals gas is independent of volume. Since the van der Waals gas tends towards the ideal gas when the volume of the system is sufficiently large, the specific heat at constant volume of the van der Waals gas, which is independent of volume, has to be equal to that of the ideal gas, i.e. $C_V=c\,NR$ [64]. Thus, we find that C_V is independent of temperature. Finally, we conclude by integrating the differential dU to obtain the following expression for the internal energy of the van der Waals gas,

$$U\left(T,V\right)=c\,NR\,T-\frac{N^2a}{V}$$

2. In a Joule expansion, we note the initial and final volumes V_i and V_f, respectively, and the initial and final temperatures T_i and T_m, respectively. The internal energy of an isolated system remains constant, i.e. $U\left(T_i,V_i\right)=U\left(T_m,V_f\right)$. Therefore, during a Joule expansion, the temperature variation is related to the volume variation by,

$$c\,NR\left(T_m-T_i\right)-N^2a\left(\frac{1}{V_f}-\frac{1}{V_i}\right)=0$$

which can be recast in the form

$$T_m-T_i=\frac{Na}{cR}\left(\frac{V_i-V_f}{V_iV_f}\right)$$

During a Joule expansion, the volume increases, i.e. $V_f>V_i$, which implies that the temperature decreases, i.e. $T_m<T_i$. We note that the introduction of an attractive force, characterised by the parameter a, is what accounts for the cooling. For a real gas, this represents a temperature variation of about $0.01\,\%$ [41].

Exercises

6.1 Melting Ice

A mixture of ice and water is heated up in such a way that the ice melts. The ice melts at a rate r and the molar latent heat of ice melting is $\ell_{s\ell}$.

a) Determine the thermal power P_Q transferred to the ice.
b) Determine the entropy rate of change $\dot{S}$.

Numerical Application:

$\Omega = 2.0 \cdot 10^{-2} \text{ mol s}^{-1}$, $\ell_{s\ell} = 6.0 \cdot 10^3 \text{ J mol}^{-1}$.

6.2 Cooling Water with Ice Cubes

Water is cooled with ice cubes (Fig. 6.10). The water and the ice cubes are considered as an isolated system. Initially, the ice cubes are at melting temperature T_0 and the water at temperature T_i. The total initial mass of ice is M' and the initial mass of water is M. The latent heat of melting of ice per unit mass is $\ell_{s\ell}^*$ and the specific heat per unit mass of water is c_V^*.

a) Determine the final temperature T_m of the water.
b) Determine the final temperature T_m of the water if melted ice (i.e. water) had been added at melting temperature T_0 instead of ice.

Numerical Application:

$M = 0.45$ kg, $M' = 0.05$ kg, $T_i = 20°C$, $T_0 = 0°C$, $\ell_{s\ell}^* = 3.33 \cdot 10^6 \text{ J kg}^{-1}$ and $c_V^* = 4.19 \cdot 10^3 \text{ J kg}^{-1} \text{ K}^{-1}$.

6.3 Wire through Ice without Cutting

A steel wire is wrapped over a block of ice with two heavy weights attached to the ends of the wire. The wire passes through a block of ice without cutting the block in two. The ice melts under the wire and the water freezes again above the wire. The wire is considered a rigid rod of negligible mass laying on the ice block with an area of contact A. The two weights of mass M each are hanging at both ends of the wire (Fig. 6.15). The entire system is at atmospheric pressure p_0 and the ice is held at temperature $T_m - \Delta T$ where T_m is the melting temperature at atmospheric pressure. The latent heat of melting of ice is $\ell_{s\ell}$, the molar volume of water v_ℓ and the molar volume of ice is v_s. Determine the minimal mass M of each weight for this experiment to succeed, i.e. for the wire to pass through the ice block.

6.4 Dupré's Law

A liquid is at equilibrium with its vapour. The vapour is assumed to be an ideal gas. The liquid has a molar latent heat of vaporisation $\ell_{\ell g}$ that depends on temperature, with $\ell_{\ell g} = A - BT$, where A and B are constants. Apply the Clausius–Clapeyron relation (6.50) and consider that the molar volume of the liquid phase is negligible compared to the vapour phase, i.e. $v_\ell \ll v_g$. Use the ideal gas law (5.47) for the

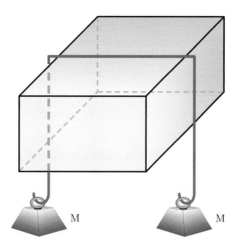

Figure 6.15 A steel wire wrapped over a block of ice with two heavy weights hanging on both sides passes through the ice, cutting the block in two.

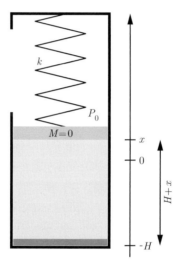

Figure 6.16 A cylinder containing a substance in liquid and gas phases is closed by a spring-loaded piston. The zero of the coordinate x is at the rest position of the spring. The mass of the piston is neglected.

vapour phase. Show that at equilibrium at a temperature T, the vapour pressure p depends on temperature according to Dupré's law,

$$\ln \left(\frac{p}{p_0} \right) = \frac{A}{R} \left(\frac{1}{T_0} - \frac{1}{T} \right) - \frac{B}{R} \ln \left(\frac{T}{T_0} \right)$$

where p_0 is the vapour pressure at T_0.

6.5 Hydropneumatic Accumulator

A container contains a substance in gaseous and liquid phases at room temperature (Fig. 6.16). The container is closed by a piston of surface area A, held back by a spring of elastic constant k. We neglect the mass of the piston. For simplicity, we neglect the volume of the liquid compared to that of the gas. The atmospheric pressure is p_0 and assumed independent of temperature.

a) Determine the temperature derivative of the gas pressure $\dfrac{dp}{dT}$ when there is no liquid phase present in the container.

b) Determine the temperature derivative of the gas pressure $\dfrac{dp}{dT}$ when liquid is present.

6.6   Positivity of Thermal Response Coefficients

To establish the positivity of the specific heat at constant pressure C_p compressibility coefficient at constant temperature κ_T (see relations (6.31)), follow the steps given here [65]:

a) Show that the Mayer relation (5.42) can be recast as,

$$C_p = C_V + \frac{\alpha^2}{\kappa_T} V T$$

where α is the thermal coefficient of expansion,

$$\alpha = \frac{1}{V} \frac{\partial V(T,p)}{\partial T} \qquad and \qquad \kappa_T = -\frac{1}{V} \frac{\partial V(T,p)}{\partial p}$$

b) Show that

$$\frac{\partial^2 F(T,V)}{\partial V^2} = \frac{\dfrac{\partial^2 U}{\partial S^2} \dfrac{\partial^2 U}{\partial V^2} - \left(\dfrac{\partial^2 U}{\partial S \partial V} \right)^2}{\dfrac{\partial^2 U}{\partial S^2}}$$

c) Conclude from these two results that $\kappa_T \geq 0$ and $C_p \geq 0$.

6.7   Heat Pipe

Heat pipes are devices used to transfer heat over a certain distance. A typical heat pipe looks like a metal rod, but modern versions, which are used for example to cool the hottest part of a phone, have a flat geometry. The principle of heat pipes is also considered in aerospace research [66]. Here, we will examine a simple model to understand the principle of a heat pipe (Fig. 6.17). The pressure difference Δp is modelled in a linear approximation, with $\Delta p = R_p \Omega$, where Ω is the rate of substance flowing down the pipe. The system is considered in a stationary state, so that the heat transfer P_Q is the same (in absolute value) on both sides. The heat of evaporation $\ell_{\ell g}$ is given and assumed independent of temperature. The temperature difference $\Delta T = T_+ - T_-$ is assumed small in order to simplify the calculations. Neglect the molar volume of the liquid v_ℓ compared to that of the vapour v_g and treat it as an ideal gas. Express the heat transfer P_Q as a function of the temperature difference ΔT.

6.8   Vapour Pressure of Liquid Droplets

Consider a cloud of droplets and assume that they all have the same diameter r. According to the Laplace formula (exercise 4.8), the pressure $p(r)$ inside the droplets of radius r is related to the vapour pressure $p_0(r)$ by,

$$p(r) = p_0(r) + \frac{2\gamma}{r}$$

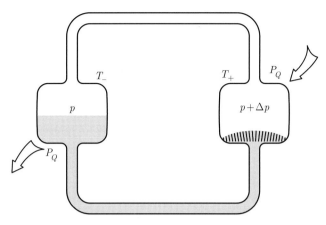

Figure 6.17 Principle of a heat pipe: At the hot side, the liquid passes through a wick and vaporises at pressure $p + \Delta p$. At the cold side, the vapour condenses at pressure p.

where γ is the surface tension. We note p_∞ the vapour pressure for an infinite radius. At temperature T, show that,

$$p_0(r) = p_\infty + \frac{2\gamma}{r}\frac{p_\infty v_\ell}{RT}$$

where v_ℓ is the molar volume of liquid in the limit where $p_\infty v_\ell \ll RT$ since the molar volume of liquid is much smaller than the molar volume of gas.

6.9   Melting Point of Nanoparticles

The surface tension modifies the melting point of particles. The effect is important when the diameter is in the nanometer range. A differential equation has to be written for $T_m(r)$, the melting temperature of particles of radius r. In order to perform this thermodynamical analysis, assume that the pressure p_s inside the particles is defined [67]. At atmospheric pressure p_0 and for infinitely large particles, the melting temperature is noted T_∞. The surface tension is γ_s for a solid particle and γ_l for a liquid one. According to exercise 4.8, the Laplace pressure $p_s(r)$ for a solid nanoparticle and the Laplace pressure $p_\ell(r)$ for a liquid nanoparticle are given by,

$$p_s(r) = \frac{2\gamma_s}{r} \quad \text{and} \quad p_\ell(r) = \frac{2\gamma_\ell}{r}$$

Determine the temperature difference $T_\infty - T_m(r)$ in terms of the latent heat of melting $\ell_{s\ell} = T_\infty(s_\ell - s_s)$ and the molar volumes v_s and v_ℓ that are both assumed to be independent of the radius r. Therefore, perform a series expansion in terms of the radius r on the chemical equilibrium condition. This result is known as the **Gibbs–Thomson equation**. For some materials, a lowering of the melting temperature can be expected, i.e. $T_m(r) < T_\infty$. This effect has been observed on individual nanoparticles by electron microscopy [68]. It is used to sinter ceramics at low temperatures [69].

6.10   Work on a Van der Waals Gas

A mole of oxygen, considered as a van der Waals gas, undergoes a reversible isothermal expansion at fixed temperature T_0 from an initial volume V_i to a final volume V_f. Determine the work W_{if} performed on the van der Waals gas in terms of the parameters a, and b.

Numerical Application:

$T_0 = 273$ K, $V_i = 22.3 \cdot 10^{-3}$ m^3, $V_f = 3V_i$, $p_0 = 1.013 \cdot 10^5$ Pa, $a = 0.14$ Pa m^6 and $b = 3.2 \cdot 10^{-6}$ m^3.

6.11 Inversion Temperature of the Joule–Thomson Process

A van der Waals gas is going through a Joule–Thomson process that keeps the enthalpy H constant (problem 4.8.3). A van der Waals gas in characterised by the following equations of state,

$$p = \frac{NRT}{V - Nb} - \frac{N^2 a}{V^2} \quad \text{and} \quad U = cNRT - \frac{N^2 a}{V}$$

and the amount of gas is constant, i.e. $N = $ const. Use the condition $dH = 0$ in order to obtain an expression for the derivative $\dfrac{dT}{dV}$. Determine the temperature T_0 at which this derivative changes sign.

6.12 Lever Rule

A phase diagram is drawn for a mixture of two substances at a fixed pressure p with a liquid phase and a gaseous phase (Fig. 6.18). The substances are labelled 1 and 2 and the diagram is shown as a function of the concentration c_1 of substance 1. There is a range of temperature for which there is coexistence of two phases. Answer the following questions, treating the concentrations c_1^A and c_1^B as given values.

a) Apply the Gibbs phase rule (6.62) to find the number of degrees of freedom when two phases coexist at a fixed pressure p.

b) We distill a substance 1 with an initial concentration c_1^A by heating the liquid up to the temperature T_C. Determine the final concentration of substance 1 after distillation.

c) A mixture with a concentration c_1^C of substance 1 is put in a container. The mixture is brought to a temperature T_C while the pressure remains at p. Establish that,

$$N_\ell \left(c_1^C - c_1^A \right) = N_g \left(c_1^B - c_1^C \right)$$

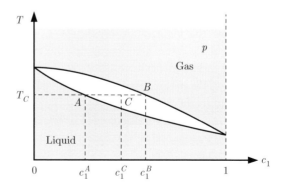

Figure 6.18   Phase diagram of a binary mixture presenting two phases and an exclusion zone (see § 6.4).

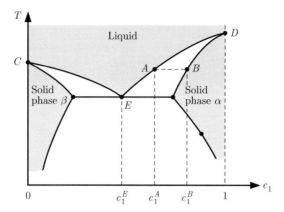

Figure 6.19 Phase diagram of a binary mixture presenting a eutectic point at E.

where N_ℓ the amount of mixture in the liquid phase, and N_g that in the gas phase. This is known as the **lever rule**.

6.13 Eutectic

A phase diagram is drawn for a mixture of two substances at a fixed pressure p with one liquid phase and two solid phases (Fig. 6.19). The substances are labelled 1 and 2 and the diagram is shown as a function of the concentration c_1 of substance 1. This diagram presents what is called a **eutectic** point. At the eutectic concentration c_1^E, the freezing temperature is the lowest. In particular, it is lower than the freezing temperatures of the pure substances (points C and D). At the eutectic, the liquid freezes into a mixture of two solid phases, the α and β phases.

a) Consider a liquid at concentration c_1^A. As the temperature is lowered, the point A is reached. Describe qualitatively what happens then.

b) Describe what happens if a liquid of composition c_1^E is cooled.

Heat Engines

Nicolas Léonard Sadi Carnot, 1796–1832

Sadi Carnot trained as an engineer at Ecole Polytechnique (Paris) under the guidance of eminent professors such as Ampere, Gay-Lussac and Poisson. He published *'Reflections on the motive power of heat'* at the age of 24. This ground-breaking treatise turned out to be his only publication, as he died at a young age during a cholera epidemic.

7.1 Historical Introduction

The first thermal machines were based on the work of Otto von Guericke (1602–1686), who built a pneumatic machine, a kind of syringe that evacuates air inside a large enclosure. This machine is said to have been presented to Emperor Ferdinand III. Von Guericke

Figure 7.1 Newcomen's pump (1712) was one of the very first heat engines. A valve lets high pressure vapour in the cylinder. A cold water entry causes the condensation of this vapour and the pressure drop causes the fall of the piston.

also created a partial vacuum inside a metallic sphere consisting of two hemispheres assembled together. Despite the poor quality of the joint, 16 horses tried to separate the two hemispheres without success. Seven years later, von Guericke built an engine based on this vacuum effect. The air was evacuated from a cylinder closed by a piston. The piston was connected to a load via a mechanical system. The motion of the piston would lift the load.

Christian Huygens and his assistant Denis Papin used the Von Guericke engine and set on fire a small amount of powder inside a cylinder. This provoked an expansion of the gas that lifted the piston. Cooling the gas created a depression that brought the piston back to its initial position. Later, Denis Papin perfected this machine by using steam and showed that the condensation of water vapour induced similar effects to the depression of the gas.

Another scientist, Thomas Newcomen (1664–1729), had the idea of injecting water in a container to condense the vapour (Fig. 7.1). Later on, a proper condenser was invented by James Watt (1736–1819). He found through careful observation that vaporisation required latent heat. However, Joseph Black (1728–1799) had established its existence before him.

Carnot spent his youth in Paris and entered the Institute of Technology at the age of 16. First, he served his time in the military, then he fully devoted his time to research. In 1824, he published his only book, a landmark in the history of thermodynamics entitled *Reflections on the Motive Power of Heat and on Machines Fitted to Develop that Power*. This 100-page-long book contained only five figures and a few equations, which appeared in the footnotes. The book was barely noticed by the scientific community until Clausius published in 1834 an article in the journal of the Institute of Technology in Paris. In this article, he showed how the ideas of Sadi Carnot could be expressed mathematically. This led to a second publication of Carnot's book that became much more appreciated. However, the book quickly went out of print. Lord Kelvin, who struggled to find a copy, eventually got it translated in 1890. From then on, the ideas of Carnot widely spread throughout the scientific community. Carnot's work remains remarkable today because it laid the foundations of thermodynamics through a reflection that began with technical issues and rose to a level of abstraction that would be an inspiration to Clausius and Lord Kelvin.

In his 'reflections', Carnot first pointed out that mechanical power resulted from thermal processes: *'Everyone knows that heat can produce motion. That it possesses vast motive power no one can doubt, in these days when the steam engine is everywhere so well known. Heat also provokes the vast movements which take place on the earth. It causes the agitations of the atmosphere, the ascension of clouds, the fall of rain and of meteors, the currents of water that span the surface of the globe, of which man has thus far employed but a small portion. Even earthquakes and volcanic eruptions are the result of heat'* [70].

Then, Carnot considered an ideal steam engine and focused on its working principle, not its actual fabrication. He assumed that the circulating heat was a fluid he called the 'caloric'. His view was that the production of motive power in steam engines was due, therefore, not to an actual consumption of caloric, but to its transportation from a warm to a cold body. *'It is not sufficient to give birth to an impelling power: it is necessary that there should also be cold; without it, the heat would be useless'.*

He made the analogy with a waterfall: *'We can compare with sufficient accuracy the motive power of heat to that of a waterfall. Each has a maximum that we cannot exceed [...]. The motive power of a waterfall depends on its height and on the quantity of the liquid; the motive power of heat depends also on the quantity of caloric used, and on what may be termed, the height of its fall, that is to say, the difference of temperature of the bodies between which the exchange of caloric is made'.* Carnot illustrated his viewpoint by considering a cycle without losses and showed *ad absurdum* that there could not be a more efficient cycle. Thus, he established that the efficiency of the machine was independent of the substance inside and that it depended only on the temperatures in between which it operated. However, the 'caloric' fluid hypothesis was unsuccessful in accounting for experimental results and Carnot later reviewed his statement and asserted that this 'caloric' fluid hypothesis had to be abandoned.

His thinking contained what would become later the first law of thermodynamics : *'Heat is simply motive power, or rather motion which has changed form [...]. We can then establish the general proposition that motive power is, in quantity, invariable in nature; that it is correctly speaking, never either produced or destroyed. It is true that it changes form [...]'.*

7.2 Thermal Machine

Before engaging in the analysis of the thermal machine originally made by Carnot, let us become acquainted with the notion of thermal machines. Formally, a system is called a *thermal machine* if it performs a heat transfer between two thermal baths, thus allowing a mechanical action form the environment on the system and vice versa. In a typical thermal machine, heat is effectively transferred between hot and cold heat reservoirs by means of a machine that passes periodically through the same states. Thus, the operation of a thermal machine is not described by a single continuous process, but by a series of distinct processes. The real mechanism can be complex. However, the basic working principle of a thermal machine can be understood by identifying the presence of hot and cold heat reservoirs, also referred to as a hot '*source*' and a cold '*sink*' in this context. In this section, we learn to recognise hot and cold reservoirs in some very simple thermal machines.

For instance, let us consider the so-called 'vacuum engine' or 'flame engine' (Fig. 7.2). The flame is the hot source of the engine and the metallic blade constitutes the cold sink. Once hot air near the flame is trapped inside the cylinder, it gets colder by contact with the blade. Consequently, its pressure decreases below atmospheric pressure and the piston acts on the flywheel, which moves the piston inside the cylinder and provokes the rotation of the flywheel. This rotation leads to the opening of the valve. When the valve opens up again, hot air rushes inside the cylinder and the cycle starts again.

Let us now consider a toy steamboat (Fig. 7.3). Here, the flame is considered the hot source and the pipes cooled by the surrounding water constitute the cold sink. The steam produced in the heated capsule expels the water contained in the pipes. This causes a thrust that gives rise to a displacement of the boat. The steam condenses as it is cooled by the pipes and the cycle restarts.

In a hand-held Stirling engine (Fig. 7.4), the hot reservoir corresponds to the palm of the hand holding it. The skin has a higher temperature than the surrounding air, which corresponds to the cold reservoir. The temperature difference between the hand and the

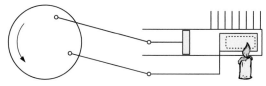

Figure 7.2   In this engine, a flame heats the air in front of a valve that opens and closes using a system of rods linked to a flywheel. The decrease of air pressure caused by the cooling fins causes a mechanical action on the piston that drives the flywheel.

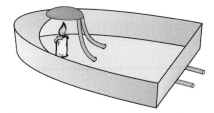

Figure 7.3   In a small toy boat, a flame brings the water, contained in a metal capsule, to boil. The resulting steam ejects the water contained in the pipes underneath the boat, providing a thrust to the boat. The steam condenses on the wall of the cold pipes, and the cycle ends by a slow inlet of water.

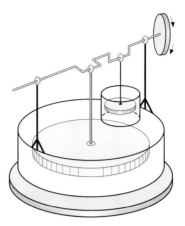

Figure 7.4 Sketch of a Stirling engine working with the heat input of a hand pressed against the metallic base of the device. Air inside the cylindrical container alternates between the top where it gets colder and the bottom where it gets hotter.

Figure 7.5 A duck keeps dipping its beak into a glass of water. The body of the duck contains a pipe filled with a liquid and its vapour. Vaporisation cools the beak, causing the liquid to rise until the duck dips to a horizontal position in which the liquid flows back down into the body, causing the duck to straighten again.

air is sufficient for this small engine to drive the rotation of a wheel. We will analyse the principle of a Stirling engine in § 7.8.

Sometimes the operation of a thermal machine can give the illusion of perpetual motion. This is due to misleading appearances. In the experiment of the rocking duck (Fig. 7.5), the beak cools because water evaporates in the room. This cooling lifts the liquid contained inside the duck until the whole duck flips towards the water again. The water is kept at room temperature by heat transfer with the atmosphere. The working principle of this heat engine can be verified by monitoring the temperature of the beak with a thermocouple mounted on it. Furthermore, it can be shown that the duck stops after a while if it is enclosed in a container, because the surrounding air becomes saturated with water vapour.

To conclude this section, we note that there are thermal machines that perform heat transfers between several thermal baths. Such is the case of the Diesel cycle which will be analysed in § 7.10.5. In the following section, we will look at machines that come into contact only with two heat reservoirs.

7.3 Carnot Cycle

The thermal processes between thermal machines and heat reservoirs are usually irreversible, in particular because the temperature of the reservoir is different from that of the system. As we showed in § 3.2, heat transfer between two regions that have different temperatures is always an irreversible process. Thus, in order for a heat transfer to be reversible, the temperature of the reservoir has to be identical to that of the system when it comes in contact with either one of them. The mechanical processes between the system and the environment may also be irreversible.

Let us now consider a thermal machine that only undergoes reversible processes. Its operation is defined by the cycle that was first put forth by Carnot (Fig. 7.6). A gas, considered as a simple system, is brought to temperature T^+ of a hot reservoir. They are then brought into contact with each other (state 1). While in contact, a heat transfer occurs and the entropy S of the gas increases from S^- to S^+. For this process to be reversible it has to happen slowly enough so that there is nearly no temperature difference between the reservoir and the gas. The gas is then detached from the hot reservoir (state 2) and undergoes a reversible adiabatic process that adjusts its temperature T to the temperature T^- of a cold reservoir. The gas is then brought into contact with the cold reservoir (state 3). While in contact, a heat transfer occurs and the entropy S of the gas decreases from S^+ to S^-. Again, for this process to be reversible, it has to happen slowly enough so that there is nearly no temperature difference between the reservoir and the gas. The gas is then detached from the cold reservoir (state 4) and undergoes an adiabatic process that adjusts its temperature T to the temperature T^+ of the hot reservoir. Thus, the system comes back to its initial state (state 1) and the cycle can start again.

In the thermal machine described by Carnot, this cycle is repeated a large number of times. Every reversible cycle consisting of two isothermal and two adiabatic processes is

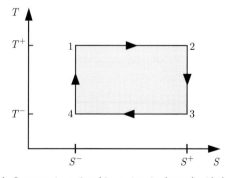

Figure 7.6 (T, S) diagram of a Carnot cycle. Processes $1 \rightarrow 2$ and $3 \rightarrow 4$ are isothermal, with the gas temperature being T^+ and T^-. Processes $2 \rightarrow 3$ and $4 \rightarrow 1$ are adiabatic and reversible, with the gas entropy being S^+ and S^-.

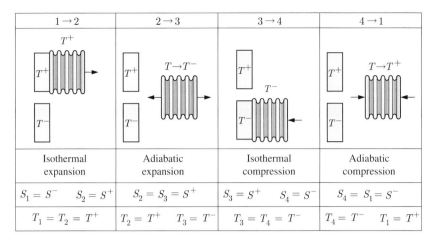

$1 \to 2$	$2 \to 3$	$3 \to 4$	$4 \to 1$
Isothermal expansion	Adiabatic expansion	Isothermal compression	Adiabatic compression
$S_1 = S^- \qquad S_2 = S^+$	$S_2 = S_3 = S^+$	$S_3 = S^+ \qquad S_4 = S^-$	$S_4 = S_1 = S^-$
$T_1 = T_2 = T^+$	$T_2 = T^+ \qquad T_3 = T^-$	$T_3 = T_4 = T^-$	$T_4 = T^- \qquad T_1 = T^+$

Figure 7.7   The four processes of the Carnot cycle performed on a gas enclosed in a bellows.

called a ***Carnot cycle*** (Fig. 7.6). Later on, we will show further that this cycle has optimal efficiency [71].

In order to emphasise the role of the heat reservoirs in the Carnot cycle (Fig. 7.7) [72], let us consider that the gas is contained in a bellows of which the base can easily be set in thermal contact with either one of the heat reservoirs. We assume that when the bellows is not in contact with a heat reservoir, the system is adiabatic and undergoes reversible processes only. Hence, the entropy is constant during the adiabatic processes. When the bellows is isolated from the heat sources, the volume is adjusted in order to reach one of the temperatures of the heat sources. The bellows is then put into contact with the hot reservoir and the volume increases under the effect of the isothermal heat transfer. Thus, according to relation (5.73), the entropy of the gas in the bellows increases at constant temperature. Then, the bellows is detached from the hot reservoir. The gas is expanded adiabatically to adjust its temperature to the cold reservoir. The bellows can then be put into contact with that reservoir and the gas can be compressed at constant temperature. Thus, according to relation (5.73), the entropy of the gas decreases at constant temperature. The bellows can then be detached and the cycle starts again.

The Carnot cycle represented on a (T, S) diagram (Fig. 7.6) allows an immediate analysis of heat transfer and mechanical action. The heat Q provided to the gas during a cycle corresponds to the sum of the heat transfers provided to the gas during the four processes,

$$Q = Q_{12} + Q_{23} + Q_{34} + Q_{41} \tag{7.1}$$

Since processes $2 \to 3$ and $4 \to 1$ are adiabatic, the heat transfers vanish,

$$Q_{23} = Q_{41} = 0 \qquad \text{(Carnot cycle)} \tag{7.2}$$

Using definition (2.25), the heat (7.1) provided to the gas during a cycle is given by,

$$Q = \int_{S_1}^{S_2} T \, dS - \int_{S_4}^{S_3} T \, dS = T^+ \left(S^2 - S^1 \right) - T^- \left(S^3 - S^4 \right)$$
$$= \left(T^+ - T^- \right) \left(S^+ - S^- \right) \qquad \text{(Carnot cycle)} \tag{7.3}$$

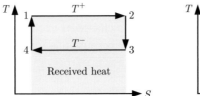 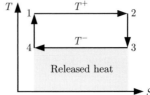

Carnot cycle: The rectangular area on the diagram on the left-hand side represents the heat received per cycle from the hot reservoir and the rectangular area on the diagram on the right-hand side represents the heat released per cycle to the cold reservoir.

where the negative sign in front of the second integral comes from the inversion of the integration limits. The first integral of equation (7.3) corresponds to the heat extracted from the hot source during a cycle. It is illustrated by the shaded rectangular area on the diagram on the left-hand side (Fig. 7.8). The second integral corresponds to the heat released to the cold source during a cycle. It is illustrated by the rectangular shaded area on the diagram on the right-hand side (Fig. 7.8). Thus, the resulting heat $Q > 0$ provided to the gas is illustrated by the area of the cycle, as it corresponds to the difference of these two areas.

The work per cycle W performed on the gas is the sum of the work performed on the gas during each process,

$$W = W_{12} + W_{23} + W_{34} + W_{41} \tag{7.4}$$

Since the initial and final states of a cycle are identical and the internal energy is a state function, the internal energy variation per cycle corresponding to the sum of the internal energy variations of the gas during the four processes vanishes,

$$\Delta U = \Delta U_{12} + \Delta U_{23} + \Delta U_{34} + \Delta U_{41} = 0 \tag{7.5}$$

According to the first law (1.44), the work performed on the gas during a cycle is the opposite of the heat provided to the system during this cycle,

$$W = -Q \tag{7.6}$$

Since the initial and final states of a cycle coincide and the entropy is a state function, the entropy variation per cycle, corresponding to the sum of the entropy variations of the gas during the four processes, vanishes,

$$\Delta S = \Delta S_{12} + \Delta S_{23} + \Delta S_{34} + \Delta S_{41} = 0 \tag{7.7}$$

Since processes $2 \to 3$ and $4 \to 1$ are adiabatic and reversible, the entropy is constant, i.e.

$$\Delta S_{23} = \Delta S_{41} = 0 \qquad \text{(Carnot cycle)} \tag{7.8}$$

Thus, relations (7.7) and (7.8) imply,

$$\Delta S_{12} = -\Delta S_{34} \qquad \text{(Carnot cycle)} \tag{7.9}$$

For a clockwise cycle on a (T, S) diagram (Fig. 7.8), the heat provided to the gas during the heat transfer between the reservoirs is positive, i.e. $Q > 0$ and the work performed on the gas by the environment is negative, i.e. $W < 0$. Such a cycle is called an ***engine cycle***. A thermal machine undergoing a clockwise cycle on the (T, S) diagram, i.e. where $Q = -W > 0$, is called a ***heat engine*** because this machine receives heat and in return, performs work on the environment.

For a counterclockwise cycle on a (T, S) diagram (Fig. 7.8), the heat provided to the gas during the heat transfer between the two heat reservoirs is negative, i.e. $Q < 0$ and the work performed on the gas by the environment is positive, i.e. $W > 0$. Such a cycle is called a ***refrigeration cycle*** or ***heat pump cycle***. A thermal machine describing a counterclockwise cycle on the (T, S) diagram, i.e. where $W = -Q > 0$, is called a ***refrigerator*** or a ***heat pump*** because work is performed on this machine by the environment. As a consequence, it extracts heat from the cold reservoir (refrigerator) or brings heat to the hot reservoir (heat pump).

7.4 Reversible Processes on an Ideal Gas

In this section, we will calculate changes in physical quantities during reversible processes performed on an ideal gas. In the next section, we will apply these results to analyse the Carnot cycle in more detail.

It will be useful to keep in mind how energy changes during the processes we will be analysing. Taking into account definitions (1.39) and (5.62), the internal energy variation ΔU_{if} during any process that brings the ideal gas from an initial state i to a final state f is given by,

$$\Delta U_{if} = \int_{U_i}^{U_f} dU = c\,NR \int_{T_i}^{T_f} dT = c\,NR\,(T_f - T_i) \tag{7.10}$$

According to definitions (4.62) and (5.65), the enthalpy variation ΔH_{if} during this process is given by,

$$\Delta H_{if} = \int_{H_i}^{H_f} dH = (c+1)\,NR \int_{T_i}^{T_f} dT = (c+1)\,NR\,(T_f - T_i) \tag{7.11}$$

We will use the word ***expansion*** to define a process during which the volume of the system increases and ***contraction*** the inverse process. Likewise, we will use the word ***compression*** to define a process that increases the pressure of the system and ***decompression*** the inverse process.

In the remainder of this section, we compute the heat Q_{if} provided to the system, the work W_{if} performed on the system and the entropy variation ΔS_{if} of the system for adiabatic, isothermal, isochoric and isobaric processes that take the system from an initial state i to a final state f.

7.4.1 Reversible Adiabatic Processes

Reversible adiabatic processes are isentropic, which means that they occur at constant entropy, i.e. $dS = 0$. Thus, the entropy variation (5.74) during a reversible adiabatic process vanishes,

$$\Delta S_{if} = \int_{S_i}^{S_f} dS = 0 \tag{7.12}$$

and the heat transfer Q_{if} provided to the system during a reversible adiabatic process vanishes as well,

$$Q_{if} = \int_i^f T dS = 0 \tag{7.13}$$

Since there is no heat transfer between the system and the exterior, the first law (1.44) implies that the work performed on the system is equal to the internal energy variation (7.10),

$$W_{if} = \Delta U_{if} = c\,NR \int_{T_i}^{T_f} dT = c\,NR\,(T_f - T_i) \tag{7.14}$$

7.4.2 Reversible Isothermal Processes

Isothermal processes occur at constant temperature, i.e. $dT = 0$. According to definitions (2.42) and (5.47), the work performed on the system during an isothermal process at constant temperature T is written as,

$$W_{if} = -\int_i^f p\,dV = -NR\,T \int_{V_i}^{V_f} \frac{dV}{V} = -NR\,T \ln\left(\frac{V_f}{V_i}\right) \tag{7.15}$$

According to (5.62), the internal energy variation at constant temperature vanishes, i.e. $\Delta U_{if} = 0$. Taking into account the first law (1.44) and equation (5.92), the heat Q_{if} provided to the system during a reversible isothermal process is given by,

$$Q_{if} = -W_{if} = \int_i^f p\,dV = NR\,T \int_{V_i}^{V_f} \frac{dV}{V} = NR\,T \ln\left(\frac{V_f}{V_i}\right) \tag{7.16}$$

According to (5.62), the internal energy differential of an ideal gas at constant temperature vanishes, i.e. $dU = 0$. Taking into account the Gibbs relation (4.1) and the ideal gas equation of state (5.47), the entropy variation ΔS_{if} of the system during a reversible isothermal process is written as,

$$\Delta S_{if} = \int_{S_i}^{S_f} dS = \int_i^f \frac{p\,dV}{T} = NR \int_{V_i}^{V_f} \frac{dV}{V} = NR \ln\left(\frac{V_f}{V_i}\right) \tag{7.17}$$

7.4.3 Reversible Isochoric Processes

Isochoric processes occur at constant volume, i.e. $dV = 0$. According to definition (2.42), the work performed on the system during a reversible isochoric process vanishes,

$$W_{if} = -\int_i^f p\,dV = 0 \tag{7.18}$$

Taking into account the first law (1.44) and definition (7.10) for the internal energy variation, the heat Q_{if} provided to the system during a reversible isochoric process is written as,

$$Q_{if} = \Delta U_{if} = \int_{U_i}^{U_f} dU = c\,NR \int_{T_i}^{T_f} dT = c\,NR\,(T_f - T_i) \tag{7.19}$$

Taking into account the Gibbs relation (4.1) under the constraint of constant volume, the entropy variation ΔS_{if} of the system during a reversible isothermal process is written as,

$$\Delta S_{if} = \int_{S_i}^{S_f} dS = c\,NR \int_{T_i}^{T_f} \frac{dT}{T} = c\,NR \ln\left(\frac{T_f}{T_i}\right) \tag{7.20}$$

7.4.4 Reversible Isobaric Processes

Isobaric processes occur at constant pressure, i.e. $dp = 0$. According to the definitions (2.42) and (5.47), the work performed on the system during a reversible isobaric process at constant pressure p is written as,

$$W_{if} = -\int_i^f p\,dV = -p \int_{V_i}^{V_f} dV = -p\,(V_f - V_i) = -NR\,(T_f - T_i) \tag{7.21}$$

Taking into account (4.61) and definition (7.11) for the enthalpy variation, the heat Q_{if} provided to the system during a reversible isobaric process is written as,

$$Q_{if} = \Delta H_{if} = \int_{H_i}^{H_f} dH = (c+1)\,NR \int_{T_i}^{T_f} dT = (c+1)\,NR\,(T_f - T_i) \tag{7.22}$$

Taking into account the enthalpy differential (4.32) under the constraint of constant pressure and the ideal gas equation of state (5.47), the entropy variation ΔS_{if} of the system during a reversible isothermal process is written as,

$$\Delta S_{if} = \int_{S_i}^{S_f} dS = \int_i^f \frac{dH}{T} = (c+1)\,NR \int_{T_i}^{T_f} \frac{dT}{T} = (c+1)\,NR \ln\left(\frac{T_f}{T_i}\right) \tag{7.23}$$

7.5 Carnot Cycle for an Ideal Gas

With the equations established in the previous section, we can now analyse in more detail the Carnot cycle performed on an ideal gas. To represent this cycle in the (p, V) diagram (Fig. 7.9), we take into account that an isothermal process is characterised by the condition $pV = \text{const}$ and an adiabatic process by the condition $pV^\gamma = \text{const}$. Since $\gamma > 1$, the

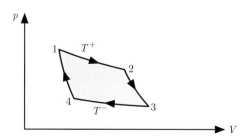

Figure 7.9 (p, V) diagram of the Carnot cycle. The slope differences are exaggerated for clarity.

absolute value of the slope of the adiabatic process is larger than that of the isothermal process.

In a (p, V) diagram, the area of a cycle is positive when the cycle is oriented in a clockwise manner and corresponds to $-W$, where W is the work performed by the environment on the system. Thus, on a (p, V) diagram, an engine cycle is oriented in a clockwise manner and a refrigeration cycle in a counterclockwise manner. Moreover, the area of a cycle in a (T, S) diagram corresponds to the heat Q provided to the system per cycle. Since $Q = -W$, the area of a cycle is the same in both diagrams. Given that $\delta Q = T \, dS$ and $\delta W = -p \, dV$, the cycle has the same orientation in the (p, V) and (T, S) diagrams.

Equation (7.10) yields the internal energy variation of an ideal gas during the isothermal processes $1 \to 2$ (at temperature T^+) and $3 \to 4$ (at temperature T^-),

$$\Delta U_{12} = \Delta U_{34} = 0 \tag{7.24}$$

Likewise, equation (7.10) yields the internal energy variation of the ideal gas during the adiabatic processes $2 \to 3$ and $4 \to 1$,

$$\Delta U_{23} = c\,NR \int_{T_2}^{T_3} dT = -c\,NR \left(T^+ - T^- \right)$$

$$\Delta U_{41} = c\,NR \int_{T_4}^{T_1} dT = c\,NR \left(T^+ - T^- \right) \tag{7.25}$$

Expressions (7.24) and (7.25) for the energy variations during the processes of the Carnot cycle satisfy (7.5), which states that the internal energy variation ΔU per cycle vanishes.

Equation (7.11) yields the enthalpy variation of an ideal gas during the isothermal processes $1 \to 2$ (at temperature T^+) and $3 \to 4$ (at temperature T^-),

$$\Delta H_{12} = \Delta H_{34} = 0 \tag{7.26}$$

Likewise, equation (7.11) yields the enthalpy variation of an ideal gas during the adiabatic processes $2 \to 3$ and $4 \to 1$,

$$\Delta H_{23} = (c+1)\,NR \int_{T_2}^{T_3} dT = -(c+1)\,NR \left(T^+ - T^- \right)$$

$$\Delta H_{41} = (c+1)\,NR \int_{T_4}^{T_1} dT = (c+1)\,NR \left(T^+ - T^- \right) \tag{7.27}$$

The results (7.26) and (7.27) show that the enthalpy variation ΔH per cycle vanishes, as it should since enthalpy is a state function.

Using definition (5.68) of the coefficient γ, relation (5.78) obtained for any adiabatic process $i \to f$ yields a relation between the volumes, i.e.

$$T_f V_f^{\gamma-1} = T_i V_i^{\gamma-1} \qquad \Rightarrow \qquad \frac{V_f}{V_i} = \left(\frac{T_i}{T_f} \right)^c \tag{7.28}$$

Identifying equation (7.28) applied to the adiabatic processes $2 \to 3$ and $4 \to 1$ yields,

$$\frac{V_3}{V_4} = \frac{V_2}{V_1} \tag{7.29}$$

Taking into account equations (7.15) and (7.29), we find the work performed on an ideal gas during the isothermal processes $1 \rightarrow 2$ (at temperature T^+) and $3 \rightarrow 4$ (at temperature T^-),

$$W_{12} = - \int_1^2 p\,dV = -NRT^+ \int_{V_1}^{V_2} \frac{dV}{V} = -NRT^+ \ln\left(\frac{V_2}{V_1}\right)$$

$$W_{34} = - \int_3^4 p\,dV = -NRT^- \int_{V_3}^{V_4} \frac{dV}{V} = NRT^- \ln\left(\frac{V_3}{V_4}\right) \tag{7.30}$$

$$= NRT^- \ln\left(\frac{V_2}{V_1}\right)$$

Equation (7.14) yields the work performed on an ideal gas during the adiabatic processes $2 \rightarrow 3$ and $4 \rightarrow 1$,

$$W_{23} = \Delta U_{23} = c\,NR \int_{T_2}^{T_3} dT = -c\,NR\left(T^+ - T^-\right)$$

$$W_{41} = \Delta U_{41} = c\,NR \int_{T_4}^{T_1} dT = c\,NR\left(T^+ - T^-\right) \tag{7.31}$$

Taking into account definition (7.4), as well as equations (7.30) and (7.31), the work W performed on an ideal gas during a cycle is written as,

$$W = -NR\left(T^+ - T^-\right) \ln\left(\frac{V_2}{V_1}\right) \tag{7.32}$$

Taking into account equations (7.16) and (7.30), we find the heat provided to an ideal gas during the isothermal processes $1 \rightarrow 2$ at the hot reservoir and $3 \rightarrow 4$ at the cold reservoir,

$$Q_{12} = -W_{12} = NRT^+ \ln\left(\frac{V_2}{V_1}\right) \tag{7.33}$$

$$Q_{34} = -W_{23} = -NRT^- \ln\left(\frac{V_3}{V_4}\right) = -NRT^- \ln\left(\frac{V_2}{V_1}\right)$$

Equation (7.13) yields the heat provided to the ideal gas during the adiabatic processes $2 \rightarrow 3$ and $4 \rightarrow 1$,

$$Q_{23} = Q_{41} = 0 \tag{7.34}$$

Taking into account definition (7.1), as well as equations (7.33) and (7.34), the heat Q provided to the system per cycle is written as,

$$Q = NR\left(T^+ - T^-\right) \ln\left(\frac{V_2}{V_1}\right) \tag{7.35}$$

We note that equations (7.32) and (7.35) satisfy the first law (7.6) applied to a cycle.

Taking into account equations (7.17) and (7.29), we find the entropy variation of an ideal gas during the isothermal processes $1 \rightarrow 2$ (at temperature T^+) and $3 \rightarrow 4$ (at temperature T^-),

$$\Delta S_{12} = \int_1^2 \frac{p\,dV}{T} = NR \int_{V_1}^{V_2} \frac{dV}{V} = NR \ln\left(\frac{V_2}{V_1}\right) \tag{7.36}$$

$$\Delta S_{34} = \int_3^4 \frac{p\,dV}{T} = NR \int_{V_3}^{V_4} \frac{dV}{V} = -NR \ln\left(\frac{V_3}{V_4}\right) = -NR \ln\left(\frac{V_2}{V_1}\right)$$

Equation (7.17) yields the entropy variation of an ideal gas during the adiabatic processes $2 \to 3$ and $4 \to 1$,

$$\Delta S_{23} = \Delta S_{41} = 0 \tag{7.37}$$

Equations (7.36) and (7.37) satisfy (7.7), that is, the entropy variation ΔS per cycle vanishes.

7.6 Efficiency and Coefficients of Performance

The **efficiency** η of a heat engine, running an engine cycle ($Q > 0$), is defined as the ratio of the work $- W$ performed by the gas on the environment and the heat Q^+ extracted from the hot reservoir at temperature T^+,

$$\eta = -\frac{W}{Q^+} \tag{7.38}$$

The negative sign is there for the work performed by the system on the environment to be positive. Taking into account relation (7.6) between the work W performed on the gas and the heat Q provided to the gas per cycle, we can write

$$\eta = \frac{Q}{Q^+} \tag{7.39}$$

where $0 < Q < Q^+$, thus the efficiency $0 < \eta < 1$.

The **heating coefficient of performance** ε^+ of a heat pump, running a refrigeration cycle ($Q < 0$), is defined as the ratio of the heat $- Q^+$ released to the hot reservoir at temperature T^+ and the work W performed on the gas by the environment,

$$\varepsilon^+ = -\frac{Q^+}{W} \tag{7.40}$$

Taking into account the efficiency (7.39) and relation (7.6) between work W performed on the gas and heat Q provided to the gas per cycle, we can write

$$\varepsilon^+ = \frac{Q^+}{Q} = \frac{1}{\eta} \tag{7.41}$$

where $Q^+ < Q < 0$, thus the heating coefficient of performance $\varepsilon^+ > 1$.

The **cooling coefficient of performance** ε^- of a refrigerator, running a refrigeration cycle ($Q < 0$), is defined as the ratio of the heat Q^- extracted from the cold reservoir at temperature T^- and the work W performed on the gas by the environment,

$$\varepsilon^- = \frac{Q^-}{W} \tag{7.42}$$

where $Q^- > 0$. Taking into account the efficiency (7.39) and relation (7.6) between work performed on the gas W and heat provided to the gas Q per cycle, we can write,

$$\varepsilon^- = -\frac{Q^-}{Q} \tag{7.43}$$

These definitions apply to any thermal machine running between two heat reservoirs. In the next two sections, we determine the efficiency and the coefficients of performance for the Carnot cycle.

7.6.1 Carnot Engine Cycle

Taking into account the results (7.33), (7.35) and (7.36), we can write the heat Q^+ extracted from the hot reservoir, the heat $-Q^-$ released to the cold reservoir and the heat Q provided to the gas during a Carnot engine cycle (Fig. 7.10) as,

$$\begin{aligned}
Q^+ &= Q_{12} = T^+ \Delta S_{12} > 0 \\
Q^- &= Q_{34} = T^- \Delta S_{34} = -T^- \Delta S_{12} < 0 \\
Q &= \left(T^+ - T^-\right) \Delta S_{12} > 0
\end{aligned} \tag{7.44}$$

Thus, for a Carnot cycle running as an engine cycle, we have indeed,

$$Q = Q^+ + Q^- > 0 \tag{7.45}$$

Applying equations (7.44) to the efficiency definition (7.39), we obtain the **Carnot efficiency** η_C,

$$\eta_C = \frac{T^+ - T^-}{T^+} = 1 - \frac{T^-}{T^+} < 1 \tag{7.46}$$

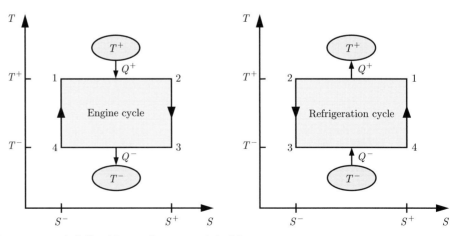

Figure 7.10   Carnot engine cycle (left) and Carnot refrigeration cycle (right).

7.6.2 Carnot Refrigeration Cycle

Taking into account the results (7.33), (7.35) and (7.36), we write the heat Q^- extracted from the cold reservoir, the heat Q^+ released to the hot reservoir and the heat Q provided to the gas during a Carnot refrigeration cycle (Fig. 7.10) as,

$$Q^- = Q_{34} = T^- \, \Delta S_{34} = - T^- \, \Delta S_{12} > 0$$
$$Q^+ = Q_{12} = T^+ \, \Delta S_{12} < 0 \tag{7.47}$$
$$Q = \left(T^+ - T^- \right) \Delta S_{12} < 0$$

Thus, for a Carnot cycle running as a refrigeration cycle, we have indeed,

$$Q = Q^- + Q^+ < 0 \tag{7.48}$$

Applying equations (7.47) to the definitions 7.41 and 7.43, we obtain expressions for the Carnot **heating coefficient of performance** ε_C^+ and the Carnot **cooling coefficient of performance** ε_C^-,

$$\varepsilon_C^+ = \frac{1}{\eta_C} = \frac{T^+}{T^+ - T^-} > 1$$
$$\varepsilon_C^- = \frac{1 - \eta_C}{\eta_C} = \frac{T^-}{T^+ - T^-} > 0 \tag{7.49}$$

7.7 Endoreversible Carnot Cycle

Until now we have considered only reversible Carnot cycles. In this section, we include irreversible heat transfers between the heat reservoirs and a system that runs a Carnot engine cycle. Thus, heat transfers are driven by temperature differences between the reservoirs and the system during the isothermal processes of a Carnot cycle (Fig. 7.11).

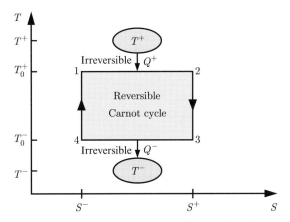

Figure 7.11   Endoreversible Carnot engine cycle : the temperatures of the reservoirs and isothermal processes are assumed to be different in order to take into account the irreversibility of heat transfers between the reservoirs and the engine.

The heat extracted from the hot reservoir is $Q^+ > 0$ and the heat released to the cold reservoir $Q^- < 0$. The temperature of the hot reservoir is T^+ and the temperature of the cold reservoir T^-. The system undergoes a Carnot cycle in which the isothermal processes occur at temperature T_0^+ when the system is in contact with the hot reservoir and T_0^- when it is in contact with the cold reservoir. The temperatures satisfy the order relation,

$$T^- < T_0^- < T_0^+ < T^+ \tag{7.50}$$

As this cycle contains a Carnot cycle within irreversible heat transfers with heat reservoirs, it is called an **endoreversible Carnot cycle**. The prefix 'endo' means 'inside' in Greek.

In order to analyse the endoreversible Carnot engine cycle and determine its efficiency, we consider that the heat transfers between the reservoirs and the system are stationary. Thus, taking into account definition (1.41) and the transport equation (3.16), we obtain,

$$Q^+ = \int_0^{\Delta t^+} P_Q \, dt = \kappa \frac{A}{\ell} \left(T^+ - T_0^+ \right) \Delta t^+ > 0$$
$$Q^- = \int_0^{\Delta t^-} P_Q \, dt = \kappa \frac{A}{\ell} \left(T^- - T_0^- \right) \Delta t^- < 0 \tag{7.51}$$

where Δt^+ and Δt^- are the durations of the heat transfers with the hot and cold reservoirs. In (7.51), we consider that the parameters κ, A and ℓ are the same for the heat transfers to both reservoirs. The duration of adiabatic processes is negligible compared to the duration of the isothermal processes. Based on this approximation and (7.51), the duration of a cycle Δt is given by,

$$\Delta t = \Delta t^+ + \Delta t^- = \frac{\ell}{\kappa A} \left(\frac{Q^+}{T^+ - T_0^+} + \frac{Q^-}{T^- - T_0^-} \right) \tag{7.52}$$

Since we have a Carnot cycle occur running between the temperatures T_0^+ and T_0^-, we can use the expressions (7.46), (7.38), (7.42) and (7.49) for the efficiency and the cooling coefficient of performance in order to obtain the relations,

$$\eta_C = \frac{T_0^+ - T_0^-}{T_0^+} = -\frac{W}{Q^+} \quad \text{and} \quad \varepsilon_C^- = \frac{T_0^-}{T_0^+ - T_0^-} = \frac{Q^-}{W} \tag{7.53}$$

Applying the expressions of Q^+ and Q^- derived from equation (7.53) to equation (7.52), we find,

$$\Delta t = -\frac{\ell W}{\kappa A} \left(\frac{T_0^+}{\left(T^+ - T_0^+ \right) \left(T_0^+ - T_0^- \right)} + \frac{T_0^-}{\left(T_0^+ - T_0^- \right) \left(T_0^- - T^- \right)} \right) \tag{7.54}$$

where $W < 0$ for an engine cycle. The average mechanical power of this engine cycle is defined as the ratio of the work performed on the environment and the cycle duration, i.e. $-W/\Delta t$. We consider that the temperatures T^+ and T^- of the two reservoirs are given. Thus, the average mechanical power is a function of the temperatures T_0^+ and T_0^- of the isothermal processes. The efficiency of the engine cycle is maximum when its average mechanical power is maximum. The average mechanical power is maximum when the partial derivatives of $-W/\Delta t$ with respect to the variables T_0^+ and T_0^- vanish. Thus, we obtain two equations from which we can deduce expressions for the temperatures T_0^+ and T_0^- in terms of the temperatures of the reservoirs T^+ and T^- [73],

$$T_0^+ = \frac{T^+}{2}\left(1 + \sqrt{\frac{T^-}{T^+}}\right) \quad \text{and} \quad T_0^- = \frac{T^-}{2}\left(1 + \sqrt{\frac{T^+}{T^-}}\right) \tag{7.55}$$

The entropy variation (7.7) per cycle vanishes, because the entropy is a state function. Using equations (7.37) and (7.44), we find,

$$\Delta S = \Delta S_{12} + \Delta S_{34} = \frac{Q^+}{T_0^+} + \frac{Q^-}{T_0^-} = 0 \tag{7.56}$$

thus,

$$\frac{Q^-}{Q^+} = -\frac{T_0^-}{T_0^+} \tag{7.57}$$

Taking into account definition (7.39) as well as equations (7.45), (7.55) and (7.57), the efficiency η_{EC} of the endoreversible Carnot engine cycle of maximum power is given by,

$$\eta_{EC} = \frac{Q}{Q^+} = 1 + \frac{Q^-}{Q^+} = 1 - \frac{T_0^-}{T_0^+} = 1 - \sqrt{\frac{T^-}{T^+}} \tag{7.58}$$

Taking into account the order relation (7.50) for the temperatures, the results (7.46) and (7.58) imply that the efficiency of the endoreversible Carnot cycle is lower than that of the Carnot cycle, i.e. $\eta_{EC} < \eta_C$.

Let us apply the efficiency formula (7.58) to the case of a nuclear power plant. We model it by an endoreversible Carnot cycle established between a hot reservoir, a nuclear reactor, and a cold reservoir, a river or some other water source. The temperature of the heat-transfer fluid at the exit of a typical nuclear reactor around $300°$ C. The cooling water temperature is about $20°$ C. The efficiency of the reversible Carnot cycle running at these temperatures is $\eta_C = 0.49$ and the maximum efficiency of the endoreversible Carnot cycle is $\eta_{EC} = 0.28$. Measured efficiencies for nuclear power plants are typically $\eta \sim 0.3$. Clearly, an irreversible cycle is much more realistic than a reversible cycle to model a working nuclear power plant.

7.8 Stirling Engine

So far we have considered an idealised machine that runs a Carnot cycle. Then we have improved the model by introducing the irreversibility that occurs during the heat transfers at the hot and cold reservoirs. Let us now look at some of the particular features of a typical heat engine. We take as an example the engine designed by reverend Stirling. By definition, a Stirling engine is one in which a fixed amount of gas undergoes cycles consisting of two isothermal and two isochoric processes. We will consider that the system is closed, that it contains an ideal gas and that the following processes are reversible:

1. The gas is heated and dilates at constant temperature T^+ (isothermal expansion).
2. The gas is cooled at constant volume V^+ by flowing along the regenerator (isochoric decompression).
3. The gas is cooled and contracts at constant temperature T^- (isothermal contraction).
4. The gas is heated at constant volume V^- by flowing along the regenerator (isochoric compression).

The Stirling engine (Fig. 7.12) contains two cylinders, a heating element, a cooling element and a 'regenerator'. Fuel is burned in a continuous manner outside the engine. Thus, this method has the advantage of allowing an optimal chemical combustion quality. Two rods are used at a dephasing angle of $90°$ between the two motions. The two disks acting on the rods are mounted so that they rotate together. In the quarter turn after the position of the wheels as shown, the gas gets into the hot zone. The volume of the gas is constant in cylinder 1. Then, the piston of cylinder 2 causes a pressure drop of the gas during a process, which is approximatively isothermal at temperature T^+, and performs work on the environment. During the isothermal contraction at temperature T^-, the environment performs a work on the gas, but it is smaller than the work performed by the gas on the environment during the isothermal expansion at temperature T^+. Therefore, during a cycle, the machine has performed work on the environment.

During the isochoric processes, the internal energy of the gas changes. Since no work is performed on the gas, there has to be a heat transfer. Typically, heat is transferred to and from a powder or a grid, called a ***regenerator***, through which the gas flows (Fig. 7.13).

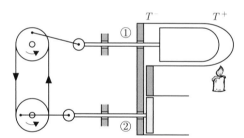

Figure 7.12 Working principle of a Stirling engine with separate cylinders.

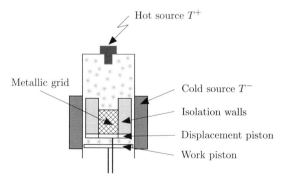

Figure 7.13 In a Stirling engine with a single cylinder, the displacement piston plays the role of the piston 1 (Fig. 7.12) and the work piston is equivalent to piston 2. The gas has to pass through a metallic grid called the regenerator.

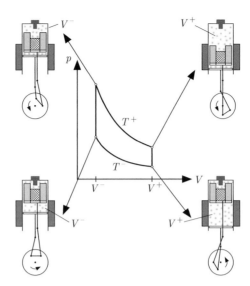

Figure 7.14 Stirling cycle in a (p, V) diagram and position of the two pistons at the four corners of the cycle.

The temperature of the regenerator varies with each passage of the gas. During isochoric processes, the thermal power characterising each heat transfer depends on the specific heat of the regenerator. If a Stirling engine is used to reach low temperatures, the regenerator specific heat decreases at low temperatures and consequently, the power of the Stirling refrigerator diminishes. In cryogenic applications, Gifford–McMahon machines are used. They are made of several cooling stages, each containing a powder that is optimised for a certain temperature range.

The Stirling engine (Fig. 7.13) operates with two pistons moving in the same cylinder. The position of the pistons during the cycle is represented in a (p, V) diagram (Fig. 7.14).

7.9 Heat Pump and Refrigerator

So far, we have analysed thermal machines in which a gas undergoes a thermodynamic cycle. Here, we briefly allude to the possibility of using a two-phase flow to run a thermal machine. In practice, heat pumps and refrigerators for example, take advantage of the latent heat of vaporisation (Fig. 7.15). The fluid is in its liquid state and at a temperature $T^- - \Delta T$ when it enters an evaporator, which is immersed in a cold reservoir at a temperature T^-. Heat is transferred from the reservoir to the liquid, thus causing it to evaporate.

The vapour at the outlet of the evaporator is compressed to higher pressure and temperature. Thus, the temperature of the gas $T^+ + \Delta T$ is higher than that of the condenser, which is immersed in a hot reservoir at temperature T^+. Heat is transferred from the gas to the hot reservoir, thus causing condensation of the gas. At the outlet of the condenser, the liquid passes through an expansion valve and reaches a lower pressure and temperature.

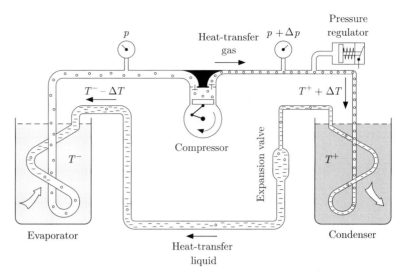

Figure 7.15 In a heat pump or refrigerator, the heat-transfer fluid turns into a liquid at the condenser and into a gas at the evaporator. With permission from Mechatronik Schule Winterthur -MSW, 8400 Winterthur, Switzerland, application note NT0028.

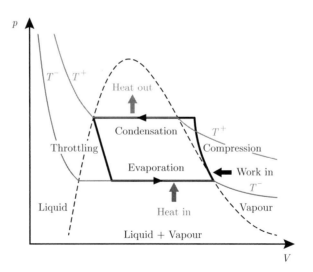

Figure 7.16 A refrigeration cycle on the (p, V) diagram of a typical biphasic fluid.

Thus, the liquid is brought back to its initial state at the inlet of the evaporator and the cycle starts again (Fig. 7.16).

In a refrigerator, the evaporator is meant to be cooled. In a heat pump, the condenser is meant to be heated up. The compressor and the expansion valve are usually electrically powered. The pressure difference has to be sufficiently high for the gas to condense on the hot side and for the liquid to evaporate on the cold side. An exercise is given to analyse the operation of a Rankine engine (§ 7.9).

7.10 Worked Solutions

7.10.1 Heat Pump

We would like to heat a building using a heat engine that performs work on a heat pump. During each cycle, the heat engine extracts an amount of heat Q_1^+ from a hot reservoir at temperature T_1^+, releases an amount of heat Q_1^- to a cold reservoir at temperature T^- and provides work W to a heat pump (Fig. 7.17). During each cycle, the heat pump uses the work W performed by the thermal machine to extract an amount of heat Q_2^- from the cold reservoir at temperature T^- and releases an amount of heat Q_2^+ to a hot reservoir at temperature T_2^+, which represents the building that is to be heated up. The temperatures satisfy the order relation $T_1^+ > T_2^+ > T^-$. We consider that the heat engine and the heat pump run reversible Carnot cycles. Show that $|Q_2^+| > |Q_1^+|$.

Solution:
Taking into account definitions (7.38) and (7.46), the efficiency of the heat engine is written,

$$\eta_C = \left| \frac{W}{Q_1^+} \right| = 1 - \frac{T^-}{T_1^+}$$

Taking into account definitions (7.40) and (7.49), the heating coefficient of performance of the heat pump is written,

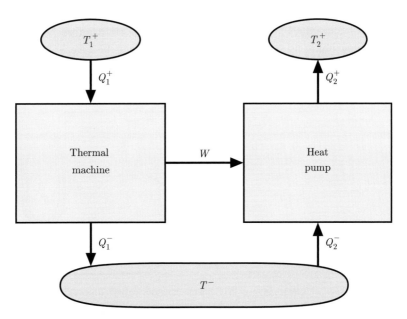

Figure 7.17   Heat pump driven by a thermal machine.

$$\varepsilon_C^+ = \left| \frac{Q_2^+}{W} \right| = \left(1 - \frac{T^-}{T_2^+} \right)^{-1}$$

Multiplying these two equations yields,

$$\eta_C \, \varepsilon_C^+ = \left| \frac{Q_2^+}{Q_1^+} \right| = \frac{1 - \dfrac{T^-}{T_1^+}}{1 - \dfrac{T^-}{T_2^+}} > 1$$

The inequality is true because $T_1^+ > T_2^+$. This means that the heat provided to the hot source at temperature T_2^+ by the heat pump is larger than the heat extracted by the heat engine from the hot source at temperature T_1^+.

For example, some buildings can be heated up by a heat pump operating according to the scheme described in this problem. Water from a lake is brought to a heat pump, where heat is extracted from the water and released the buildings. The water is returned to the lake at a temperature $2°\,C$ less than that of the incoming water.

7.10.2 Clément–Desormes Experiment

It is possible to determine the γ coefficient of an ideal gas by measuring the pressures obtained during a sequence of processes known as the Clément–Desormes experiment [74]. Contrary to its common implementation, here the gas remains inside an enclosure of variable volume (Fig. 7.18).

The U-shaped tube provides a way to measure the gas pressure with respect to that of the atmosphere by a displacement of the liquid inside the tube. The volume of the tube is negligible with respect to the volume V of the sphere. At first, the valve is open and the pressure p_0 is the atmospheric pressure, the temperature T_0 is the room temperature and the volume V_0 is the total volume of gas inside the sphere and the syringe. Then, the valve

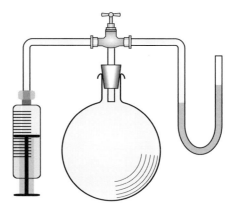

Figure 7.18   In a special implementation of the Clément–Desormes method to measure the γ coefficient of a gas, a U-shaped tube is used to measure pressure variations, and a syringe ensures that the processes occur with a fixed amount of gas.

is closed and the gas in the syringe is slowly injected into the sphere. This process is an isothermal compression. The pressure difference Δp_1 between the intermediate pressure p_1 and the initial pressure p_0 is then measured. Finally, the piston of the syringe is pulled out as fast as possible in order to bring the pressure of the gas in the sphere back to its initial value p_0. This process is an adiabatic decompression. At this point, the volume of gas in the sphere and the syringe is V_2 and the system reaches thermal equilibrium during an isochoric compression. The pressure difference Δp_2 between the final pressure p_2 and the initial pressure p_0 is measured. Show that the measured pressure differences can be used to determine the γ coefficient according to the formula,

$$\gamma \simeq \frac{\Delta p_1}{\Delta p_1 - \Delta p_2}$$

The approximation holds in the limit where $\Delta p_1 \ll p_0$ and $\Delta p_2 \ll p_0$. Use series expansion to first-order in $\Delta p_1/p_0$ and $\Delta p_2/p_0$ to find this result.

Solution:
In the initial state, the ideal gas has a volume V_0, a pressure p_0 and a temperature T_0. According to the equation of state (5.47) of an ideal gas,

$$p_0 V_0 = NR\,T_0$$

At the end of the isothermal compression process at temperature T_0, the ideal gas has a pressure p_1 and it is entirely contained in the sphere of volume V. According to the ideal gas equation of state (5.47),

$$p_1 V = NR\,T_0$$

At the end of the adiabatic decompression process, the pressure of the ideal gas is equal to the initial pressure p_0 and the total volume of gas in the sphere and the syringe is V_2. According to equation (5.90),

$$p_1 V^\gamma = p_0 V_2^\gamma$$

At the end of the isochoric compression process, the gas occupies a volume V_2 at pressure p_2 and its temperature is equal to the initial temperature T_0. According to equation (5.47),

$$p_2 V_2 = NR\,T_0$$

The previous equations imply that,

$$\frac{p_1}{p_0} = \left(\frac{V_2}{V}\right)^\gamma = \left(\frac{NR\,T_0}{p_2}\,\frac{p_1}{NR\,T_0}\right)^\gamma = \left(\frac{p_1}{p_0}\,\frac{p_0}{p_2}\right)^\gamma$$

Using definitions $p_1 = p_0 + \Delta p_1$ and $p_2 = p_0 + \Delta p_2$, the previous expression becomes,

$$1 + \frac{\Delta p_1}{p_0} = \left(1 + \frac{\Delta p_1}{p_0}\right)^\gamma \left(1 + \frac{\Delta p_2}{p_0}\right)^{-\gamma}$$

Since the pressure variations are small compared to the initial pressure, i.e. $\Delta p_1 \ll p_0$ and $\Delta p_2 \ll p_0$, the first-order series expansion in $\Delta p_1/p_0$ and $\Delta p_2/p_0$ are written,

$$1 + \frac{\Delta p_1}{p_0} \simeq \left(1 + \gamma\,\frac{\Delta p_1}{p_0}\right)\left(1 - \gamma\,\frac{\Delta p_2}{p_0}\right)$$

or alternatively,

$$1 + \frac{\Delta p_1}{p_0} \simeq 1 + \gamma \left(\frac{\Delta p_1}{p_0} - \frac{\Delta p_2}{p_0} \right) - \gamma^2 \frac{\Delta p_1}{p_0} \frac{\Delta p_2}{p_0}$$

where the second term is a second-order term that can be neglected. Thus,

$$\Delta p_1 \simeq \gamma \left(\Delta p_1 - \Delta p_2 \right)$$

which implies that,

$$\gamma \simeq \frac{\Delta p_1}{\Delta p_1 - \Delta p_2}$$

7.10.3 Brayton Engine Cycle

An ideal gas undergoes four reversible processes known as the Brayton engine cycle (Fig. 7.19):

   $1 \rightarrow 2$ adiabatic compression,
   $2 \rightarrow 3$ isobaric expansion,
   $3 \rightarrow 4$ adiabatic decompression,
   $4 \rightarrow 1$ isobaric contraction.

The pressures p_1 and p_2 and the volumes V_1 and V_3 are given.

1. Determine W_{34}, the work performed during the adiabatic decompression $3 \rightarrow 4$.
2. Determine Q_{23}, the heat provided during the isobaric expansion $2 \rightarrow 3$.
3. Determine ΔS_{41}, the entropy variation during the isobaric contraction $4 \rightarrow 1$.
4. Draw the (T, S) diagram of the cycle.

Solution:

1. *According to expression* (7.14) *for the work performed during an adiabatic process,*

$$W_{34} = \Delta U_{34} = c\,NR \int_{T_3}^{T_4} dT = c\,NR\,(T_4 - T_3)$$

Taking into account the ideal gas equation of state (5.47),

$$p_3 V_3 = NR\,T_3 \qquad and \qquad p_4 V_4 = NR\,T_4$$

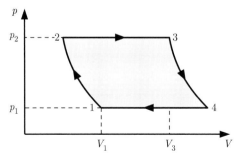

Figure 7.19 (p, V) diagram of a Brayton engine cycle.

the properties of isobaric processes,

$$p_3 = p_2 \quad \text{and} \quad p_4 = p_1$$

and the property (5.90) of adiabatic processes,

$$p_3 V_3^\gamma = p_4 V_4^\gamma \quad \Rightarrow \quad V_4 = \left(\frac{p_3}{p_4}\right)^{1/\gamma} V_3 = \left(\frac{p_2}{p_1}\right)^{1/\gamma} V_3$$

the work W_{34} can be recast as,

$$W_{34} = c\,(p_4 V_4 - p_3 V_3) = c\,(p_1 V_4 - p_2 V_3) = cV_3 \left(p_1 \left(\frac{p_2}{p_1}\right)^{1/\gamma} - p_2\right)$$

2. According to expression (7.22), the heat provided during an isobaric process is,

$$Q_{23} = \Delta H_{23} = (c+1)\,NR \int_{T_2}^{T_3} dT = (c+1)\,NR\,(T_3 - T_2)$$

Taking into account the ideal gas equation of state (5.47),

$$p_2 V_2 = NR\,T_2 \quad \text{and} \quad p_3 V_3 = NR\,T_3$$

the property of an isobaric process,

$$p_3 = p_2$$

and the property (5.90) of adiabatic processes,

$$p_1 V_1^\gamma = p_2 V_2^\gamma \quad \Rightarrow \quad V_2 = \left(\frac{p_1}{p_2}\right)^{1/\gamma} V_1$$

the heat Q_{23} can be recast as,

$$Q_{23} = (c+1)\,(p_3 V_3 - p_2 V_2) = (c+1)\,p_2\,(V_3 - V_2)$$

$$= (c+1)\,p_2 \left(V_3 - \left(\frac{p_1}{p_2}\right)^{1/\gamma} V_1\right)$$

3. According to definition (7.23) of the entropy variation during an isobaric process, the ideal gas equation of state (5.47) and the property of adiabatic processes stated above, we have,

$$\Delta S_{41} = \int_4^1 \frac{dH}{T} = (c+1)\,NR \int_{T_4}^{T_1} \frac{dT}{T} = (c+1)\,NR \ln\left(\frac{T_1}{T_4}\right)$$

$$= (c+1)\,NR \ln\left(\frac{V_1}{V_4}\right) = (c+1)\,NR \ln\left(\left(\frac{p_1}{p_2}\right)^{1/\gamma} \frac{V_1}{V_3}\right)$$

4. According to equation (7.12), the adiabatic processes are vertical lines on a (T, S) diagram. Solving equation (7.23) for the temperature shows that isobaric processes are exponentials on a (T, S) diagram (Fig. 7.20).

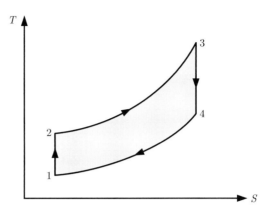

Figure 7.20 (T, S) diagram of the Brayton engine cycle.

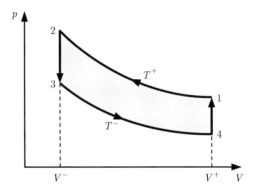

Figure 7.21 (p, V) diagram of the Stirling refrigeration cycle.

7.10.4 Stirling Refrigeration Cycle

An ideal gas undergoes four reversible processes known as the Stirling refrigeration cycle (Fig. 7.21):

   $1 \rightarrow 2$ isothermal compression,
   $2 \rightarrow 3$ isochoric decompression,
   $3 \rightarrow 4$ isothermal decompression,
   $4 \rightarrow 1$ isochoric compression.

The minimum volume V^-, the maximum volume V^+, the temperatures of the hot reservoir T^+ and that of the cold reservoir T^- are given.

1. Determine the work values W_{12} and W_{34} performed during the isothermal compression $1 \rightarrow 2$ and decompression $3 \rightarrow 4$.
2. Determine the heat values Q_{12}, Q_{23}, Q_{34} and Q_{41} during all the processes and deduce the heat Q provided per cycle.
3. Determine the cooling coefficient of performance ε^- of the refrigeration cycle.
4. Draw the (T, S) diagram of the cycle.

Solution:

1. *According to definition (7.15), the work performed during the isothermal processes at temperatures T^+ and T^- are given by,*

$$W_{12} = -\int_1^2 p\,dV = -NR\,T^+ \int_{V^+}^{V^-} \frac{dV}{V} = -NR\,T^+ \ln\left(\frac{V^-}{V^+}\right)$$

$$W_{34} = -\int_3^4 p\,dV = -NR\,T^- \int_{V^-}^{V^+} \frac{dV}{V} = -NR\,T^- \ln\left(\frac{V^+}{V^-}\right)$$

2. *According to definition (7.16), the heat provided during the isothermal processes at temperatures T^+ and T^- are given by,*

$$Q_{12} = -W_{12} = NR\,T^+ \ln\left(\frac{V^-}{V^+}\right)$$

$$Q_{34} = -W_{34} = NR\,T^- \ln\left(\frac{V^+}{V^-}\right)$$

According to definition (7.19), the heat provided during the isochoric processes at volumes V^+ and V^- are given by,

$$Q_{23} = \Delta\,U_{23} = c\,NR \int_{T^+}^{T^-} dT = c\,NR\left(T^- - T^+\right)$$

$$Q_{41} = \Delta\,U_{41} = c\,NR \int_{T^-}^{T^+} dT = c\,NR\left(T^+ - T^-\right)$$

According to definition (7.1), the heat provided per cycle is,

$$Q = Q_{12} + Q_{23} + Q_{34} + Q_{41} = -NR\left(T^+ - T^-\right)\ln\left(\frac{V^+}{V^-}\right) < 0$$

3. *Taking into account the fact that the heat extracted from the cold reservoir at temperature T^- is $Q^- = Q_{34}$, the cooling coefficient of performance (7.43) is given by,*

$$\varepsilon^- = -\frac{Q_{34}}{Q} = \frac{NR\,T^- \ln\left(\dfrac{V^+}{V^-}\right)}{NR\left(T^+ - T^-\right)\ln\left(\dfrac{V^+}{V^-}\right)} = \frac{T^-}{T^+ - T^-}$$

which is equal to the cooling coefficient of performance (7.49) of the Carnot cycle.

4. *By definition, the isothermal processes are horizontal lines on the (T, S) diagram. Solving equation (7.20) for temperature, we find that the isochoric processes are exponentials on a (T, S) diagram (Fig. 7.22).*

7.10.5 Diesel Engine Cycle

An ideal gas undergoes four reversible processes known as the Diesel cycle (Fig. 7.23):

  1 $\rightarrow$ 2 adiabatic compression,
  2 $\rightarrow$ 3 isobaric decompression,

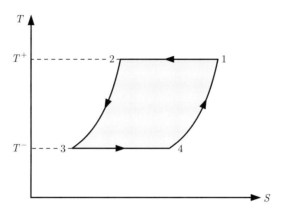

Figure 7.22 (T, S) diagram of the Stirling refrigeration cycle.

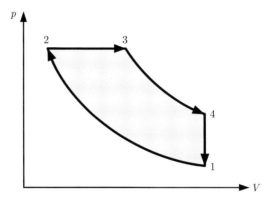

Figure 7.23 (p, V) diagram of the Diesel engine cycle.

$3 \rightarrow 4$ adiabatic decompression,

$4 \rightarrow 1$ isobaric compression.

The pressure p_1 and the volumes V_1, V_2 and V_3 are given.

1. Determine the pressures p_2 and p_4.
2. Determine the entropy variations ΔS_{12}, ΔS_{23}, ΔS_{34} and ΔS_{41} during all four processes and deduce the entropy variation ΔS per cycle.
3. Determine the internal energy variations ΔU_{12}, ΔU_{23}, ΔU_{34} and ΔU_{41} during all four processes and deduce the internal energy variation ΔU per cycle.

Solution:

1. *Taking into account the property (5.90) of an adiabatic process, the pressure p_2 is written as,*

$$p_1 V_1^{\gamma} = p_2 V_2^{\gamma} \qquad \Rightarrow \qquad p_2 = p_1 \left(\frac{V_1}{V_2} \right)^{\gamma}$$

Moreover, an isobaric process is characterised by equal pressures, i.e. $p_2 = p_3$, and an isochoric process by equal volumes, i.e. $V_4 = V_1$. Thus, the pressure p_4 is written as,

$$p_3 V_3^\gamma = p_4 V_4^\gamma \quad \Rightarrow \quad p_4 = p_3 \left(\frac{V_3}{V_4}\right)^\gamma = p_2 \left(\frac{V_3}{V_1}\right)^\gamma = p_1 \left(\frac{V_3}{V_2}\right)^\gamma$$

2. *Using definitions (5.68), (7.12), (7.20) and (7.23), the entropy variations* ΔS_{12}, ΔS_{23}, ΔS_{34} *and* ΔS_{41} *during all four processes are given by,*

$$\Delta S_{12} = 0$$

$$\Delta S_{23} = \int_2^3 \frac{dH}{T} = (c+1) NR \int_{T_2}^{T_3} \frac{dT}{T} = (c+1) NR \ln \left(\frac{T_3}{T_2}\right)$$

$$= (c+1) NR \ln \left(\frac{V_3}{V_2}\right)$$

$$\Delta S_{34} = 0$$

$$\Delta S_{41} = \int_4^1 \frac{dU}{T} = c NR \int_{T_4}^{T_1} \frac{dT}{T} = c NR \ln \left(\frac{T_1}{T_4}\right)$$

$$= c NR \ln \left(\frac{p_1}{p_4}\right)$$

This implies that the entropy variation ΔS *per cycle vanishes,*

$$\Delta S = \Delta S_{12} + \Delta S_{23} + \Delta S_{34} + \Delta S_{41} = 0$$

which is a consequence of the fact that the entropy S *is a state function.*

3. *Taking into account the ideal gas equation of state (5.47) and definition (7.10), the internal energy variation* ΔU_{if} *during a reversible process* $i \to f$ *is given by,*

$$\Delta U_{if} = c NR \int_{T_i}^{T_f} dT = c NR (T_f - T_i) = c (p_f V_f - p_i V_i)$$

Thus the internal energy variations ΔU_{12}, ΔU_{23}, ΔU_{34} *and* ΔU_{41} *during all four processes are given by,*

$$\Delta U_{12} = c (p_2 V_2 - p_1 V_1) = c p_1 \left(\left(\frac{V_1}{V_2}\right)^\gamma V_2 - V_1\right)$$

$$\Delta U_{23} = c (p_3 V_3 - p_2 V_2) = c p_1 \left(\frac{V_1}{V_2}\right)^\gamma (V_3 - V_2)$$

$$\Delta U_{34} = c (p_4 V_4 - p_3 V_3) = c p_1 \left(\left(\frac{V_3}{V_2}\right)^\gamma V_1 - \left(\frac{V_1}{V_2}\right)^\gamma V_3\right)$$

$$\Delta U_{41} = c (p_1 V_1 - p_4 V_4) = c p_1 \left(V_1 - \left(\frac{V_3}{V_2}\right)^\gamma V_1\right)$$

This implies that the internal energy variation ΔU *per cycle vanishes,*

$$\Delta U = \Delta U_{12} + \Delta U_{23} + \Delta U_{34} + \Delta U_{41} = 0$$

which is a consequence of the fact that the internal energy density U *is a state function.*

Exercises

7.1   Refrigerator

A thermoelectric refrigerator becomes cold by expelling heat into the environment at a temperature T^+. The power supplied to the device is P_W and the thermal power corresponding to the rejected heat is P_Q. Determine the lowest temperature T^- that the system can reach if it had an optimal efficiency.

Numerical Application:
$P_W = 100\,\text{W}$, $P_Q = 350\,\text{W}$ and $T^+ = 25°\text{C}$.

7.2   Power Plant Cooled by a River

A power plant operates between a hot reservoir consisting of a combustion chamber or a nuclear reactor and a cold reservoir consisting of the water of a river. It is modelled as a thermal machine operating between the hot reservoir at temperature T^+ and the cold reservoir at temperature T^-. Analyse this power plant by using the following instructions:

a) Determine the maximum efficiency η_C of this power plant and the thermal power P_{Q^+}, describing the heat exchange with the combustion chamber.
b) Assume that its real efficiency is $\eta = k\eta_C$ and find the thermal power P_{Q^-}, describing the heat exchange with the river.
c) Determine the temperature difference ΔT of the water flowing at a rate $\dot{V}$ down the river. The water has a density m and a specific heat at constant pressure per unit of mass c_p^*.

Numerical Application:
$P_W = -750\,\text{MW}$, $T^+ = 300°\text{C}$, $T^- = 19°\text{C}$, $k = 60\,\%$, $\dot{V} = 200\,\text{m}^3/\text{s}$, $m = 1{,}000\,\text{kg/m}^3$ and $c_p^* = 4{,}181\,\text{J/kg\,K}$.

7.3   Braking Cycle

A system is made up of a vertical cylinder which is sealed at the top and closed by a piston at the bottom. A valve A controls the intake of gas at the top and an exhaust valve B (also at the top) is held back by a spring that exerts a constant pressure p_2 on the valve. The system goes through the following processes:

- $0 \longrightarrow 1$: the piston is at the top of the cylinder; valve A opens up and the piston is lowered into it so that some of the gas at atmospheric pressure $p_0 = p_1$ is added to the cylinder. The gas is at room temperature T_1. Valve B is closed. The maximum volume occupied by the incoming gas is V_1.
- $1 \longrightarrow 2$: Valve A is now closed and the piston moves upward, fast enough so that the process can be considered adiabatic. Valve B remains closed as long as the pressure during the rise of the piston is lower than p_2. As the piston continues in its rise, the gas reaches pressure $p_2 = 10p_1$, at a temperature T_2 in a volume V_2. Assume a reversible adiabatic process for which equations (5.90) and (5.83) apply.

- 2 $\longrightarrow$ 3: As the piston keeps moving up, valve B opens up, the pressure is $p_3 = p_2$ and the gas is released in the environment while valve A still remains closed until the piston reaches the top, where $V_3 = V_0 = 0$.
- 3 $\longrightarrow$ 0: Valve B closes and valve A opens up. The system is ready to start over again.

Analyse this cycle by using the following instructions:

a) Draw the (p, V) diagram for the three processes that the system is undergoing.
b) Determine the temperature T_2 and the volume V_2.
c) Find the work W performed per cycle.

Numerical Application:
$V_0 = V_3 = 0$, $p_0 = p_1 = 10^5$ Pa, $V_1 = 0.25$ l, $T_1 = 27°$ C and $\gamma = 1.4$.

7.4 Lenoir Cycle

The Lenoir cycle is a model for the operation of a combustion engine patented by Jean Joseph Etienne Lenoir in 1860 (Figs. 7.2 and 7.3). This idealised cycle is defined by three reversible processes:

- 1 $\longrightarrow$ 2 isochoric compression
- 2 $\longrightarrow$ 3 adiabatic expansion
- 3 $\longrightarrow$ 1 isobaric contraction

Assume that the cycle is performed on an ideal gas characterised by the coefficient c found in relation (5.62). The following values of some state variables of the gas are assumed to be known: the pressure p_1, volumes V_1 and V_3, temperature T_1 and the number of moles of gas N. Analyse this cycle by using the following instructions :

a) Draw the (p, V) and (T, S) diagrams of the cycle.
b) Determine the entropy variation ΔS_{12} of the gas during the isochoric process $1 \longrightarrow 2$.
c) Express the temperature T_2 in terms of the heat exchanged Q_{12} during the isochoric process $1 \longrightarrow 2$.
d) Determine the pressure p_2 in terms of the pressure p_1, the volume V_1 and the heat exchanged Q_{12}.
e) Determine the pressure p_3 in terms of the pressure p_2 and volumes V_2 and V_3.
f) Determine the work W_{23} performed during the adiabatic process $2 \longrightarrow 3$ and the heat Q_{23} exchanged during this process.
g) Find the work W_{31} performed during the isobaric process $3 \longrightarrow 1$ and the heat Q_{31} exchanged during this process.
h) Find the efficiency of the cycle η_L defined in conformity with relation (7.38) as,

$$\eta_L = -\frac{W_{23} + W_{31}}{Q_{12}}$$

Express the efficiency η_L in terms of the temperatures T_1, T_2 and T_3.

7.5 Otto Cycle

The Otto cycle is a model for a spark ignition engine and represents the mode of operation of most non-diesel car engines. It consists of four processes when the system is closed, and of two additional isobaric processes when the system is open, corresponding to air intake and exhaust. Thus, we have,

- $0 \longrightarrow 1$ isobaric air intake
- $1 \longrightarrow 2$ adiabatic compression
- $2 \longrightarrow 3$ isochoric heating
- $3 \longrightarrow 4$ adiabatic expansion
- $4 \longrightarrow 1$ isochoric cooling
- $1 \longrightarrow 0$ isobaric gas exhaust

Assume that the adiabatic processes are reversible and that the gas is an ideal gas characterised by the coefficient c found in relation (5.62) and coefficient $\gamma = (c + 1)/c$. The following values of state variables are assumed to be known: the pressure p_1, the volumes $V_1 = V_4$ and $V_2 = V_3$, the temperature T_3, and the number of moles N of air at the intake. Analyse this cycle by using the following instructions:

a) Draw the (p, V) and (T, S) diagrams of the cycle. On the (p, V) diagram, show also the intake and exhaust processes.
b) Describe what the engine does in each of the processes.
c) Explain why an exchange of air with the exterior is needed.
d) On the (p, V) and (T, S) diagrams determine the relation between the area enclosed in the cycles and the work W and the heat Q per cycle.
e) Determine all the state variables at points 1, 2, 3 and 4 of the cycle, i.e. find p_2, p_3, p_4, T_2 and T_4.
f) Compute the work W performed per cycle and the heat Q exchanged during a cycle.
g) Determine the efficiency of the Otto cycle,

$$\eta_O = -\frac{W}{Q^+}$$

where $Q^+ = Q_{23}$.

7.6 Atkinson Cycle

James Atkinson was a British engineer who designed several combustion engines. The thermodynamic cycle bearing his name is a modification of the Otto cycle intended to improve its efficiency. The trade-off in achieving higher efficiency is a decrease in the work performed per cycle. The idealised Atkinson cycle consists of the following reversible processes:

- $1 \longrightarrow 2$: adiabatic compression
- $2 \longrightarrow 3$: isochoric heating
- $3 \longrightarrow 4$: isobaric heating
- $4 \longrightarrow 5$: adiabatic expansion
- $5 \longrightarrow 6$: isochoric cooling
- $6 \longrightarrow 1$: isobaric cooling

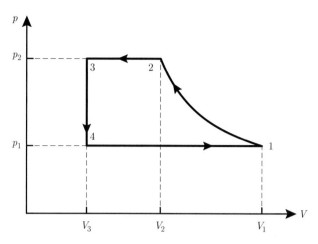

Figure 7.24 (p, V) diagram of a refrigeration cycle

Assume that the adiabatic processes are reversible and that the cycle is operated on an ideal gas characterised by,

$$pV = NRT \qquad U = cNRT \qquad \gamma = \frac{c+1}{c}$$

The following physical quantities that characterise the cycle are assumed to be known: volumes V_1, V_2 and V_6, pressures p_1 and p_3, temperature T_5, and the number of moles N of gas. Analyse this cycle by using the following instructions:

a) Draw the (p, V) diagram of the Atkinson cycle.
b) Determine the pressures p_2, p_4, p_5, p_6, the volumes V_3, V_4, V_5 and temperatures T_1, T_2, T_3, T_4, T_6, in terms of the known physical quantities.
c) Find the works W_{12}, W_{23}, W_{34}, W_{45}, W_{56}, W_{61} and the work W performed per cycle.
d) Find the heat transfers Q_{12}, Q_{23}, Q_{34}, Q_{45}, Q_{56}, Q_{61} and the heat $Q^+ = Q_{23} + Q_{34}$ provided to the gas.
e) Determine the efficiency of the Atkinson cycle,

$$\eta_A = -\frac{W}{Q^+}$$

7.7 Refrigeration Cycle

An ideal gas characterised by the coefficient c found in relation (5.62) and the coefficient $\gamma = (c+1)/c$ undergoes a refrigeration cycle consisting of four reversible processes (Fig. 7.24):

- $1 \longrightarrow 2$: adiabatic compression
- $2 \longrightarrow 3$: isobaric compression
- $3 \longrightarrow 4$: isochoric cooling
- $4 \longrightarrow 1$: isobaric expansion

Analyse this cycle by using the following instructions:

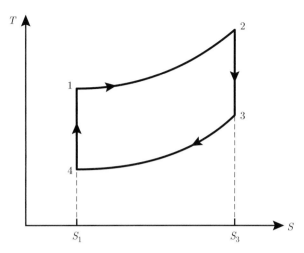

Figure 7.25 (T, S) diagram of a Rankine cycle operated on an ideal gas

a) Determine the volume V_2 in terms of the volumes V_1 and V_3 and the pressures p_1 and p_2.

b) Find the entropy variation ΔS_{23} during the isobaric compression.

c) Determine the heat exchanged Q_{23} during the isobaric compression.

d) Assume now that instead of an ideal gas a fluid is used, which is entirely in a gaseous state at point 2 and completely in a liquid state at point 3. The isobaric compression $2 \longrightarrow 3$ is then a phase transition occurring at temperature T and characterised by the molar latent heat of vaporisation $\ell_{\ell g}$. Determine the entropy variation ΔS_{23} during the phase transition in terms of the number of moles N of fluid, the volume V_2, the pressure p_2 and the molar latent heat of vaporisation $\ell_{\ell g}$, assuming that $p V = NR T$ in the gas phase.

7.8   Rankine Cycle

An ideal gas characterised by the coefficient c found in relation (5.62) and the coefficient $\gamma = (c + 1)/c$ undergoes a Rankine engine cycle consisting of four reversible processes:

- $1 \longrightarrow 2$: isobaric expansion
- $2 \longrightarrow 3$: adiabatic expansion
- $3 \longrightarrow 4$: isobaric compression
- $4 \longrightarrow 1$: adiabatic compression

Thus, the cycle is represented by a rectangle in a (T, S) diagram (Fig. 7.25).

Analyse this cycle by using the following instructions:

a) Draw the (p, V) diagram of a Rankine cycle for an ideal gas.

b) Determine the works performed W_{12}, W_{23}, W_{34} and W_{41} and the work performed per cycle W in terms of the enthalpies H_1, H_2, H_3 and H_4.

c) Find the heat provided by the hot reservoir $Q^+ = Q_{12}$ in terms of the enthalpies H_1, H_2, H_3 and H_4.

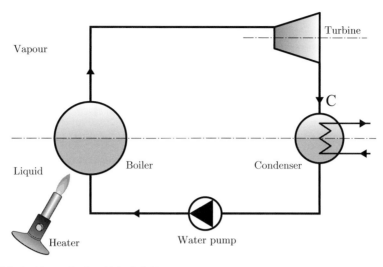

Figure 7.26 Diagram of the Rankine engine for a biphasic fluid.

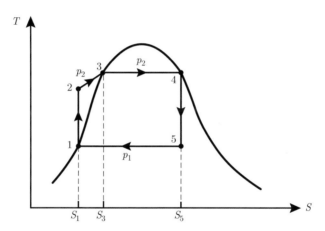

Figure 7.27 (T, S) diagram of the Rankine cycle for a biphasic fluid.

d) Determine the efficiency of the Rankine cycle for an ideal fluid defined as,

$$\eta_R = -\frac{W}{Q^+}$$

7.9 Rankine Cycle for a Biphasic Fluid

An engine consists of a boiler, a condenser, a turbine and a water pump (Fig. 7.26). This engine is operating a Rankine cycle on a biphasic fluid (Fig. 7.27). The cycle consists of five processes:

- $1 \longrightarrow 2$: The fluid coming out of the turbine is completely condensed (1). The liquid goes then through an isentropic compression from an initial pressure p_1 to a final pressure p_2.

- $2 \longrightarrow 3$: The liquid is heated up at constant pressure p_2 by the boiler. It undergoes an isobaric heating until it reaches the vaporisation temperature (3).
- $3 \longrightarrow 4$: The liquid is vaporised at constant pressure p_2. It goes through a phase transition until saturation is reached (4).
- $4 \longrightarrow 5$: The fluid undergoes an isentropic expansion from an initial pressure p_2 to a final pressure p_1.
- $5 \longrightarrow 1$: The fluid is condensed at constant pressure p_1. It goes through a phase transition until full condensation is reached (1).

Analyse this cycle by using the following instructions:

a) Determine the heat provided by the boiler $Q^+ = Q_{23} + Q_{34}$, the heat released at the condenser $Q^- = Q_{51}$ in terms of the enthalpies per unit mass h_1^*, h_2^*, h_4^* and h_5^* and the M of fluid undergoing this cycle (Fig. 7.27).
b) Find the work performed by the pump W_{12} and the work performed on the turbine W_{45} in terms of the enthalpies per unit mass h_1^*, h_2^* and h_5^* and the mass M by using the results obtained for the open system presented in § 4.11 and assuming that the mechanical power is due to the chemical power P_C of the fluid flowing through the pump and the turbine, i.e. $P_W = P_C$.
c) Determine the efficiency of the Rankine cycle for a biphasic fluid defined as,

$$\eta_R = -\frac{W}{Q^+}$$

| 8 | **Chemistry and Electrochemistry** |

Josiah Willard Gibbs, 1839–1903

J. W. Gibbs received the first engineering PhD degree of the USA from Yale University. After spending 13 years throughout Europe, he came back to Yale as a professor of theoretical physics. He published in 1876 his work on reactive mixtures and, in 1902, a book in which he laid the foundations of statistical physics.

8.1 Historical Introduction

As we will see in this chapter, thermodynamics is also well suited to describe chemical reactions and electrochemical processes. In order to appreciate how thermodynamics has contributed to the field of chemistry, we have to be aware of the long journey undertaken by generations of scientists before chemical concepts could be expressed naturally in the framework of thermodynamics.

Newton tried to interpret the apparent affinity among substances as resulting from mutual attractions between particles. The ***phlogistic*** principle was meant to give an understanding of chemical interactions that was distinct from the mechanistic views of seventeenth-century Newtonian physics. For *Georg Stahl* (1660–1734), chemist and personal doctor of the king of Prussia, phlogistic was a very subtle matter emanating from the combustion of coal or the corrosion of metals. The more a material would contain phlogistic, the more likely it would burn. The theory was proved wrong, in particular when it was shown that magnesium gained weight while burning whereas it had to be assumed that it was loosing some phlogistic. It was *Antoine Lavoisier* (1743–1794) who brought to light the role played by oxygen in combustion, thus solving the paradox concerning magnesium. He set the foundations for a calorific understanding of combustion.

In the eighteenth century, chemists were seeking a unifying concept that would play a role in chemistry analogous to that of force in Newtonian mechanics [75]. The concept that emerged in the nineteenth century was affinity [76]. It was the catalog of observations collected during the sixteenth and seventeenth centuries that allowed chemists to define the notion of *affinity*: a tendency of substances to form bonds. Famous scientists such as Berthollet, Davy, Ampère and Faraday undertook the task of generating an ordered list of relative affinities of substances. A Swiss chemist named *Germain Henri Hess* (1802–1850) found also that the heat of a reaction was the same whether the process occurred in one or more intermediate steps.

At the end of the nineteenth century, thermodynamics had systematised the notions of internal energy and entropy, and described how systems exchange energy in the form of heat and work. Thermodynamic quantities that would account for chemical reactions had yet to be introduced. This was achieved by J. W. Gibbs, a physicist, chemist and mathematician. In order to do so, he introduced the notion of *chemical potential*.

As we will see in this chapter, the notion of osmotic pressure identified by *Jacobus Henricus van't Hoff* (1852–1911), is closely related to that of chemical potential. For this discovery, he was awarded in 1901 the first Nobel Prize in Chemistry. In 1922, *Théophile de Donder* (1872–1957) gave the mathematical definition for the affinity of a chemical reaction, which is used to this day, in terms of chemical potentials and stoichiometric coefficients.

8.2 Chemical Reactions

In the previous chapters, we described the thermodynamics of simple systems or subsystems in the absence of chemical reactions. Here, we formalise the notion of chemical reaction and in the next section we define the affinity of a chemical reaction. A *chemical reaction* is a matter transformation whereby chemical substances are modified. The initial chemical substances are called the *reactants* and the final substances, the *products*.

For instance, let us consider the chemical reaction that transforms molecular hydrogen H_2 and molecular oxygen O_2 into molecular water H_2O and vice versa,

$$2\,H_2 + O_2 \ \rightleftarrows\ 2\,H_2O \tag{8.1}$$

where the arrows denote the directions of the reactions. The chemical reaction (8.1) can be recast by moving all the terms on the left-hand side to the right-hand side and by changing their signs,

$$0 \rightleftarrows -2\,H_2 - O_2 + 2\,H_2O \tag{8.2}$$

The coefficients -2, -1, 2 appearing in equation (8.2) are called the **stoichiometric coefficients**. This reformulation is a convention. For any chemical reaction, the chemical symbols can be represented by $(X_1, X_2, X_3) = (H_2, O_2, H_2O)$ and the stoichiometric coefficients of the chemical reaction (8.2), labelled a, by $(\nu_{a1}, \nu_{a2}, \nu_{a3}) = (-2, -1, 2)$. Thus, the equation (8.2) associated to the chemical reaction a is written,

$$0 \rightleftarrows \sum_{A=1}^{3} \nu_{aA}\, X_A \tag{8.3}$$

8.2.1 Extent of Reaction

Let us now consider a chemical reaction a in a system consisting of r chemical substances labelled by the index $A = 1, \ldots, r$. We denote the numbers of moles of chemical substances by $\{N_A\} \equiv \{N_1, \ldots, N_r\}$. In a chemical reaction a, the infinitesimal variation of the number of moles of each substance is in proportion to their stoichiometric coefficient, that is,

$$\frac{dN_1}{\nu_{a1}} = \frac{dN_2}{\nu_{a2}} = \ldots = \frac{dN_r}{\nu_{ar}} = d\xi_a \tag{8.4}$$

This can be recast as,

$$dN_A = \nu_{aA}\, d\xi_a \qquad \forall\, A = 1, \ldots, r \tag{8.5}$$

where the dimensionless quantity ξ_a is called the **extent** of the chemical reaction a. If $d\xi_a > 0$, the chemical reaction a occurs from left to right and if $d\xi_a < 0$, it occurs from right to left. Integration of (8.5) yields,

$$N_A(t) = N_A(0) + \nu_{aA}\, \xi_a(t) \tag{8.6}$$

where $\xi_a(0) = 0$. According to expression (8.6), for a chemical reaction a, the only parameter that governs the time evolution of the number of moles N_A of a chemical substance A is the extent ξ_a of this reaction. The chemical **reaction rate** Ω_a associated to the chemical reaction a is defined as the time derivative of the extent of the reaction (8.5) [77],

$$\Omega_a = \dot{\xi}_a \tag{8.7}$$

The time derivative deduced from (8.5) yields the variation rate of the numbers of moles of substance A,

$$\dot{N}_A = \nu_{aA}\, \Omega_a \tag{8.8}$$

We can think of the extent ξ_a of the chemical reaction a as the analog in chemistry to the position $\boldsymbol{r}$ of a point mass in mechanics, and the chemical reaction rate Ω_a as the analog of the velocity $\boldsymbol{v}$. Indeed, the chemical reaction rate is sometimes called the chemical reaction velocity.

In chemistry, it is common for some substances to take part in several chemical reactions. Let us consider that in addition to the chemical reaction (8.1), there is a reaction that transforms carbon monoxide CO into carbon dioxide CO_2 and vice versa,

$$2\,CO + O_2 \; \rightleftarrows \; 2\,CO_2 \tag{8.9}$$

Molecular oxygen O_2 takes part in both chemical reactions (8.1) and (8.9). Hence, they are **coupled reactions**. The variation of the number of moles of oxygen is equal to the sum of the variations due to both chemical reactions (8.1) and (8.9). Thus, for a system consisting of r chemical substances $A = 1, \ldots, r$ coupled by n chemical reactions $a = 1, \ldots, n$, the differential (8.5) becomes,

$$dN_A = \sum_{a=1}^{n} \nu_{aA}\, d\xi_a \tag{8.10}$$

Integration of equation (8.10) yields,

$$N_A\,(t) = N_A\,(0) + \sum_{a=1}^{n} \nu_{aA}\, \xi_a\,(t) \tag{8.11}$$

where $\xi_a\,(0) = 0 \; \forall\, a = 1, \ldots, n$. Taking into account definition (8.7), the time derivative of equation (8.10) is written as,

$$\dot{N}_A = \sum_{a=1}^{n} \nu_{aA}\, \Omega_a \tag{8.12}$$

8.2.2 Affinity

Chemical reactions often occur in an environment where the temperature T and the pressure p are constant. This means that the system environment can be considered as a heat and work reservoir. To describe the thermodynamics of a chemical system in such an environment, we choose as thermodynamical potential the Gibbs free energy $G\,(T, p, \{N_A\})$. According to definitions (4.39) and (8.10), the differential of the Gibbs free energy is written as,

$$dG = -S\,dT + V\,dp + \sum_{a=1}^{n} \left(\sum_{A=1}^{r} \mu_A\, \nu_{aA} \right) d\xi_a \tag{8.13}$$

where the sums with respect to the chemical reactions a and substances A commute. The last term of the differential describes the infinitesimal variation of the Gibbs free energy due to chemical reactions a.

For a system coupled to a heat reservoir at temperature T and to a work reservoir at pressure p, the chemical equilibrium is given by,

$$\frac{\partial G}{\partial \xi_a} = 0 \qquad \forall\, a = 1, \ldots, n \qquad \text{(chemical equilibrium)} \tag{8.14}$$

Taking into account the differential (8.13), we can write equation (8.14) in terms of the chemical potentials μ_A and stoichiometric coefficients ν_{aA},

$$\sum_{A=1}^{r} \mu_A\, \nu_{aA} = 0 \qquad \forall\, a = 1, \ldots, n \qquad \text{(chemical equilibrium)} \tag{8.15}$$

It is common practice in chemistry to characterise a chemical reaction a at constant temperature and pressure by the **Gibbs free energy of reaction** defined as,

$$\Delta_a G \equiv \frac{\partial G}{\partial \xi_a} = \sum_{A=1}^{r} \mu_A \nu_{aA} \tag{8.16}$$

The Gibbs free energy is minimum at equilibrium, as stated in equation (4.58). Therefore, the sign of $\Delta_a G$ determines the direction in which a chemical reaction a occurs at constant temperature and pressure:

1. $\Delta_a G < 0$: reaction a evolves spontaneously from left to right,
2. $\Delta_a G = 0$: reaction a has reached equilibrium,
3. $\Delta_a G > 0$: reaction a evolves spontaneously from right to left.

Alternatively, a chemical reaction a is characterised by its **affinity**, defined as the opposite of the the Gibbs free energy of reaction,

$$\mathcal{A}_a = - \frac{\partial G}{\partial \xi_a} \tag{8.17}$$

Taking into account equation (8.16), the affinity reads,

$$\mathcal{A}_a = - \sum_{A=1}^{r} \mu_A \nu_{aA} \tag{8.18}$$

The chemical notion affinity allows us to draw a parallel with the mechanical notion of force. We can assimilate the extent of a chemical reaction a to a coordinate ξ_a, as one assimilates a vector $\boldsymbol{r}$ to a position in mechanics. Then, according to equation (8.17), the affinity $\mathcal{A}_a$ of the chemical reaction a is the opposite of the derivative of the Gibbs free energy G, just like a conservative mechanical force $\boldsymbol{F}$ is the opposite of the derivative of the potential energy U,

$$\boldsymbol{F} = - \boldsymbol{\nabla} U \equiv - \frac{\partial U}{\partial \boldsymbol{r}} \tag{8.19}$$

Chemical reactions can also occur in an environment where the temperature T is variable and the pressure p is constant. This means that the environment can be considered as a work reservoir. To describe the thermodynamics of a chemical system in such an environment, we choose as thermodynamical potential the enthalpy $H\left(S\left(T, p, \{N_A\}\right), p, \{N_A\}\right)$. According to relation (8.10), the variation of the number of moles of substance A is entirely determined by the extent ξ_a of the chemical reaction a, the enthalpy can be written as $H\left(S\left(T, p, \{N_A\left(\xi_a\right)\}\right), p, \{N_A\left(\xi_a\right)\}\right)$. The enthalpy variation with respect to the extent of the chemical reaction a at constant pressure is written as,

$$\frac{dH}{d\xi_a} = \sum_{A=1}^{N} \left(\frac{\partial H}{\partial S} \frac{\partial S}{\partial N_A} \frac{\partial N_A}{\partial \xi_a} + \frac{\partial H}{\partial N_A} \frac{\partial N_A}{\partial \xi_a} \right) \tag{8.20}$$

Applying Schwarz theorem (4.67) to the Gibbs free energy $G\left(T, p, \{N_A\}\right)$ we obtain,

$$\frac{\partial}{\partial T} \left(\frac{\partial G}{\partial N_A} \right) = \frac{\partial}{\partial N_A} \left(\frac{\partial G}{\partial T} \right) \tag{8.21}$$

Using definition (4.40) for entropy $S\left(T, p, \{N_A\}\right)$ and definition (4.42) for chemical potential $\mu_A\left(T, p, \{N_A\}\right)$, the differential equation (8.21) yields the Maxwell relation,

$$\frac{\partial \mu_A}{\partial T} = -\frac{\partial S}{\partial N_A} \tag{8.22}$$

Moreover, using expressions (4.33) for temperature T, (4.35) for chemical potential μ_A, (8.10) for the extent of the reaction and the Maxwell relation (8.22), we can rewrite equation (8.20) as,

$$\frac{dH}{d\xi_a} = -T\frac{\partial}{\partial T}\left(\sum_{A=1}^{N} \mu_A \nu_{aA}\right) + \sum_{A=1}^{N} \mu_A \nu_{aA} \tag{8.23}$$

We have taken into account the fact that the stoichiometric coefficients ν_{aA} are constant. In the neighbourhood of a chemical equilibrium, the second term is negligible with respect to the first. Thus, equation (8.23) reduces to,

$$\frac{dH}{d\xi_a} = -T\frac{\partial}{\partial T}\left(\sum_{A=1}^{N} \mu_A \nu_{aA}\right) \tag{8.24}$$

It is common practice in chemistry to characterise a chemical reaction a at pressure p by the **enthalpy of reaction** defined as,

$$\Delta_a H \equiv \frac{dH}{d\xi_a} \tag{8.25}$$

In view of equations (8.16) and (8.25), the enthalpy of reaction $\Delta_a H$ can be determined if the Gibbs free energy of reaction $\Delta_a G$ is known for different temperatures.

As expressed by equation (4.55), enthalpy is minimum at equilibrium, which implies that the sign of the enthalpy of reaction determines the direction in which a reaction at constant pressure occurs:

1. $\Delta_a H < 0$: it evolves spontaneously from left to right,
2. $\Delta_a H = 0$: it has reached equilibrium,
3. $\Delta_a H > 0$: it evolves spontaneously from right to left.

According to equation (4.61), the enthalpy variation ΔH_{if} during an isobaric process from an initial state i to a final state f is equal to the heat Q_{if} provided to the system, i.e. $\Delta H_{if} = Q_{if}$. Thus, the reaction is said to be **exothermic** when $\Delta_a H < 0$ and **endothermic** when $\Delta_a H > 0$.

8.3 Matter Balance and Chemical Dissipation

In this section, we expand on the general results established in Chapter 2 for a simple system consisting of r chemical substances, in order to determine the dissipation due to n chemical reactions occurring in a simple open system. The causes of the variation of the amount of matter of every chemical substance A are the chemical power P_C and the chemical reactions a occurring in the system. Thus, in view of equation (8.12), the matter balance equation implies,

$$\sum_{A=1}^{r} \mu_A \dot{N}_A = P_C + \sum_{A=1}^{r} \mu_A \left(\sum_{a=1}^{n} \nu_{aA} \Omega_a \right) \quad \text{(open system)} \tag{8.26}$$

Using the definition (8.17) of the chemical affinity, the matter balance equation (8.26) implies,

$$P_C = \sum_{A=1}^{r} \mu_A \dot{N}_A + \sum_{a=1}^{n} \mathcal{A}_a \Omega_a \quad \text{(open system)} \tag{8.27}$$

after taking into account the fact that the sums commute. Expression (8.27) for the chemical power P_C is a generalisation of expression (2.17) for a reversible matter transfer in the presence of chemical reactions. Matter transfers are irreversible processes, as we will see later. The general expression (8.27) will be established from a description of continuous media in § 10.3.5.

When a simple system is subjected to a reversible mechanical action, then according to relation (2.26), i.e. $P_W = -p \dot{V}$, the entropy production rate (2.15) becomes,

$$\Pi_S = \frac{1}{T} \left(P_C - \sum_{A=1}^{r} \mu_A \dot{N}_A \right) \geq 0 \tag{8.28}$$

In view of equation (8.27), equation (8.28) reduces to,

$$\Pi_S = \frac{1}{T} \sum_{a=1}^{n} \mathcal{A}_a \Omega_a \geq 0 \tag{8.29}$$

Thus, the entropy production rate Π_S of the system is entirely due to the chemical reactions occurring inside the system.

In the particular case of a closed system, i.e. $P_C = 0$, the matter balance equation (8.27) is reduced to,

$$\sum_{A=1}^{r} \mu_A \dot{N}_A = - \sum_{a=1}^{n} \mathcal{A}_a \Omega_a \quad \text{(closed system)} \tag{8.30}$$

8.4 Molar Volume, Entropy and Enthalpy

8.4.1 Molar Volume

Let us consider the volume $V(T, p, \{N_A\})$ as a function of temperature T, pressure p and number of moles $\{N_A\}$ of chemical substances $A = 1, \ldots, r$. The volume $V(T, p, \{N_A\})$ and each number of moles N_A are extensive quantities, whereas temperature T and pressure p are intensive quantities. Thus, the volume $V(T, p, \{N_A\})$ multiplied by any real number λ must satisfy the identity,

$$\lambda V(T, p, \{N_A\}) = V(T, p, \{\lambda N_A\}) \tag{8.31}$$

Differentiating equation (8.31) with respect to λ and evaluating it for $\lambda = 1$, we obtain,

$$V(T, p, \{N_A\}) = \sum_{A=1}^{r} \frac{\partial V}{\partial N_A} N_A \qquad (8.32)$$

The ***molar volume*** defined as,

$$v_A(T, p, \{N_A\}) = \frac{\partial V(T, p, \{N_A\})}{\partial N_A} \qquad (8.33)$$

Since the molar volume $v_A(T, p, \{N_A\})$ is an intensive quantity, it must satisfy the identity,

$$v_A(T, p, \{N_A\}) = v_A(T, p, \{\lambda N_A\}) \qquad (8.34)$$

The concentration of substance A in the system is defined as,

$$c_A = \frac{N_A}{\displaystyle\sum_{B=1}^{r} N_B} \qquad (8.35)$$

Taking into account relation (8.35) and evaluating identity (8.34) for $\lambda = \left(\sum_{B=1}^{r} N_B\right)^{-1}$, we find,

$$v_A(T, p, \{N_A\}) = v_A(T, p, \{c_A\}) \qquad (8.36)$$

In view of relations (8.33) and (8.36), the expression (8.32) for the volume can be written as,

$$V(T, p, \{N_A\}) = \sum_{A=1}^{r} v_A(T, p, \{c_A\}) N_A \qquad (8.37)$$

8.4.2 Molar Entropy

Let us consider the entropy $S(T, p, \{N_A\})$ as a function of temperature T, pressure p and number of moles $\{N_A\}$ of chemical substances $A = 1, \dots, r$. The entropy $S(T, p, \{N_A\})$ and the number of moles N_A of each chemical substance A are extensive quantities, whereas temperature T and pressure p are intensive quantities. Thus, the entropy $S(T, p, \{N_A\})$ multiplied by λ must satisfy the identity,

$$\lambda S(T, p, \{N_A\}) = S(T, p, \{\lambda N_A\}) \qquad (8.38)$$

Differentiating equation (8.38) with respect to λ and evaluating it for $\lambda = 1$, we obtain,

$$S(T, p, \{N_A\}) = \sum_{A=1}^{r} \frac{\partial S}{\partial N_A} N_A \qquad (8.39)$$

The ***molar entropy*** is defined as,

$$s_A(T, p, \{N_A\}) = \frac{\partial S(T, p, \{N_A\})}{\partial N_A} \qquad (8.40)$$

Since the molar entropy $s_A(T, p, \{N_A\})$ is an intensive quantity, it must satisfy the identity,

$$s_A(T, p, \{N_A\}) = s_A(T, p, \{\lambda N_A\}) \qquad (8.41)$$

Taking into account relation (8.35) and evaluating identity (8.41) for $\lambda = \left(\sum_{B=1}^{r} N_B \right)^{-1}$, we find,

$$s_A \left(T, p, \{N_A\} \right) = s_A \left(T, p, \{c_A\} \right) \tag{8.42}$$

In view of relations (8.40) and (8.42), the expression (8.39) for the entropy can be written as,

$$S \left(T, p, \{N_A\} \right) = \sum_{A=1}^{r} s_A \left(T, p, \{c_A\} \right) N_A \tag{8.43}$$

8.4.3 Molar Enthalpy

In order to define the molar enthalpy, we first consider the enthalpy $H \left(S \left(T, p, \{N_A\} \right), p, \{N_A\} \right)$ as a function of temperature T, pressure p and number of moles $\{N_A\}$ of chemical substances $A = 1, \ldots, r$. The enthalpy $H \left(S \left(T, p, \{N_A\} \right), p, \{N_A\} \right)$, the entropy $S \left(T, p, \{N_A\} \right)$ and the number of moles N_A are extensive quantities. Temperature T and pressure p are intensive quantities. Thus, taking into account equation (8.38), the enthalpy $H \left(S \left(T, p, \{N_A\} \right), p, \{N_A\} \right)$ multiplied by λ must satisfy the identity,

$$\lambda H \left(S \left(T, p, \{N_A\} \right), p, \{N_A\} \right) = H \left(S \left(T, p, \{\lambda N_A\} \right), p, \{\lambda N_A\} \right) \tag{8.44}$$

Differentiating equation (8.44) with respect to λ and evaluating it for $\lambda = 1$, we obtain,

$$H \left(S \left(T, p, \{N_A\} \right), p, \{N_A\} \right) = \sum_{A=1}^{r} \frac{dH}{dN_A} N_A \tag{8.45}$$

where the pressure p is kept constant while performing the partial derivative with respect to N_A,

$$\frac{dH}{dN_A} = \frac{\partial H}{\partial S} \frac{\partial S}{\partial N_A} + \frac{\partial H}{\partial N_A} \tag{8.46}$$

The ***molar enthalpy*** is defined as,

$$h_A \left(s_A \left(T, p, \{N_A\} \right), p, \{N_A\} \right) = \frac{dH}{dN_A} \tag{8.47}$$

Since the molar enthalpy $h_A \left(s_A \left(T, p, \{N_A\} \right), p, \{N_A\} \right)$ is an intensive quantity, it must satisfy the identity,

$$h_A \left(s_A \left(T, p, \{N_A\} \right), p, \{N_A\} \right) = h_A \left(s_A \left(T, p, \{\lambda N_A\} \right), p, \{\lambda N_A\} \right) \tag{8.48}$$

Taking into account relation (8.35) and evaluating identity (8.48) for $\lambda = \left(\sum_{B=1}^{r} N_B \right)^{-1}$, we find,

$$h_A \left(s_A \left(T, p, \{N_A\} \right), p, \{N_A\} \right) = h_A \left(s_A \left(T, p, \{c_A\} \right), p, \{c_A\} \right) \tag{8.49}$$

In view of relations (8.47) and (8.49), the expression (8.45) for the volume can be written as,

$$H\left(S\left(T,p,\{N_A\}\right),p,\{N_A\}\right) = \sum_{A=1}^{r} h_A\left(s_A\left(T,p,\{c_A\}\right),p,\{c_A\}\right) N_A \qquad (8.50)$$

Using expressions (4.33) for temperature, (4.35) for chemical potential, (8.40) for molar entropy and (8.47) for molar enthalpy in equation (8.46), we obtain the following expression for the chemical potential,

$$\mu_A = h_A - T s_A \qquad (8.51)$$

8.4.4 Enthalpy of Formation

In order to compute the enthalpy of a **chemical compound**, let us consider a chemical reaction a between $r-1$ reactants, $X_1, \ldots, X_{r-1}$, that yields as a single product the compound C,

$$-\nu_{a1} X_1 - \ldots - \nu_{ar-1} X_{r-1} \to C \qquad (8.52)$$

The stoichiometric coefficients ν_{aA} are negative since the substances $A = 1, \ldots, r-1$ are reactants. The difference between the molar enthalpy h_C of the compound C and the sum of the molar enthalpies h_A of its constituents $X_1, \ldots, X_{r-1}$ is called the **enthalpy of formation**,

$$\Delta h^0 = h_C + \sum_{A=1}^{r-1} \nu_{aA} h_A \qquad (8.53)$$

where $\nu_{aA} < 0$ ($\forall A = 1, \ldots, r-1$). In practice, enthalpies of formation are calculated by invoking **Hess's law** that states that the enthalpy of a global chemical reaction is independent of the existence of intermediate chemical reactions. The enthalpy of a reaction can then be calculated by splitting this reaction in a series of reactions that can effectively occur, or not. The enthalpy of the global reaction is therefore the sum of the enthalpies of the intermediate reactions [78]. Hess's law is a consequence of the fact that enthalpy is a state function, which implies that its variation depends only on the initial and final states [79].

8.5 Mixture of Ideal Gases

In this section, we establish an expression for the chemical potential of a substance that belongs to a mixture of several substances. We will use the simplest possible model, according to which, the substances are independent from one another and each one can be considered as an ideal gas.

8.5.1 Chemical Potential of a Pure Ideal Gas

First, we need to establish the expression for the chemical potential $\mu(T,p)$ of a pure ideal gas as a function of pressure p. Applying Schwarz theorem (4.67) to the Gibbs free energy $G(T,p,N)$, we find,

$$\frac{\partial}{\partial p}\left(\frac{\partial G}{\partial N}\right) = \frac{\partial}{\partial N}\left(\frac{\partial G}{\partial p}\right) \tag{8.54}$$

Using definitions (4.41) for the volume $V(T,p,N)$ and (4.42) for the chemical potential $\mu(T,p,N)$ of the ideal gas, the differential equation (8.54) leads to the Maxwell relation,

$$\frac{\partial \mu}{\partial p} = \frac{\partial V}{\partial N} \tag{8.55}$$

Applying the equation of state (5.47) of the pure ideal gas to the Maxwell relation (8.55), we find the differential equation,

$$\frac{\partial \mu}{\partial p} = \frac{\partial}{\partial N}\left(\frac{NRT}{p}\right) = \frac{RT}{p} \tag{8.56}$$

When integrating the differential equation (8.56) from a pressure p_0 to a pressure p while keeping temperature T constant we obtain,

$$\int_{\mu(T,p_0)}^{\mu(T,p)} d\mu = RT \int_{p_0}^{p} \frac{dp'}{p'} \tag{8.57}$$

This yields the expression for the chemical potential of an ideal gas,

$$\mu(T,p) = \mu(T,p_0) + RT\ln\left(\frac{p}{p_0}\right) \tag{8.58}$$

8.5.2 Chemical Potential of an Ideal Gas Mixture

Now, we would like to determine the chemical potential of a mixture of r ideal gases $A = 1,\ldots,r$ in a system of volume V. According to the ideal gas equation of state (5.47), we can define the partial pressure p_A of each gas as,

$$p_A = \frac{N_A RT}{V} \tag{8.59}$$

The pressure p of the gas mixture is the sum of the partial pressures,

$$p = \sum_{A=1}^{r} p_A = \left(\sum_{A=1}^{r} N_A\right)\frac{RT}{V} \tag{8.60}$$

This means that the pressure p is entirely determined by the set $\{p_A\}$ of partial pressures. Applying Schwarz theorem (4.67) to the Gibbs free energy $G\left(T,p\left(\{p_A\}\right),\{N_A\}\right)$ we find,

$$\frac{\partial}{\partial p_A}\left(\frac{\partial G}{\partial N_A}\right) = \frac{\partial}{\partial N_A}\left(\frac{\partial G}{\partial p_A}\right) \tag{8.61}$$

Taking into account (8.60), we have,

$$\frac{\partial G}{\partial p_A} = \frac{\partial p}{\partial p_A}\frac{\partial G}{\partial p} = \frac{\partial G}{\partial p} \tag{8.62}$$

Using the definitions (4.41) for the volume $V(T,p(\{p_A\}),\{N_A\})$ and (4.42) for the chemical potential $\mu_A(T,p_A,N_A)$ of the ideal gas A and in view of identity (8.62), Schwarz theorem (8.61) leads to the Maxwell relation,

$$\frac{\partial \mu_A}{\partial p_A} = \frac{\partial V}{\partial N_A} \tag{8.63}$$

Applying expression (8.59) for the partial pressure p_A of an ideal gas A to the Maxwell relation (8.63), we find the differential equation,

$$\frac{\partial \mu_A}{\partial p_A} = \frac{\partial}{\partial N_A} \left(\frac{N_A R T}{p_A} \right) = \frac{R T}{p_A} \tag{8.64}$$

When integrating the differential equation (8.64) from a pressure p to a pressure p_A while keeping the temperature T constant we obtain,

$$\int_{\mu_A(T,p)}^{\mu_A(T,p_A)} d\mu_A = R T \int_{p}^{p_A} \frac{dp'_A}{p'_A} \tag{8.65}$$

This yields the expression for the chemical potential,

$$\mu_A(T,p_A) = \mu_A(T,p) + R T \ln \left(\frac{p_A}{p} \right) \tag{8.66}$$

Taking into account definitions (8.59), (8.60) and (8.35) for the partial pressure p_A, the total pressure p and the concentration c_A, their ratio is written as,

$$\frac{p_A}{p} = \frac{N_A}{\displaystyle\sum_{B=1}^{r} N_B} = c_A \tag{8.67}$$

which means that the partial pressure p_A is a function of the pressure p and of the concentration c_A. Thus, the chemical potential in equation (8.66) becomes a function of T, p and c_A,

$$\mu_A(T,p,c_A) = \mu_A(T,p) + R T \ln(c_A) \tag{8.68}$$

where $\mu_A(T,p)$ is the chemical potential of the pure ideal gas A. According to equation (8.68), when the ideal gas A is pure, i.e. $c_A = 1$, the chemical potential $\mu_A(T,p) = \mu_A(T,p,1)$.

In practice, many mixtures can often be described in a first approximation by making the drastic assumption that each constituent of the mixture is an ideal gas. The **ideal mixture** is defined as the mixture where the chemical potential of each constituent satisfies relation (8.68) [80]. Notice we adopt the convention according to which a property of a substance is that of the pure substance when the concentration is not found in the list of state variables of that property.

8.5.3 Molar Entropy of an Ideal Gas Mixture

Using definition (8.40) for molar entropy, the Maxwell relation (8.22) is written as,

$$s_A(T,p,c_A) = -\frac{\partial \mu_A(T,p,c_A)}{\partial T} \qquad \text{(mixture)}$$

$$s_A(T,p) = -\frac{\partial \mu_A(T,p)}{\partial T} \qquad \text{(pure)} \tag{8.69}$$

The partial derivative with respect to temperature T of the expression (8.68) for the chemical potential yields,

$$\frac{\partial}{\partial T} \mu_A(T,p,c_A) = \frac{\partial}{\partial T} \mu_A(T,p) + R \ln(c_A) \tag{8.70}$$

Taking into account the identities (8.69), equation (8.70) is recast as,

$$s_A(T,p,c_A) = s_A(T,p) - R\ln(c_A) \tag{8.71}$$

where $s_A(T,p)$ is the molar entropy of the pure ideal gas A. The second term on the right-hand side of expression (8.71) is called the **entropy of mixing**. This result is known as the 'Gibbs theorem' [81].

A **colligative property** of a chemical solution is a property that is due to the mixing of constituents, but is independent of the nature of these constituents. The entropy of mixing (8.71) is an example of such a property.

8.5.4 Molar Enthalpy of an Ideal Gas Mixture

According to expression (8.51) for the molar enthalpy, we have,

$$h_A\left(s_A(T,p,c_A),p,c_A\right) = \mu_A(T,p,c_A) + T s_A(T,p,c_A) \qquad \text{(mixture)}$$

$$h_A\left(s_A(T,p),p\right) = \mu_A(T,p) + T s_A(T,p) \qquad \text{(pure)} \tag{8.72}$$

Applying expressions (8.68) for the chemical potential $\mu_A(T,p,c_A)$ and (8.71) for the molar entropy $s_A(T,p,c_A)$, we can combine the first and second equations of (8.72) to write,

$$h_A\left(s_A(T,p,c_A),p,c_A\right) = h_A\left(s_A(T,p),p\right) \tag{8.73}$$

This means that the molar enthalpy of an ideal gas in an ideal mixture is independent of its concentration c_A.

8.5.5 Law of Mass Action

Taking into account equations (4.37) and (8.68), the Gibbs free energy $G(T,p,\{N_A\})$ can be written as,

$$G(T,p,\{N_A\}) = \sum_{A=1}^{r} N_A\,\mu_A(T,p,c_A) = \sum_{A=1}^{r} N_A\left(\mu_A(T,p) + RT\ln(c_A)\right) \tag{8.74}$$

Calculating the partial derivative of the Gibbs free energy with respect to the extent ξ_a of the chemical reaction a, taking into account definition (8.35) for the concentration c_A and using equation (8.10), we can recast dN_A in terms of $d\xi_a$ and find,

$$\frac{\partial G}{\partial \xi_a} = \sum_{A=1}^{r} \nu_{aA}\,\mu_A(T,p) + RT\sum_{A=1}^{r} \nu_{aA}\ln(c_A)$$

$$+ RT\sum_{A=1}^{r} N_A \frac{\partial}{\partial \xi_a}\ln\left(\frac{N_A}{\sum_{B=1}^{r} N_B}\right) \tag{8.75}$$

The second term on the right-hand side of equation (8.75) can in turn be recast as,

$$RT\sum_{A=1}^{r} \nu_{aA}\ln(c_A) = RT\ln\left(\prod_{A=1}^{r} c_A^{\nu_{aA}}\right) \tag{8.76}$$

Taking into account equation (8.10), the third term on the right-hand side of (8.75) can be written as,

$$R T \sum_{A=1}^{r} N_A \frac{\partial}{\partial \xi_a} \left(\ln(N_A) - \ln \left(\sum_{B=1}^{r} N_B \right) \right)$$

$$= R T \sum_{A=1}^{r} N_A \left(\frac{\nu_{aA}}{N_A} - \frac{\sum_{C=1}^{r} \nu_{aC}}{\sum_{B=1}^{r} N_B} \right) = 0 \tag{8.77}$$

At chemical equilibrium, equation (8.14) holds. Thus, the right-hand side of equation (8.75) vanishes. Taking into account identities (8.76) and (8.77), the chemical equilibrium is therefore written as,

$$\sum_{A=1}^{r} \nu_{aA} \mu_A (T,p) + R T \ln \left(\prod_{A=1}^{r} c_A^{\nu_{aA}} \right) = 0 \qquad \text{(chemical equilibrium)} \tag{8.78}$$

The **equilibrium constant** K_a of the chemical reaction a is defined as,

$$K_a = \exp \left(- \frac{1}{R T} \sum_{A=1}^{r} \nu_{aA} \mu_A (T,p) \right) \tag{8.79}$$

Substituting this definition into equation (8.78), we find what is known as the **mass action law**,

$$K_a = \prod_{A=1}^{r} c_A^{\nu_{aA}} \qquad \text{(chemical equilibrium)} \tag{8.80}$$

This is also called the **Guldberg and Waage law**. It can be used to estimate the Gibbs free energy of a mixture from measurements of its concentrations at equilibrium.

8.6 Osmosis

Using the expression we established for the chemical potential of a substance in a mixture as a function of its concentration, we can now explain the phenomenon known as osmosis, which we define as follows. Consider a system consisting of two simple subsystems separated by a fixed permeable membrane. One subsystem contains a solvent and the other the solvent and a solute. When the membrane has the particular property that it lets the solvent pass through it, but not the solute, it is said to be an **osmotic membrane**. The spontaneous transfer across the membrane of solvent from the side where it is pure to the side which has the solute is called **osmosis**.

Let us assume that one subsystem is a large container of pure water and that a cell containing salt water is immersed in the pure water. The cell is separated from the container of pure water by an osmotic membrane (Fig. 8.1). The salt water in the cell and the pure

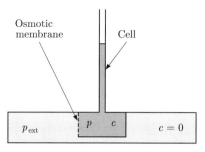

Figure 8.1 An osmotic cell is immersed in a container filled with pure water. The left wall of the cell is a fixed osmotic membrane that lets pure water pass through, but not the ions in solution in the cell. The osmotic cell contains salt water with a salt concentration c. The pressures is p_{ext} in the container and p in the cell.

water in the container can be considered as two simple subsystems. The container is large enough with respect to the size of the cell that the pure water in the container can be considered as a heat and work reservoir at temperature T and pressure p_{ext}. The cell comprises a long and vertical open-ended tube. The cell is supposed to be shallow enough with respect to the height of the tube so that we can assume that the pressure p is uniform in the part of the cell where the membrane is located. The tube is thin enough so that the volume of liquid in the tube remains low compared to the volume of liquid in the cell. Thus, when the liquid rises in the tube, it can be assumed that the concentration c of salt in the cell does not change significantly.

The transfer of pure water across the membrane stops when the two subsystems reach the chemical equilibrium (3.43) characterised by the equality of the chemical potentials of the solvent on either side of the osmotic membrane,

$$\mu\left(T, p_{ext}\right) = \mu\left(T, p, 1 - c\right) \qquad \text{(chemical equilibrium)} \tag{8.81}$$

In the approximation of an ideal mixture, equation (8.68) implies that the chemical potential is written as,

$$\mu\left(T, p, 1 - c\right) = \mu\left(T, p\right) + RT \ln\left(1 - c\right) \tag{8.82}$$

where $1 - c$ is the concentration of water in the cell. As the salt concentration is low, i.e. $c \ll 1$, we can use the following series expansion in c,

$$\ln\left(1 - c\right) = -c + \mathcal{O}\left(c^2\right) \tag{8.83}$$

Thus, at chemical equilibrium, to first-order in c, relations (8.81) and (8.82) imply that,

$$\mu\left(T, p_{ext}\right) = \mu\left(T, p\right) - cRT \qquad \text{(chemical equilibrium)} \tag{8.84}$$

We need to work out how the chemical potential of pure water depends on pressure. Since pure water is incompressible, its molar volume v is independent of pressure. The definition (8.33) of the molar volume and Maxwell relation (8.55) imply that for pure water,

$$\mu\left(T, p\right) - \mu\left(T, p_{ext}\right) = \int_{p_{ext}}^{p} \frac{\partial \mu}{\partial p}\, dp = v \int_{p_{ext}}^{p} dp = v\left(p - p_{ext}\right) \tag{8.85}$$

Comparing relations (8.84) and (8.85), we find the law first established by van't Hoff [82],

$$v\,(p - p_{\text{ext}}) = c\,R\,T \qquad \text{(chemical equilibrium)} \tag{8.86}$$

The pressure difference $p - p_{\text{ext}}$ between the cell and the container is called the **osmotic pressure**. The osmotic pressure $p - p_{\text{ext}}$ is equal to the hydrostatic pressure of the water column in the tube.

It can also be interpreted as the partial pressure of an ideal gas consisting of N_s moles in the volume V of the cell. Indeed, using definition (8.35), the salt concentration $c \ll 1$ in the cell can be written in terms of the number N_s of salt moles and the number N_e of water moles, where $N_s \ll N_e$,

$$c = \frac{N_s}{N_e + N_s} \simeq \frac{N_s}{N_e} \tag{8.87}$$

The pure water molar volume is written as,

$$v \simeq \frac{V}{N_e} \tag{8.88}$$

Taking into account equations (8.87) and (8.88), equation (8.86) can be expressed in terms of the number of moles N_s of salt as,

$$p - p_{\text{ext}} = \frac{N_s\,R\,T}{V} \tag{8.89}$$

Notice that the equality of the chemical potential of the water on both sides of the membrane implies the equality of the partial pressure $p - N_s\,R\,T/V$ of water inside the cell and the pressure p_{ext} of the surrounding pure water.

8.7 Electrochemistry

Until now, we considered that the system consisted of substances without net electric charge. In this section, we extend our thermodynamic description of chemical systems by introducing electrically charged substances. The study of the interplay between chemical and electrical phenomena is called **electrochemistry**. It is an important research field since energy storage and electricity production using electrochemical cells could potentially play an important role in regard to the energy crisis.

8.7.1 Electrochemical potential

When electrically charged substances are in equilibrium, the quantity that characterises equilibrium is a generalisation of the chemical potential called the **electrochemical potential** $\bar{\mu}$. It is defined as the sum of the chemical potential μ and an electrostatic term, which is the product of the molar electric charge $z\,F_F$ of the substance and the **electrostatic potential** φ,

$$\bar{\mu} = \mu + z\,F_F\,\varphi \tag{8.90}$$

F_F is the **Faraday constant**, defined as the electric charge of a mole of electrons,

$$F_F = -\mathcal{N}_A e = 96,487 \, C \, mol^{-1} \tag{8.91}$$

where $\mathcal{N}_A$ is the Avogadro number, $e = -1.602 \cdot 10^{-19} \, C$ is the charge of an electron.

In equation (8.90), z is the **electrovalence**, defined as the number of electrons to add ($z > 0$) or take away ($z < 0$) from each ion to make the substance electrically neutral. Equation (8.90) will be established on a rigorous basis in Chapter 9. It may be helpful to remember that the electrochemical potential $\bar{\mu}$ and the electrostatic potential φ are intensive quantities.

8.7.2 Oxidation and Reduction

Electrochemical reactions transform neutral substances into electrically charged substances and vice versa. For instance, let us consider the chemical reaction between copper atoms Cu, zinc atoms Zn, and their ions Cu^{2+} and Zn^{2+},

$$Zn + Cu^{2+} \rightarrow Zn^{2+} + Cu \tag{8.92}$$

This electrochemical reaction takes place spontaneously in the indicated direction. In order to benefit from such a reaction, two electrodes consisting of copper and zinc respectively, are immersed into electrolytic baths, which are solutions containing the ions of both metals. The electrochemical reaction is either an oxidation or a reduction. An oxidation produces electrons and consumes electrons. Equation (8.92) is an example of such a reaction, which is called a redox reaction. In this reaction, the Zn atom undergoes an **oxidation** and transforms into a Zn^{2+} ion giving two electrons 2 e,

$$Zn \rightarrow Zn^{2+} + 2\,e \tag{8.93}$$

and the Cu^{2+} ion undergoes a **reduction** and transforms into a Cu atom by consuming two electrons 2 e,

$$Cu^{2+} + 2\,e \rightarrow Cu \tag{8.94}$$

The electrode where the oxidation occurs is called the **anode** and the electrode where the reduction occurs is called the **cathode**.

In order to analyse the equilibrium of an oxidation or reduction reaction, the atoms, ions and electrons are treated as distinct 'chemical' substances. By replacing the chemical potential by the electrochemical potential in the equilibrium condition (8.15), we obtain the equilibrium of the zinc oxidation reaction (8.93),

$$-\bar{\mu}_{Zn} + \bar{\mu}_{Zn^{2+}} + 2\,\bar{\mu}_e = 0 \tag{8.95}$$

where the stoichiometric coefficients are determined by the oxidation (8.93). Similarly, the equilibrium condition for the copper reduction (8.94) is given by,

$$\bar{\mu}_{Cu} - \bar{\mu}_{Cu^{2+}} - 2\,\bar{\mu}_e = 0 \tag{8.96}$$

where the stoichiometric coefficients are determined by the reduction (8.94).

8.7.3 Nernst Potential

We are now in a position to analyse what is happening in a battery. In order to keep our reasoning and the notation clear, we follow two steps. First, we consider a hypothetical battery where the electrolytes on the anode and cathode sides differ only by their concentration of Cu ions. This allows us to introduce the notion of Nernst potential. Then, we will consider a battery with both Cu and Zn ions.

Let us first consider two separate cells, which both consist of a Cu electrode in an electrolytic bath containing water, Cu^{2+} ions and SO_4^{2-} ions. The two electrodes are identical, and likewise, the electrolyte pressure and temperature is the same on both sides. The electrolytic baths of both cells are connected by a salt bridge that only lets the SO_4^{2-} ions pass. This ensures electric neutrality of the electrolytic baths, except, as we will see, at the interfaces with the electrodes (Fig. 8.2) [83].

The concentration c^+ of ions Cu^{2+} is assumed higher in the $(+)$ cell than the concentration c^- of ions Cu^{2+} in the $(-)$ cell. The chemical reactions between water and ions and between the water and the copper electrodes can be ignored because their influence on the equilibrium state is negligible. At equilibrium, an electrostatic potential difference $\Delta\varphi$ is measured between the two electrodes. We show now how this electrostatic potential difference $\Delta\varphi$ can be expressed in terms of the concentrations c^+ and c^-.

At equilibrium, the reduction of copper (8.94) at the $(+)$ and $(-)$ electrodes are expressed as,

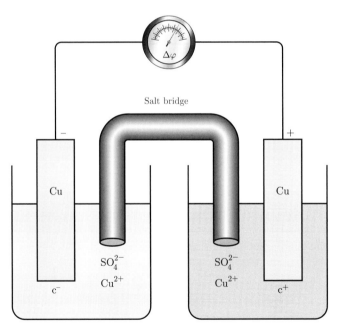

Figure 8.2   An electrostatic potential difference $\Delta\varphi$ is measured by a voltmeter between two electrodes of Cu immersed into electrolytic baths of Cu^{2+} ions with concentrations c^- (left) and c^+ (right), with $c^+ > c^-$. The salt bridge only allows the SO_4^{2-} ions to pass from one electrolytic bath to the other.

$$\bar{\mu}_{Cu}^{(+)} - \bar{\mu}_{Cu^{2+}}^{(+)} - 2\,\bar{\mu}_e^{(+)} = 0$$
$$\bar{\mu}_{Cu}^{(-)} - \bar{\mu}_{Cu^{2+}}^{(-)} - 2\,\bar{\mu}_e^{(-)} = 0 \qquad (8.97)$$

Using definition (8.90), the electrochemical potentials can be expressed in terms of the chemical potentials and the electrostatic potentials. The copper atoms of the two identical electrodes are neutral. Thus, their electrostatic potential energy can be taken to be zero and their chemical potentials are equal,

$$\bar{\mu}_{Cu}^{(+)} = \mu_{Cu} \quad \text{and} \quad \bar{\mu}_{Cu}^{(-)} = \mu_{Cu} \qquad (8.98)$$

The electrochemical potentials of the ions are given by,

$$\bar{\mu}_{Cu^{2+}}^{(+)} = \mu_{Cu^{2+}}^{(+)} + 2\,F_F\,\varphi_{Cu^{2+}}^{(+)} \quad \text{and} \quad \bar{\mu}_{Cu^{2+}}^{(-)} = \mu_{Cu^{2+}}^{(-)} + 2\,F_F\,\varphi_{Cu^{2+}}^{(-)} \qquad (8.99)$$

The chemical potential of the electrons μ_e is the same for the two electrodes since they consist of the same metal,

$$\bar{\mu}_e^{(+)} = \mu_e - F_F\,\varphi_e^{(+)} \quad \text{and} \quad \bar{\mu}_e^{(-)} = \mu_e - F_F\,\varphi_e^{(-)} \qquad (8.100)$$

Taking into account relations (8.98), (8.99) and (8.100), equations (8.97) become,

$$\mu_{Cu} - \mu_{Cu^{2+}}^{(+)} - 2\,\mu_e + 2\,F_F\left(\varphi_e^{(+)} - \varphi_{Cu^{2+}}^{(+)}\right) = 0$$
$$\mu_{Cu} - \mu_{Cu^{2+}}^{(-)} - 2\,\mu_e + 2\,F_F\left(\varphi_e^{(-)} - \varphi_{Cu^{2+}}^{(-)}\right) = 0 \qquad (8.101)$$

The presence of electrically charged SO_4^{2-} ions in the electrolytic bath largely 'neutralises' the electrostatic fields generated by the ions around the electrodes. This phenomenon is called **electrostatic screening**. Thus, the electrostatic potentials of the ions in the baths are negligible with respect to the electrostatic potentials of the conduction electrons on the electrodes,

$$\varphi_{Cu^{2+}}^{(+)} \ll \varphi_e^{(+)} \quad \text{and} \quad \varphi_{Cu^{2+}}^{(-)} \ll \varphi_e^{(-)} \qquad (8.102)$$

According to equation (8.68), the chemical potential of the copper ions $\mu_{Cu^{2+}}^{(+)}$ and $\mu_{Cu^{2+}}^{(-)}$ depends on their respective concentrations c^+ and c^- in the electrolytic baths,

$$\mu_{Cu^{2+}}^{(+)} = \mu_{Cu^{2+}} + R\,T\ln\left(c^+\right)$$
$$\mu_{Cu^{2+}}^{(-)} = \mu_{Cu^{2+}} + R\,T\ln\left(c^-\right) \qquad (8.103)$$

where $\mu_{Cu^{2+}}$ is the chemical potential of the pure copper ions. Using relations (8.103) in equations (8.101) and taking into account the screening (8.102) of the electrostatic potential of the ions, we obtain,

$$\mu_{Cu} - \mu_{Cu^{2+}} - R\,T\ln\left(c^+\right) - 2\,\mu_e + 2\,F_F\,\varphi_e^{(+)} = 0$$
$$\mu_{Cu} - \mu_{Cu^{2+}} - R\,T\ln\left(c^-\right) - 2\,\mu_e + 2\,F_F\,\varphi_e^{(-)} = 0 \qquad (8.104)$$

Subtracting both equations (8.104) from one another yields the electrostatic potential difference between the two electrodes,

$$\Delta\varphi \equiv \varphi_e^{(+)} - \varphi_e^{(-)} = \frac{R\,T}{2\,F_F}\,\ln\left(\frac{c^+}{c^-}\right) \qquad (8.105)$$

This concentration-dependent potential is called the ***Nernst potential***. It can be understood as follows: in the cell containing the highest concentration c^+ of Cu^{2+} ions, the reduction is more important than in the cell containing the lowest concentration c^- of Cu^{2+} ions. Thus, the copper electrode in the cell with a concentration c^+ has a larger electric charge than the electrode in the other cell. This asymmetry induces the Nernst potential.

8.7.4 Electric Battery

In this section, we would like to describe the working of a particular electric battery, called the Daniell cell, that consists of a Cu electrode and a Zn electrode. The Daniell cell is an electrochemical system that is analogous to the system presented in § 8.7.3 in which the left Cu electrode has been replaced by a Zn electrode (Fig. 8.3).

The oxidation (8.93) that occurs at the Zn anode implies that this electrode consumes electrons. The reduction (8.94) that occurs at the Cu cathode implies that this electrode releases electrons into the electrolytic bath. The left electrolytic bath, where the Zn anode is immersed, contains Zn^{2+} and SO_4^{2-} ions, and the right electrolytic bath, where the Cu cathode is immersed, contains Cu^{2+} and SO_4^{2-} ions. At equilibrium, an electrostatic potential difference $\Delta\varphi$ is measured between the cathode and the anode. It can be used by closing the electrical circuit outside the cell.

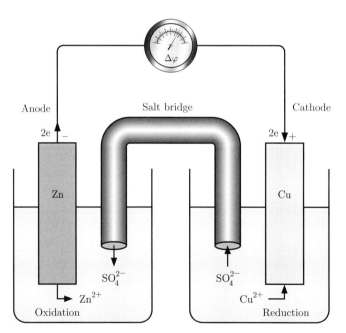

Figure 8.3 An electrostatic potential difference $\Delta\varphi$ is measured by a voltmeter between a Cu cathode and a Zn anode. The cathode is immersed in an electrolytic bath containing Cu^{2+} and SO_4^{2-} ions. The anode is immersed in an electrolytic bath containing Zn^{2+} and SO_4^{2-} ions. The salt bridge only allows SO_4^{2-} ions to pass from one electrolytic bath to the other.

At electrochemical equilibrium, the copper reduction at the cathode $(+)$ yields the condition (8.101). Taking into account the screening (8.102), the condition (8.101) is written as,

$$\mu_{Cu} - \mu_{Cu^{2+}}^{(+)} - 2\,\mu_e + 2\,F_F\,\varphi_e^{(+)} = 0 \tag{8.106}$$

Replacing the copper by zinc in equation (8.106) yields the equilibrium condition at the anode $(-)$,

$$\mu_{Zn} - \mu_{Zn^{2+}}^{(-)} - 2\,\mu_e + 2\,F_F\,\varphi_e^{(-)} = 0 \tag{8.107}$$

Subtracting equations (8.106) and (8.107) yields the electrostatic potential difference between the two electrodes,

$$\Delta\varphi \equiv \varphi_e^{(+)} - \varphi_e^{(-)} = \frac{1}{2\,F_F}\left(\left(\mu_{Cu^{2+}}^{(+)} - \mu_{Cu}\right) - \left(\mu_{Zn^{2+}}^{(-)} - \mu_{Zn}\right) \right) \tag{8.108}$$

When the electric circuit is closed outside the battery, the electrons (of negative electric charge e) move in the outer circuit from the anode $(-)$ to the cathode $(+)$. By convention, the electric current is defined for positive charges. The signs $(+)$ and $(-)$ with which the electrodes are labeled correspond to this convention according to which an electric current (of positive charges) goes from the anode $(+)$ to the cathode $(-)$ through the external electric circuit. In order for the charge circuit to form a closed loop, a current of SO_4^{2-} ions of negative electric charge must run through the salt bridge connecting the electrolytic baths. The SO_4^{2-} ions move from the cathode $(+)$ side to the anode $(-)$ side of the Daniell cell.

8.8 Worked Solutions

8.8.1 Dissipation of an Ideal Gas

Consider an isolated system consisting of two subsystems, labelled 1 and 2, containing an ideal gas. The two subsystems are separated by a fixed diathermal and permeable wall (Fig. 3.3). The system is kept at a constant temperature T. Assume that the pressure p_2 of subsystem 2 is slightly larger than the pressure p_1 of the subsystem 1, i.e. $\Delta p = p_2 - p_1 \ll p_2$. Determine the entropy production rate Π_S of the system in terms of Δp.

Solution:

At constant temperature T, the entropy production rate (3.44) is written

$$\Pi_S = -\frac{1}{T}\left(\mu_1 - \mu_2\right)\dot{N}_1 \geq 0$$

Equation (8.58) is written in terms of the chemical potentials $\mu_1 \equiv \mu(T,p_1)$ and $\mu_2 \equiv \mu(T,p_2)$ as,

$$\mu(T,p_1) = \mu(T,p_2) + R\,T \ln\left(\frac{p_1}{p_2}\right) = \mu(T,p_2) + R\,T \ln\left(1 - \frac{\Delta p}{p_2}\right)$$

Since $\Delta p \ll p_2$ the following series expansion can be used,

$$\ln\left(1 - \frac{\Delta p}{p_2}\right) = -\frac{\Delta p}{p_2} + \mathcal{O}\left(\frac{\Delta p^2}{p_2}\right)$$

Thus, to first order in $\Delta p / p_2$,

$$\mu_1 - \mu_2 = -RT\frac{\Delta p}{p_2}$$

which implies that,

$$\Pi_S = R\frac{\Delta p}{p_2}\dot{N}_1 \geq 0$$

This implies also that $\dot{N}_1 > 0$. Thus, the gas moves from the subsystem 2 towards the subsystem 1.

8.8.2 Isothermal Mixture

An isolated system of volume V consists of two simple subsystems of volumes V_1 and V_2 initially separated by a fixed diathermal wall. The temperature T and the pressure p are the same in both subsystems. The temperature T is constant and the total volume V is fixed. Initially, there are N_1 moles of an ideal gas of substance 1 in one of the subsystems and N_2 moles of an ideal gas of substance 2 in the other. The wall is removed and the ideal gases diffuse and mix [84]. There is no chemical reaction between the ideal gases.

1. Determine the internal energy variation ΔU of the system during the mixing.
2. Determine the enthalpy variation ΔH of the system during the mixing.
3. Determine the entropy variation ΔS of the system during the mixing in terms of the ideal gas concentrations c_1 and c_2 in the mixture and show that it is positive.
4. Determine the Gibbs free energy variation ΔG of the system during the mixing in terms of the ideal gas concentrations c_1 and c_2 in the mixture and show that it is negative.

Solution:

1. *Since the system is isolated, the internal energy variation (1.30) during the mixing vanishes,*

$$\Delta U = 0$$

2. *According to expression (5.65), the enthalpy variation of an ideal gas during an isothermal process vanishes. Thus, the total enthalpy variation during the mixing vanishes,*

$$\Delta H = 0$$

3. *Taking into account the expression (7.17) of the entropy variation for an isothermal process, the entropy variation of the system is written,*

$$\Delta S = \Delta S_1 + \Delta S_2 = N_1 R \ln\left(\frac{V}{V_1}\right) + N_2 R \ln\left(\frac{V}{V_2}\right)$$

Before the mixing, the ideal gases satisfy the equations of state (5.47),

$$V_1 = \frac{N_1 R T}{p} \quad and \quad V_2 = \frac{N_2 R T}{p}$$

Summing the two previous equations yields an expression for the total volume V,

$$V = V_1 + V_2 = \frac{(N_1 + N_2) R T}{p}$$

Substituting the two previous equations in the expression of ΔS yields,

$$\Delta S = N_1 R \ln\left(\frac{N_1 + N_2}{N_1}\right) + N_2 R \ln\left(\frac{N_1 + N_2}{N_2}\right)$$

According to expression (8.35), the ideal gas concentrations are defined as,

$$c_1 = \frac{N_1}{N_1 + N_2} \quad and \quad c_2 = \frac{N_2}{N_1 + N_2}$$

Substituting the two previous equations in the expression of ΔS yields,

$$\Delta S = -N_1 R \ln(c_1) - N_2 R \ln(c_2) = -R \ln\left(c_1^{N_1} c_2^{N_2}\right)$$

Since the concentrations $0 < c_1 < 1$ and $0, < c_2 < 1$,

$$\ln\left(c_1^{N_1} c_2^{N_2}\right) < 0 \quad \Rightarrow \quad \Delta S > 0$$

Thus, the entropy increases during the mixing.

4. *According to equations (4.31) and (4.38) for an isothermal process, the Gibbs free energy variation ΔG is related to the enthalpy variation ΔH*

$$\Delta G = \Delta H - T \Delta S - S \Delta T = -T \Delta S = R T \ln\left(c_1^{N_1} c_2^{N_2}\right) < 0$$

Thus, the Gibbs free energy variation during an isothermal mixing is negative. Moreover, the previous inequality means that the mixture at constant pressure and temperature is a spontaneous process.

8.8.3 Chemical Equilibrium

In a cylinder, there are N_{N_2} moles of molecular nitrogen N_2 and N_{H_2} moles of molecular hydrogen H_2 that can be considered as ideal gases. The system is closed by a piston of negligible weight. The ideal gases are initially separated by an impermeable wall of negligible mass (Fig. 8.4). They are kept at constant temperature T and at constant pressure p. When the wall is removed, a chemical reaction, denoted a, occurs due to a catalyst that can be ignored in the analysis. The chemical reaction produces ammonia NH_3 that can be considered as an ideal gas. We assume that temperature and pressure are constant during the chemical reaction.

1. Define the chemical reaction a and determine the stoichiometric coefficients ν_{aN_2}, ν_{aH_2} and ν_{aNH_3}.
2. In the particular case where the system consists initially of N moles of molecular nitrogen and $3N$ moles of molecular hydrogen that transform entirely into ammonia, determine the ratio of the volumes $V_{NH_3} / (V_{N_2} + V_{H_2})$.

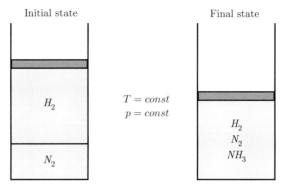

Figure 8.4 Initially, molecular hydrogen H_2 and molecular nitrogen N_2 are separated by a wall in a cylinder closed by a piston. After removing the wall, a chemical reaction produces ammonia NH_3.

3. Determine the equilibrium condition.
4. Express this equilibrium condition in terms of the molecular nitrogen concentration c_{N_2} and the molecular hydrogen concentration c_{H_2}.

Solution:

1. *The stoichiometric coefficients of the chemical reaction a are* $\nu_{aN_2} = -1$, $\nu_{aH_2} = -3$ *and* $\nu_{aNH_3} = 2$,

$$N_2 + 3\,H_2 \rightleftarrows 2\,NH_3$$

2. *In the particular case where the chemical reaction transforms entirely the molecular nitrogen and hydrogen into ammonia, the initial volumes of ideal gases* V_{N_2}, V_{H_2} *and the final volume of ideal gas* V_{NH_3} *are given by,*

$$V_{N_2} = \frac{NR\,T}{p} \quad and \quad V_{H_2} = \frac{3\,NR\,T}{p} \quad and \quad V_{NH_3} = \frac{2\,NR\,T}{p}$$

Thus, the volume ratio of the ideal gas is given by,

$$\frac{V_{NH_3}}{V_{N_2} + V_{H_2}} = \frac{1}{2}$$

This implies that the chemical reaction a causes the decrease of the gas volume during the production of ammonia.

3. *Taking into account the explicit expression of the stoichiometric coefficients, the equilibrium condition* (8.15) *of the chemical potentials is written as,*

$$-\mu_{N_2} - 3\,\mu_{H_2} + 2\,\mu_{NH_3} = 0$$

4. *According to equation* (8.68),

$$\mu_{N_2}\,(T,p,c_{N_2}) = \mu_{N_2}\,(T,p) + R\,T\,\ln\,(c_{N_2})$$
$$\mu_{H_2}\,(T,p,c_{H_2}) = \mu_{N_2}\,(T,p) + R\,T\,\ln\,(c_{H_2})$$
$$\mu_{NH_3}\,(T,p,c_{NH_3}) = \mu_{NH_3}\,(T,p) + R\,T\,\ln\,(c_{NH_3})$$

and since the ammonia concentration is related to the molecular nitrogen and hydrogen concentrations by,

$$c_{NH_3} = 1 - c_{N_2} - c_{H_2}$$

the chemical equilibrium condition is recast as,

$$- \mu_{N_2}(T,p) - 3\,\mu_{H_2}(T,p) + 2\,\mu_{NH_3}(T,p) = R\,T\ln \left(\frac{c_{N_2}\,c_{H_2}^3}{\left(1 - c_{N_2} - c_{H_2}\right)^2} \right)$$

8.8.4 Osmotic Cell

Determine the time evolution of an osmotic cell by considering a system consisting of two subsystems labelled 1 and 2, separated by an osmotic wall of area A. Subsystem 1 is an osmotic cell that contains salt water with a low initial salt concentration, i.e. $c \ll 1$. Subsystem 2 contains only pure water and constitutes a heat and work reservoir. The number of moles of pure water in the osmotic cell is N_1 and the number of moles of salt is N_s. The two subsystems are kept at constant temperature T and pressure p.

1. Determine the time evolution of the number of moles of salt $N_s(t)$ in the osmotic cell.
2. Determine the time evolution of the number of moles of pure water $N_1(t)$ in the osmotic cell, assuming that the time derivative of this quantity is proportional to the chemical potential difference of water in the two subsystems.

Solution:

1. *The osmotic wall does not let salt pass. Thus, the number of moles of salt in the salt water is constant, i.e.*

$$N_s = \text{const}$$

2. *According to relation (3.45), the rate of change of the number of moles of pure water in the osmotic cell $\dot{N}_1$ is given by,*

$$\dot{N}_1 = F\frac{A}{\ell}\,(\mu_2 - \mu_1)$$

where F is Fick's diffusion constant and ℓ is the characteristic diffusion length. The chemical potentials of water in the cell μ_1 and in the reservoir μ_2 are related by equation (8.82),

$$\mu_1 = \mu_2 + R\,T\ln(1 - c)$$

Since the salt concentration in the salt water is low, i.e. $c \ll 1$, the following series expansion in terms of c can be used,

$$\ln(1 - c) = -c + \mathcal{O}\left(c^2\right)$$

It implies that,

$$\mu_2 - \mu_1 = R\,T\,c$$

and, for a low salt concentration,

$$c = \frac{N_s}{N_1}$$

Thus, the evolution equation can be recast as,

$$\dot{N}_1 = \frac{1}{N_1 \, \tau}$$

where the characteristic time τ is given by,

$$\tau = \frac{\ell}{A \, F \, N_s \, R \, T}$$

The integral of the evolution equation reads,

$$\int_{N_1(0)}^{N_1(t)} N_1 \, dN_1 = \frac{1}{\tau} \int_0^t dt'$$

which implies that the solution is given by,

$$\frac{1}{2} \left(N_1(t)^2 - N_1(0)^2 \right) = \frac{t}{\tau}$$

Thus,

$$N_1(t) = \sqrt{N_1(0)^2 + \frac{2 \, t}{\tau}}$$

8.8.5 Thermogalvanic Cell

Consider an isolated system consisting of two simple subsystems labelled 1 and 2 separated by a diathermal, permeable and fixed wall of area A (Fig. 3.5). Both subsystems contain the same electrically charged substance. In view of § 3.6.5, the thermal power $P_Q^{(21)}$ and the chemical power $P_C^{(21)}$ exerted by subsystem 2 on subsystem 1 and the rate $\Omega^{(21)}$ of substance entering subsystem 1 are given by the coupled system of equations,

$$\begin{pmatrix} P_Q^{(21)} + P_C^{(21)} \\ \Omega^{(21)} \end{pmatrix} = \begin{pmatrix} L_{QQ} & L_{QN} \\ L_{NQ} & L_{NN} \end{pmatrix} \begin{pmatrix} \frac{1}{T_1} - \frac{1}{T_2} \\ \frac{\mu_2}{T_2} - \frac{\mu_1}{T_1} \end{pmatrix}$$

1. Show that the entropy production rate can be written as a positive quadratic form, i.e.

$$\Pi_S = x^T \mathsf{L} x > 0 \quad \text{where} \quad x \neq 0$$

   Deduce from this inequality that $\det(\mathsf{L}) > 0$ and $\text{Tr}(\mathsf{L}) > 0$. Show that the eigenvalues λ_1 and λ_2 of matrix L are positive.
2. When the wall is impermeable and diathermal, show that the system of coupled equations predicts the existence of the 'discrete' **Fourier law** (3.16) and express the thermal conductivity κ in terms of the empirical coefficients.
3. When the temperature T is uniform and the system contains a neutral substance, show that the coupled system of equations predicts the existence of a 'discrete' **Fick law** (3.45) and express the diffusion coefficient F in terms of the empirical coefficients.

4. When the temperature difference between the two subsystems is kept constant, i.e. $T_1 = T - \Delta T/2$ and $T_2 = T + \Delta T/2$ ($\Delta T \ll T$), and when the substance consists of electrons of temperature-independent chemical potential μ_e and electrochemical potentials $\bar{\mu}_1 = \mu_e - q_e \Delta\varphi/2$ and $\bar{\mu}_2 = \mu_e + q_e \Delta\varphi/2$, show that in the stationary regime ($\dot{N}_1 = 0$), the electrostatic potential difference $\Delta\varphi$ is given by,

$$\Delta\varphi = -\varepsilon \Delta T$$

The existence of this potential difference is called the **Seebeck effect**. Find an expression for the Seebeck coefficient ε in terms of the empirical coefficients, in the limit where $\mu_e \ll L_{NQ}/L_{NN}$.

5. When the electrochemical potential depends on temperature, i.e. $\bar{\mu}_1 = \mu - q\Delta\varphi/2 - s\Delta T/2$ and $\bar{\mu}_2 = \mu + q\Delta\varphi/2 + s\Delta T/2$ where μ is the temperature-independent chemical potential and $s = -\partial\mu/\partial T$ the molar entropy of the substance, find under the condition $\dot{N}_1 = 0$ an expression for the Seebeck coefficient ε in terms of s, in the limit where $\mu \ll Ts$ and $L_{NQ}/L_{NN} \ll Ts$.

Solution:

1. *Since the wall is fixed, the volume of subsystem 1 is constant, i.e. $\dot{V}_1$, and the mechanical power exerted by subsystem 2 on subsystem 1 vanishes, i.e. $P_W^{(21)} = 0$. Thus, the internal energy variation rate is equal to the sum of the thermal power and the chemical power exerted by subsystem 2 on subsystem 1, i.e. $\dot{U}_1 = P_Q^{(21)} + P_C^{(21)}$. Expression (3.50) for the entropy variation rate $\dot{S}$ of the isolated system, where the chemical potentials are replaced by the electrochemical potentials, is equal to the entropy production rate,*

$$\Pi_S = \left(\frac{1}{T_1} - \frac{1}{T_2}\right)\left(P_Q^{(21)} + P_C^{(21)}\right) + \left(\frac{\bar{\mu}_2}{T_2} - \frac{\bar{\mu}_1}{T_1}\right)\Omega^{(21)} > 0$$

This can be expressed as a scalar product, i.e.

$$\Pi_S = \left(\frac{1}{T_1} - \frac{1}{T_2}, \frac{\bar{\mu}_2}{T_2} - \frac{\bar{\mu}_1}{T_1}\right)\cdot\begin{pmatrix} P_Q^{(21)} \\ \Omega^{(21)} \end{pmatrix} > 0$$

Using the coupled system of equations, the entropy production rate can be written as a positive definite quadratic form,

$$\Pi_S = \left(\frac{1}{T_1} - \frac{1}{T_2}, \frac{\bar{\mu}_2}{T_2} - \frac{\bar{\mu}_1}{T_1}\right)\begin{pmatrix} L_{QQ} & L_{QN} \\ L_{NQ} & L_{NN} \end{pmatrix}\begin{pmatrix} \frac{1}{T_1} - \frac{1}{T_2} \\ \frac{\bar{\mu}_2}{T_2} - \frac{\bar{\mu}_1}{T_1} \end{pmatrix} > 0$$

Defining vector $\boldsymbol{x}$ and matrix L as,

$$\boldsymbol{x} = \begin{pmatrix} \frac{1}{T_1} - \frac{1}{T_2} \\ \frac{\bar{\mu}_2}{T_2} - \frac{\bar{\mu}_1}{T_1} \end{pmatrix} \quad \text{and} \quad \mathsf{L} = \begin{pmatrix} L_{QQ} & L_{QN} \\ L_{NQ} & L_{NN} \end{pmatrix}$$

the positive definite quadratic form is written as,

$$\Pi_S = \boldsymbol{x}^T \mathsf{L}\boldsymbol{x} > 0$$

where the exponent T denotes the transpose obtained by transforming rows into columns and vice versa. In order to determine the determinant and the trace of the positive definite matrix L, *we express this matrix using an invertible* 2×2 *matrix,*

$$M = \begin{pmatrix} M_{11} & M_{12} \\ M_{21} & M_{22} \end{pmatrix}$$

in the following manner [85],

$$x^T L x = x^T M^T M x = (M x)^T \cdot (M x) = \|M x\|^2 > 0 \quad \forall x \neq 0$$

Then, we show that the determinant of the matrix L *is positive, i.e.*

$$\det(L) = \det(M^T M) = \det(M)^2 > 0$$

Writing the positive definite matrix $L = M^T M$ *explicitly in components yields,*

$$\begin{pmatrix} L_{QQ} & L_{QN} \\ L_{NQ} & L_{NN} \end{pmatrix} = \begin{pmatrix} M_{11}^2 + M_{21}^2 & M_{11} M_{12} + M_{21} M_{22} \\ M_{11} M_{12} + M_{21} M_{22} & M_{12}^2 + M_{22}^2 \end{pmatrix}$$

Thus, the empirical coefficients satisfy the following conditions,

$$L_{QQ} = M_{11}^2 + M_{21}^2 > 0$$
$$L_{NN} = M_{12}^2 + M_{22}^2 > 0$$
$$L_{QN} = L_{NQ} = M_{11} M_{12} + M_{21} M_{22}$$

Hence, the trace of the matrix L *is positive, i.e.*

$$\text{Tr}(L) = L_{QQ} + L_{NN} > 0$$

Since the determinant is the product of the eigenvalues and the trace is the sum of the eigenvalues, we find,

$$\det(L) = \lambda_1 \lambda_2 > 0 \qquad \text{Tr}(L) = \lambda_1 + \lambda_2 > 0$$

This implies that the eigenvalues are positive, i.e. $\lambda_1 > 0$ *and* $\lambda_2 > 0$.

2. *For a system consisting of an impermeable diathermal wall, i.e.* $P_C^{(21)} = 0$, *the amount of matter is constant in each subsystem, i.e.* $\Omega^{(21)} = 0$. *Thus, the system of coupled equations reduces to,*

$$P_Q^{(21)} = L_{QQ} \left(\frac{1}{T_1} - \frac{1}{T_2} \right) + L_{QN} \left(\frac{\bar{\mu}_2}{T_2} - \frac{\bar{\mu}_1}{T_1} \right)$$

$$0 = L_{NQ} \left(\frac{1}{T_1} - \frac{1}{T_2} \right) + L_{NN} \left(\frac{\bar{\mu}_2}{T_2} - \frac{\bar{\mu}_1}{T_1} \right)$$

The second equation implies that,

$$\left(\frac{\bar{\mu}_2}{T_2} - \frac{\bar{\mu}_1}{T_1} \right) = - \frac{L_{NQ}}{L_{NN}} \left(\frac{1}{T_1} - \frac{1}{T_2} \right)$$

Substituting this relation into the first equation yields the equation,

$$P_Q^{(21)} = \frac{L_{QQ} L_{NN} - L_{QN} L_{NQ}}{L_{NN}} \left(\frac{1}{T_1} - \frac{1}{T_2} \right)$$

This corresponds to the 'discrete' Fourier law (3.16),

$$P_Q^{(21)} = \kappa \frac{A}{\ell} (T_2 - T_1)$$

where the thermal conductivity κ is expressed in terms of the empirical coefficients of the wall area A and characteristic length ℓ as,

$$\kappa = \frac{\ell}{A} \left(\frac{L_{QQ} L_{NN} - L_{QN} L_{NQ}}{T_1 T_2 L_{NN}} \right) = \frac{\ell}{A} \frac{\det(L)}{T_1 T_2 L_{NN}} > 0$$

3. *For a system at uniform temperature T, i.e. $T_1 = T_2 = T$, consisting of an electrically neutral substance, i.e. $\bar{\mu}_1 = \mu_1$ and $\bar{\mu}_2 = \mu_2$, the second equation of the system of coupled equations reduces to,*

$$\Omega^{(21)} = \dot{N}_1 = \frac{L_{NN}}{T} (\mu_2 - \mu_1)$$

which corresponds to the 'discrete' Fick law (3.45),

$$\dot{N}_1 = F \frac{A}{\ell} (\mu_2 - \mu_1)$$

where the diffusion coefficient F is expressed in terms of the empirical coefficients of the wall area A and characteristic length ℓ as,

$$F = \frac{\ell}{A} \frac{L_{NN}}{T} > 0$$

4. *The series expansion of $1/T_1$ and $1/T_2$ to first order in $\Delta T/T$ is given by,*

$$\frac{1}{T_1} = \frac{1}{T} \left(1 - \frac{\Delta T}{2T} \right)^{-1} = \frac{1}{T} \left(1 + \frac{\Delta T}{2T} \right) + \mathcal{O} \left(\frac{\Delta T^2}{T} \right)$$

$$\frac{1}{T_2} = \frac{1}{T} \left(1 + \frac{\Delta T}{2T} \right)^{-1} = \frac{1}{T} \left(1 - \frac{\Delta T}{2T} \right) + \mathcal{O} \left(\frac{\Delta T^2}{T} \right)$$

Thus, to first order in $\Delta T/T$,

$$\frac{1}{T_1} - \frac{1}{T_2} = \frac{\Delta T}{T^2} \quad and \quad \frac{\bar{\mu}_2}{T_2} - \frac{\bar{\mu}_1}{T_1} = -\mu_e \frac{\Delta T}{T^2} + q_e \frac{\Delta\varphi}{T}$$

Substituting these relations into the equation of point 2 yields the equation,

$$-\mu_e \frac{\Delta T}{T} + q_e \Delta\varphi = -\frac{L_{NQ}}{L_{NN}} \frac{\Delta T}{T}$$

*that describes the **Seebeck effect**,*

$$\Delta\varphi = -\frac{1}{q_e T} \left(\frac{L_{NQ}}{L_{NN}} - \mu_e \right) \Delta T \equiv -\varepsilon \Delta T$$

where in the limit $\mu_e \ll L_{NQ}/L_{NN}$ the Seebeck coefficient ε is expressed in terms of the empirical coefficients as,

$$\varepsilon = \frac{1}{q_e T} \frac{L_{NQ}}{L_{NN}}$$

5. *To first order in $\Delta T / T$,*

$$\frac{1}{T_1} - \frac{1}{T_2} = \frac{\Delta T}{T^2} \qquad and \qquad \frac{\bar{\mu}_2}{T_2} - \frac{\bar{\mu}_1}{T_1} = -\mu \frac{\Delta T}{T^2} + q \frac{\Delta \varphi}{T} + s \frac{\Delta T}{T}$$

Substituting these relations into the equation of point 2 yields the equation,

$$-\mu \frac{\Delta T}{T} + q \Delta \varphi + s \Delta T = -\frac{L_{NQ}}{L_{NN}} \frac{\Delta T}{T}$$

that describes the **Seebeck effect**,

$$\Delta \varphi = -\frac{1}{qT} \left(\frac{L_{NQ}}{L_{NN}} - \mu + Ts \right) \Delta T$$

where in the limit $\mu \ll Ts$ and $L_{NQ}/L_{NN} \ll Ts$ the Seebeck coefficient ε is expressed in terms of the coefficient s as,

$$\varepsilon = \frac{s}{q}$$

and represents the thermogalvanic effect (§ 8.13). This effect is used in thermogalvanic cells [86, 87, 88].

Exercises

8.1 Oxidation of Ammonia

The chemical reaction of ammonia oxidation reads,

$$4\,NH_3 + 5\,O_2 \;\rightarrow\; 4\,NO + 6\,H_2O$$

Consider that initially this reaction is taking place with $N_{NH_3}(0)$ moles of NH_3 and $N_{O_2}(0)$ moles of O_2. Find the amount of NH_3, O_2, NO and H_2O at the end of the reaction.

Numerical Application:

$N_{NH_3}(0) = 2\,mol$, $N_{O_2}(0) = 2\,mol$, $N_{NO}(0) = 0\,mol$ and $N_{H_2O}(0) = 0\,mol$.

8.2 Acetylene Lamp

Acetylene (C_2H_2) can be produced through a chemical reaction between water (H_2O) and calcium carbide (CaC_2):

$$CaC_2\,(s) + 2\,H_2O\,(l) \;\rightarrow\; C_2H_2\,(g) + Ca(OH)_2\,(s)$$

where (s) and (l) indicate whether the substance is solid or liquid. A cave explorer considers using an acetylene torch, known to consume this gas at a volume rate of $\dot{V}$ (at standard conditions of temperature and pressure). As the expedition is due to last a time t, find the amount of calcium carbide that the explorer would need if he chose this type of light source. Determine the amount of water used by this torch during this time.

Numerical Application:

$T = 0°C, p = 10^5$ atm, $\dot{V} = 10$ l/h and $t = 8$ h.

8.3 Coupled Chemical Reactions

The oxidation of methane can take place according to either one of the following reactions:

$$CH_4 + 2O_2 \xrightarrow{1} CO_2 + 2H_2O$$
$$2CH_4 + 3O_2 \xrightarrow{2} 2CO + 4H_2O$$

When the reactions stop at time t_f because all the methane is burned, the total mass of the products (CO_2, CO, H_2O) is

$$m(t_f) = m_{CO_2}(t_f) + m_{CO}(t_f) + m_{H_2O}(t_f)$$

Determine the initial mass of methane $m_{CH_4}(0)$ in terms of the total mass of the products $m(t_f)$ and the mass of water $m_{H_2O}(t_f)$.

Numerical Application:

$m(t_f) = 24.8$ g and $m_{H_2O}(t_f) = 12.6$ g.

8.4 Variance

The variance v is the number of degrees of freedom of a system consisting of r substances in m phases taking part in n chemical reactions. The variance v is obtained by subtracting n constraints from the number of degrees of freedom f determined by the Gibbs phase rule (6.62),

$$v = f - n = r - m - n + 2$$

The pressure p and the temperature T are not fixed. Otherwise, there are additional constraints to fix p and T. Apply this concept to the following situation.

a) Determine the variance v of methane cracking described by the chemical reaction:

$$CH_4(g) \leftrightarrows C(g) + 2H_2(g)$$

b) A system consists of three phases and there is one chemical reaction between the substances. The variance is known to be two. Determine the number r of substances in the system.

c) A system is at a fixed temperature and consists of three phases. Its variance is known to be two and there are two chemical reactions between the substances. Determine the number r of substances in the system.

8.5 Enthalpy of Formation

a) There are two isomers of butane: butane (C_4H_{10}) and isobutane (methylpropane) (iso-C_4H_{10}). Determine the standard enthalpy of isomerisation $\Delta h°$ of butane to isobutane in terms of the enthalpies of formation of the two isomers, $h_{C_4H_{10}}$ and $h_{\text{iso-}C_4H_{10}}$.

b) The lunar module 'Eagle' of the Apollo mission was propelled using the energy released by the reaction:

$$H_2NN(CH_3)_2(l) + 2\,N_2O_4(l) \rightarrow 3\,N_2(g) + 2\,CO_2(g) + 4\,H_2O(g)$$

Determine the molar enthalpy Δh° of this exothermic reaction in terms of the enthalpies of formation of the reactants, $h_{H_2NN(CH_3)_2(l)}$, $h_{N_2O_4(l)}$ and of the products $h_{N_2(g)}$ $h_{CO_2(g)}$, h_{H_2O}.

c) The combustion of acetylene (C_2H_2) is described by the chemical reaction:

$$C_2H_2(g) + \frac{5}{2}O_2(g) \rightarrow 2\,CO_2(g) + H_2O(l)$$

Determine the enthalpy of formation $h_{C_2H_2}$ of acetylene (C_2H_2) in terms of the molar enthalpies $h_{O_2(g)}$, $h_{CO_2(g)}$, $h_{H_2O(g)}$, the molar enthalpy of the reaction Δh° and the vaporisation molar enthalpy of water h_{vap}.

Numerical Application:

a) $h_{C_4H_{10}} = -2{,}877\,kJ/mol$, $h_{i\text{-}C_4H_{10}} = -2{,}869\,kJ/mol$,
b) $h_{H_2O(g)} = -242\,kJ/mol$, $h_{CO_2(g)} = -394\,kJ/mol$,
$h_{N_2(g)} = 0\,kJ/mol$, $h_{N_2O_4(l)} = 10\,kJ/mol$, $h_{H_2NN(CH_3)_2(l)} = 52\,kJ/mol$
c) $\Delta h^\circ = -1{,}300\,kJ/mol$, $h_{vap} = 44\,kJ/mol$, $h_{O_2(g)} = 0\,kJ/mol$,

8.6   Work and Heat of a Chemical Reaction

Steel wool is placed inside a cylinder filled with molecular oxygen O_2, considered as an ideal gas. A piston ensures a constant pressure of the gas. The steel wool reacts with the molecular oxygen to form iron rust Fe_2O_3,

$$2\,Fe + \frac{3}{2}\,O_2 \rightarrow Fe_2O_3$$

The reaction is slow, so that the gas remains at ambient temperature T_0. Determine the heat Q_{if}, the work W_{if} and the internal energy variation ΔU_{if} in terms of the enthalpy of reaction ΔH_{if} for a reaction involving two moles of iron.

Numerical Application:

$\Delta H_{if} = -830\,kJ$, $T_0 = 25°C$.

8.7   Mass Action Law: Esterification

The Fischer esterification reaction is given by,

$$R\text{-}(C=O)\text{-}OH + R\text{-}OH \rightleftarrows R\text{-}(C=O)\text{-}OR + H_2O$$

Determine the equilibrium constant K of this reaction in terms of the concentrations of the reactants $c_{R\text{-}(C=O)\text{-}OH}$, $c_{R\text{-}OH}$ and of the products $c_{R-(C=O)-OR}$ and c_{H_2O} at equilibrium.

Numerical Application:

$c_{R\text{-}(C=O)\text{-}OH} = 1/3$, $c_{R\text{-}OH} = 1/3$, $c_{R-(C=O)-OR} = 2/3$, $c_{H_2O} = 2/3$.

8.8   Mass Action Law: Carbon Monoxide

In a reactor of volume V_0, initially empty, solid carbon is introduced in an excess amount together with $N_{CO_2(g)}(0)$ moles of carbon dioxide. The reactor is brought to temperature T_0 and the system reaches a chemical equilibrium,

$$CO_2(g) + C(s) \rightleftarrows 2\,CO(g)$$

At equilibrium, which occurs at time $t = t_f$, the density of the gases relative to air is δ. Determine:

a) the pressure $p(t_f)$ in the reactor.
b) the equilibrium constant K.
c) the variance v as defined in exercise (8.4).

Numerical Application:

$V_0 = 1\,l$, $T_0 = 1,000°C$, $N_{CO_2(g)}(0) = 0.1\,\text{mol}$, $\delta = 1.24$, $M_{air} = 29\,g$.

8.9   Entropy of Mixing

A gas container of fixed volume V is divided into two compartments by an impermeable fixed wall. One compartment contains ideal gas 1, the other ideal gas 2. Both sides are at pressure p and temperature T. When the wall is removed, the system reaches equilibrium. During this process, going from an initial state i to a final state f, the system is held at constant temperature T. There is no chemical interaction between the two gases. Therefore, the mixture is an ideal gas also.

a) Determine the internal energy variation ΔU_{if} during this process.
b) Show that the total entropy variation ΔS_{if} is given by (Fig. 8.5),

$$\Delta S_{if} = -\,(N_1 + N_2)\,R \sum_{A=1}^{2} c_A \ln(c_A)$$

8.10   Raoult's Law

A container at a pressure p and temperature T contains two substances 1 and 2, present both in liquid and gas phases. Estimate the partial pressure p_A of substance

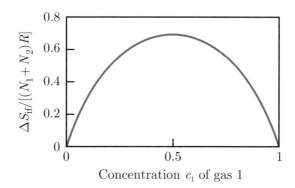

Figure 8.5   Entropy of mixing, as a function of the concentration c_1 of gas 1.

A in the gas phase ($A = 1, 2$) as a function of the concentrations $c_1^{(\ell)}$ and $c_2^{(\ell)}$ of substances 1 and 2 in the liquid phase. Raoult's law relates the partial pressure p_A of substance A to the saturation pressure p_A°,

$$p_A = p_A^\circ c_A^{(\ell)}$$

where the saturation pressure p_A° is the pressure that the pure substance A would have in the gas phase in equilibrium with the liquid phase at temperature T. Establish Raoult's law by assuming that the liquid and gas mixtures can be treated as ideal mixtures (§ 8.5.2) and by considering that molar volumes in the liquid phase are negligible compared to molar volumes in the gaseous phase.

8.11   Boiling Temperature of Salt Water

Consider a mixture of water and salt with a low salt concentration. Use the ideal mixture law (8.68) to evaluate the chemical potential of water in the salt solution. Recall that according to relation (8.51), for any substance A in any given phase, $\mu_A(T) = h_A - T s_A$. Assume that near the boiling temperature T_0 of pure water, the molar enthalpy h_A and the molar entropy s_A of the liquid and vapour phases do not depend on temperature. Determine the boiling temperature variation $T - T_0$ as a function of salt concentration c_A.

8.12   Battery Potential

Apply the general definition of the battery potential,

$$\Delta \varphi = - \frac{1}{z F_F} \sum_A \nu_{aA} \mu_A$$

to the Daniell cell (§ 8.7.4) and show that it yields relation (8.108). Show that the battery potential can be written as,

$$\Delta \varphi = \frac{\mathcal{A}_a}{z F_F} = - \frac{\Delta_a G}{z F_F} = - \frac{\Delta_a H - T \Delta_a S}{z F_F}$$

where

$$\Delta_a H = \sum_A \nu_{aA} h_A \qquad \text{and} \qquad \Delta_a S = \sum_A \nu_{aA} s_A$$

8.13   Thermogalvanic Cell

Consider an electrochemical cell made of two half-cells that are identical except that they are maintained at different temperatures. This is called a thermogalvanic cell. Determine the thermogalvanic coefficient,

$$\alpha = \frac{\partial \Delta \varphi}{\partial T} = \frac{\Delta_a S}{z F_F}$$

using the definition of the battery potential introduced in exercise (8.12).

8.14   Gas Osmosis

An isolated system consists of two rigid subsystems of volumes V_1 and V_2 separated by a rigid and porous membrane. Helium (He) can diffuse through the membrane, but oxygen (O_2) cannot. We label the gases as A for helium and B for oxygen. The whole system is in thermal equilibrium at all times. Each gas can be considered an ideal

Initial   Final

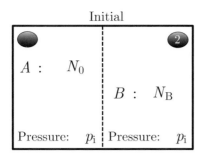
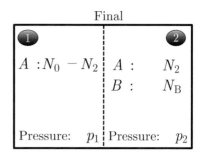

Figure 8.6 A system is divided into two by an osmotic membrane which lets substance A diffuse through it, but not substance B.

gas, i.e. they satisfy the equations of state (5.46) and (5.47), namely, $pV = NRT$ and $U = cNRT$. The gas mixture obeys the ideal mixture relation (8.68), that is,

$$\mu_A(T,p,c_A) = \mu_A(T,p) + RT\ln(c_A)$$
$$\mu_B(T,p,c_B) = \mu_B(T,p) + RT\ln(c_B)$$

where $\mu_A(T,p)$ and $\mu_B(T,p)$ are the chemical potentials of substances A and B when they are pure, c_A and c_B are the concentrations of A and B. Initially, the system contains N_0 moles of helium in subsystem 1, and N_B moles of oxygen in subsystem 2 (Fig. 8.6). The numbers of moles N_0 and N_B are chosen so that the initial pressure p_i is the same in both subsystems. At all times, each subsystem is assumed to be homogeneous. Designate by N_1 and N_2 the number of moles of helium in subsystems 1 and 2, respectively.

a) At equilibrium, show that $\mu_A(T,p_1) = \mu_A(T,p_2,c_A)$.
b) Deduce from the previous result a relation between the pressures p_1 and p_2 when the two sub-systems reach equilibrium. Express c_A, p_1 and p_2 in terms of N_2. Determine p_1 and p_2 in terms of the initial pressure p_i under the condition of equal volume, i.e. $V_1 = V_2 = V_0$.

8.15 Osmosis Power Plant

At sea level, water from the outlet of a river is diverted to a power plant that operates on the principle of osmosis. A turbine is installed in the pipe that brings the river water to an osmotic membrane separating the clear water from the salt water of the sea. The sea water at the location of the membrane is assumed to have a constant low salt concentration c, i.e. $c \ll 1$. The pure water pressure in the river and in the sea is p_0. Because of osmosis, water is driven from the river through the turbine and then across the osmotic membrane into the sea. Just after the turbine and before the membrane the pressure is $p_1 = p_0 - \Delta p$. Calculate the mechanical power of the water flowing through the turbine,

$$P_W = \Delta p \, \dot{N} v$$

where v is the molar volume of water and $\dot{N}$ is the number of moles per unit time flowing through the osmotic membrane. The hydrodynamics of the turbine is such that we can assume,

$$\Delta p = \frac{R_H \dot{N}}{v}$$

so that $P_W = R_H \dot{N}^2$, i.e. it is similar to the form of the Joule power for electrical heating. Use the ideal mixture relation (8.68) to determine the mechanical power P_W. Since the salt concentration c is low enough, we can assume at ambient temperature that $\Delta\mu \gg RTc$. Show that the mechanical power is given by,

$$P_W = \Delta\mu \dot{N}$$

where $\Delta\mu$ is the chemical potential drop between the river and the sea water.

Part III

CONTINUOUS MEDIA

9

Matter and Electromagnetic Fields

Joseph Larmor, 1857–1942

J. Larmor was appointed professor at the University of Cambridge in 1885. In 1897, while doing research on ether, he derived what are now called the Lorentz transformations of space-time coordinates, two years before Lorentz did. Larmor was particularly interested in the interactions between matter and electromagnetic fields. In 1896, he established that a magnetic dipole density subjected to a magnetic induction field undergoes a precession that bears his name. In 1903, he became 'Lucasian Professor of Mathematics', 234 years after Sir Isaac Newton and 76 years before Stephen Hawking.

9.1 Historical Introduction

Before engaging in a formal thermodynamic description of matter interacting with electromagnetic fields, a historical perspective will illustrate the significance of this topic in our modern world. First, we describe certain aspects of magnetic and dielectric materials. Then, we retrace the development of the concept of energy associated with materials in electromagnetic fields.

According to Chinese writings dating back to 4000 B.C., the first compasses were made of meteoric iron [89]. They were shaped as spoons and could freely rotate on a polished copper surface. These experiments were the first observations of the **orientation of a magnetised rod** in a magnetic induction field. William Gilbert published in 1600 a treatise on magnetism, a remarkable example of the emerging modern experimental sciences [90]. Gilbert observed and analysed the mutual attraction forces exerted between two magnetic objects.

The conception of a scale used to measure magnetisation is attributed to **Faraday**. The operating principle of this scale is based on the force exerted on a magnetic object immersed in an inhomogeneous magnetic field, a subject we will address in this chapter. While nowadays it is more common to measure the field induced by a magnetic sample, there are instruments of very high sensitivity that expand on the concept of Faraday. For example, some magnetometers detect the oscillations of an elastic blade holding a sample under test and resonating at the frequency of an applied field gradient. There are also devices that measure the torsion that a magnetic sample exerts on its support as a result of its interaction with an applied field [91].

In 1889, the German Georg Quincke observed that the meniscus of a magnetic liquid changed shape in the presence of an applied magnetic field. The Frenchman Louis-Georges Gouy derived an expression for the interaction force exerted on a cylinder containing a paramagnetic liquid immersed in a uniform magnetic field. His considerations are at the core of a susceptibility measurement method that bears his name. Following on the same ideas, nowadays magnetic fluids called ferro-fluids are used to visualise magnetic domains at the surface of a magnetic solid.

The **Stern and Gerlach** experiment, performed for the first time in 1922, tested and motivated the fundamental concepts of quantum mechanics. In that experiment, the interaction force between an atomic magnetic dipole and a magnetic induction field gradient was measured by observing the deviation of a beam of magnetic dipoles passing in between magnetic poles.

Let us now turn our attention to dielectric materials. The dielectric properties of materials have a more limited history and remain further away from daily-life experience. The ancient Greeks already observed **pyroelectricity** in tourmaline, i.e. the appearance of electric charges when heating a solid. In the nineteenth century, Pierre Curie and his brother Jacques observed that the charge of a pyroelectric material is not the same depending on whether the sample had a uniform or a non-uniform temperature. They realised that this effect was due to **piezoelectricity**, that is, the coupling between the electrical and mechanical properties of a solid.

In 1920 Valasek discovered **ferroelectricity**, that is, the spontaneous alignment of electric polarisation analogous to that of magnetisation in ferromagnetism. He observed this in a hydrated sodium potassium tartrate crystal, called nowadays Rochelle salt. This substance remained the only known example of ferroelectricity until professors Scherrer and Busch, at the Federal Institute of Technology in Zurich, discovered ferroelectricity in numerous other substances [92].

Dielectrophoresis refers to the means of moving objects that rely on the force exerted by an electric field gradient on electric dipoles. It was around 1950 that this effect was discovered by H. Pohl [93]. Since biological cells have dielectric properties, this effect can be used in medical applications.

It is also possible to manipulate dielectric objects with intense laser beams which are referred to as ***optical tweezers*** in this context. Depending on the respective refraction indices of the object and the medium in which it is immersed, the object is attracted or repelled by an optical beam. This technique was described in detail in an article in 1986 [94]. One of the authors, Steven Chu, showed later that the method could trap and cool neutral atoms. In 1997, he was awarded the Nobel Prize for his work with Claude Cohen-Tannoudji and William D. Philips.

As we will see in this chapter, the notion of energy associated with matter in electromagnetic fields is complex. To get a historical perspective on its development, let us go back to the nineteenth century. In order to ensure the continuity of the electric current between the plates of a capacitor, James Clerk ***Maxwell*** [95] introduced a displacement current term to the magnetostatic equation of Ampère. The theoretical justification of the displacement current was actually based on a thermodynamic argument. Indeed, this current ensures the continuity of the electromagnetic energy flux through a dielectric medium. Using this current, Maxwell established a set of equations that bear his name and describe the evolution of electromagnetic fields in interaction with matter. He noted that these equations predicted the existence of electromagnetic waves that propagate in vacuum at the speed of light. The existence of these waves was experimentally demonstrated by Heinrich Hertz in 1887.

Based on the works of Maxwell, John Henry ***Poynting*** established in 1884 the electromagnetic energy balance equation in a theorem named after him [96]. Nikola Tesla, a great inventor, was concerned with the transport of electromagnetic energy.

Peter ***Debye*** [97] was interested in the dielectric properties of matter and completed Poynting's description by adding a term to the expression of the electromagnetic energy density in order to account for the interaction between electric polarisation of matter and electric fields. His analysis was based on the works of Clausius [98] and Mossotti [99], who independently had established a relationship between the dielectric properties of vacuum and matter, thus characterising the electric polarisability of matter.

Similarly, Joseph ***Larmor*** [100] introduced an additional term in the energy density which accounted for the interaction between magnetisation and magnetic fields.

This term implies the Larmor precession that dominates nuclear magnetisation dynamics. The Bloch equations [101] describe this and predict magnetic resonance, a phenomenon that enables the development of numerous technological applications, notably, magnetic resonance imaging.

9.2 Insulators and Electromagnetic Fields

9.2.1 Electromagnetic Fields and State Fields

In this section, we would like to describe the thermodynamics of a material system consisting of a homogeneous rigid sample, made of r electrically neutral chemical substances A, in the presence of electromagnetic fields. In a thermodynamic description of electromagnetic phenomena, all electric charge carriers, such as conduction electrons in a metal, have to be treated as distinct substances. Since here the substances are neutral, the material is an ***electric insulator***.

Electromagnetism is a local theory in the sense that the fields at a position x and a time t are defined for local systems of infinitesimal volume centred around a spatial point x. The volume of a local system is actually not physically infinitesimal but, compared to the macroscopic scale of the whole system, it is small enough to be considered as such. On this macroscopic scale, the thermodynamic system is treated as a continuous medium consisting of local systems of infinitesimal volume.

The description that we will adopt involves six different fields, namely: ***electric field E***, ***magnetic field H***, ***electric polarisation P***, ***magnetisation M***, ***electric displacement field D*** and ***magnetic induction field B***. Electromagnetic fields E and B are intensive quantities and electromagnetic fields D, H, P and M are specific quantities [102].

In general, electric polarisation P and magnetisation M consist of two parts: one part that is induced by intensive electromagnetic fields E and B and a permanent part. In order to have constitutive relations that are bijective, we do not treat systems with electric or magnetic hysteresis. In other words, we consider only the part that is induced by the fields. Furthermore, we assume that there are no magnetoelectric effects. Thus, the electric polarisation P of matter is induced by the electric field E [103],

$$P = \varepsilon_0 \, \chi_e \cdot E \tag{9.1}$$

where χ_e is the symmetric electric susceptibility tensor of matter and ε_0 is the electric permittivity of vacuum. Likewise, the magnetisation M of matter is induced by the magnetic induction field B [104],

$$M = \mu_0^{-1} \, \chi_m \cdot B \tag{9.2}$$

where χ_m is the symmetric magnetic susceptibility tensor of matter and μ_0^{-1} is the inverse of the magnetic permeability of vacuum. In this chapter, we consider only the case of linear electromagnetism. Then, the electric susceptibility tensor χ_e and the magnetic susceptibility tensor χ_m are independent of the electromagnetic fields E and B.

The electric displacement field D represents the electric dipole density induced by the electric field E. Since the field D is a specific quantity, it can be considered as the sum of the electric dipole densities in vacuum $\varepsilon_0 \, E$ and in matter P, which are induced by the electric field E [105],

$$D = \varepsilon_0 \, E + P \tag{9.3}$$

We proceed now in a similar fashion to determine a relationship between fields that characterise magnetism, i.e. we take into account the intensive or specific character of these fields. The opposite of the magnetic field $-H$ represents the magnetic dipole density induced by the magnetic induction field B. Since the field $-H$ is a specific quantity, it can be considered as the sum of the magnetic dipole densities in vacuum $-\mu_0^{-1} \, B$ and in matter M, both being induced by the magnetic induction field B [106],

$$-H = -\mu_0^{-1} \, B + M \tag{9.4}$$

The negative signs appearing in equation (9.4) are a consequence of the historical developments that led to the definition of the magnetisation M [107]. Equations (9.3) and (9.4) express a relation between electric fields D and E, and magnetic fields H and B. They are called the ***constitutive relations*** of electromagnetism.

The thermodynamic variables of the local system are chosen in order to account for the local properties of matter. In order to describe the thermal, electrical and magnetic properties of matter, we choose as local state fields the entropy density s, the set of densities $\{n_A\}$ of the chemical substances A, the electric polarisation P and the magnetisation M. The intensive electromagnetic fields E and B cannot be chosen as state variables of the local system since they do not necessarily vanish in the absence of matter. In view of the local description explained above, the local state fields are specific quantities that are functions of position x and time t.

9.2.2 Energy

We would like to establish an expression for the internal energy density $u\,(s, \{n_A\}, P, M)$ of the local material system in terms of the state variables we have chosen. In order to do so, we need to use some results from the theory of electromagnetism. The state is determined by the state fields D and B for the following reasons. There is no need to provide energy to the system in order to maintain a distribution of electric charges that is related to the field D through Gauss' equation. Thus, field D can be taken as the electric state field. To the contrary, energy needs to be provided in order to maintain a distribution of electric currents that is related to the field H through Ampère's equation. Therefore, we need to choose field B, that is, the field conjugated to the field H, as the magnetic state field. It can be shown that the total internal energy density $u^{\text{tot}}\,(s, \{n_A\}, D, B)$ of the local material system and of the electromagnetic fields is given by [108, 109],

$$u^{\text{tot}}\,(s, \{n_A\}, D, B) = Ts + \sum_{A=1}^{r} \mu_A n_A + \frac{1}{2}\, E \cdot D + \frac{1}{2}\, H \cdot B \qquad (9.5)$$

The first term in the right-hand side of equation (9.5) corresponds to the thermal energy density. The second term corresponds to the chemical energy density. The third and fourth terms represent the total electromagnetic energy density of the local material system interacting with the electromagnetic fields.

Moreover, it can be shown that the internal energy density $u^{\text{em}}\,(E, B)$ of the electromagnetic fields in vacuum is given by,

$$u^{\text{em}}\,(E, B) = \frac{1}{2}\, \varepsilon_0\, E^2 + \frac{1}{2}\, \frac{B^2}{\mu_0} \qquad (9.6)$$

The internal energy density $u^{\text{em}}\,(E, B)$ of the electromagnetic fields in vacuum does not belong to the local material system and has to be subtracted from the total internal energy density $u^{\text{tot}}\,(s, \{n_A\}, D, B)$ in order to conform to our choice of local thermodynamic system.

It is apparent that we need to be able to change state variables and we will do so following the general procedure established in § 4.3. By analogy with enthalpy, which is a state function of the intensive field pressure p, we define a magnetic enthalpy as a state function of the magnetic induction field $\boldsymbol{B}$, which is an intensive field, whereas the conjugated field, the magnetisation $\boldsymbol{M}$, is a specific field. Thus, we define the magnetic enthalpy density $h_m\left(s, \{n_A\}, \boldsymbol{P}, \boldsymbol{B}\right)$ as the difference between the total internal energy density $u^{\text{tot}}\left(s, \{n_A\}, \boldsymbol{D}, \boldsymbol{B}\right)$ and the internal energy density $u^{\text{em}}\left(\boldsymbol{E}, \boldsymbol{B}\right)$ of the electromagnetic fields in vacuum, i.e.

$$h_m\left(s, \{n_A\}, \boldsymbol{P}, \boldsymbol{B}\right) = u^{\text{tot}}\left(s, \{n_A\}, \boldsymbol{D}, \boldsymbol{B}\right) - u^{\text{em}}\left(\boldsymbol{E}, \boldsymbol{B}\right) \tag{9.7}$$

Using the constitutive equations (9.3) and (9.4) of electromagnetism, we obtain an expression for the magnetic enthalpy density $h_m\left(s, \{n_A\}, \boldsymbol{P}, \boldsymbol{B}\right)$,

$$h_m\left(s, \{n_A\}, \boldsymbol{P}, \boldsymbol{B}\right) = Ts + \sum_{A=1}^{r} \mu_A\, n_A + \frac{1}{2}\, \boldsymbol{P} \cdot \boldsymbol{E} - \frac{1}{2}\, \boldsymbol{M} \cdot \boldsymbol{B} \tag{9.8}$$

The internal energy density $u\left(s, \{n_A\}, \boldsymbol{P}, \boldsymbol{M}\right)$ of the material system is obtained by Legendre transformation of the magnetic enthalpy density $h_m\left(s, \{n_A\}, \boldsymbol{P}, \boldsymbol{B}\right)$ with respect to the magnetic induction field $\boldsymbol{B}$, i.e.

$$u\left(s, \{n_A\}, \boldsymbol{P}, \boldsymbol{M}\right) = h_m\left(s, \{n_A\}, \boldsymbol{P}, \boldsymbol{B}\right) - \frac{\partial h_m}{\partial \boldsymbol{B}} \cdot \boldsymbol{B} \tag{9.9}$$

This procedure introduces a derivative of a scalar field with respect to a vector field. We define the partial derivative of an arbitrary scalar field $S\left(\boldsymbol{V}, \ldots\right)$ with respect to an arbitrary vector field $\boldsymbol{V} = V_x\, \hat{\boldsymbol{x}} + V_y\, \hat{\boldsymbol{y}} + V_z\, \hat{\boldsymbol{z}}$ as,

$$\frac{\partial S}{\partial \boldsymbol{V}} = \frac{\partial S}{\partial V_x}\, \hat{\boldsymbol{x}} + \frac{\partial S}{\partial V_y}\, \hat{\boldsymbol{y}} + \frac{\partial S}{\partial V_z}\, \hat{\boldsymbol{z}}$$

Taking into account the magnetic constitutive relation (9.2), the partial derivative of the magnetic enthalpy density $h_m\left(s, \{n_A\}, \boldsymbol{P}, \boldsymbol{B}\right)$ is given by,

$$\frac{\partial h_m}{\partial \boldsymbol{B}} = -\boldsymbol{M}$$

Thus, the internal energy density $u\left(s, \{n_A\}, \boldsymbol{P}, \boldsymbol{M}\right)$ of the material system is given by,

$$u\left(s, \{n_A\}, \boldsymbol{P}, \boldsymbol{M}\right) = Ts + \sum_{A=1}^{r} \mu_A\, n_A + \frac{1}{2}\, \boldsymbol{P} \cdot \boldsymbol{E} + \frac{1}{2}\, \boldsymbol{M} \cdot \boldsymbol{B} \tag{9.10}$$

where fields $\boldsymbol{E}$ and $\boldsymbol{B}$ are respectively, functions of fields $\boldsymbol{P}$ and $\boldsymbol{M}$ that can be deduced from equations (9.1) and (9.2). The third term in equation (9.10) is the interaction energy density between the local electric dipole density $\boldsymbol{P}$ and the electric field $\boldsymbol{E}$. The last term is the interaction energy density between the local magnetic dipole density $\boldsymbol{M}$ and the magnetic induction field $\boldsymbol{B}$.

The internal energy density of the system $u\left(s, \{n_A\}, \boldsymbol{P}, \boldsymbol{M}\right)$ is the state function we initially wanted to determine. Since it is a function of the local state fields of the system s, $\{n_A\}$, $\boldsymbol{P}$ and $\boldsymbol{M}$, its differential $du\left(s, \{n_A\}, \boldsymbol{P}, \boldsymbol{M}\right)$ can be expressed as,

$$du\left(s, \{n_A\}, \boldsymbol{P}, \boldsymbol{M}\right) = \frac{\partial u}{\partial s}\, ds + \sum_{A=1}^{r} \frac{\partial u}{\partial n_A}\, dn_A$$
$$+ \frac{\partial u}{\partial \boldsymbol{P}} \cdot d\boldsymbol{P} + \frac{\partial u}{\partial \boldsymbol{M}} \cdot d\boldsymbol{M} \tag{9.11}$$

The state fields s, $\{n_A\}$, $\boldsymbol{P}$ and $\boldsymbol{M}$ are specific quantities. The intensive fields conjugated to the state fields are given by,

$$\frac{\partial u}{\partial s} = T, \qquad \frac{\partial u}{\partial n_A} = \mu_A, \qquad \frac{\partial u}{\partial \boldsymbol{P}} = \boldsymbol{E}, \qquad \frac{\partial u}{\partial \boldsymbol{M}} = \boldsymbol{B} \tag{9.12}$$

Applying definitions (9.12) to the differential (9.11), we obtain,

$$du\left(s, \{n_A\}, \boldsymbol{P}, \boldsymbol{M}\right) = T\, ds + \sum_{A=1}^{r} \mu_A\, dn_A + \boldsymbol{E} \cdot d\boldsymbol{P} + \boldsymbol{B} \cdot d\boldsymbol{M} \tag{9.13}$$

The infinitesimal variation of the internal energy density (9.10) is obtained by differentiation of equation (9.10),

$$du = T\, ds + s\, dT + \sum_{A=1}^{r} \left(\mu_A\, dn_A + n_A\, d\mu_A\right) + \boldsymbol{P} \cdot d\boldsymbol{E} + \boldsymbol{E} \cdot d\boldsymbol{P} + \boldsymbol{B} \cdot d\boldsymbol{M} + \boldsymbol{M} \cdot d\boldsymbol{B} \tag{9.14}$$

Taking into account equation (9.13), equation (9.14) requires that,

$$s\, dT + \sum_{A=1}^{r} n_A\, d\mu_A + \boldsymbol{P} \cdot d\boldsymbol{E} + \boldsymbol{M} \cdot d\boldsymbol{B} = 0 \tag{9.15}$$

which is a generalisation of the volume-specific Gibbs–Duhem equation at constant pressure.

Having established the above results for a local system, the expression for the internal energy of a homogeneous uniform system can now be determined. The state variables of the system are the entropy S, the number of moles N_A of chemical elements A, the induced electric dipole $\boldsymbol{p}$ and the induced magnetic dipole $\boldsymbol{m}$. These variables are defined as the integral over the volume of the state fields, which are specific quantities,

$$S \equiv \int_V dV\, s, \qquad N_A \equiv \int_V dV\, n_A, \qquad \boldsymbol{p} \equiv \int_V dV\, \boldsymbol{P}, \qquad \boldsymbol{m} \equiv \int_V dV\, \boldsymbol{M} \tag{9.16}$$

The internal energy $U\left(S, \{N_A\}, \boldsymbol{p}, \boldsymbol{m}\right)$ of the system is defined as the integral of the internal energy density $u\left(s, \{n_A\}, \boldsymbol{P}, \boldsymbol{M}\right)$ over the volume V of the system, i.e.

$$U\left(S, \{N_A\}, \boldsymbol{p}, \boldsymbol{m}\right) \equiv \int_V dV\, u\left(s, \{n_A\}, \boldsymbol{P}, \boldsymbol{M}\right) \tag{9.17}$$

Since temperature T and chemical potential μ_A are homogeneous intensive state functions and electric field $\boldsymbol{E}$ and magnetic induction field $\boldsymbol{B}$ are uniform intensive state functions, then the integral of the internal energy density (9.10) reads,

$$\int_V dV\, u\left(s, \{n_A\}, \boldsymbol{P}, \boldsymbol{M}\right) = T \int_V dV\, s + \sum_{A=1}^{r} \mu_A \int_V dV\, n_A$$
$$+ \frac{1}{2}\left(\int_V dV\, \boldsymbol{P}\right) \cdot \boldsymbol{E} + \frac{1}{2}\left(\int_V dV\, \boldsymbol{M}\right) \cdot \boldsymbol{B} \tag{9.18}$$

Using definitions (9.16) and (9.17), the internal energy of the homogeneous uniform system is expressed as,

$$U\left(S,\{N_A\},\boldsymbol{p},\boldsymbol{m}\right) = TS + \sum_{A=1}^{r} \mu_A N_A + \frac{1}{2}\boldsymbol{p}\cdot\boldsymbol{E} + \frac{1}{2}\boldsymbol{m}\cdot\boldsymbol{B} \qquad (9.19)$$

9.2.3 Free Energy, Enthalpies and Free Enthalpies

We now define several thermodynamic potentials, obtained from the energy density $u\left(s,\boldsymbol{P},\boldsymbol{M}\right)$ by various Legendre transformations (see § 4.3).

Using relations (9.10) and (9.12), the free energy density $f(T,\{n_A\},\boldsymbol{P},\boldsymbol{M})$ is defined as,

$$\begin{aligned} f(T,\{n_A\},\boldsymbol{P},\boldsymbol{M}) &\equiv u - \frac{\partial u}{\partial s}s = u - Ts \\ &= \sum_{A=1}^{r} \mu_A n_A + \frac{1}{2}\boldsymbol{P}\cdot\boldsymbol{E} + \frac{1}{2}\boldsymbol{M}\cdot\boldsymbol{B} \end{aligned} \qquad (9.20)$$

Relations (9.13) and (9.20) imply that the differential of the free energy density is given by,

$$df(T,\{n_A\},\boldsymbol{P},\boldsymbol{M}) = -s\,dT + \sum_{A=1}^{r} \mu_A\,dn_A + \boldsymbol{E}\cdot d\boldsymbol{P} + \boldsymbol{B}\cdot d\boldsymbol{M} \qquad (9.21)$$

Using relations (9.10) and (9.12), the electric enthalpy density $h_e\left(s,\{n_A\},\boldsymbol{E},\boldsymbol{M}\right)$ is defined as,

$$\begin{aligned} h_e\left(s,\{n_A\},\boldsymbol{E},\boldsymbol{M}\right) &\equiv u - \frac{\partial u}{\partial \boldsymbol{P}}\cdot\boldsymbol{P} = u - \boldsymbol{E}\cdot\boldsymbol{P} \\ &= Ts + \sum_{A=1}^{r} \mu_A n_A - \frac{1}{2}\boldsymbol{P}\cdot\boldsymbol{E} + \frac{1}{2}\boldsymbol{M}\cdot\boldsymbol{B} \end{aligned} \qquad (9.22)$$

Relations (9.13) and (9.22) imply that the differential of the electric enthalpy density is given by,

$$dh_e\left(s,\{n_A\},\boldsymbol{E},\boldsymbol{M}\right) = T\,ds + \sum_{A=1}^{r} \mu_A\,dn_A - \boldsymbol{P}\cdot d\boldsymbol{E} + \boldsymbol{B}\cdot d\boldsymbol{M} \qquad (9.23)$$

Using relations (9.10) and (9.12), the magnetic enthalpy density $h_m\left(s,\{n_A\},\boldsymbol{P},\boldsymbol{B}\right)$ is expressed as,

$$\begin{aligned} h_m\left(s,\{n_A\},\boldsymbol{P},\boldsymbol{B}\right) &= u - \frac{\partial u}{\partial \boldsymbol{M}}\cdot\boldsymbol{M} = u - \boldsymbol{B}\cdot\boldsymbol{M} \\ &= Ts + \sum_{A=1}^{r} \mu_A n_A + \frac{1}{2}\boldsymbol{P}\cdot\boldsymbol{E} - \frac{1}{2}\boldsymbol{M}\cdot\boldsymbol{B} \end{aligned} \qquad (9.24)$$

This corresponds, as it should, to the definition given in equation (9.8). Relations (9.13) and (9.24) imply that the differential of the magnetic enthalpy density is given by,

$$dh_m\left(s,\{n_A\},\boldsymbol{P},\boldsymbol{B}\right) = T\,ds + \sum_{A=1}^{r} \mu_A\,dn_A + \boldsymbol{E}\cdot d\boldsymbol{P} - \boldsymbol{M}\cdot d\boldsymbol{B} \qquad (9.25)$$

Free enthalpies can be defined similarly by performing Legendre transformations of the enthalpies above with respect to entropy density s. Using relations (9.12) and (9.22), the electric free enthalpy density $g_e\left(T,\{n_A\},\boldsymbol{E},\boldsymbol{M}\right)$ is defined as,

$$g_e\left(T,\{n_A\},\boldsymbol{E},\boldsymbol{M}\right) \equiv h_e - \frac{\partial h_e}{\partial s}\, s = h_e - Ts$$

$$= \sum_{A=1}^{r} \mu_A\, n_A - \frac{1}{2}\,\boldsymbol{P}\cdot\boldsymbol{E} + \frac{1}{2}\,\boldsymbol{M}\cdot\boldsymbol{B} \qquad (9.26)$$

Relations (9.22) and (9.23) imply that the differential of the electric free enthalpy density is given by,

$$dg_e\left(T,\{n_A\},\boldsymbol{E},\boldsymbol{M}\right) = -s\,dT + \sum_{A=1}^{r} \mu_A\, dn_A - \boldsymbol{P}\cdot d\boldsymbol{E} + \boldsymbol{B}\cdot d\boldsymbol{M} \qquad (9.27)$$

Using relations (9.12) and (9.24), the magnetic free enthalpy density $g_m\left(T,\{n_A\},\boldsymbol{P},\boldsymbol{B}\right)$ is defined as,

$$g_m\left(T,\{n_A\},\boldsymbol{P},\boldsymbol{B}\right) \equiv h_m - \frac{\partial h_m}{\partial s}\, s = h_m - Ts$$

$$= \sum_{A=1}^{r} \mu_A\, n_A + \frac{1}{2}\,\boldsymbol{P}\cdot\boldsymbol{E} - \frac{1}{2}\,\boldsymbol{M}\cdot\boldsymbol{B} \qquad (9.28)$$

Relations (9.13) and (9.24) imply that the differential of the magnetic free enthalpy density is given by,

$$dg_m\left(T,\{n_A\},\boldsymbol{P},\boldsymbol{B}\right) = -s\,dT + \sum_{A=1}^{r} \mu_A\, dn_A + \boldsymbol{E}\cdot d\boldsymbol{P} - \boldsymbol{M}\cdot d\boldsymbol{B} \qquad (9.29)$$

9.2.4 Force Densities

The numerous functions defined in the previous section may become useful in several contexts. Here, we determine electromagnetic forces based on the spatial variation of these functions. Then, in § 9.5, we will discuss in detail an example of homogeneous and uniform systems in terms of these functions.

In order to determine the force density $\boldsymbol{f}_p$ exerted by the electric polarisation $\boldsymbol{P}$ of the system in the presence of an electric field $\boldsymbol{E}$, we consider that only the electric field $\boldsymbol{E}$ varies spatially. In this case, fields s, $\{n_A\}$ and $\boldsymbol{M}$ are fixed. Taking into account the differential (9.23) of the electric enthalpy density, the electric enthalpy density gradient $\boldsymbol{\nabla} h_e\left(s,\{n_A\},\boldsymbol{E},\boldsymbol{M}\right)$ reduces to,

$$\boldsymbol{\nabla}\, h_e\left(\boldsymbol{E}\right) = \frac{\partial h_e}{\partial \boldsymbol{E}}\,\boldsymbol{\nabla}\,\boldsymbol{E} = -\boldsymbol{P}\,\boldsymbol{\nabla}\,\boldsymbol{E} \qquad (9.30)$$

We use the notation $\boldsymbol{V}\,\boldsymbol{\nabla}\,\boldsymbol{W}$ to represent a vector field oriented in the direction of the gradient $\boldsymbol{\nabla}$ and defined in terms of a scalar product between the vector fields $\boldsymbol{V} = V_x\,\hat{\boldsymbol{x}} + V_y\,\hat{\boldsymbol{y}} + V_z\,\hat{\boldsymbol{z}}$ and $\boldsymbol{W} = W_x\,\hat{\boldsymbol{x}} + W_y\,\hat{\boldsymbol{y}} + W_z\,\hat{\boldsymbol{z}}$,

$$\boldsymbol{V}\,\boldsymbol{\nabla}\,\boldsymbol{W} = V_x\,\boldsymbol{\nabla}\,W_x + V_y\,\boldsymbol{\nabla}\,W_y + V_z\,\boldsymbol{\nabla}\,W_z$$

The force density $f_p(E)$ is defined as the opposite of the gradient of the electric enthalpy density, where the electric enthalpy density plays the role of the potential energy density [110],

$$f_p(E) \equiv -\nabla h_e(E) = P \nabla E \qquad (9.31)$$

The resulting force $F_p(E)$ exerted on the system is defined as the integral of the force density $f_p(E)$ over the volume V of the system,

$$F_p(E) \equiv \int_V dV f_p(E) = \int_V dV (P \nabla E) \qquad (9.32)$$

In order to determine the force density f_m exerted by the magnetisation M of the system in the presence of a magnetic induction field B, we consider that only the magnetic induction field B varies spatially. In this case, the fields s, $\{n_A\}$ and P are fixed. Taking into account the differential (9.25) of the magnetic enthalpy density, the magnetic enthalpy density gradient $\nabla h_m(s, \{n_A\}, P, B)$ reduces to,

$$\nabla h_m(B) = \frac{\partial h_m}{\partial B} \nabla B = -M \nabla B \qquad (9.33)$$

The force density $f_m(B)$ is defined as the opposite of the magnetic enthalpy density gradient, where the magnetic enthalpy density plays the role of the potential energy density [111],

$$f_m(B) \equiv -\nabla h_m(B) = M \nabla B \qquad (9.34)$$

The resulting force $F_m(B)$ exerted on the system is defined as the integral of the force density $f_m(B)$ over the volume V of the system,

$$F_m(B) \equiv \int_V dV f_m(B) = \int_V dV (M \nabla B) \qquad (9.35)$$

9.3 Conductors and Electromagnetic Fields

So far, we have described electric insulators consisting of electrically neutral substances in the presence of electromagnetic fields. Now, we consider electrically charged substances that can move in the system. Thus, the system is a **conductor**. In view of the laws of electromagnetism, when dealing with a conductor, we need to include the electromagnetic fields in the system in order to treat properly the interactions between matter and electromagnetic fields. Thus, the system consists now of electrically charged matter and electromagnetic fields. We choose as local state fields the entropy density s, the set of densities $\{n_A\}$ of the chemical substances A, the electric displacement field D and the magnetic induction field B.

In general, the choice of state fields of a thermodynamic system depends on the description to be made. For instance, for an insulator, we chose the state fields s, $\{n_A\}$, P, M that describe only the properties of matter. In the case of an electric conductor, given the laws of electromagnetism, the state fields s, $\{n_A\}$, D, B are chosen to describe the properties of matter, of the electromagnetic fields and of their interaction.

The electric constitutive relation (9.3) and the linear relation (9.1) yield,

$$\boldsymbol{D} = \boldsymbol{\varepsilon} \cdot \boldsymbol{E} \tag{9.36}$$

where the electric permittivity tensor $\boldsymbol{\varepsilon}$ is related to the electric susceptibility tensor $\boldsymbol{\chi}_e$ by [103],

$$\boldsymbol{\varepsilon} = \varepsilon_0 \left(\mathbb{1} + \boldsymbol{\chi}_e\right) \tag{9.37}$$

Similarly, the magnetic constitutive relation (9.4) and the linear relation (9.2) yield,

$$\boldsymbol{H} = \boldsymbol{\mu}^{-1} \cdot \boldsymbol{B} \tag{9.38}$$

where the inverse of the magnetic permeability tensor $\boldsymbol{\mu}^{-1}$ is related to the magnetic susceptibility tensor $\boldsymbol{\chi}_m$ by [104],

$$\boldsymbol{\mu}^{-1} = \mu_0^{-1} \left(\mathbb{1} - \boldsymbol{\chi}_m\right) \tag{9.39}$$

For a homogeneous sample, the electric permittivity tensor $\boldsymbol{\varepsilon} = \varepsilon \mathbb{1}$ and the inverse of the magnetic permeability tensor $\boldsymbol{\mu}^{-1} = \mu^{-1} \mathbb{1}$ are multiples of the identity tensor $\mathbb{1}$. The scalars ε and μ^{-1} are the electric permittivity and the inverse of the magnetic permeability of the matter. In this case, the electromagnetic constitutive relations (9.36) and (9.38) reduce to [112],

$$\boldsymbol{D} = \varepsilon \boldsymbol{E} \tag{9.40}$$

$$\boldsymbol{H} = \mu^{-1} \boldsymbol{B} \tag{9.41}$$

The internal energy density of the system $u\left(s, \{n_A\}, \boldsymbol{D}, \boldsymbol{B}\right)$ is a state function, namely a function of the local state fields of the system s, $\{n_A\}$, $\boldsymbol{D}$ and $\boldsymbol{B}$. For a local system consisting of matter and electromagnetic fields, the internal energy density is given by the expression (9.5). Therefore, the total energy density $u^{\text{tot}}\left(s, \{n_A\}, \boldsymbol{D}, \boldsymbol{B}\right)$ is renamed here $u\left(s, \{n_A\}, \boldsymbol{D}, \boldsymbol{B}\right)$. Differentiating this expression with respect to the state fields and taking into account the constitutive relations (9.40) and (9.41) yields the internal energy density differential $du\left(s, \{n_A\}, \boldsymbol{D}, \boldsymbol{B}\right)$ of the local system that consists of matter and electromagnetic fields,

$$du\left(s, \{n_A\}, \boldsymbol{D}, \boldsymbol{B}\right) = T ds + \sum_A \mu_A \, dn_A + \boldsymbol{E} \cdot d\boldsymbol{D} + \boldsymbol{H} \cdot d\boldsymbol{B} \tag{9.42}$$

where the electric field $\boldsymbol{E}$ and the magnetic field $\boldsymbol{H}$ are fields conjugated to the electromagnetic state fields of the system $\boldsymbol{D}$ and $\boldsymbol{B}$. This is mathematically expressed as,

$$\boldsymbol{E} = \frac{\partial u}{\partial \boldsymbol{D}} \quad \text{and} \quad \boldsymbol{H} = \frac{\partial u}{\partial \boldsymbol{B}} \tag{9.43}$$

Taking into account the constitutive equations (9.40) and (9.41), the internal energy density $u\left(s, \{n_A\}, \boldsymbol{D}, \boldsymbol{B}\right)$ that satisfies the integrability relations found in (9.43) is given by,

$$u\left(s, \{n_A\}, \boldsymbol{D}, \boldsymbol{B}\right) = Ts + \sum_{A=1}^{r} \mu_A \, n_A + \frac{D^2}{2\varepsilon} + \frac{B^2}{2\mu} \tag{9.44}$$

The first term on the right-hand side of equation (9.44) corresponds to the thermal energy density. The second is the chemical energy density. The last two terms represent the energy density of the electromagnetic fields and their interaction with matter. In contrast to expression (9.10) for the internal energy density, equation (9.44) contains the energy density of the electromagnetic fields, because in this section the electromagnetic fields are an integral part of the system.

9.3.1 Conductors and Electrostatic Fields

Let us consider here the particular case of an isolated system which consists of a single homogeneous electric conductor of volume V_{int}, maintained in an enclosure of volume $V = V_{int} + V_{ext}$. Electrostatic fields are generated by the electric charges of the system itself (Fig. 9.1). The enclosure does not belong to the system but the field inside it does. The volume V_{ext} outside the conductor is so large that the electrostatic potential is negligible on the walls of the enclosure, as they are located very far away from the conductor.

In electrostatics, the electric charges do not move in the reference frame of the system and there is no electric current. Thus, the system does not generate a magnetic induction field. Therefore, the internal energy density (9.44) reduces to,

$$u\left(s, \{n_A\}, \boldsymbol{D}\right) = Ts + \sum_{A=1}^{r} \mu_A\, n_A + \frac{\boldsymbol{D}^2}{2\varepsilon} \qquad (9.45)$$

Using the electric constitutive relation (9.40), the internal energy density (9.45) can be expressed as,

$$u\left(s, \{n_A\}, \boldsymbol{D}\right) = Ts + \sum_{A=1}^{r} \mu_A\, n_A + \frac{1}{2}\,\boldsymbol{E} \cdot \boldsymbol{D} \qquad (9.46)$$

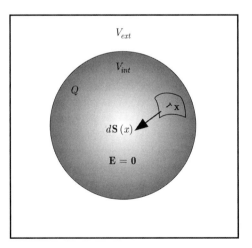

Figure 9.1   An electric conductor of volume V_{int} and electric charge Q is located in an enclosure. The electric field $\boldsymbol{E}$ vanishes inside the conductor. Any volume integral over the outside volume V_{ext} can be transformed into a surface integral expressed in terms of the infinitesimal surface element $d\boldsymbol{S}$ that is defined positively when oriented towards the interior of the conductor.

where the last term is the electrostatic energy density. In electrostatics, the electric field E is the opposite of the electrostatic potential gradient φ [113],

$$E = -\nabla\varphi \tag{9.47}$$

Thus, the electrostatic energy density can be expressed as [114],

$$\frac{1}{2}E \cdot D = -\frac{1}{2}(\nabla\varphi) \cdot D = \frac{1}{2}\varphi(\nabla \cdot D) - \frac{1}{2}\nabla \cdot (\varphi D) \tag{9.48}$$

where the second equality results from a vector analysis identity. The integral of equation (9.48) over the volume V of the system gives the electrostatic energy of the system. It reads,

$$\frac{1}{2}\int_V dV\,(E \cdot D) = \frac{1}{2}\int_V dV\left(\varphi(\nabla \cdot D)\right) - \frac{1}{2}\int_V dV\left(\nabla \cdot (\varphi D)\right) \tag{9.49}$$

The differential expression of Gauss' electrostatic law provides a relationship between the electric displacement field D and the electric charge density q [115],

$$\nabla \cdot D = q \tag{9.50}$$

The electric field vanishes inside an electric conductor (Fig. 9.1) [116],

$$E = 0 \qquad \text{(inside conductor)} \tag{9.51}$$

Taking into account the electric constitutive relation (9.40) for a homogeneous conductor, as well as equations (9.50) and (9.51), we find that the electric charge density vanishes inside the electric conductor,

$$q = 0 \qquad \text{(inside conductor)} \tag{9.52}$$

This means that the moving electric charges are on the surface of the conductor (Fig. 9.1).

Equations (9.51), (9.40) and (9.52) imply that the volume integration in equation (9.49) vanishes when it is carried out over the volume V_{int} of the electric conductor. The non-vanishing contributions to the volume integrals (9.49) stem from integration over the volume V_{ext} outside the electric conductor. In vacuum, there are no electric charges, in other words, the electric charge density vanishes outside the electric conductor. Taking into account Gauss' electrostatic law (9.50), the first integral on the right-hand side of equation (9.49) vanishes when integrated over the volume V_{ext} outside the electric conductor. Using the divergence theorem, the second integral over the volume V_{ext} outside the electric conductor can be expressed as the sum of an integral over the internal surface of the outside volume (i.e. the surface S of the conductor) [117] and over the external surface of the outside volume (i.e. the walls of the enclosure). Since the electrostatic potential is negligible on the walls of the enclosure, the divergence theorem reduces to,

$$\frac{1}{2}\int_{V_{ext}} dV\,(E \cdot D) = -\frac{1}{2}\int_{V_{ext}} dV\left(\nabla \cdot (\varphi D)\right) = -\frac{1}{2}\int_S dS \cdot (\varphi D) \tag{9.53}$$

where the infinitesimal surface element dS is oriented towards the exterior of the volume V_{ext}, i.e. towards the inside of the electric conductor (Fig. 9.1).

Taking into account condition (9.51), the electrostatic potential φ is constant on the surface of the conductor [118],

$$\varphi = \text{const} \qquad \text{(surface conductor)} \tag{9.54}$$

Then, the right-hand side of (9.53) can be deduced using the integral expression of Gauss' electrostatic law, which states [119],

$$-\int_S d\boldsymbol{S} \cdot \boldsymbol{D} = Q \tag{9.55}$$

The negative sign is present here because the infinitesimal surface element appearing in Gauss' electrostatic law is defined positively when oriented towards the exterior of the conductor. Thus, taking into account Gauss' electrostatic law (9.55) and property (9.54), we conclude that the electrostatic energy (9.53) reduces to,

$$\frac{1}{2}\int_{V_{\text{ext}}} dV\,(\boldsymbol{E} \cdot \boldsymbol{D}) = \frac{1}{2}\,Q\,\varphi \tag{9.56}$$

where φ is the electrostatic potential at the surface of the conductor. The internal energy $U(S, \{N_A\}, Q)$ of the system can be obtained from the internal energy density by integration over the volume V of the system,

$$U(S, \{N_A\}, Q) = \int_V dV\,u\,(s, \{n_A\}, \boldsymbol{D}) \tag{9.57}$$

The entropy S and the number of moles N_A of the substance A are obtained by integrating the respective densities s and n_A over the volume V_{int} of the electric conductor,

$$S = \int_{V_{\text{int}}} dV\,s \qquad \text{and} \qquad N_A = \int_{V_{\text{int}}} dV\,n_A \tag{9.58}$$

As the conductor is homogeneous, temperature T and chemical potential μ_A of substance A are homogeneous intensive fields. When integrating over the volume V of the system the internal energy density $u\,(s, \{n_A\}, \boldsymbol{D})$ given by equation (9.46) and taking into account equations (9.56), (9.57) and (9.58), we find the internal energy of the system,

$$U(S, \{N_A\}, Q) = TS + \sum_{A=1}^{r} \mu_A N_A + \frac{1}{2}\,\varphi\,Q \tag{9.59}$$

In linear electrostatics, the electric charge Q is a linear function of the electrostatic potential φ. Thus, the internal energy differential (9.59) is given by,

$$dU(S, \{N_A\}, Q) = T\,dS + \sum_{A=1}^{r} \mu_A\,dN_A + \varphi\,dQ \tag{9.60}$$

The electrostatic enthalpy of the system $H_e\,(S, \{N_A\}, \varphi)$ is a function of the intensive variable φ. It is obtained by a Legendre transformation with respect to the extensive variable Q of the internal energy $U(S, \{N_A\}, Q)$ given by (9.59),

$$\begin{aligned} H_e\,(S, \{N_A\}, \varphi) &\equiv U - \frac{\partial U}{\partial Q}\,Q = U - \varphi\,Q \\ &= TS + \sum_{A=1}^{r} \mu_A N_A - \frac{1}{2}\,Q\,\varphi \end{aligned} \tag{9.61}$$

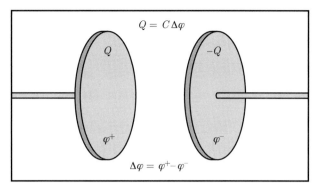

Figure 9.2   A parallel plate capacitor has an electric charge Q on the plate at electrostatic potential φ^+ and an electric charge $-Q$ on the plate at electrostatic potential φ^-.

The differential of the electrostatic enthalpy (9.61) is given by,

$$dH_e\left(S, \{N_A\}, \varphi\right) = T\,dS + \sum_{A=1}^{r} \mu_A\,dN_A - Q\,d\varphi \tag{9.62}$$

Based on the above results, we can now analyse a capacitor that consists of two separate conductors. The electric charges held by each one of them generate an electrostatic potential difference. The electrostatic potential of the positively charged conductor is φ^+ and its electric charge is $Q^+ > 0$. Likewise, the electrostatic charge of the negatively charged conductor is φ^- and its electric charge is $Q^- < 0$. The electric charges Q^+ and Q^- are opposite,

$$Q \equiv Q^+ = -Q^- > 0 \tag{9.63}$$

and the potential difference $\Delta\varphi$ between the conductors is defined as,

$$\Delta\varphi = \varphi^+ - \varphi^- > 0 \tag{9.64}$$

In order to determine the expression for the electrostatic energy $U(Q)$ and the electrostatic enthalpy $H_e(\Delta\varphi)$ of this capacitor, we consider that it consists of two simple subsystems, which are the two conductors. Taking into account definitions (9.59), (9.61), (9.63) and (9.64), we find that the electrostatic energy and the electrostatic enthalpy are [120],

$$U(Q) = \frac{1}{2}\left(\varphi^+ Q^+ + \varphi^- Q^-\right) = \frac{1}{2}\Delta\varphi\,Q \tag{9.65}$$

$$H_e(\Delta\varphi) = -\frac{1}{2}\left(Q^+ \varphi^+ + Q^- \varphi^-\right) = -\frac{1}{2}Q\,\Delta\varphi \tag{9.66}$$

The electric charge Q is proportional to the electrostatic potential difference $\Delta\varphi$ between the conductors and the **electrostatic capacity** called the **capacitance** C,

$$Q = C\,\Delta\varphi \tag{9.67}$$

The capacitance C is proportional to the electric permittivity ε of the material between the plates and depends upon the geometric shape of the conductors. Using equation (9.67), the electrostatic energy (9.65) and the electrostatic enthalpy (9.66) become,

$$U(Q) = \frac{1}{2}\frac{Q^2}{C} \tag{9.68}$$

$$H_e(\Delta\varphi) = -\frac{1}{2}C\Delta\varphi^2 \tag{9.69}$$

9.3.2 Conductors and Magnetostatic Fields

As a second example of conductors and electromagnetic fields, we consider here a system consisting of a fixed, electrically conducting coil in an enclosure of volume V. A magnetostatic field is generated by the electric current that flows through the coil. This current is driven by a current source located outside the enclosure. The enclosure and the current source do not belong to the system (Fig. 9.3). The coil consists of enough turns per unit length that the field can be considered uniform inside and vanishingly small outside the coil. In magnetostatics, the electric current is stationary. In view of the conservation of electric charges, the electric current entering the conductor is equal to the electric current flowing through and coming out of it. Thus, the system does not generate variations of the electric displacement field. Therefore, the internal energy density (9.44) reduces to,

$$u(s, \{n_A\}, \boldsymbol{B}) = Ts + \sum_{A=1}^{r}\mu_A n_A + \frac{\boldsymbol{B}^2}{2\mu} \tag{9.70}$$

Using the magnetic constitutive relation (9.41), the internal energy density (9.70) can be expressed as,

$$u(s, \{n_A\}, \boldsymbol{B}) = Ts + \sum_{A=1}^{r}\mu_A n_A + \frac{1}{2}\boldsymbol{H}\cdot\boldsymbol{B} \tag{9.71}$$

where the third term is the magnetostatic energy density. The magnetic induction field $\boldsymbol{B}$ is the curl of the magnetostatic potential $\boldsymbol{A}$,

$$\boldsymbol{B} = \boldsymbol{\nabla}\times\boldsymbol{A} \tag{9.72}$$

Thus, the magnetostatic energy density can be recast as [121],

$$\frac{1}{2}\boldsymbol{H}\cdot\boldsymbol{B} = \frac{1}{2}\boldsymbol{H}\cdot(\boldsymbol{\nabla}\times\boldsymbol{A}) = \frac{1}{2}(\boldsymbol{\nabla}\times\boldsymbol{H})\cdot\boldsymbol{A} + \frac{1}{2}\boldsymbol{\nabla}\cdot(\boldsymbol{A}\times\boldsymbol{H}) \tag{9.73}$$

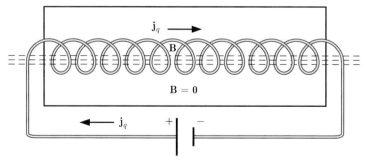

Figure 9.3   A stationary electric current density $\boldsymbol{j}_q$ flows through a coil located in an enclosure. The magnetic induction field $\boldsymbol{B}$ is constant inside the coil and negligible outside it.

where the second equality is a vector analysis identity. The integral of equation (9.73) over the volume V of the system reads,

$$\frac{1}{2} \int_V dV \, (\boldsymbol{H} \cdot \boldsymbol{B}) = \frac{1}{2} \int_V dV \left((\boldsymbol{\nabla} \times \boldsymbol{H}) \cdot \boldsymbol{A} \right) + \frac{1}{2} \int_V dV \left(\boldsymbol{\nabla} \cdot (\boldsymbol{A} \times \boldsymbol{H}) \right) \qquad (9.74)$$

The differential expression of Ampère's magnetostatic law links the magnetic field $\boldsymbol{H}$ and the electric current density $\boldsymbol{j}_q$. It is given by [122],

$$\boldsymbol{\nabla} \times \boldsymbol{H} = \boldsymbol{j}_q \qquad (9.75)$$

Due to the stationary electric current flowing through the coil, the magnetic induction field $\boldsymbol{B}$, assumed to be long enough and tightly wound, is constant, uniform and oriented along the axis of symmetry of the coil inside the coil (Fig. 9.3). It is negligibly small outside of it.

$$\boldsymbol{B} = \boldsymbol{0} \qquad \text{(outside coil)} \qquad (9.76)$$

Using the divergence theorem, the second integral on the right-hand side of relation (9.74) can be replaced by an integral over the enclosure of the system [117]. According to the magnetic constitutive relation (9.41) and in view of relation (9.76), the integrand of the second integral on the right-hand side of relation (9.74) vanishes on those parts of the enclosure that are outside the coil. Moreover, the magnetic induction field $\boldsymbol{B}$ inside the coil is orthogonal to the plane of the wire loops whereas the magnetostatic vector potential $\boldsymbol{A}$ is oriented in that plane. Thus, the magnetic constitutive relation (9.41) implies that the vector $\boldsymbol{A} \times \boldsymbol{H}$ is oriented in the plane of the wire loops inside the coil. Since the vector corresponding to an infinitesimal surface element is orthogonal to the enclosure, the integrand of the second integral on the right-hand side of relation (9.74) vanishes on the parts of the enclosure which are inside the coil. Taking into account Ampère's magnetostatic law (9.75), relation (9.74) becomes,

$$\frac{1}{2} \int_V dV \, (\boldsymbol{H} \cdot \boldsymbol{B}) = \frac{1}{2} \int_V dV \, (\boldsymbol{j}_q \cdot \boldsymbol{A}) \qquad (9.77)$$

We now make use of the fact that the integral on the right-hand side of relation (9.77) brings a non-vanishing contribution only in the volume occupied by the wires through which the constant electric current density $\boldsymbol{j}_q$ flows. An infinitesimal volume element dV of the wire is the scalar product of an infinitesimal surface element $d\boldsymbol{S}$ oriented along the wire and an infinitesimal displacement vector oriented along the wire also (Fig. 9.4),

$$dV = d\boldsymbol{S} \cdot d\boldsymbol{r} \qquad (9.78)$$

Thus, relation (9.77) can be expressed as the product of current density integrated over the wire section S and an integral of the magnetostatic potential vector $\boldsymbol{A}$ (Fig. 9.4) on the contour C defined by one turn of the coil,

$$\frac{1}{2} \int_V dV \, (\boldsymbol{H} \cdot \boldsymbol{B}) = \frac{1}{2} \int_S d\boldsymbol{S} \cdot \boldsymbol{j}_q \int_C d\boldsymbol{r} \cdot \boldsymbol{A} \qquad (9.79)$$

The electric current I flowing through the wire of the coil is given by the electric current density $\boldsymbol{j}_q$ integrated over the section S of the wire [123],

$$I = \int_S d\boldsymbol{S} \cdot \boldsymbol{j}_q \qquad (9.80)$$

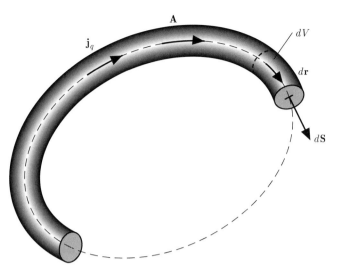

A stationary electric current density j_q flows through one of the conducting loops of a coil. At any point in the wire, the electric current density j_q, the magnetostatic potential vector A, the infinitesimal displacement element dr and the infinitesimal surface element dS associated to the wire are all collinear, parallel to the electric current density, which itself is orthogonal to the wire section at that location.

The magnetostatic flux Φ that flows through the coil is given by the integral on the contour C of the magnetostatic potential vector A [124],

$$\Phi = \int_C dr \cdot A \tag{9.81}$$

Here, the magnetostatic potential vector A is tangent to the surface of the coil and orthogonal to the axis of symmetry of the coil. Thus, by taking into account definitions (9.80) and (9.81), the magnetostatic energy (9.79) can be reduced to,

$$\frac{1}{2} \int_V dV (H \cdot B) = \frac{1}{2} I \Phi \tag{9.82}$$

The internal energy $U(S, \{N_A\}, \Phi)$ of the system is deduced from the internal energy density by an integral over the volume V of the system,

$$U(S, \{N_A\}, \Phi) = \int_V dV u (s, \{n_A\}, B) \tag{9.83}$$

Since the electric current that flows through the coil is stationary, the temperature T and chemical potential μ_A of substance A are intensive fields that have the same value everywhere. When integrating over the volume V of the system the internal energy density $u(s, \{n_A\}, B)$ given by equation (9.71) and taking into account equations (9.82), (9.83) and (9.58), we find the internal energy of the system,

$$U(S, \{N_A\}, \Phi) = TS + \sum_{A=1}^{r} \mu_A N_A + \frac{1}{2} I \Phi \tag{9.84}$$

In linear magnetostatics, the magnetostatic flux Φ is a linear function of the electric current I. Thus, the differential of the internal energy (9.84) is given by,

$$dU\left(S, \{N_A\}, \Phi\right) = T\, dS + \sum_{A=1}^{r} \mu_A\, dN_A + I\, d\Phi \tag{9.85}$$

The magnetostatic enthalpy of the local system $H_m\left(S, \{N_A\}, I\right)$ depends on the extensive variable I. It is obtained by a Legendre transformation of expression (9.84) for the internal energy $U\left(S, \{N_A\}, \Phi\right)$ with respect to the intensive variable Φ, i.e.

$$H_m\left(S, \{N_A\}, I\right) \equiv U - \frac{\partial U}{\partial \Phi}\Phi = U - I\Phi$$
$$= TS + \sum_{A=1}^{r} \mu_A\, N_A - \frac{1}{2}\Phi I \tag{9.86}$$

The differential of the magnetostatic enthalpy (9.86) is given by,

$$dH_m\left(S, \{N_A\}, I\right) = T\, dS + \sum_{A=1}^{r} \mu_A\, dN_A - \Phi\, dI \tag{9.87}$$

Taking into account definitions (9.84) and (9.86), the magnetostatic energy and the magnetostatic enthalpy of the coil can be written as [125],

$$U\left(\Phi\right) = \frac{1}{2}I\Phi \tag{9.88}$$

$$H_m\left(I\right) = -\frac{1}{2}\Phi I \tag{9.89}$$

The electric current I flowing through the coil is linked to the magnetostatic flux Φ by the *magnetostatic inductance* L [124],

$$\Phi = LI \tag{9.90}$$

where L is proportional to the magnetic permeability μ and depends on the shape of the coil. Using equation (9.67), the magnetostatic energy (9.88) and the magnetostatic enthalpy (9.89) become,

$$U\left(\Phi\right) = \frac{1}{2}\frac{\Phi^2}{L} \tag{9.91}$$

$$H_m\left(I\right) = -\frac{1}{2}LI^2 \tag{9.92}$$

9.4　Conductor and External Electromagnetic Fields

In this section, we would like to describe the thermodynamics of an electric conductor subjected to external electromagnetic fields that are much larger than the electromagnetic fields generated by the system. Therefore, the latter can be neglected. We first consider electrostatic, then magnetostatic fields.

9.4.1 Conductor and External Electrostatic Fields

We will consider here a system that is analogous to the one described in § 9.3.1, that is, it consists of a single homogeneous electric conductor, fixed in an enclosure. However, here the sphere is subjected to a constant uniform external electric field E. This external electric field is strong enough so that the electric field generated by the electric charges of the system can be neglected. In this case, the external electric field E is independent of the electric displacement field D. Therefore, in view of the conditions (9.43), the electrostatic energy density is $E \cdot D$ and the internal energy density is given by,

$$u\left(s, \{n_A\}, D\right) = Ts + \sum_{A=1}^{r} \mu_A n_A + E \cdot D \tag{9.93}$$

Comparing the internal energy density (9.93) with the internal energy density (9.46), we notice that these expressions differ by a factor of $1/2$. The presence of this factor depends on whether the electric field E is an internal field that is a function of the electric displacement field D or an external field that is independent of D. Following an approach analogous to the one in § 9.3.1, the analog of expression (9.59) for the internal energy of the system is readily found to be,

$$U\left(S, \{N_A\}, Q\right) = TS + \sum_{A=1}^{r} \mu_A N_A + Q\varphi \tag{9.94}$$

where φ represents the external electrostatic potential. The electric charge Q is the sum of the electric charges of the chemical substances A, i.e.

$$Q = \sum_{A=1}^{r} q_A N_A \tag{9.95}$$

where q_A represents the electric charge of a mole of chemical substance A. Equation (9.95) implies that the internal energy (9.94) of the local system can be expressed as,

$$U\left(S, \{N_A\}, Q\right) = TS + \sum_{A=1}^{r} \bar{\mu}_A N_A \tag{9.96}$$

where the electrochemical potential $\bar{\mu}_A$ is defined as,

$$\bar{\mu}_A \equiv \mu_A + q_A \varphi \tag{9.97}$$

We now consider an external electric field E applied between the two conductors of a parallel plate capacitor. This field generates an electrostatic potential difference $\Delta\varphi$. Using equations (9.63), (9.64) and (9.67), the electrostatic energy (9.94) generated by the external field is found to be,

$$U(Q) = Q\,\Delta\varphi = C\,\Delta\varphi^2 \tag{9.98}$$

To conclude this section, we consider a system consisting of a substance A with permanent electric dipoles p_A subjected to an external electric field E. In that case, the electrochemical potential is written as $\bar{\mu}_A \equiv \mu_A - p_A \cdot E$ where the role played by q_A and φ in equation (9.97) is now played by p_A and $-E$ [126].

9.4.2 Conductor and External Magnetostatic Fields

We will consider here a system that is analogous to the one described in § 9.3.2, that is, it consists of a homogeneous conducting coil held fixed in an electrically neutral enclosure. The coil is subjected to a uniform and constant external magnetic induction field $\boldsymbol{B}$. We consider that the external magnetic induction field is strong enough so that the magnetic induction field generated by the stationary electric current in the system can be neglected. Therefore, the external magnetic field $\boldsymbol{H}$ is independent of the magnetic induction field $\boldsymbol{B}$. Taking into account conditions (9.43), the magnetostatic energy density is then $\boldsymbol{H} \cdot \boldsymbol{B}$, and the internal energy density is given by,

$$u\,(s, \{n_A\}, \boldsymbol{B}) = Ts + \sum_{A=1}^{r} \mu_A\, n_A + \boldsymbol{H} \cdot \boldsymbol{B} \tag{9.99}$$

Comparing the internal energy density (9.99) with the internal energy density (9.71), we notice that these expressions differ by a factor of $1/2$. The presence of this factor depends on whether the magnetic field $\boldsymbol{H}$ is an internal field that is a function of the magnetic induction field $\boldsymbol{B}$ or an external field independent of $\boldsymbol{B}$. Thus, following an approach analogous to § 9.3.2, the analog of expression (9.84) for the internal energy of the system can be written as,

$$U\,(S, \{N_A\}, \Phi) = TS + \sum_{A=1}^{r} \mu_A\, N_A + I\,\Phi \tag{9.100}$$

where Φ represents the external magnetostatic flux. The electric current I is the sum of the electric currents of the chemical substances A, i.e.

$$I = \sum_{A=1}^{r} I_A\, N_A \tag{9.101}$$

where I_A represents the electric current due to the motion of the chemical substance A. Equation (9.101) implies that the internal energy (9.100) of the local system can be expressed as,

$$U\,(S, \{N_A\}, \Phi) = TS + \sum_{A=1}^{r} \bar{\mu}_A\, N_A \tag{9.102}$$

where the magnetochemical potential $\bar{\mu}_A$ is defined as,

$$\bar{\mu}_A \equiv \mu_A + I_A\, \Phi \tag{9.103}$$

The external magnetic induction field $\boldsymbol{B}$ applied inside the coil generates a magnetostatic flux Φ. Using the expressions (9.90) and (9.100), the magnetostatic energy generated by the external field yields,

$$U\,(\Phi) = \Phi\,I = L\,I^2 \tag{9.104}$$

To conclude this section, we consider a system consisting of a substance A with permanent magnetic dipoles $\boldsymbol{m}_A$ subjected to an external magnetic induction field $\boldsymbol{B}$. In that case, the electrochemical potential is written as $\bar{\mu}_A \equiv \mu_A - \boldsymbol{m}_A \cdot \boldsymbol{B}$ where the role played by q_A and φ in equation (9.97) is now played by $\boldsymbol{m}_A$ and $-\boldsymbol{B}$ [126].

9.5 Adiabatic Demagnetisation

Adiabatic demagnetisation can be used as a cooling method for samples to reach very low temperatures. In this section, we expand on the notion of thermal response coefficients introduced in §5.2 in order to give a thermodynamic description of this cooling method. We will use the methods of Chapter 4 to infer the behaviour of entropy $S(T, \boldsymbol{B})$ as a function of temperature T and magnetic induction field $\boldsymbol{B}$. An example of such a function $S(T, \boldsymbol{B})$ (Fig. 9.5) will help us keep track of the meaning of the successive results obtained in this section.

Let us first describe the principle of adiabatic demagnetisation and its technological relevance. In order to cool a system, it is first brought to state 'A' (Fig. 9.5) and is held in contact with a thermal bath. Then, while keeping this contact with a thermal bath, a magnetic induction field $\boldsymbol{B}$ is applied to the system until it reaches state 'B'. The contact with the thermal bath is then removed and the magnetic induction field is cancelled. This brings the system to state 'C'. In view of the qualitative features of function $S(T, \boldsymbol{B})$ (Fig. 9.5), we expect a temperature drop.

This method was originally suggested by Debye [127]. It was used to cool a paramagnetic salt, before applying a magnetic field all the while keeping the salt temperature constant. Finally, the salt was thermally isolated and the magnetic field was reduced to zero, causing a drop of the salt temperature. Using this method, physicists can reach temperatures as low as a mK ! Recent research has led to the discovery of new materials that are performant at temperatures useful for commercial applications, like refrigerators [128]. For industrial magnetic cooling, numerous cycles have to be performed which consist of two adiabatic and two isothermal processes. It is of course necessary to design devices that subject these magnetic substances to these processes [129].

In order to understand the principle of adiabatic demagnetisation, we will consider a sample consisting of a paramagnetic, uniform, electrically neutral and insulating single

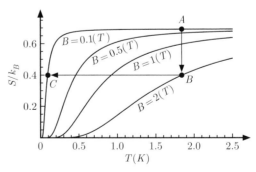

Figure 9.5 The entropy per magnetic moment is expressed as a function of temperature T and magnetic induction field $\boldsymbol{B}$. The numerical values are obtained for a fictitious magnetic moment of 8 Bohr magnetons μ_B that can only be either parallel or antiparallel to the magnetic induction field $\boldsymbol{B}$. The temperature is assumed to be low enough for the contribution of crystal vibrations to be negligible.

chemical substance A in the presence of a magnetic induction field $\boldsymbol{B}$. First, we would like to determine the variation of the entropy density s when the magnetic induction field $\boldsymbol{B}$ increases and the temperature T is held constant (i.e. $T = T_0$). This corresponds to the path (A, B) (Fig. 9.5). Second, we will examine the temperature variation when the magnetic induction field $\boldsymbol{B}$ decreases and the entropy density is held constant (i.e. $s = s_0$), which corresponds to the path (B, C) (Fig. 9.5).

For a homogeneous solid (i.e. $dn_A = 0 \; \forall \, A$), in the absence of electric polarisation (i.e. $\boldsymbol{P} = \boldsymbol{0}$), the differential expression (9.13) for internal energy density $du\,(s, \{n_A\}, \boldsymbol{P}, \boldsymbol{M})$ reduces to,

$$du\,(s, \boldsymbol{M}) = T ds + \boldsymbol{B} \cdot d\boldsymbol{M} \tag{9.105}$$

A paramagnetic material is characterised by the magnetisation $\boldsymbol{M}$ induced by a magnetic induction field $\boldsymbol{B}$. We consider here the Curie magnetic equation of state,

$$\boldsymbol{M} = \frac{C}{\mu_0 \, T} \boldsymbol{B} \tag{9.106}$$

where $C > 0$ is the Curie constant [130].

9.5.1 Isothermal Process

We now examine how the entropy density s depends on the induced magnetisation $\boldsymbol{M}$ when the temperature T is held constant (i.e. $T = T_0$). For a homogeneous solid (i.e. $dn_A = 0$), in absence of electric polarisation (i.e. $\boldsymbol{P} = \boldsymbol{0}$), the expression (9.21) for the free energy density $df\,(T, n_A, \boldsymbol{P}\boldsymbol{M})$ reduces to,

$$df\,(T, \boldsymbol{M}) = -s\,dT + \boldsymbol{B} \cdot d\boldsymbol{M} \tag{9.107}$$

Applying the Schwarz theorem (4.67) to the free energy density $f\,(T, \boldsymbol{M})$ yields,

$$\frac{\partial}{\partial \boldsymbol{M}} \left(\frac{\partial f}{\partial T} \right) = \frac{\partial}{\partial T} \left(\frac{\partial f}{\partial \boldsymbol{M}} \right) \tag{9.108}$$

Using Curie's law (9.106), the differential (9.107) and Schwarz theorem (9.108), we can find the following Maxwell relation,

$$-\frac{\partial s}{\partial \boldsymbol{M}} = \frac{\partial \boldsymbol{B}}{\partial T} = \frac{\mu_0}{C} \boldsymbol{M} \tag{9.109}$$

The entropy density $s\,(T, \boldsymbol{M})$ is a function of two state fields: the temperature T and the induced magnetisation $\boldsymbol{M}$. Thus, the differential $ds\,(T, \boldsymbol{M})$ is given by,

$$ds\,(T, \boldsymbol{M}) = \frac{\partial s}{\partial T} dT + \frac{\partial s}{\partial \boldsymbol{M}} \cdot d\boldsymbol{M} \tag{9.110}$$

At constant temperature T_0, equations (9.110) and (9.109) imply that,

$$ds\,(T_0, \boldsymbol{M}) = -\frac{\mu_0}{C} \boldsymbol{M} \cdot d\boldsymbol{M} \tag{9.111}$$

When integrating the differential (9.111) with respect to $\boldsymbol{M}$ from the initial state $s_i\,(T_0, \boldsymbol{M}_i)$ to the final state $s_f\,(T_0, \boldsymbol{M}_f)$ yields,

$$s_f\,(T_0, \boldsymbol{M}_f) - s_i\,(T_0, \boldsymbol{M}_i) = -\frac{\mu_0}{2\,C}\left(\boldsymbol{M}_f^2 - \boldsymbol{M}_i^2\right) \tag{9.112}$$

To express the entropy density s as a function of the magnetic induction field $\boldsymbol{B}$, we use Curie's law (9.106) and we find,

$$s_f\,(T_0, \boldsymbol{B}_f) - s_i\,(T_0, \boldsymbol{B}_i) = -\frac{C}{2\,\mu_0\,T_0^2}\left(\boldsymbol{B}_f^2 - \boldsymbol{B}_i^2\right) \tag{9.113}$$

Thus, during an isothermal process at temperature T_0, the entropy density s decreases (i.e. $s_f < s_i$) when the norm of the magnetic induction field $\|\boldsymbol{B}\|$ increases (i.e. $\|\boldsymbol{B}_f\| > \|\boldsymbol{B}_i\|$).

9.5.2 Adiabatic Process

Now, let us determine the temperature variation when the magnetic induction field $\boldsymbol{B}$ decreases and the entropy density s is held constant (i.e. $s = s_0$). To obtain the corresponding thermal response coefficient, we apply the cyclic identity of partial derivatives (4.81) to the state functions s, T and $\boldsymbol{B}$ and find,

$$\frac{\partial s}{\partial T}\frac{\partial T}{\partial \boldsymbol{B}} \cdot \frac{\partial \boldsymbol{B}}{\partial s} = -1 \qquad \Rightarrow \qquad \frac{\partial T}{\partial \boldsymbol{B}} = -\frac{\partial T}{\partial s}\frac{\partial s}{\partial \boldsymbol{B}} \tag{9.114}$$

For a homogeneous solid (i.e. $dn_A = 0$), in the absence of electric polarisation (i.e. $\boldsymbol{P} = \boldsymbol{0}$), expression (9.29) for the magnetic free enthalpy density differential $dg_m(T, n_A, \boldsymbol{P}, \boldsymbol{B})$ reduces to,

$$dg_m\,(T, \boldsymbol{B}) = -s\,dT - \boldsymbol{M} \cdot d\boldsymbol{B} \tag{9.115}$$

Applying the Schwarz theorem (4.67) to the magnetic free enthalpy density $g_m\,(T, \boldsymbol{B})$ yields,

$$\frac{\partial}{\partial \boldsymbol{B}}\left(\frac{\partial g_m}{\partial T}\right) = \frac{\partial}{\partial T}\left(\frac{\partial g_m}{\partial \boldsymbol{B}}\right) \tag{9.116}$$

Using Curie's law (9.106), the differential (9.115) and the Schwarz theorem (9.116), we obtain the following Maxwell relation,

$$-\frac{\partial s}{\partial \boldsymbol{B}} = -\frac{\partial \boldsymbol{M}}{\partial T} = \frac{C\,\boldsymbol{B}^2}{\mu_0\,T^2} \tag{9.117}$$

In order to complete our analysis, we will need to determine in expression (9.114) the derivative of T with respect to s when $\boldsymbol{B}$ is constant. This quantity is closely related to the specific heat at constant magnetic induction field, as we shall now see.

For the same homogeneous solid, expression (9.25) for the magnetic enthalpy density differential $dh_m(s, n_A, \boldsymbol{P}, \boldsymbol{B})$ reduces to,

$$dh_m\,(s, \boldsymbol{B}) = T\,ds - \boldsymbol{M} \cdot d\boldsymbol{B} \tag{9.118}$$

The internal energy density differential (9.105) and the magnetic enthalpy density differential (9.118) can be expressed in terms of thermal response coefficients, i.e.

$$du = c_M \, dT + \boldsymbol{L_M} \cdot d\boldsymbol{M} \qquad (9.119)$$

$$dh_{\mathrm{m}} = c_B \, dT + \boldsymbol{L_B} \cdot d\boldsymbol{B} \qquad (9.120)$$

where the thermal response coefficients are defined as,

$$c_M \left(T, \boldsymbol{M} \right) \equiv \left. \frac{\partial u}{\partial T} \right|_{\boldsymbol{M}} = T \frac{\partial s \left(T, \boldsymbol{M} \right)}{\partial T} \,, \qquad \boldsymbol{L_M} \left(T, \boldsymbol{M} \right) \equiv \left. \frac{\partial u}{\partial \boldsymbol{M}} \right|_T \qquad (9.121)$$

$$c_B \left(T, \boldsymbol{B} \right) \equiv \left. \frac{\partial h_{\mathrm{m}}}{\partial T} \right|_{\boldsymbol{B}} = T \frac{\partial s \left(T, \boldsymbol{B} \right)}{\partial T} \,, \qquad \boldsymbol{L_B} \left(T, \boldsymbol{B} \right) \equiv \left. \frac{\partial h_{\mathrm{m}}}{\partial \boldsymbol{B}} \right|_T \qquad (9.122)$$

Applying the Schwarz theorem (4.67) to the internal energy density $u\left(s \left(T, \boldsymbol{M} \right), \boldsymbol{M} \right)$ yields,

$$\frac{\partial}{\partial \boldsymbol{M}} \left(\left. \frac{\partial u}{\partial T} \right|_{\boldsymbol{M}} \right) = \frac{\partial}{\partial T} \left(\left. \frac{du}{d\boldsymbol{M}} \right|_T \right) \qquad (9.123)$$

Using Curie's law (9.106), the differential (9.119) and the Schwarz theorem (9.123), we get the following Maxwell relation,

$$\frac{\partial c_M}{\partial \boldsymbol{M}} = \frac{\partial \boldsymbol{L_M}}{\partial T} \qquad (9.124)$$

Using definition (9.121) for the specific heat density c_M, Maxwell relation (9.109) and Curie's law (9.106), we can express the product of entropy density differential (9.110) and temperature as,

$$T ds = T \frac{\partial s}{\partial T} \, dT + T \frac{\partial s}{\partial \boldsymbol{M}} \cdot d\boldsymbol{M} = c_M \, dT - \boldsymbol{B} \cdot d\boldsymbol{M} \qquad (9.125)$$

Substituting this result in expression (9.105) for $du \left(s, \boldsymbol{M} \right)$, the latter reduces to,

$$du = c_M \, dT \qquad (9.126)$$

Comparing equations (9.119) and (9.126), we find that $\boldsymbol{L_M} = \boldsymbol{0}$. The identity (9.124) implies that the specific heat c_M is only a function of T, i.e. $c_M = c_M \left(T \right)$.

The determination of $c_M \left(T \right)$ requires a statistical approach that goes beyond the framework of thermodynamics. For a temperature so low that the magnetic contribution to the specific heat dominates, we can assume that the specific heat density $c_M \left(T \right)$ has a temperature dependence given by,

$$c_M \left(T \right) = \frac{a}{\mu_0 \, T^2} \qquad (9.127)$$

where $a > 0$ is a constant.

In order to obtain a relation between specific heat densities c_M and c_B, we take the partial derivative of the enthalpy density (9.24) with respect to temperature T and make use of Curie's law (9.106) to obtain,

$$\left. \frac{\partial h_{\mathrm{m}}}{\partial T} \right|_{\boldsymbol{B}} = \left. \frac{\partial u}{\partial T} \right|_{\boldsymbol{M}} - \frac{\partial}{\partial T} \left(\frac{C \, \boldsymbol{B}^2}{\mu_0 \, T} \right)$$

Using the definitions of specific heat densities (9.121) and (9.122), this relation amounts to an extension to magnetism of Mayer's relation,

$$c_B - c_M = \frac{C\,\boldsymbol{B}^2}{\mu_0\,T^2} \tag{9.128}$$

Using equations (9.127) and (9.128), we deduce that c_B is given by [131],

$$c_B\,(T, \boldsymbol{B}) = \frac{a + C\,\boldsymbol{B}^2}{\mu_0\,T^2}$$

The definition (9.122) of c_B implies that,

$$\frac{\partial s}{\partial T} = \frac{c_B}{T} = \frac{a + C\,\boldsymbol{B}^2}{\mu_0\,T^3} \qquad \Rightarrow \qquad \frac{\partial T}{\partial s} = \frac{\mu_0\,T^3}{a + C\,\boldsymbol{B}^2} \tag{9.129}$$

Substituting the partial derivatives (9.129) and (9.117) in equation (9.114) yields,

$$\frac{\partial T}{\partial \boldsymbol{B}} = \frac{C\,T\,\boldsymbol{B}}{a + C\,\boldsymbol{B}^2} \qquad \Rightarrow \qquad \frac{dT}{T} = \frac{1}{2}\frac{d\left(a + C\,\boldsymbol{B}^2\right)}{a + C\,\boldsymbol{B}^2} \tag{9.130}$$

When integrating relation (9.130) with respect to T from the initial state $T_i\,(s_0, \boldsymbol{B}_i)$ to the final state $T_f\,(s_0, \boldsymbol{B}_f)$, we finally obtain,

$$\frac{T_f(s_0, \boldsymbol{B}_f)}{T_i(s_0, \boldsymbol{B}_i)} = \sqrt{\frac{a + C\,\boldsymbol{B}_f^2}{a + C\,\boldsymbol{B}_i^2}} \tag{9.131}$$

Therefore, during a reversible adiabatic process, the temperature T decreases (i.e. $T_f < T_i$) when the norm of the magnetic induction field $\|\boldsymbol{B}\|$ decreases (i.e. $\|\boldsymbol{B}_f\| < \|\boldsymbol{B}_i\|$). Thus, we established the results necessary to understand cooling by adiabatic demagnetisation.

9.6 Worked Solutions

9.6.1 Mayer Relations for a Paramagnetic System

This exercise illustrates the systematic method presented in § 5.4. We apply it to a paramagnetic system. We seek to find the relation between the specific heat at constant magnetic field and the specific heat at constant magnetisation. In order to obtain this result, we need to know the following about this system:

- The magnetisation $\boldsymbol{M}$ is related to the magnetic induction field $\boldsymbol{B}$ through the Curie magnetic equation of state (9.106),

$$\boldsymbol{M} = \frac{C}{\mu_0\,T}\,\boldsymbol{B}$$

  where $C > 0$ and $\mu_0 > 0$ and T is the temperature of the system.
- The internal energy density is a function of the entropy density s and the magnetisation $\boldsymbol{M}$, i.e. $u\,(s, \boldsymbol{M})$ according to relation (9.105),

$$du\,(s, \boldsymbol{M}) = T\,ds + \boldsymbol{B} \cdot d\boldsymbol{M}$$

- The magnetic enthalpy density $h_m(s, \boldsymbol{B})$ is obtained by performing a Legendre transformation of internal energy density with respect to the magnetisation, i.e. $h_m(s, \boldsymbol{B}) = u - \boldsymbol{B} \cdot \boldsymbol{M}$, so that according to relation (9.118),

$$dh_{\mathrm{m}}(s, \boldsymbol{B}) = T\,ds - \boldsymbol{M} \cdot d\boldsymbol{B}$$

- The free energy density $f(T, \boldsymbol{M})$ is obtained by performing a Legendre transformation of internal energy density with respect to the temperature, i.e. $f(T, \boldsymbol{M}) = u - Ts$,

$$df(T, \boldsymbol{M}) = -s\,dT + \boldsymbol{B} \cdot d\boldsymbol{M}$$

Use the method detailed in § 5.4 to determine the difference between the values of the specific heat per unit volume at fixed $\boldsymbol{B}$, respectively at fixed $\boldsymbol{M}$, which are defined by,

$$c_B = \frac{\partial h_m}{\partial T}\bigg|_B \qquad \text{and} \qquad c_M = \frac{\partial u}{\partial T}\bigg|_M$$

Solution:

The magnetic enthalpy density is written as,

$$h_m\Big(s(T, \boldsymbol{B}), \boldsymbol{B}\Big) = u\Big(s\big(T, \boldsymbol{M}(T, \boldsymbol{B})\big), \boldsymbol{M}(T, \boldsymbol{B})\Big) - \boldsymbol{B} \cdot \boldsymbol{M}(T, \boldsymbol{B})$$

The temperature derivative of the magnetic enthalpy density is given by,

$$\frac{\partial h_m}{\partial s}\frac{\partial s}{\partial T} = \frac{\partial u}{\partial s}\frac{ds}{dT} + \frac{\partial u}{\partial \boldsymbol{M}} \cdot \frac{\partial \boldsymbol{M}}{\partial T} - \boldsymbol{B} \cdot \frac{\partial \boldsymbol{M}}{\partial T}$$

where,

$$\frac{ds}{dT} = \frac{\partial s}{\partial T} + \frac{\partial s}{\partial \boldsymbol{M}} \cdot \frac{\partial \boldsymbol{M}}{\partial T} \qquad \text{and} \qquad \boldsymbol{B} = \frac{\partial u}{\partial \boldsymbol{M}}$$

This implies that,

$$\frac{\partial h_m}{\partial s}\frac{\partial s}{\partial T} = \frac{\partial u}{\partial s}\frac{\partial s}{\partial T} + \frac{\partial u}{\partial s}\frac{\partial s}{\partial \boldsymbol{M}} \cdot \frac{\partial \boldsymbol{M}}{\partial T}$$

Since,

$$\frac{\partial h_m}{\partial T}\bigg|_B = \frac{\partial h_m}{\partial s}\frac{\partial s}{\partial T} \qquad \text{and} \qquad \frac{\partial u}{\partial T}\bigg|_M = \frac{\partial u}{\partial s}\frac{\partial s}{\partial T} \qquad \text{and} \qquad T = \frac{\partial u}{\partial s}$$

the previous relation is recast as,

$$\frac{\partial h_m}{\partial T}\bigg|_B = \frac{\partial u}{\partial T}\bigg|_M + T\frac{\partial s}{\partial \boldsymbol{M}} \cdot \frac{\partial \boldsymbol{M}}{\partial T}$$

Therefore,

$$c_B = c_M + T\frac{\partial s}{\partial \boldsymbol{M}} \cdot \frac{\partial \boldsymbol{M}}{\partial T}$$

Applying the Schwarz theorem (4.67) to the free energy density $f(T, \boldsymbol{M})$ we obtain,

$$\frac{\partial}{\partial T}\left(\frac{\partial f}{\partial \boldsymbol{M}}\right) = \frac{\partial}{\partial \boldsymbol{M}}\left(\frac{\partial f}{\partial T}\right)$$

which yields the Maxwell relation,

$$\frac{\partial \boldsymbol{B}}{\partial T} = -\frac{\partial s}{\partial \boldsymbol{M}}$$

and implies that,

$$c_B = c_M - T \frac{\partial \boldsymbol{B}}{\partial T} \cdot \frac{\partial \boldsymbol{M}}{\partial T}$$

According to the Curie magnetic equation of state (9.106),

$$\frac{\partial \boldsymbol{B}}{\partial T} = \frac{\boldsymbol{B}}{T} \qquad \text{and} \qquad \frac{\partial \boldsymbol{M}}{\partial T} = -\frac{C \boldsymbol{B}}{\mu_0 T^2}$$

we finally obtain,

$$c_B = c_M + \frac{C \boldsymbol{B}^2}{\mu_0 T^2}$$

9.6.2 Parallel Plate Capacitor

Consider a capacitor consisting of two metallic parallel plates. The plates of the capacitor carry an electric charge Q and the electrostatic potential difference between the plates is $\Delta\varphi$. The capacitance of the empty capacitor is C_0. Between the plates, we insert a dielectric material of uniform and constant electric susceptibility χ_e. The capacitance of the capacitor then becomes $C = (1 + \chi_e) C_0$.

1. If the electric charge Q of the plates of the capacitor is constant (Fig. 9.6), determine the work W_Q performed in order to insert the dielectric material into the capacitor.
2. If the electrostatic potential difference $\Delta\varphi$ between the plates of the capacitor is maintained constant by an external power supply (Fig. 9.6), determine the work W_φ performed in order to insert the dielectric material into the capacitor.
3. Determine the work W^{ext} performed by the power supply in order to maintain the constant electrostatic potential difference $\Delta\varphi$ between the plates of the capacitor.

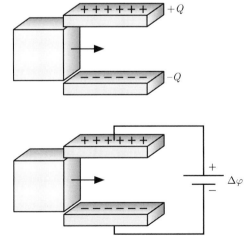

Figure 9.6   A parallel plate capacitors can be characterised either by a constant electric charge Q (top) or a constant electrostatic potential difference $\Delta\varphi$ (bottom).

Solution:

1. *The initial capacitance C_i is that of the capacitor in the absence of a dielectric material and the final capacitance C_f is that of the capacitor in the presence of a dielectric material,*

$$C_i = C_0 \quad and \quad C_f = C_0 (1 + \chi_e)$$

When the electric charge between the plates of the capacitor is kept constant, i.e. $Q = const$, the work W_Q performed in order to insert the dielectric material in the capacitor is equal to the electrostatic energy variation ΔU_{if} between the initial state i and the final state f. Taking into account equation (9.68), the work can be written as,

$$W_Q = \Delta U_{if} = \frac{1}{2} \frac{Q^2}{C_f} - \frac{1}{2} \frac{Q^2}{C_i} = -\frac{1}{2} \frac{\chi_e}{1 + \chi_e} \frac{Q^2}{C_0} < 0$$

Thus, the capacitor attracts the dielectric material and minimises its electrostatic energy.

2. *When the electrostatic potential difference between the plates of the capacitor is kept constant, i.e. $\Delta\varphi = const$, the work W_φ performed in order to insert the dielectric material into the capacitor is equal to the electrostatic enthalpy variation $(\Delta H_e)_{if}$ between the initial state i and the final state f. Taking into account equation (9.69), the work is found to be,*

$$W_\varphi = (\Delta H_e)_{if} = -\frac{1}{2} C_f \Delta\varphi^2 + \frac{1}{2} C_i \Delta\varphi^2 = -\frac{\chi_e}{2} C_0 \Delta\varphi^2 < 0$$

Again, the capacitor attracts the dielectric material to minimise its electrostatic enthalpy.

3. *The work W^{ext} performed by the external power supply to keep constant the electrostatic potential difference $\Delta\varphi$ when the dielectric material is inserted into the capacitor is equal to the variation of the electrostatic energy between the initial state i and the final state f. Using equation (9.98), the work can be written as,*

$$W^{ext} = C_f \Delta\varphi^2 - C_i \Delta\varphi^2 = \chi_e C_0 \Delta\varphi^2 > 0$$

9.6.3 Coil

Consider a coil consisting of a metallic wire of perfect conductivity (no electric resistance). An electric current I flows through the wire of the coil. The magnetic inductance of the empty coil is L_0. We insert into the coil a paramagnetic material of uniform and constant magnetic susceptibility χ_m. Then, the magnetic inductance of the coil becomes $L = (1 - \chi_m)^{-1} L_0$.

1. If the magnetostatic flux Φ is constant in the coil (Fig. 9.7), determine the work W_Φ performed in order to insert the paramagnetic material in the coil.
2. If the electric current I flowing through the coil is kept constant (Fig. 9.7) thanks to an external power supply, determine the work W_I performed in order to insert the paramagnetic material in the coil.
3. Determine the work W^{ext} performed by the external power supply in order to generate a constant electric current I.

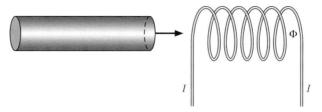

Figure 9.7 An electric current I flows through a coil generating a magnetostatic flux Φ.

Solution:

1. *The initial inductance L_i is that of the coil in absence of a paramagnetic material and the final inductance L_f is that of the coil when the paramagnetic material fills the coil,*

$$L_i = L_0 \qquad and \qquad L_f = \frac{L_0}{1 - \chi_m}$$

When the magnetostatic flux in the coil is kept constant, i.e. $\Phi = const$, the work W_Φ performed in order to insert the paramagnetic material in the coil is equal to the magnetostatic energy variation ΔU_{if} between the initial state i and the final state f. Taking into account equation (9.91), the work can be written as,

$$W_\Phi = \Delta U_{if} = \frac{1}{2}\frac{\Phi^2}{L_f} - \frac{1}{2}\frac{\Phi^2}{L_i} = -\frac{1}{2}\frac{\chi_m}{L_0}\Phi^2 < 0$$

Thus, the coil attracts the paramagnetic material and minimises its magnetostatic energy.

2. *When the electric current flowing through the coil is kept constant, i.e. $I = const$, the work W_I performed in order to insert the paramagnetic material into the coil is equal to the magnetostatic enthalpy variation $(\Delta H_m)_{if}$ between the initial state i and the final state f. Taking into account equation (9.92), the work is found to be,*

$$W_I = (\Delta H_m)_{if} = -\frac{1}{2}L_f I^2 + \frac{1}{2}L_i I^2 = -\frac{1}{2}\left(\frac{\chi_m}{1 - \chi_m}\right)L_0 I^2 < 0$$

Here again, the coil attracts the paramagnetic material to minimise its magnetostatic energy.

3. *The work W^{ext} performed by the external power supply to keep constant the electric current I flowing through the wire when the paramagnetic material is inserted into the coil is equal to the variation of the magnetostatic energy between the initial state i and the final state f. Taking into account equation (9.104), this work is written as,*

$$W^{ext} = L_f I^2 - L_i I^2 = \frac{\chi_m}{1 - \chi_m}L_0 I^2 > 0$$

9.6.4 Rise of a Dielectric Liquid

Consider a dielectric liquid of density ρ, of uniform susceptibility χ_e located in a U-shaped tube. The left end of the liquid is placed between the plates of a capacitor (Fig. 9.8). Part of

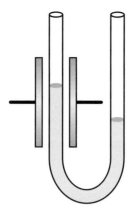

Figure 9.8 A dielectric liquid rises in a *U*-shaped tube under the effect of an inhomogeneous applied electric field.

the liquid is then subjected to an inhomogeneous external electric field $\boldsymbol{E}$ that generates an induced electric polarisation $\boldsymbol{P} = \varepsilon_0 \, \chi_e \, \boldsymbol{E}$.

1. Express the force $\boldsymbol{F}_p$ as an integral on the surface of the liquid.
2. At equilibrium, determine the level difference Δh of the surface of the liquid in the two branches.

Solution:

1. *The force is equal to the integral of the force density* (9.31) *over the volume of the liquid,*

$$\boldsymbol{F}_p = \int_V \boldsymbol{f}_p \, dV = \int_V (\boldsymbol{P} \boldsymbol{\nabla} \, \boldsymbol{E}) \, dV = \frac{\varepsilon_0}{2} \, \chi_e \int_V \left(\boldsymbol{\nabla} \, \boldsymbol{E}^2 \right) \, dV$$

A vector analysis theorem states that the integral over the volume V of a gradient of a scalar field is equal to the integral of this scalar field over the surface S of the solid. Thus,

$$\boldsymbol{F}_p = \frac{\varepsilon_0}{2} \, \chi_e \int_S \boldsymbol{E}^2 \, d\boldsymbol{S} = \frac{1}{2} \int_S (\boldsymbol{P} \cdot \boldsymbol{E}) \, d\boldsymbol{S} = \frac{1}{2} \, (\boldsymbol{P} \cdot \boldsymbol{E}) \, \boldsymbol{S}$$

where d$\boldsymbol{S}$ is the vector normal to an infinitesimal surface element and $\boldsymbol{S}$ is the vector normal to the surface element of the liquid positively defined upwards.

2. *At equilibrium, the gravitation potential energy of the liquid in the left part of the tube compared to its value in the right part of the tube is equal to the density of electric dipolar energy* (9.10),

$$\rho g \Delta h = \frac{1}{2} \boldsymbol{P} \cdot \boldsymbol{E}$$

where g is the acceleration of the gravitational field. Thus,

$$\Delta h = \frac{\boldsymbol{P} \cdot \boldsymbol{E}}{2 \rho g}$$

In passing, we note that we could describe the rise of a paramagnetic liquid in a coil wound around one branch of the U-tube where,

$$\Delta h = \frac{\boldsymbol{M} \cdot \boldsymbol{B}}{2\rho g}$$

9.6.5 Paramagnetic Scale

Consider a cylindrical paramagnetic solid of height h, section S and uniform magnetic susceptibility χ_m. The cylinder is placed between the poles of a magnet shaped in order to deliberately create an inhomogeneous of magnetic field. Thus, the solid is subjected to an inhomogeneous magnetic external field $\boldsymbol{B}$ generating an induced magnetisation $\boldsymbol{M} = \mu_0^{-1} \chi_m \boldsymbol{B}$. The force $\boldsymbol{F_m}$ acting on the cylinder is counterbalanced by a weight on a scale with beams of equal lengths (Fig. 9.9).

1. Express the force $\boldsymbol{F_m}$ as an integral on the surface of the solid.
2. At equilibrium, determine the excess mass m of the counterweight of the scale that compensates the magnetic force.

Solution:

1. *The force is equal to the integral of the force density* (9.34) *over the volume of the solid,*

$$\boldsymbol{F_m} = \int_V \boldsymbol{f_m} \, dV = \int_V (\boldsymbol{M} \boldsymbol{\nabla} \boldsymbol{B}) \, dV = \frac{1}{2\mu_0} \chi_m \int_V (\boldsymbol{\nabla} \boldsymbol{B}^2) \, dV$$

A vector analysis theorem states that the integral on the volume V of the gradient of a scalar field is equal to the integral of this scalar field over the surface S of the solid. Thus,

$$\boldsymbol{F_m} = \frac{1}{2\mu_0} \chi_m \int_S \boldsymbol{B}^2 d\boldsymbol{S} = \frac{1}{2} \int_S (\boldsymbol{M} \cdot \boldsymbol{B}) \, d\boldsymbol{S}$$

$$= \frac{1}{2} (\boldsymbol{M_+} \cdot \boldsymbol{B_+} - \boldsymbol{M_-} \cdot \boldsymbol{B_-}) \, \boldsymbol{S}$$

where $d\boldsymbol{S}$ is the vector normal to an infinitesimal surface element, $\boldsymbol{S}$ is the vector normal to the top surface element of the solid, $\boldsymbol{B_+}$ and $\boldsymbol{B_-}$ are the values of the magnetic

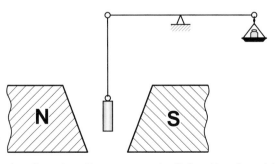

Figure 9.9   A scale is used to measure the pull experienced by a paramagnetic cylinder subjected to an inhomogeneous magnetic field.

induction field on the top and bottom horizontal surfaces of the solid, M_+ and M_- are the values of the magnetisation at the top and bottom horizontal surfaces of the solid.

2. *At equilibrium, the force F_m is equal and opposite to the weight of the counterweight of mass m,*

$$mg = \frac{1}{2}\left(M_+ \cdot B_+ - M_- \cdot B_-\right) S$$

where g is the acceleration of the gravitational field. Thus,

$$m = \frac{\left(M_+ \cdot B_+ - M_- \cdot B_-\right) S}{2g}$$

Exercises

9.1 Vapour Pressure of a Paramagnetic Liquid

At constant temperature T, show that the vapour pressure $p(T, B)$ of a paramagnetic liquid has the following dependence on the magnetic induction field B,

$$p(T, B) - p(T, 0) = -p(T, 0)\,\frac{c\,B^2}{2\,\mu_0\,R\,T^2}$$

in the small field limit, i.e. $cB \ll \mu_0 R T^2$. The liquid phase satisfies the Curie equation of state (9.106),

$$m_\ell = \frac{c\,B}{\mu_0\,T}$$

where m_ℓ is the molar magnetisation. The molar volume v_ℓ of the liquid phase can be neglected compared to the molar volume v_g of the gaseous phase, i.e. $v_\ell \ll v_g$.

9.2 Magnetic-Field Induced Adsorption or Desorption

We consider a substance that can be either in a gaseous phase inside a rigid container or in an absorbed phase on the surface of a substrate inside it. We analyse the equilibrium reached by the substance in the gaseous and adsorbed phases. The entire system is at a fixed temperature T. We assume that when the substance is adsorbed, it acquires a magnetisation M. The reason whether and why this might happen has been the subject of much research [132]. The gas has no magnetisation. Use relation (8.58) for the pressure dependence of the chemical potential of the gaseous phase and assume that the chemical potential of the adsorbed phase is independent of pressure. Find the dependence on magnetic induction field B of the pressure p.

9.3 Magnetic Battery

A U-shaped tube contains a solution of paramagnetic salt, such as $CoSO_4$. One side of the U-shaped tube, considered as a subsystem 1, is in a magnetic induction field B, the other side, considered as a subsystem 2, is at $B = 0$. The initial concentration of paramagnetic Co^{++} salt is c_0 and the ideal mixture relation applies. Find the electric potential difference $\Delta\varphi$ in terms of the magnetic induction field B and the molar

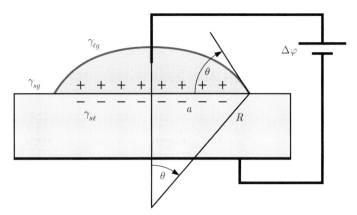

Figure 9.10 Contact angle θ for a drop for an applied electrostatic potential difference $\Delta\varphi$.

magnetisation $\boldsymbol{m}_1$ of electrodes made of the same material dipped in the the salt solution. Determine also the concentration ratio of paramagnetic Co^{++} salt in both subsystems at equilibrium in terms of the magnetic induction field $\boldsymbol{B}$.

9.4 Electrocapilarity

Electrowetting is a means of driving microfluidic motion by modifying the surface tension [133]. Here, we determine the effect of an electrostatic potential difference $\Delta\varphi$ applied between the solid and liquid phases on the surface tension. Since the liquid is now electrically charged a double layer forms at the interface between the solid and the liquid phases. The capacitance per unit area c of this double layer is supposed to be known. Determine the electrically charged surface tension $\bar{\gamma}_{s\ell}$ in terms of the neutral surface tension $\gamma_{s\ell}$.

9.5 Magnetic Clausius–Clapeyron Equation

Some magnetic materials undergo a first-order phase transition when a magnetic induction field $\boldsymbol{B}$ is applied. These materials are of interest because some of them have a strong magnetocaloric effect, which could be used for refrigeration applications [134]. In these materials, at any given temperature T, there is a critical magnetic induction field $\boldsymbol{B}$ in which the two phases coexist. The two phases, labelled 1 and 2, are characterised by their molar magnetisation $\boldsymbol{m}_1$ and $\boldsymbol{m}_2$ and their molar entropies per unit volume s_1 and s_2. Show that the temperature derivative of the magnetic induction field $\boldsymbol{B}$ satisfies the magnetic Clausius–Clapeyron equation,

$$\frac{dB}{dT} = -\left(s_1 - s_2\right)\left(\boldsymbol{m}_1 - \boldsymbol{m}_2\right)^{-1}$$

which is analogous to the Clausius–Clapeyron equation (6.49).

9.6 Magnetocaloric Effect

The magnetocaloric effect is defined by the temperature variation observed in a material under adiabatic conditions when the applied magnetic induction field $\boldsymbol{B}$ varies. The effect becomes particularly large when the material undergoes a phase transition due to the magnetic induction field variation. An application of this effect was

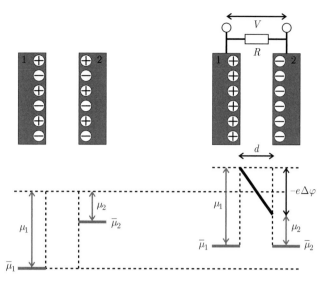

Figure 9.11 A parallel plate capacitor is made of metals 1 and 2 which have electrochemical potentials $\bar{\mu}_1$ and $\bar{\mu}_2$ and chemical potentials μ_1 and μ_2. On the left, the plates are not connected and the electrochemical potentials are not equal. On the right, when conduction electrons of charge e (sign included) are allowed to flow, the electrochemical potentials are equal. If the spacing d varies over time, electric charges flow through the resistance R and a voltage V can be measured.

considered for space exploration [135]. The applied magnetic induction field may be produced by permanent magnets [136]. Show that the infinitesimal temperature variation dT during an adiabatic process characterised by an infinitesimal magnetic induction field variation $d\boldsymbol{B}$ at constant pressure is given by,

$$dT = -\frac{T}{c_B}\frac{\partial \boldsymbol{M}}{\partial T}\cdot d\boldsymbol{B}$$

where c_B is the specific heat per unit volume at constant magnetic induction field $\boldsymbol{B}$.

9.7 Kelvin Probe

Two parallel plates are made of two metals 1 and 2 inside a vacuum chamber (Fig. 9.11). The chemical potentials of electrons in metals 1 and 2 are μ_1 and μ_2. When contact is established between the two metals, conduction electrons flow until equilibrium is reached. As a result of this electron flow, there is a charge Q and $-Q$ at the surface of the electrodes. These electric charges produce an electric field $\boldsymbol{E}$ between the plates. According to relation (9.67) the corresponding electrostatic potential difference $\Delta\varphi$ is related to the charge Q by,

$$Q = C\,\Delta\varphi$$

where C is the capacitance of the capacitor. The plates have a surface area A and they are separated by a distance d. It can be shown that the capacitance C is given by,

$$C = \frac{\varepsilon_0 A}{d}$$

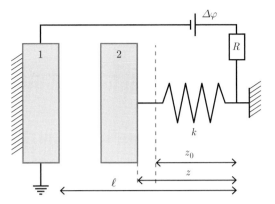

Figure 9.12   A parallel plate capacitor is made of metals 1 and 2. The metallic plate 1 is fixed mechanically. The metallic plate 2 vibrates under the effect of a spring of elastic constant k. When it vibrates, conduction electrons flow through the resistance R and the spring. The electrostatic potential difference $\Delta\varphi$ is maintained constant by the power supply.

where ε_0 is the electric permittivity of vacuum. For practical applications, the flow of electric charges is too difficult to detect. It is easier to make the separation distance d oscillate in time, which causes C to vary, hence Q oscillates in time, which implies that an electric current of intensity I flows through the resistance R. Determine the voltage V detected between the plates as a function of $\dot{d}$, μ_1 and μ_2.

9.8   Electromechanical Circuit

In the electromechanical system presented here (Fig. 9.12), a metallic plate 1 is fixed mechanically and a metallic plate 2 vibrates under the effect of a spring of elastic constant k, of length z and of natural length z_0. The distance between the plates is $\ell - z$ and their surface area is A. The capacitance of the parallel plate capacitor is,

$$C = \frac{\varepsilon_0 A}{\ell - z}$$

The electrostatic potential difference $\Delta\varphi$ between the plates is maintained constant by a power supply, which generates a flow of conduction electrons through the circuit. The chemical potential of the conduction electrons on plate 1 and 2 are μ_1 and μ_2. At equilibrium, we assume that the deviation of the plate $\Delta z = z - z_0$ is small enough, i.e. $\Delta z \ll z_0$. Determine this deviation Δz to first-order.

10

Thermodynamics of Continuous Media

Carl Henry Eckart, 1902–1973

C. H. Eckart obtained a PhD in physics from the University of Princeton in 1925 and became a professor at the University of Chicago in 1928. He was mainly interested in the mathematical symmetries of quantum physics and developed with Eugene Wigner a theorem named after them both that relates symmetry groups and conservation laws.

10.1 Historical Introduction

The development of thermodynamics began during the industrial revolution. It was driven by the necessity to improve the efficiency of steam engines. Ever since the pioneering work of Carnot [137] in 1824 and up to the 1950s, thermodynamics was a theory essentially restricted to the description of equilibrium states and transformations between them [138]. Hence, thermodynamics could have been called thermostatics.

After the pioneering work undertaken by Eckart in 1940 [139, 140], Stückelberg [14] reformulated thermodynamics in the 1950s as a theory based on two fundamental laws and formulated in such a way that he could express the time evolution of a thermodynamic system in terms of a set of first-order differential equations. Thereby, he extended the thermostatic theory of equilibrium states and obtained a genuine thermodynamic theory, that is, a theory describing the time evolution of thermodynamic systems and their behaviour in the neighbourhood of equilibrium states.

Furthermore, he applied quite systematically the principle of Galilean invariance. According to this principle, in classical physics, the laws governing the time evolution of a physical system are independent of the choice of inertial frame of reference. In the limit where the relative velocity between two inertial frames of reference is low compared to the speed of light in vacuum, the transformations relating the space-time coordinates of a physical event described with respect to two frames of reference are the Galilean transformations. Therefore, the laws of classical physics are invariant under a Galilean transformation. This is called the ***Galilean invariance principle***. This invariance principle has many consequences. For example, we will see in this chapter how it leads to the mass conservation that Lavoisier postulated.

We will follow Stückelberg's approach and establish a thermodynamic description of continuous media by applying the first and second laws of thermodynamics locally. In order to do this, we will introduce the notion of continuity equations for thermodynamic fields, and we will derive differential equations that describe the time evolution of local systems. However, we will not consider any rotation of local systems on themselves.

The key result of this chapter is that the entropy production rate can be expressed as a sum of terms, which each consist of the product of a current and a generalised force. The rather demanding formal approach presented here lays the foundation for the next chapter, where transport phenomena are analysed.

10.2 Continuity Equations

10.2.1 Definition

In Chapter 1, we defined the state of a thermodynamic system by a set of variables. Now, we consider here a system with no intrinsic rotation, hence with no angular velocity around its centre of mass, i.e. $\omega(t) = \mathbf{0}$. For such a system, the first law requires the existence of the state function momentum $\boldsymbol{P}(t)$ and the second law requires the existence of the state function entropy $S(t)$. We choose to take $\boldsymbol{P}(t)$ and $S(t)$ as state variables. Thus, the state of the system is defined by the variables momentum $\boldsymbol{P}(t)$, entropy $S(t)$ and by a finite set of n extensive time dependent variables $X_1(t)$, $X_2(t)$, ..., $X_n(t)$ [109]. This means that the state of the thermodynamic system is formally defined by a set of extensive state variables, i.e. $\{\boldsymbol{P}(t), S(t), X_1(t), X_2(t), \ldots, X_n(t)\}$.

Certain physical quantities $F(t)$ are 'state functions', i.e.

$$F(t) \equiv F\Big(\boldsymbol{P}(t), S(t), X_1(t), X_2(t), \ldots, X_n(t)\Big) \tag{10.1}$$

Although a piece of matter consists of a large finite number of elementary microscopic constituents, it can be considered in an empirical approach as a continuous medium. We describe the dynamics of this continuous medium with respect to an inertial frame of reference [141, 142]. Thus, the enclosure of the thermodynamic system is defined in terms of spatial points and not material points. Spatial points are points that belong to the inertial frame of reference and material points correspond to infinitesimal elements of matter that may move with respect to this frame of reference. In this description, the volume of a thermodynamic system $V(t)$ and the surface of the enclosure $S(t)$ are time dependent. At any given time t, the neighbourhood of every spatial point $x \in V(t)$ constitutes a local thermodynamic system. This local system contains a large enough number of elementary constituents to be insensitive to statistical fluctuations, but it is small enough to be considered as infinitesimal. Thus, we consider local infinitesimal systems which are simple systems that can be described as material points of variable mass. We restrict our analysis to continuous media devoid of shear and vorticity.

The thermodynamic state of the system is characterised by the momentum density field $p(x, t)$, the entropy density field $s(x, t)$ and by a set of n fields $x_1(x, t), x_2(x, t), \ldots, x_n(x, t)$ that are specific quantities. The local thermodynamic state is formally defined by a set of state fields, i.e. $\{p(x, t), s(x, t), x_1(x, t), \ldots, x_n(x, t)\}$.

Since the momentum $P(t)$ is an extensive state variable, it is related to the momentum density $p(x, t)$ by an integral over the volume $V(t)$, i.e.

$$P(t) = \int_{V(t)} dV(x)\, p(x, t) \tag{10.2}$$

where the volume of the infinitesimal local system $dV(x)$ is time independent, because it is defined with respect to the inertial frame of reference. Similarly, the entropy $S(t)$ is an extensive state variable related to the entropy density $s(x, t)$ by an integral over the volume $V(t)$, i.e.

$$S(t) = \int_{V(t)} dV(x)\, s(x, t) \tag{10.3}$$

The extensive state variables $X_i(t)$, where $i \in \{1, \ldots, n\}$, are related to the corresponding densities $x_i(x, t)$ by an integral over the volume $V(t)$, i.e.

$$X_i(t) = \int_{V(t)} dV(x)\, x_i(x, t) \tag{10.4}$$

For an arbitrary extensive scalar physical quantity described by the scalar state function $F(t)$, the field $f(x, t)$ describing the density of this quantity on the scale of the local system is a function of the state fields,

$$f(x, t) \equiv f\Big(p(x, t), s(x, t), x_1(x, t), \ldots, x_n(x, t)\Big) \tag{10.5}$$

It is related to $F(t)$ by an integral over the volume $V(t)$, i.e.

$$F(t) = \int_{V(t)} dV(x)\, f(x, t) \tag{10.6}$$

A **continuity equation** of an extensive state function $F(t)$ is the equation that describes the local time evolution of the corresponding state function density $f(x, t)$. In order to obtain the continuity equation of the extensive scalar state function $F(t)$, we perform the mathematical variation of it. The mathematical notion of variation is a generalisation of differentiation in the sense that it is done with respect to a function instead of a variable. The integral expression (10.6) of the extensive state function $F(t)$ is a function of the state field function $f(x, t)$ and of the infinitesimal integration volume $dV(x)$, which are functions of space x and time t coordinates. The infinitesimal element $\delta F(t)$ is the sum of an integral over the volume $V(t)$ and an integral over the surface $S(t)$ of the system,

$$\delta F(t) = \int_{V(t)} dV(x)\, \delta f(x, t) + \int_{S(t)} \delta dV(x, t)\, f(x, t) \tag{10.7}$$

The first integrand represents the local variation of the state field function $\delta f(x, t)$ and the second integrand represents the infinitesimal local volume variation $\delta dV(x, t)$ on the surface $S(t)$. The latter is the scalar product of the displacement vector field $\delta r(x, t)$ and the surface element vector field $dS(x)$,

$$\delta dV(x, t) = dS(x) \cdot \delta r(x, t) \tag{10.8}$$

The time derivative of the extensive state function $\dot{F}(t)$, the partial derivative with respect to time of the state field function $\partial_t f(x, t)$ and the material velocity field $v(x, t)$ are defined as,

$$\dot{F}(t) \equiv \frac{dF(t)}{dt} = \lim_{\delta t \to 0} \frac{\delta F(t)}{\delta t}$$
$$\partial_t f(x, t) \equiv \frac{\partial f(x, t)}{\partial t} = \lim_{\delta t \to 0} \frac{\delta f(x, t)}{\delta t} \tag{10.9}$$
$$v(x, t) \equiv \lim_{\delta t \to 0} \frac{\delta r(x, t)}{\delta t}$$

Thus, the time derivative of the extensive state function $\dot{F}(t)$ is given by,

$$\dot{F}(t) = \int_{V(t)} dV(x)\, \partial_t f(x, t) + \int_{S(t)} dS(x) \cdot v(x, t) f(x, t) \tag{10.10}$$

Using the divergence theorem [143], this time derivative (10.10) can be expressed as,

$$\dot{F}(t) = \int_{V(t)} dV(x) \left[\partial_t f(x, t) + \nabla \cdot \Big(f(x, t)\, v(x, t) \Big) \right] \tag{10.11}$$

The **divergence** is defined as,

$$\nabla \cdot \Big(f(x, t)\, v(x, t) \Big) \equiv \partial_i \Big(f(x, t)\, v^i(x, t) \Big)$$

where we used the compact notation $\partial_i \equiv \partial/\partial x^i$ for the partial derivative with respect to the spatial direction x^i and the Einstein summation convention (10.12). The Einstein summation convention consists in removing the summation symbol $\sum$ when writing a sum of vectorial or tensorial components and replacing it by identical alternated subscript and exponent. The scalar product of two vectors x and y is then written in components as,

$$\boldsymbol{x} \cdot \boldsymbol{y} = \sum_{i=1}^{3} x_i y^i \equiv x_i y^i \tag{10.12}$$

where the coefficients x_i correspond to the components of a row vector and the coefficients y^i to the components of a column vector. The sum of the product of these components yields a scalar. The component j of the vector obtained by the scalar product of a vector $\boldsymbol{x}$ and a tensor y (i.e. a mathematical object represented by a 3×3 matrix) is written in components as,

$$(\boldsymbol{x} \cdot \mathsf{y})^j = \sum_{i=1}^{3} x_i y^{ij} \equiv x_i y^{ij} \tag{10.13}$$

The trace of the scalar product of two tensors x and y is written in components as,

$$\mathsf{x} : \mathsf{y} = \sum_{i,j=1}^{3} x_{ij} y^{ji} \equiv x_{ij} y^{ji} \tag{10.14}$$

where the double sum of the product of these components yields a scalar.

Previously in equation (10.11), the time derivative of the state function $F(t)$ has been entirely established on a mathematical level. We now have to describe the physical causes of the variation rate of the state function $F(t)$, of which there are two. The first is the flux of $F(t)$ that flows through the surface of the enclosure of the system. This flux is described locally by a current density field $\boldsymbol{j}_f(\boldsymbol{x}, t)$. The second is a production or destruction process of $F(t)$ inside the system. It is described locally by the source density field $\pi_f(\boldsymbol{x}, t)$.

The source density field $\pi_f(\boldsymbol{x}, t)$ can have internal or external origins. In the next section, we shall see for example, that when considering the continuity equation for the amount of a chemical substance A, the source density (10.25) has an internal origin. However, for the energy, the corresponding source density (10.36) must have an external origin.

By convention, the current density field $\boldsymbol{j}_f(\boldsymbol{x}, t)$ points in the direction in which the extensive quantity is transported, the infinitesimal surface element $d\boldsymbol{S}(\boldsymbol{x})$ points outwards of the enclosure (Fig. 10.1) and the source density field $\pi_f(\boldsymbol{x}, t)$ is defined positively when it contributes to an increase of $F(t)$. Thus, from the standpoint of physical processes, the integral expression for the time derivative of the state function $F(t)$ is given by,

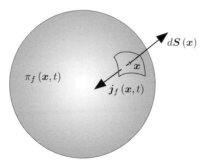

Figure 10.1 A schematic representation of the global thermodynamic system, of the source density $\pi_f(\boldsymbol{x}, t)$, of the orientation of the infinitesimal surface element $d\boldsymbol{S}(\boldsymbol{x})$ and of the current density $\boldsymbol{j}_f(\boldsymbol{x}, t)$.

$$\dot{F}(t) = \int_{V(t)} dV(\mathbf{x}) \, \pi_f(\mathbf{x},t) - \int_{S(t)} d\mathbf{S}(\mathbf{x}) \cdot \mathbf{j}_f(\mathbf{x},t) \tag{10.15}$$

Using the divergence theorem, this time derivative (10.15) can be expressed as,

$$\dot{F}(t) = \int_{V(t)} dV(\mathbf{x}) \left[\pi_f(\mathbf{x},t) - \boldsymbol{\nabla} \cdot \mathbf{j}_f(\mathbf{x},t) \right] \tag{10.16}$$

By equating the mathematical expression (10.11) and the physical expression (10.16) for the time derivative of the state function $\dot{F}(t)$, we find that $f(\mathbf{x},t)$ must obey the local continuity equation given by,

$$\partial_t f(\mathbf{x},t) + \boldsymbol{\nabla} \cdot \Big(f(\mathbf{x},t)\, \mathbf{v}(\mathbf{x},t) \Big) = \pi_f(\mathbf{x},t) - \boldsymbol{\nabla} \cdot \mathbf{j}_f(\mathbf{x},t) \tag{10.17}$$

The term $f(\mathbf{x},t)\,\mathbf{v}(\mathbf{x},t)$ is the **convective** current density field and $\mathbf{j}_f(\mathbf{x},t)$ is the **conductive** current density field.

Note that the continuity equation (10.17) is expressed in an inertial frame of reference. We now wish to obtain a continuity equation which is expressed in a frame of reference located on the centre of mass of the local system. For this puprose, we introduce the time derivative of the state field function $\dot{f}(\mathbf{x},t)$ defined as,

$$\dot{f}(\mathbf{x},t) = \partial_t f(\mathbf{x},t) + (\mathbf{v}(\mathbf{x},t) \cdot \boldsymbol{\nabla}) f(\mathbf{x},t) \tag{10.18}$$

where $\mathbf{v}(\mathbf{x},t) \cdot \boldsymbol{\nabla} \equiv v^i(\mathbf{x},t)\, \partial_i$. The time derivative $\dot{f}(\mathbf{x},t)$, also called the **material derivative** [144], is invariant under the action of a Galilean transformation (§ 10.4.1). It corresponds to the time derivative in the frame of reference of the centre of mass of the local system. Using the definition (10.18) and the vectorial identity,

$$\boldsymbol{\nabla} \cdot \Big(f(\mathbf{x},t)\, \mathbf{v}(\mathbf{x},t) \Big) = f(\mathbf{x},t)\, \boldsymbol{\nabla} \cdot \mathbf{v}(\mathbf{x},t) + (\mathbf{v}(\mathbf{x},t) \cdot \boldsymbol{\nabla}) f(\mathbf{x},t) \tag{10.19}$$

the continuity equation (10.17) can be expressed in the local frame of reference as,

$$\dot{f}(\mathbf{x},t) + (\boldsymbol{\nabla} \cdot \mathbf{v}(\mathbf{x},t))\, f(\mathbf{x},t) + \boldsymbol{\nabla} \cdot \mathbf{j}_f(\mathbf{x},t) = \pi_f(\mathbf{x},t) \tag{10.20}$$

which is the continuity equation for a scalar state function $F(t)$. The generalisation to a vectorial state function $\mathbf{F}(t)$ is straightforward. The continuity equation for such a state function is given by,

$$\dot{\mathbf{f}}(\mathbf{x},t) + (\boldsymbol{\nabla} \cdot \mathbf{v}(\mathbf{x},t))\, \mathbf{f}(\mathbf{x},t) + \boldsymbol{\nabla} \cdot \mathrm{j}_f(\mathbf{x},t) = \boldsymbol{\pi}_f(\mathbf{x},t) \tag{10.21}$$

where $\mathbf{f}(\mathbf{x},t)$ is the density of the vectorial state field function, $\mathrm{j}_f(\mathbf{x},t)$ is the tensorial current density field and $\boldsymbol{\pi}_f(\mathbf{x},t)$ is the vectorial source density field. The divergence of the tensorial current density field $\boldsymbol{\nabla} \cdot \mathrm{j}_f(\mathbf{x},t)$ is a vector, the components of which are given by,

$$\big(\boldsymbol{\nabla} \cdot \mathrm{j}_f(\mathbf{x},t) \big)^i = \partial_k j_f^{ki}(\mathbf{x},t) \tag{10.22}$$

where we used the implicit Einstein summation convention. It is important to note that the continuity equations (10.20) and (10.21) have this structure even if the corresponding state field functions depend on the choice of frame of reference.

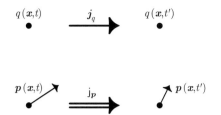

Figure 10.2 Time variation of the scalar $q\,(\textbf{\textit{x}}, t)$ caused by the vector $\textbf{\textit{j}}_q\,(\textbf{\textit{x}}, t)$ and time variation of the norm and direction of the vector $\textbf{\textit{p}}\,(\textbf{\textit{x}}, t)$ caused by the tensor $\mathrm{j}_{\textbf{\textit{p}}}\,(\textbf{\textit{x}}, t)$.

To illustrate the notion of current density, we examine now two physical examples where the velocity field is uniform, i.e. $\nabla \cdot \textbf{\textit{v}}\,(\textbf{\textit{x}}, t) = 0$, but not constant. This means that the norm and the orientation of the velocity field can change globally over time.

First, we consider the electric charge, an extensive scalar physical quantity and its corresponding density $q\,(\textbf{\textit{x}}, t)$. In the absence of a source density, i.e. $\pi_q\,(\textbf{\textit{x}}, t) = 0$, the continuity equation (10.20) implies that the time variation of the scalar $q\,(\textbf{\textit{x}}, t)$, due to the matter displacement, is caused physically by the vector $\textbf{\textit{j}}_q\,(\textbf{\textit{x}}, t)$ (Fig. 10.2). In an arbitrary frame of reference, the electric current density $\textbf{\textit{j}}_q\,(\textbf{\textit{x}}, t)$ is represented by a 3-component vector written in components as $j_q^i\,(\textbf{\textit{x}}, t)$.

Second, we consider the momentum, an extensive vectorial physical quantity, and its corresponding density $\textbf{\textit{p}}\,(\textbf{\textit{x}}, t)$. In the absence of a source density, i.e. $\pi_{\textbf{\textit{p}}}\,(\textbf{\textit{x}}, t) = \textbf{0}$, the continuity equation (10.21) implies that the time variation of the vector $\textbf{\textit{p}}\,(\textbf{\textit{x}}, t)$, due to the matter displacement, is caused physically by the tensor $\mathrm{j}_{\textbf{\textit{p}}}\,(\textbf{\textit{x}}, t)$ (Fig. 10.2). In an arbitrary frame of reference, the momentum current density $\mathrm{j}_{\textbf{\textit{p}}}\,(\textbf{\textit{x}}, t)$ is represented by a 3×3 matrix written in components as $j_p^{ki}\,(\textbf{\textit{x}}, t)$.

10.2.2 Thermodynamic State Field

The state of a thermodynamic system is defined by a set of state variables. In a continuous system, the local state is defined by a set of state fields. Thus, when considering a continuous medium consisting of electrically charged chemical substances interacting with an external electrostatic field, we assume that the local state is entirely defined by four types of state fields which are represented by the following specific quantities:

1. the momentum density $\textbf{\textit{p}}\,(\textbf{\textit{x}}, t)$,
2. the entropy density $s\,(\textbf{\textit{x}}, t)$,
3. the set of densities $\{n_A\,(\textbf{\textit{x}}, t)\}$ of r chemical substances
     where $A = 1, \ldots, r$,
4. the electric charge density $q\,(\textbf{\textit{x}}, t)$.

In other words, the thermodynamic state of the continuous medium is defined by a set of state fields $\{\textbf{\textit{p}}\,(\textbf{\textit{x}}, t)\,, s\,(\textbf{\textit{x}}, t)\,, \{n_A\,(\textbf{\textit{x}}, t)\}, q\,(\textbf{\textit{x}}, t)\}$. The other physical fields that describe the thermodynamics of the continuous medium are defined locally and are functions of these state fields. Certain fields are of particular interest in the following developments:

1. the velocity $v\,(p, s, \{n_A\}, q)$,
2. the mass density $m\,(p, s, \{n_A\}, q)$,
3. the energy density $e\,(p, s, \{n_A\}, q)$,
4. the internal energy density $u\,(p, s, \{n_A\}, q)$,

In order to simplify the notation, we do not mention explicitly the dependence of the state fields p, s, $\{n_A\}$ and q on space x and time t. In the remainder of this chapter, we will not explicitly mention this dependence. However, it remains implicit.

10.2.3 State Field Continuity Equations

The analytical form of the continuity equations for the extensive physical quantities is given by equation (10.20) for scalar fields and by equation (10.21) for vector fields. We shall now apply these equations to several thermodynamic state fields and characterise their corresponding source densities.

In order to satisfy the evolution condition of the second law (2.1) applied locally to our continuous system, the entropy source density π_s must be positive definite, i.e.

$$\pi_s \geqslant 0 \tag{10.23}$$

The entropy continuity equation is then given by,

$$\dot{s} + (\nabla \cdot v)\, s + \nabla \cdot j_s = \pi_s \geqslant 0 \tag{10.24}$$

where j_s is the entropy current density.

The source density π_A of a chemical substance A participating in chemical reactions a is proportional to the stoichiometric coefficients ν_{aA},

$$\pi_A = \sum_{a=1}^{n} \omega_a \nu_{aA} \tag{10.25}$$

where ω_a is the local reaction rate density of chemical reaction a. The sum extends to the n possible chemical reactions in the system. If the reaction rate density ω_a is positive, then the substances with a positive stoichiometric coefficient ν_{aA} are produced and those with a negative stoichiometric coefficient ν_{aA} are consumed. It is the opposite if the reaction rate density is negative. The continuity equation for the chemical substance A is given by,

$$\dot{n}_A + (\nabla \cdot v)\, n_A + \nabla \cdot j_A = \sum_{a=1}^{n} \omega_a \nu_{aA} \tag{10.26}$$

where j_A is the current density of the chemical substance A.

The electric charge continuity equation is given by,

$$\dot{q} + (\nabla \cdot v)\, q + \nabla \cdot j_q = 0 \tag{10.27}$$

where j_q is the conductive electric current. It is related to the total electric current density j by,

$$j = j_q + q\,v \tag{10.28}$$

Here, $q\,v$ represents the convective electric current density. A convective current is due to the displacement of the centre of mass of a local fluid element and a conductive current is a matter displacement within the local centre of mass frame of reference. The momentum source density is due to the forces associated to the external fields only, i.e.

$$\pi_p = f^{\text{ext}} \tag{10.29}$$

where f^{ext} is the force density of the external fields. The momentum current j_p is a tensor (i.e. a mathematical object represented by a 3×3 matrix) that is defined as the opposite of the stress tensor τ accounting for mechanical effects [145],

$$j_p = -\tau \tag{10.30}$$

The momentum continuity equation is then given by,

$$\dot{p} + (\nabla \cdot v)\, p - \nabla \cdot \tau = f^{\text{ext}} \tag{10.31}$$

where the components of the divergence of the symmetric stress tensor $\nabla \cdot \tau$ are given by,

$$(\nabla \cdot \tau)^i = \frac{1}{2}\, \partial_j \left(\tau^{ji} + \tau^{ij} \right)$$

The conservation of angular momentum, of which the density is defined as $\ell = r \times p$, implies that the stress tensor τ is represented by a 3×3 symmetric matrix [109] (§ 10.4.3).

The constitutive relation of mechanics is written as,

$$p = m\left(s, \{n_A\}, q\right) v\left(p, s, \{n_A\}, q\right) \tag{10.32}$$

Taking into account equation (10.32), the momentum continuity equation (10.31) becomes,

$$m\dot{v} + \left(\dot{m} + (\nabla \cdot v)\, m \right) v = f^{\text{ext}} + \nabla \cdot \tau \tag{10.33}$$

and depends on the choice of inertial frame of reference. However, in classical physics, the equations of motion are the same in all the inertial frames of reference. This is a consequence of Galilean invariance. In order for the equation of motion (10.33) to be independent of the choice of frame of reference, the expression in the brackets of (10.33) must vanish. This yields the mass continuity equation,

$$\dot{m} + (\nabla \cdot v)\, m = 0 \tag{10.34}$$

The mass continuity equation (10.34) implies that there is no mass source density and no conductive mass current density as well. This means that in classical physics, mass is a 'strongly conserved' quantity.

Taking into account the mass continuity equation (10.34), the momentum continuity (10.31) reduces to Newton's 2nd law applied to a continuous medium,

$$m\dot{v} = f^{\text{ext}} + \nabla \cdot \tau \tag{10.35}$$

The term $\nabla \cdot \tau$ is a force density which is related to the constraints exerted at the surface of the fluid.

In order to satisfy the first law (1.11) applied locally to the continuous system, the energy source density π_e, which does not result from an energy current flowing through the enclosure, is due to the action of external forces on the system. Thus, the energy source density π_e that corresponds to the external power density is expressed as,

$$\pi_e = \boldsymbol{v} \cdot \boldsymbol{f}^{\text{ext}} \tag{10.36}$$

The energy continuity equation is then given by,

$$\dot{e} + (\boldsymbol{\nabla} \cdot \boldsymbol{v})\, e + \boldsymbol{\nabla} \cdot \boldsymbol{j}_e = \boldsymbol{v} \cdot \boldsymbol{f}^{\text{ext}} \tag{10.37}$$

where $\boldsymbol{j}_e$ is the energy current density.

The internal energy density of the system $u\,(s, \{n_A\}, q)$ is defined as the energy density with respect to the frame of reference where the infinitesimal material element is at rest. Thus, the energy density $e\,(\boldsymbol{p}, s, \{n_A\}, q)$ can be decomposed into a kinetic energy and an internal energy,

$$e\,(\boldsymbol{p}, s, \{n_A\}, q) = \frac{\boldsymbol{p}^2}{2\, m\,(s, \{n_A\}, q)} + u\,(s, \{n_A\}, q) \tag{10.38}$$

The structure of equation (10.38) is a consequence of the fact that the velocity field $\boldsymbol{v}$ is the field conjugated to the momentum density field, i.e. $\boldsymbol{v} = \partial e / \partial \boldsymbol{p}$ and of the constitutive relation (10.32). Applying (10.32), the time derivative of equation (10.38) can be written as,

$$\dot{e} = \boldsymbol{v} \cdot \dot{\boldsymbol{p}} - \frac{1}{2}\, \dot{m}\, \boldsymbol{v}^2 + \dot{u} \tag{10.39}$$

Using the mass continuity equation (10.34), the momentum continuity equation (10.31) and the constitutive relation (10.32), equation (10.39) can be recast as,

$$\dot{e} = \boldsymbol{v} \cdot \left(-\,(\boldsymbol{\nabla} \cdot \boldsymbol{v})\, m\, \boldsymbol{v} + \boldsymbol{\nabla} \cdot \boldsymbol{\tau} + \boldsymbol{f}^{\text{ext}} \right) + \frac{1}{2}\, m\, \boldsymbol{v}^2\,(\boldsymbol{\nabla} \cdot \boldsymbol{v}) + \dot{u} \tag{10.40}$$

Taking into account the expression (10.38) for the energy density e and the constitutive relation (10.32), equation (10.40) can be written as,

$$\dot{e} = \dot{u} + (\boldsymbol{\nabla} \cdot \boldsymbol{v})\,(u - e) + \boldsymbol{v} \cdot (\boldsymbol{\nabla} \cdot \boldsymbol{\tau}) + \boldsymbol{f}^{\text{ext}} \cdot \boldsymbol{v} \tag{10.41}$$

Given that the stress tensor $\boldsymbol{\tau}$ is symmetric (§ 10.4.3), in order to obtain a continuity equation for the internal energy, we use the vectorial identity,

$$\boldsymbol{v} \cdot (\boldsymbol{\nabla} \cdot \boldsymbol{\tau}) = \boldsymbol{\nabla} \cdot \frac{1}{2} \left(\boldsymbol{v} \cdot \boldsymbol{\tau} + (\boldsymbol{v} \cdot \boldsymbol{\tau})^T \right) - \frac{1}{2} \left(\boldsymbol{\nabla} \boldsymbol{v} + (\boldsymbol{\nabla} \boldsymbol{v})^T \right) : \boldsymbol{\tau} \tag{10.42}$$

where the exponent ^{T} indicates that we take the transpose of the tensorial expression. The components of the symmetric vector $\boldsymbol{v} \cdot \boldsymbol{\tau} + (\boldsymbol{v} \cdot \boldsymbol{\tau})^T$ and of the symmetric tensor $(\boldsymbol{\nabla} \boldsymbol{v} + (\boldsymbol{\nabla} \boldsymbol{v})^T$ are given by,

$$\left(\boldsymbol{v} \cdot \boldsymbol{\tau} + (\boldsymbol{v} \cdot \boldsymbol{\tau})^T \right)^i = v_j \left(\tau^{ji} + \tau^{ij} \right)$$

$$\left(\boldsymbol{\nabla} \boldsymbol{v} + (\boldsymbol{\nabla} \boldsymbol{v})^T \right)_{ij} = \partial_i v_j + \partial_j v_i$$

Substituting for $\dot{e}$ in the energy continuity equation (10.37) its expression (10.41) and using the vectorial identity (10.42), we obtain the internal energy continuity equation,

$$\dot{u} + (\boldsymbol{\nabla} \cdot \boldsymbol{v})\, u + \boldsymbol{\nabla} \cdot \boldsymbol{j}_u = \pi_u \tag{10.43}$$

where the expressions for the internal energy current density j_u and the internal energy source density π_u are found to be,

$$j_u = j_e + \frac{1}{2}\left(v \cdot \tau + (v \cdot \tau)^T\right) \tag{10.44}$$

$$\pi_u = \frac{1}{2}\left(\nabla v + (\nabla v)^T\right) : \tau \tag{10.45}$$

The energy continuity equation (10.37) reduces to the internal energy continuity equation (10.43) if the continuous medium is locally at rest (i.e. $v = 0$), as it should be in view of the definitions of energy e and internal energy u.

10.3 Evolution Equations

In this section, we will use our thermodynamic approach to obtain conservation laws for mass and electric charge. These are usually presented as results from the molecular structure of matter. We will also show that the velocity field v represents the velocity of the centre of mass of the local infinitesimal element of fluid.

10.3.1 Conservation of Mass

The time derivative of the mass density $m\left(s, \{n_A\}, q\right)$ is given by,

$$\dot{m} = \frac{\partial m}{\partial s}\dot{s} + \sum_{A=1}^{r}\frac{\partial m}{\partial n_A}\dot{n}_A + \frac{\partial m}{\partial q}\dot{q} \tag{10.46}$$

Using this equation and the entropy continuity equations for entropy (10.24), chemical substance A (10.26) and electric charge (10.27), we can recast the mass continuity equation (10.34) as,

$$\begin{aligned}
&\frac{\partial m}{\partial s}\left(\pi_s - (\nabla \cdot v)s - \nabla \cdot j_s\right) \\
&+ \sum_{A=1}^{r}\frac{\partial m}{\partial n_A}\left(\sum_{a=1}^{n}\omega_a \nu_{aA} - (\nabla \cdot v)n_A - \nabla \cdot j_A\right) \\
&+ \frac{\partial m}{\partial q}\left(-(\nabla \cdot v)q - \nabla \cdot j_q\right) + (\nabla \cdot v)m = 0
\end{aligned} \tag{10.47}$$

The evolution condition of the second law requires that the entropy source density be positive definite, i.e. $\pi_s \geqslant 0$ and this, independently of the value or of the signs of the other thermodynamic fields. Thus, the differential factor multiplying π_s has to vanish, i.e.

$$\frac{\partial m}{\partial s} = 0 \tag{10.48}$$

Moreover, the mass and electric charge are two independent physical properties. This implies that,

$$\frac{\partial m}{\partial q} = 0 \tag{10.49}$$

Thus, the mass density is only a function of the chemical substance densities $\{n_A\}$, i.e. $m = m(\{n_A\})$. The dynamic equation (10.47) can then be reduced to,

$$\left(m - \sum_{A=1}^{r} \frac{\partial m}{\partial n_A} n_A\right)(\boldsymbol{\nabla} \cdot \boldsymbol{v}) + \sum_{A=1}^{r} \frac{\partial m}{\partial n_A}\left(\sum_{a=1}^{n} \omega_a \nu_{aA} - \boldsymbol{\nabla} \cdot \boldsymbol{j}_A\right) = 0 \tag{10.50}$$

This equation has to be satisfied no matter what the velocity field $\boldsymbol{v}$ is like. This imposes the condition,

$$m - \sum_{A=1}^{r} \frac{\partial m}{\partial n_A} n_A = 0 \tag{10.51}$$

The partial derivative of equation (10.51) with respect to the density n_B of chemical substance B is given by,

$$\frac{\partial m}{\partial n_B} - \sum_{A=1}^{r}\left(\frac{\partial^2 m}{\partial n_A \partial n_B} n_A + \frac{\partial m}{\partial n_A} \delta_{AB}\right) = 0 \tag{10.52}$$

Since the first and third terms of equation (10.52) cancel each other out, (10.52) implies that,

$$\frac{\partial^2 m}{\partial n_A \partial n_B} = 0 \quad \forall A = 1, \dots, r \tag{10.53}$$

and thus the partial derivatives,

$$m_A \equiv \frac{\partial m}{\partial n_A} = \text{const} \tag{10.54}$$

are independent constants. They represent the mass of a mole of chemical substance A. Thus, the mass density m is the sum of the mass densities of the different chemical substances,

$$m = \sum_{A=1}^{r} n_A m_A \tag{10.55}$$

The definition (10.54) and equations (10.50) and (10.51) imply that,

$$\sum_{A=1}^{r} m_A\left(\sum_{a=1}^{n} \omega_a \nu_{aA} - \boldsymbol{\nabla} \cdot \boldsymbol{j}_A\right) = 0 \tag{10.56}$$

Since the mass m_A of a mole of chemical substance A is a constant, that is,

$$\boldsymbol{\nabla} m_A = \mathbf{0} \tag{10.57}$$

equation (10.56) can be recast as,

$$\sum_{a=1}^{n} \omega_a\left(\sum_{A=1}^{r} m_A \nu_{aA}\right) - \boldsymbol{\nabla} \cdot \left(\sum_{A=1}^{r} m_A \boldsymbol{j}_A\right) = 0 \tag{10.58}$$

Since equation (10.58) must be satisfied for any substance current density $\boldsymbol{j}_A$, we must have,

$$\sum_{A=1}^{r} m_A \boldsymbol{j}_A = \mathbf{0} \tag{10.59}$$

This is a consequence of the conductive nature of j_A. Moreover, equation (10.58) has to be satisfied for every chemical reaction a. This condition leads to the mass conservation law of Lavoisier,

$$\sum_{A=1}^{r} m_A \, \nu_{aA} = 0 \tag{10.60}$$

The condition (10.59) leads to a self-consistent physical interpretation of the velocity v as the velocity of the center of mass of the local system. In order to establish this, we introduce the velocity v_A of the chemical substance A that satisfies the relation,

$$v_A \, n_A = v \, n_A + j_A \tag{10.61}$$

Equations (10.55), (10.59) and (10.61) imply that the material momentum density p can be expressed in terms of the v_A,

$$p = \sum_{A=1}^{r} m_A \, n_A \, v_A \tag{10.62}$$

Thus, the material velocity v is given by,

$$v = \frac{\displaystyle\sum_{A=1}^{r} m_A \, n_A \, v_A}{\displaystyle\sum_{A=1}^{r} m_A \, n_A} \tag{10.63}$$

This expression is consistent with the definition of the centre of mass velocity of the substances contained in the local thermodynamic system.

10.3.2 Conservation of Electric Charge

The electric charge is a property of the elementary constituents of matter. Thus, it is a function of the chemical substance densities $\{n_A\}$, i.e. $q = q\left(\{n_A\}\right)$.

The time derivative of the charge density $q\left(\{n_A\}\right)$ is given by,

$$\dot{q} = \sum_{A=1}^{r} \frac{\partial q}{\partial n_A} \, \dot{n}_A \tag{10.64}$$

Using this relation and the continuity equation (10.26) for chemical substance A, the continuity equation (10.27) for the electric charge becomes,

$$\sum_{A=1}^{r} \frac{\partial q}{\partial n_A} \left(\sum_{a=1}^{n} \omega_a \, \nu_{aA} - (\nabla \cdot v)\, n_A - \nabla \cdot j_A \right) + (\nabla \cdot v)\, q + \nabla \cdot j_q = 0 \tag{10.65}$$

The dynamic equation (10.65) can be recast as,

$$\left(q - \sum_{A=1}^{r} \frac{\partial q}{\partial n_A} n_A \right) (\nabla \cdot v) + \sum_{A=1}^{r} \frac{\partial q}{\partial n_A} \left(\sum_{a=1}^{n} \omega_a \, \nu_{aA} - \nabla \cdot j_A \right) + \nabla \cdot j_q = 0 \tag{10.66}$$

This equation has to be satisfied no matter what the velocity field $\boldsymbol{v}$ is like. This imposes the condition,

$$q - \sum_{A=1}^{r} \frac{\partial q}{\partial n_A} n_A = 0 \tag{10.67}$$

The partial derivative of equation (10.67) with respect to the density n_B of chemical substance B is given by,

$$\frac{\partial q}{\partial n_B} - \sum_{A=1}^{r} \left(\frac{\partial^2 q}{\partial n_A \partial n_B} n_A + \frac{\partial q}{\partial n_A} \delta_{AB} \right) = 0 \tag{10.68}$$

Since the first and third terms of equation (10.68) cancel each other out, this equation implies that,

$$\frac{\partial^2 q}{\partial n_A \partial n_B} = 0 \quad \forall A = 1, \ldots, r \tag{10.69}$$

and thus the partial derivatives,

$$q_A \equiv \frac{\partial q}{\partial n_A} = \text{const} \tag{10.70}$$

are independent constants. They represent the electric charge of a mole of chemical substance A. Thus, the electric charge density q is the sum of the electric charge densities of the different chemical substances,

$$q = \sum_{A=1}^{r} n_A q_A \tag{10.71}$$

In our phenomenological approach, matter consists of three electric charge carriers that can be considered as different chemical substances A:

1. anions, for which $q_A < 0$,
2. cations, for which $q_A > 0$,
3. neutral substances, for which $q_A = 0$.

Definition (10.70) and equations (10.66) and (10.67) imply that,

$$\sum_{A=1}^{r} q_A \left(\sum_{a=1}^{n} \omega_a \nu_{aA} - \boldsymbol{\nabla} \cdot \boldsymbol{j}_A \right) + \boldsymbol{\nabla} \cdot \boldsymbol{j}_q = 0 \tag{10.72}$$

Since the electric charge q_A of a mole of chemical substance A is a constant,

$$\boldsymbol{\nabla} \, q_A = \boldsymbol{0}$$

equation (10.72) can be recast as,

$$\sum_{a=1}^{n} \omega_a \left(\sum_{A=1}^{r} q_A \nu_{aA} \right) + \boldsymbol{\nabla} \cdot \left(\boldsymbol{j}_q - \sum_{A=1}^{r} q_A \boldsymbol{j}_A \right) = 0 \tag{10.73}$$

Equation (10.73) has to be satisfied whatever the field of substance current density $\boldsymbol{j}_A$ looks like. This implies an expression for the conductive electric current density $\boldsymbol{j}_q$ in terms of the chemical substance current densities $\boldsymbol{j}_A$, namely,

$$j_q = \sum_{A=1}^{r} q_A j_A \tag{10.74}$$

Moreover, equation (10.73) has to be satisfied for every chemical reaction a. This implies the electric charge conservation law,

$$\sum_{A=1}^{r} q_A \nu_{aA} = 0 \tag{10.75}$$

10.3.3 Energy Balance

The local energy balance is given by applying the first law locally. We seek to establish thermodynamic evolution equations that are independent of the choice of frame of reference. Therefore, we will work out an energy balance for the internal energy as it is, by definition of internal energy, independent of the choice of frame of reference. We begin by computing the time derivative of the internal energy density $u\,(s, \{n_A\}, q)$. This gives,

$$\dot{u} = \frac{\partial u}{\partial s}\,\dot{s} + \sum_{A=1}^{r} \frac{\partial u}{\partial n_A}\,\dot{n}_A + \frac{\partial u}{\partial q}\,\dot{q} \tag{10.76}$$

The temperature T, the chemical potential μ_A and the external electrostatic potential φ are intensive fields defined as fields that are conjugate of the entropy density s, the chemical substance densities $\{n_A\}$ and the electric charge density q, respectively. Thus, we have,

$$T \equiv \frac{\partial u}{\partial s}\,, \qquad \mu_A \equiv \frac{\partial u}{\partial n_A} \qquad \text{and} \qquad \varphi \equiv \frac{\partial u}{\partial q} \tag{10.77}$$

Using definitions (10.77), the time derivative of the internal energy density (10.76) becomes,

$$\dot{u} = T\dot{s} + \sum_{A=1}^{r} \mu_A \dot{n}_A + \varphi \dot{q} \tag{10.78}$$

The internal energy balance is locally given by the internal energy density continuity equation (10.43). Applying the continuity equations for entropy (10.24) and chemical substance A (10.26), together with the time derivative of the internal energy density (10.78), the internal energy density continuity equation (10.43) can be recast as,

$$T\left(\pi_s - (\boldsymbol{\nabla} \cdot \boldsymbol{v})\,s - \boldsymbol{\nabla} \cdot \boldsymbol{j}_s\right) + \sum_{A=1}^{r} \mu_A \left(\sum_{a=1}^{n} \omega_a \nu_{aA} - (\boldsymbol{\nabla} \cdot \boldsymbol{v})\,n_A - \boldsymbol{\nabla} \cdot \boldsymbol{j}_A\right)$$
$$+ \varphi\left(- (\boldsymbol{\nabla} \cdot \boldsymbol{v})\,q - \boldsymbol{\nabla} \cdot \boldsymbol{j}_q\right) + (\boldsymbol{\nabla} \cdot \boldsymbol{v})\,u + \boldsymbol{\nabla} \cdot \boldsymbol{j}_u - \pi_u = 0 \tag{10.79}$$

Moreover, using equation (10.45) and applying the vectorial identities,

$$T(\boldsymbol{\nabla} \cdot \boldsymbol{j}_s) = \boldsymbol{\nabla} \cdot (T\boldsymbol{j}_s) - \boldsymbol{j}_s \cdot \boldsymbol{\nabla} T\,,$$

$$\sum_{A=1}^{r} \mu_A\,(\boldsymbol{\nabla} \cdot \boldsymbol{j}_A) = \boldsymbol{\nabla} \cdot \left(\sum_{A=1}^{r} \mu_A \boldsymbol{j}_A\right) - \sum_{A=1}^{r} \boldsymbol{j}_A \cdot \boldsymbol{\nabla} \mu_A\,,$$

$$\varphi\,(\boldsymbol{\nabla} \cdot \boldsymbol{j}_q) = \boldsymbol{\nabla} \cdot (\varphi \boldsymbol{j}_q) - \boldsymbol{j}_q \cdot \boldsymbol{\nabla} \varphi\,,$$

the internal energy balance equation (10.79) becomes,

$$
\left(u - Ts - \sum_{A=1}^{r} \mu_A n_A - q\varphi \right) (\boldsymbol{\nabla} \cdot \boldsymbol{v})
$$

$$
+ \boldsymbol{\nabla} \cdot \left(\boldsymbol{j}_u - T\boldsymbol{j}_s - \sum_{A=1}^{r} \mu_A \boldsymbol{j}_A - \varphi \boldsymbol{j}_q \right) + T\pi_s + \sum_{A=1}^{r} \mu_A \left(\sum_{a=1}^{n} \omega_a \nu_{aA} \right) \tag{10.80}
$$

$$
+ \boldsymbol{j}_s \cdot \boldsymbol{\nabla} T + \sum_{A=1}^{r} \boldsymbol{j}_A \cdot \boldsymbol{\nabla} \mu_A + \boldsymbol{j}_q \cdot \boldsymbol{\nabla} \varphi - \frac{1}{2} \left(\boldsymbol{\nabla} \boldsymbol{v} + (\boldsymbol{\nabla} \boldsymbol{v})^T \right) : \boldsymbol{\tau} = 0
$$

Here, we consider a medium without shear or vorticity. Thus, the stress tensor is given by,

$$
\boldsymbol{\tau} = \left(\tau^{\mathrm{fr}} - p \right) \mathbb{1} \tag{10.81}
$$

where $\mathbb{1}$ is the identity tensor respresented by the 3×3 identity matrix and p is the pressure. The pressure is the reversible part of the stress tensor. The irreversible part τ^{fr} is due to internal viscous friction. The pressure p and the internal viscous friction τ^{fr} are isotropic in the local frame of reference in which matter is at rest, that is, the forces associated with them have the same norm in every direction.

To generalise the approach presented in this chapter and introduce shear forces, it would be necessary to specify the geometric structure of the local system by including the deformation tensor as one of the state fields [107, 146, 147]. Moreover, to describe vorticity, it would be essential to specify the state of rotation of the local system around its centre of mass by defining a pseudo-vector angular velocity as a state field [148]. Finally, we note that in equation (10.81), we also neglected the friction due to the relative displacement of different chemical substances.

Using the decomposition (10.81), the last term of (10.80) reads,

$$
\frac{1}{2} \left(\boldsymbol{\nabla} \boldsymbol{v} + (\boldsymbol{\nabla} \boldsymbol{v})^T \right) : \boldsymbol{\tau} = (\boldsymbol{\nabla} \cdot \boldsymbol{v}) \left(\tau^{\mathrm{fr}} - p \right) \tag{10.82}
$$

In equation (10.80), let us introduce the affinity $\mathcal{A}_a$ of a chemical reaction a defined as,

$$
\mathcal{A}_a = - \sum_{A=1}^{r} \mu_A \nu_{aA} \tag{10.83}
$$

Using the definition (10.83), the double sum on the substances and the chemical reactions in the balance equation (10.80) reduces to,

$$
\sum_{A=1}^{r} \mu_A \left(\sum_{a=1}^{n} \omega_a \nu_{aA} \right) = - \sum_{a=1}^{n} \omega_a \mathcal{A}_a \tag{10.84}
$$

Using identities (10.82) and (10.84) and using expressions (10.71) and (10.74) for the electric charge density q and the conductive electric current density $\boldsymbol{j}_q$, we find that the internal energy balance equation (10.80) becomes,

$$\left(u - Ts + p - \sum_{A=1}^{r} \left(\mu_A + q_A \varphi \right) n_A \right) \left(\boldsymbol{\nabla} \cdot \boldsymbol{v} \right)$$

$$+ \boldsymbol{\nabla} \cdot \left(\boldsymbol{j}_u - T\boldsymbol{j}_s - \sum_{A=1}^{r} \left(\mu_A + q_A \varphi \right) \boldsymbol{j}_A \right) \tag{10.85}$$

$$+ T\pi_s - \sum_{a=1}^{n} \omega_a \, \mathcal{A}_a - \tau^{\mathrm{fr}} \left(\boldsymbol{\nabla} \cdot \boldsymbol{v} \right) + \boldsymbol{j}_s \cdot \boldsymbol{\nabla} T + \sum_{A=1}^{r} \boldsymbol{j}_A \cdot \left(\boldsymbol{\nabla} \mu_A + q_A \boldsymbol{\nabla} \varphi \right) = 0$$

The terms on the last line of equation (10.85) account for irreversible processes and thus change sign under time reversal (§ Time reversal).

10.3.4 Thermostatic and Thermodynamic Equations

The internal energy balance equation (10.85) has to be satisfied for any velocity field. Thus, the expression in brackets on the first line on the left-hand side of (10.85) has to vanish. This yields an explicit expression for the internal energy density,

$$u = Ts - p + \sum_{A=1}^{r} \left(\mu_A + q_A \varphi \right) n_A \tag{10.86}$$

This is a ***thermostatic equilibrium equation*** since it is the volume-specific Euler equation. Moreover, the internal energy balance equation (10.85) has to be satisfied for any energy current density field. Thus, the expression in brackets on the second line on the left-hand side of (10.85) has to vanish. This yields an explicit expression for the internal energy current density,

$$\boldsymbol{j}_u = T\boldsymbol{j}_s + \sum_{A=1}^{r} \left(\mu_A + q_A \varphi \right) \boldsymbol{j}_A \tag{10.87}$$

This is a ***reversible thermodynamic evolution equation***.

The thermostatic equation (10.86) and the reversible thermodynamic equation (10.87) allow us to deduce from the balance equation (10.85) an expression for the entropy source density,

$$\pi_s = \frac{1}{T} \left\{ \sum_{a=1}^{n} \omega_a \, \mathcal{A}_a + \tau^{\mathrm{fr}} \left(\boldsymbol{\nabla} \cdot \boldsymbol{v} \right) \right.$$

$$\left. + \boldsymbol{j}_s \cdot \left(- \boldsymbol{\nabla} T \right) + \sum_{A=1}^{r} \boldsymbol{j}_A \cdot \left(- \boldsymbol{\nabla} \mu_A - q_A \boldsymbol{\nabla} \varphi \right) \right\} \tag{10.88}$$

This is an ***irreversible thermodynamic evolution equation***. It will be the foundation of the developments of the next chapter.

10.3.5 Link between Local and Global Homogeneous Systems

Having now obtained these important results relative to local systems, we would like to establish their connection to the thermodynamics of a global system as we presented it at the beginning of this book. To do so, we consider a simple global system (i.e. homogeneous) where the centre of mass is at rest. A simple system where the centre of mass is at rest can change volume. The volume variation rate is given by the divergence of the velocity field. Thus, the global simple system is homogeneous but not uniform. This means that matter is distributed homogeneously inside the system, but the velocity field does not have the same value everywhere.

First, we determine the equilibrium equation of the homogeneous global system based on the equilibrium equation of the local homogeneous systems. The internal energy of the system $U(S, V, \{N_A\}, Q)$, a function of the state variables entropy S, volume V, numbers of moles $\{N_A\}$ of the different chemical substances and electric charge Q defined for the global system. The function $U(S, V, \{N_A\}, Q)$ can be deduced from the internal energy density $u(s, \{n_A\}, q)$ by an integral over the volume V of the global system,

$$U(S, V, \{N_A\}, Q) = \int_V dV\, u(s, \{n_A\}, q) \tag{10.89}$$

Similarly, the extensive state variables S, $\{N_A\}$ and Q of the global system are also related to their respective densities by integrals over the volume V, i.e.

$$S = \int_V dV\, s, \qquad N_A = \int_V dV\, n_A, \qquad Q = \int_V dV\, q \tag{10.90}$$

Since the system is homogeneous, the intensive fields temperature T, chemical potential μ_A and electrostatic potential φ, are independent of position. Thus, the integral of the local thermostatic equation (10.86) over the volume V of the system is given by,

$$\int_V dV\, u = T \int_V dV\, s - p \int_V dV + \sum_{A=1}^{r} (\mu_A + q_A \varphi) \int_V dV\, n_A \tag{10.91}$$

Using definitions (10.89) and (10.90) for the global state functions and state variables, the global thermostatic equilibrium equation (10.91) can be written as,

$$U(S, V, \{N_A\}, Q) = TS - pV + \sum_{A=1}^{r} (\mu_A + q_A \varphi)\, N_A \tag{10.92}$$

Second, we establish the irreversible evolution equation of the global simple system. The homogeneity of the local system implies that the gradient of the intensive scalar fields vanish, i.e. $\boldsymbol{\nabla}\, T = \boldsymbol{0}$, $\boldsymbol{\nabla}\, \mu_A = \boldsymbol{0}$ and $\boldsymbol{\nabla}\, \varphi = \boldsymbol{0}$. Thus, the local irreversible thermodynamic evolution equation (10.88) reduces to,

$$T\pi_s = \tau^{\text{fr}} (\boldsymbol{\nabla} \cdot \boldsymbol{v}) + \sum_{a=1}^{n} \mathcal{A}_a \omega_a \tag{10.93}$$

The entropy production rate Π_S, the chemical reaction rate Ω_a and the volume variation rate $\dot{V}$ are related to their corresponding densities by,

$$\Pi_S = \int_V dV\, \pi_s, \qquad \Omega_a = \int_V dV\, \omega_a, \qquad \dot{V} = \int_V dV\, (\boldsymbol{\nabla} \cdot \boldsymbol{v}) \tag{10.94}$$

The last expression is obtained from the integral relation (10.11), noting that the volume density is simply unity. The mechanical work source density π_W is defined as the internal energy source density π_u,

$$\pi_u \equiv \pi_W \tag{10.95}$$

The mechanical power P_W is expressed as an integral of the internal energy source density (10.95),

$$P_W = \int_V dV \, \pi_W \tag{10.96}$$

We can now find an expression for the mechanical power (10.96) that makes explicit the irreversible contribution in it. By taking into account expression (10.45) for the internal energy source density π_u, expression (10.81) for the stress tensor, equation (10.94) and in view of the fact that the intensive fields pressure p and τ^{fr} are the same everywhere, the mechanical power (10.96) can be written as,

$$P_W = \int_V dV \left(\frac{1}{2} \left(\boldsymbol{\nabla} \boldsymbol{v} + (\boldsymbol{\nabla} \boldsymbol{v})^T \right) : \boldsymbol{\tau} \right)$$
$$= \left(\tau^{\text{fr}} - p \right) \int_V dV \left(\boldsymbol{\nabla} \cdot \boldsymbol{v} \right) = \left(\tau^{\text{fr}} - p \right) \dot{V} \tag{10.97}$$

Furthermore, by comparing expressions (2.36) and (10.97) for the mechanical power, we conclude that for a homogeneous system,

$$\tau^{\text{fr}} = p - p_{\text{ext}} \tag{10.98}$$

Taking into account relation (10.98) and the fact that the intensive fields temperature T, chemical affinity $\mathcal{A}_a$, pressures p and p_{ext} are the same everywhere, the integral of the irreversible local evolution equation (10.88) over the volume V of the global system is given by,

$$T \int_V dV \, \pi_s = (p - p_{\text{ext}}) \int_V dV \left(\boldsymbol{\nabla} \cdot \boldsymbol{v} \right) + \sum_{a=1}^n \mathcal{A}_a \int_V dV \, \omega_a \tag{10.99}$$

Using definitions (2.36) and (10.94), the global irreversible thermodynamic evolution equation (10.99) is written as,

$$T \Pi_S = P_W + p \, \dot{V} + \sum_{a=1}^n \mathcal{A}_a \, \Omega_a \tag{10.100}$$

in agreement with relation (2.15) since we have the relations (10.84) and (10.94).

Third, we establish the evolution equation of the first law in the centre of mass frame of reference of the global system. The internal energy current density $\boldsymbol{j}_u$ is the sum of the heat current density $\boldsymbol{j}_Q$ and the matter current density $\boldsymbol{j}_C$, i.e.

$$\boldsymbol{j}_u = \boldsymbol{j}_Q + \boldsymbol{j}_C \tag{10.101}$$

where, taking into account equation (10.87), the current densities are respectively defined as,

$$\boldsymbol{j}_Q \equiv T \boldsymbol{j}_s \tag{10.102}$$

$$\boldsymbol{j}_C \equiv \sum_{A=1}^r \left(\mu_A + q_A \varphi \right) \boldsymbol{j}_A \tag{10.103}$$

The thermal power P_Q is expressed as the opposite of the integral over the surface S of the heat current density $\boldsymbol{j}_Q$, i.e.

$$P_Q = -\int_S d\boldsymbol{S} \cdot \boldsymbol{j}_Q = -\int_V dV \left(\boldsymbol{\nabla} \cdot \boldsymbol{j}_Q\right) \tag{10.104}$$

where the last equality is obtained by applying the divergence theorem. The chemical power P_C is expressed as the opposite of the integral over the surface S of the matter current density $\boldsymbol{j}_C$, i.e.

$$P_C = -\int_S d\boldsymbol{S} \cdot \boldsymbol{j}_C = -\int_V dV \left(\boldsymbol{\nabla} \cdot \boldsymbol{j}_C\right) \tag{10.105}$$

where the last equality is obtained by applying the divergence theorem. Since the vector $d\boldsymbol{S}$ is defined positively when oriented outwards and the vectors $\boldsymbol{j}_Q$ and $\boldsymbol{j}_C$ contribute to an increase of their corresponding extensive quantities. The thermal power P_Q and the chemical power P_C are defined positively when heat and matter enter the system.

It is worth noting that the thermal power P_Q is due to the heat flow through the enclosure of the global system. Consequently, it is expressed as an integral over the surface of a heat current density $\boldsymbol{j}_Q$. Likewise, the chemical power P_C is due to a matter flow through the enclosure of the global system and it is expressed as an integral over the surface of a matter current density $\boldsymbol{j}_C$. The mechanical power P_W is due to a work performed by the environment on the global system. However, it is expressed, according to the formalism of continuous media, as an integral over the volume of a mechanical work source density π_W.

Taking into account relations (10.11), (10.18) and (10.19), we can write the internal energy density variation rate of the global system as,

$$\dot{U} = \int_V dV \left(\dot{u} + (\boldsymbol{\nabla} \cdot \boldsymbol{v})\, u\right) \tag{10.106}$$

The integral of the internal energy continuity equation (10.43) over the volume of the global system is expressed as,

$$\int_V dV \left(\dot{u} + (\boldsymbol{\nabla} \cdot \boldsymbol{v})\, u\right) = -\int_V dV \left(\boldsymbol{\nabla} \cdot \boldsymbol{j}_u\right) + \int_V dV\, \pi_u \tag{10.107}$$

Using relations (10.95) and (10.101), the integral (10.107) is recast as,

$$\int_V dV \left(\dot{u} + (\boldsymbol{\nabla} \cdot \boldsymbol{v})\, u\right) = -\int_V dV \left(\boldsymbol{\nabla} \cdot \boldsymbol{j}_Q\right) - \int_V dV \left(\boldsymbol{\nabla} \cdot \boldsymbol{j}_C\right) + \int_V dV\, \pi_W \tag{10.108}$$

Taking into account definitions (10.96), (10.104), (10.105) and (10.106), we find that the integral equation (10.108) is the expression (1.29) for the first law in the frame of reference where the homogeneous system is at rest,

$$\dot{U} = P_Q + P_W + P_C \quad \text{(open system)} \tag{10.109}$$

Fourth, we establish the evolution equation of the second law in the frame of reference of the centre of mass of the global system. Taking into account relations (10.11), (10.18) and (10.19), we can write the variation rate of the entropy of the global system as,

$$\dot{S} = \int_V dV \left(\dot{s} + (\boldsymbol{\nabla} \cdot \boldsymbol{v})\, s\right) \tag{10.110}$$

Using relations (10.102) and (10.104) and in view of the fact that the temperature T is homogeneous, we obtain the identity,

$$\frac{P_Q}{T} = -\frac{1}{T} \int_V dV \, (\boldsymbol{\nabla} \cdot \boldsymbol{j}_Q) = -\int_V dV \, (\boldsymbol{\nabla} \cdot \boldsymbol{j}_s) \tag{10.111}$$

The integral of the entropy continuity equation (10.43) over the volume V of the global system is expressed as,

$$\int_V dV \left(\dot{s} + (\boldsymbol{\nabla} \cdot \boldsymbol{v}) \, s \right) = -\int_V dV (\boldsymbol{\nabla} \cdot \boldsymbol{j}_s) + \int_V dV \, \pi_s \tag{10.112}$$

Taking into account the first relation (10.94) and relations (10.110) and (10.111), we find that the integral equation (10.112) is the expression (2.18) of the second law in the frame of reference where the homogeneous system is at rest,

$$\dot{S} = \frac{P_Q}{T} + \Pi_S \tag{10.113}$$

Fifth, we establish the matter balance equation in the frame of reference of the centre of mass of the global system for electrically neutral substances, i.e. $q_A = 0$. Taking into account relations (10.11), (10.18) and (10.19), we obtain the following identity,

$$\sum_{A=1}^{r} \mu_A \dot{N}_A = \sum_{A=1}^{r} \mu_A \int_V dV \left(\dot{n}_A + (\boldsymbol{\nabla} \cdot \boldsymbol{v}) \, n_A \right) \tag{10.114}$$

For electrically neutral substances, relations (10.103) and (10.105) imply that,

$$P_C = -\sum_{A=1}^{r} \mu_A \int_V dV \, (\boldsymbol{\nabla} \cdot \boldsymbol{j}_A) \tag{10.115}$$

Using definition (8.18) for the chemical affinity $\mathcal{A}_a$ and the second integral expression (10.94), we have,

$$\sum_{a=1}^{n} \mathcal{A}_a \Omega_a = -\sum_{A=1}^{N} \mu_A \int_V dV \left(\sum_{a=1}^{n} \omega_a \nu_{aA} \right) \tag{10.116}$$

The integral of the continuity equation (10.26) of the substance A over the volume V of the global system is expressed as,

$$\int_V dV \left(\dot{n}_A + (\boldsymbol{\nabla} \cdot \boldsymbol{v}) \, n_A \right) = -\int_V dV (\boldsymbol{\nabla} \cdot \boldsymbol{j}_A) + \int_V dV \left(\sum_{a=1}^{n} \omega_a \nu_{aA} \right) \tag{10.117}$$

Taking into account relations (10.114), (10.115) and (10.116) equation (10.117) yields the matter balance equation for the electrically neutral substances in the frame of reference where the homogeneous system is at rest, which can be written as,

$$\sum_{A=1}^{r} \mu_A \dot{N}_A = P_C - \sum_{a=1}^{n} \mathcal{A}_a \Omega_a \tag{10.118}$$

10.4 Worked Solutions

10.4.1 Galilean Invariance

Consider an extensive scalar state function $F(t)$ characterising a physical property of a thermodynamic system. The state function density $f(x, t)$ is described with respect to an inertial frame of reference $\mathcal{R}$. The local spatial coordinates are the components of the vector x and the time coordinate is t.

1. In case the system is moving with a uniform velocity v_0 with respect to the inertial frame of reference $\mathcal{R}$, we can define a new inertial frame of reference $\mathcal{R}'$ where the system is at rest. The Galilean transformation relates the coordinates of the frames of reference $\mathcal{R}$ and $\mathcal{R}'$,

$$x' = x - v_0 t$$
$$t' = t$$

   The local spatial coordinates expressed with respect to $\mathcal{R}'$ are the components of the vector x' and the time coordinate is t'. Determine the transformation law between the time derivative of the state function density $\partial_t f(x, t)$ expressed with respect to the frame of reference $\mathcal{R}$ and the time derivative of the state function density $\partial_{t'} f(x', t')$ expressed with respect to the frame of reference $\mathcal{R}'$. Determine the transformation law between the spatial derivative of the state function density $\nabla f(x, t)$ expressed with respect to the frame of reference $\mathcal{R}$ and the spatial derivative of the state function density $\nabla' f(x', t')$ expressed with respect to the frame of reference $\mathcal{R}'$.

2. Show that the time derivative of the state function density $\dot{f}(x, t)$ given by equation (10.18) is a Galilean invariant by performing a Galilean transformation of uniform velocity v_0 between the inertial frames of reference $\mathcal{R}$ and $\mathcal{R}'$, where the frame of reference $\mathcal{R}'$ is an arbitrary inertial frame of reference.

Solution:

1. *Taking into account relation (10.11) and the uniformity of the velocity v_0, the time derivative of the extensive state function $F(t)$ in the inertial frame of reference $\mathcal{R}$ is written as,*

$$\dot{F}(t) = \int_{V(t)} dV(x) \left(\partial_t f(x, t) + (v_0 \cdot \nabla) f(x, t) \right)$$

   According to relation (10.11), the time derivative of the extensive state function $F(t')$ in the inertial frame of reference $\mathcal{R}'$ where the system is at rest is expressed as,

$$\dot{F}(t') = \int_{V(t')} dV(x') \, \partial_{t'} f(x', t')$$

   Since the time coordinate is invariant under a Galilean transformation, $\dot{F}(t) = \dot{F}(t')$, which implies that the integrands of the two previous equations are equal. This equality gives rise to the transformation law,

$$\partial_{t'} f(\boldsymbol{x}', t') = \partial_t f(\boldsymbol{x}, t) + (\boldsymbol{v}_0 \cdot \boldsymbol{\nabla}) f(\boldsymbol{x}, t)$$

Since the inertial frame of reference $\mathcal{R}'$ moves at velocity $\boldsymbol{v}_0$ with respect to the frame of reference $\mathcal{R}$, the frame of reference $\mathcal{R}$ moves with velocity $-\boldsymbol{v}_0$ with respect to the frame $\mathcal{R}'$. If the spatial coordinates are permuted, i.e. $\boldsymbol{x} \leftrightarrow \boldsymbol{x}'$, and the time coordinates are permuted as well, i.e. $t \leftrightarrow t'$, then $\boldsymbol{v}_0 \to -\boldsymbol{v}_0$, which implies that,

$$\partial_t f(\boldsymbol{x}, t) = \partial_{t'} f(\boldsymbol{x}', t') - (\boldsymbol{v}_0 \cdot \boldsymbol{\nabla}') f(\boldsymbol{x}', t')$$

Adding the two previous equations yields the transformation law,

$$\boldsymbol{\nabla}' f(\boldsymbol{x}', t') = \boldsymbol{\nabla} f(\boldsymbol{x}, t)$$

2. *The time derivative of the Galilean transformation of the spatial coordinates yields the Galilean transformation of the velocities,*

$$\boldsymbol{v}' = \boldsymbol{v} - \boldsymbol{v}_0$$

Taking into account definition (10.18), the transformation law for velocities and the spatial and time derivatives, we have,

$$\partial_{t'} f(\boldsymbol{x}', t') + (\boldsymbol{v}' \cdot \boldsymbol{\nabla}') f(\boldsymbol{x}', t') = \partial_t f(\boldsymbol{x}, t) + (\boldsymbol{v} \cdot \boldsymbol{\nabla}) f(\boldsymbol{x}, t)$$

This establishes the Galilean invariance of the state function density $\dot{f}(\boldsymbol{x}, t)$,

$$\dot{f}(\boldsymbol{x}', t') = \dot{f}(\boldsymbol{x}, t)$$

10.4.2 Dynamics of a Homogeneous and Uniform System

Establish the expression of Newton's 2nd law for a system consisting of a homogeneous and uniform fluid of mass M in a rectilinear motion.

Solution:

1. *For a uniform system, the divergence of the stress tensor $\boldsymbol{\tau}$ vanishes,*

$$\boldsymbol{\nabla} \cdot \boldsymbol{\tau} = \boldsymbol{0}$$

Thus, Newton's 2nd law (10.35) for a uniform fluid in a rectilinear motion reduces to,

$$\boldsymbol{f}^{\text{ext}} = m\dot{\boldsymbol{v}}$$

Since the acceleration $\boldsymbol{a} \equiv \dot{\boldsymbol{v}}$ is an intensive quantity, this equation can be integrated over the volume V of the system,

$$\int_V dV \boldsymbol{f}^{\text{ext}} = \boldsymbol{a} \int_V dV m$$

which can be expressed as,

$$\boldsymbol{F}^{\text{ext}} = M\boldsymbol{a}$$

where $\boldsymbol{F}^{\text{ext}}$ is the resulting external force acting on the system.

10.4.3 Stress Tensor

Using the angular momentum continuity equation, show that the stress tensor $\boldsymbol{\tau}$ can be represented by a 3×3 symmetric matrix, i.e. $\tau^{jk} = \tau^{kj}$.

Solution:

The angular momentum continuity equation is expressed as,

$$\dot{\boldsymbol{\ell}} + (\boldsymbol{\nabla} \cdot \boldsymbol{v})\,\boldsymbol{\ell} + \boldsymbol{\nabla} \cdot \mathrm{j}_{\boldsymbol{\ell}} = \boldsymbol{\pi}_{\boldsymbol{\ell}}$$

where $\boldsymbol{\ell} = \boldsymbol{r} \times \boldsymbol{p}$ is the angular momentum density. Taking into account equation (10.31), the derivative of the angular momentum can be written as,

$$\dot{\boldsymbol{\ell}} = \dot{\boldsymbol{r}} \times \boldsymbol{p} + \boldsymbol{r} \times \dot{\boldsymbol{p}} = \boldsymbol{r} \times \dot{\boldsymbol{p}} = -\,(\boldsymbol{\nabla} \cdot \boldsymbol{v})\,\boldsymbol{r} \times \boldsymbol{p} + \boldsymbol{r} \times (\boldsymbol{\nabla} \cdot \boldsymbol{\tau}) + \boldsymbol{r} \times \boldsymbol{f}^{\mathrm{ext}}$$

which is recast as,

$$\dot{\boldsymbol{\ell}} + (\boldsymbol{\nabla} \cdot \boldsymbol{v})\,\boldsymbol{\ell} - \boldsymbol{r} \times (\boldsymbol{\nabla} \cdot \boldsymbol{\tau}) = \boldsymbol{r} \times \boldsymbol{f}^{\mathrm{ext}}$$

Thus, the vectorial divergence of the tensorial angular momentum current density $\mathrm{j}_{\boldsymbol{\ell}}$ is given by,

$$\boldsymbol{\nabla} \cdot \mathrm{j}_{\boldsymbol{\ell}} = -\,\boldsymbol{r} \times (\boldsymbol{\nabla} \cdot \boldsymbol{\tau})$$

and the angular momentum source density $\boldsymbol{\pi}_{\boldsymbol{\ell}}$ is identified as the external torque density,

$$\boldsymbol{\pi}_{\boldsymbol{\ell}} = \boldsymbol{r} \times \boldsymbol{f}^{\mathrm{ext}}$$

According to equation (10.29), the external force density $\boldsymbol{f}^{ext}$ is the momentum source density $\boldsymbol{\pi}_{\boldsymbol{p}}$. Thus, the angular momentum source density can be expressed as,

$$\boldsymbol{\pi}_{\boldsymbol{\ell}} = \boldsymbol{r} \times \boldsymbol{\pi}_{\boldsymbol{p}}$$

According to equation (10.30), the stress tensor $\boldsymbol{\tau}$ is the opposite of the momentum current density $\mathrm{j}_{\boldsymbol{p}}$. Likewise, the angular momentum current density can be written as,

$$\mathrm{j}_{\boldsymbol{\ell}} = \boldsymbol{r} \times \mathrm{j}_{\boldsymbol{p}} = -\,\boldsymbol{r} \times \boldsymbol{\tau}$$

Hence, the vectorial divergence of the tensorial angular momentum current density $\mathrm{j}_{\boldsymbol{\ell}}$ is written as,

$$\boldsymbol{\nabla} \cdot \mathrm{j}_{\boldsymbol{\ell}} = -\,\boldsymbol{\nabla} \cdot (\boldsymbol{r} \times \boldsymbol{\tau})$$

The identification of the two expressions for $\boldsymbol{\nabla} \cdot \mathrm{j}_{\boldsymbol{\ell}}$ yields,

$$\boldsymbol{r} \times (\boldsymbol{\nabla} \cdot \boldsymbol{\tau}) = \boldsymbol{\nabla} \cdot (\boldsymbol{r} \times \boldsymbol{\tau})$$

The terms on each side of this equation are written in components as,

$$\boldsymbol{r} \times (\boldsymbol{\nabla} \cdot \boldsymbol{\tau})_i = \varepsilon_{ijk}\, r^j\, \partial_\ell\, \tau^{\ell k}$$
$$\boldsymbol{\nabla} \cdot (\boldsymbol{r} \times \boldsymbol{\tau})_i = \varepsilon_{ijk}\, \partial_\ell\, \left(r^j\, \tau^{\ell k} \right) = \varepsilon_{ijk}\, \tau^{jk} + \varepsilon_{ijk}\, r^j\, \partial_\ell\, \tau^{\ell k}$$

where we used the Einstein summation convention for alternated indices and ε_{ijk} is the antisymmetric Levi-Civita tensor obtained by permutation of the indices 1, 2 and 3 of the tensor $\varepsilon_{123} = 1$. *Thus, the components of the stress tensor have to satisfy the following condition,*

$$\varepsilon_{ijk}\, \tau^{jk} = 0$$

which can be written as,

$$\varepsilon_{ijk}\, \tau^{jk} = \frac{1}{2}\left(\varepsilon_{ijk}\, \tau^{jk} + \varepsilon_{ikj}\, \tau^{kj}\right) = \frac{1}{2}\,\varepsilon_{ijk}\left(\tau^{jk} - \tau^{kj}\right) = 0$$

since $\varepsilon_{ikj} = -\varepsilon_{ijk}$. This means that the antisymmetric part of the stress tensor vanishes. Thus, the stress tensor is represented by a 3×3 symmetric matrix, i.e.

$$\tau^{jk} = \tau^{kj}$$

10.4.4 Accelerated Fluid Container

A container with vertical walls and a rectangular base is subjected to a constant acceleration *a* oriented towards the right. The liquid, of mass density m, inside the container is assumed to be at equilibrium with respect to the container. Friction is negligible.

a) Determine the pressure inside the liquid as a function of the horizontal coordinate x and the vertical coordinate z.
b) Show that the surface of the liquid is tilted backwards with a uniform inclination angle α (Fig. 10.3). Determine the expression for α.

Solution:

a) *In the absence of viscous friction, the only external force density exerted on a small volume of liquid is its weight density,*

$$\boldsymbol{f}^{\text{ext}} = m\,\boldsymbol{g}$$

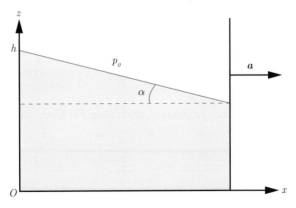

Figure 10.3   A container filled with a liquid is subjected to a constant acceleration. In a stationary state, the water surface is tilted backwards with a uniform inclination angle α.

Since there is no shear, no vorticity and no friction, i.e. $\tau^{fr} = 0$, according to equation (10.81) the divergence of the stress tensor reduces to the opposite of the pressure gradient,

$$\nabla \cdot \tau = -\nabla p$$

Thus, Newton's 2nd law (10.35) can be recast as,

$$\nabla p = -m\,\boldsymbol{a} + m\,\boldsymbol{g}$$

The pressure gradient ∇p is expressed in Cartesian coordinates as,

$$\nabla p = \frac{\partial p}{\partial x}\,\hat{\boldsymbol{x}} + \frac{\partial p}{\partial z}\,\hat{\boldsymbol{z}}$$

and the acceleration $\boldsymbol{a}$ and the gravitational field $\boldsymbol{g}$ are written as,

$$\boldsymbol{a} = a\,\hat{\boldsymbol{x}} \qquad \text{and} \qquad \boldsymbol{g} = -g\,\hat{\boldsymbol{z}}$$

which implies that,

$$\frac{\partial p}{\partial x} = -m\,a \qquad \text{and} \qquad \frac{\partial p}{\partial z} = -m\,g$$

The pressure differential can be expressed as,

$$dp\,(x,z) = \frac{\partial p}{\partial x}\,dx + \frac{\partial p}{\partial z}\,dz = -m\,a\,dx - m\,g\,dz$$

The pressure $p\,(x,z)$ is obtained by integration over the spatial coordinates x and z,

$$p\,(x,z) = -m\,a\,x - m\,g\,z + p\,(0,0)$$

where the integration constant $p\,(0,0)$ corresponds to the pressure at the origin O of the coordinate frame. The pressure $p\,(0,0)$ is the sum of the atmospheric pressure p_0 and the hydrostatic pressure of the column of liquid $m\,g\,h$,

$$p\,(0,0) = p_0 + m\,g\,h$$

Thus, the pressure at point (x,z) inside the liquid is given by,

$$p\,(x,z) = -m\,a\,x - m\,g\,z + p_0 + m\,g\,h$$

b) *The previous equation can be recast as,*

$$z = -\frac{a}{g}\,x + h - \frac{p - p_0}{m\,g}$$

At the surface of the liquid, the pressure is simply the atmospheric pressure, i.e. $p = p_0$. Thus, the previous relation reduces to,

$$z = -\frac{a}{g}\,x + h \qquad \text{(at the surface)}$$

which is a straight line with a negative slope since a, g and h are positive constants. Thus, the tilt angle α is determined by computing the ratio of the coordinates,

$$\tan \alpha = -\frac{z}{x} = \frac{a}{g} \qquad \text{thus} \qquad \alpha = \arctan\left(\frac{a}{g}\right)$$

10.4.5 Archimedes' Force Accelerometer

Consider a homogeneous cork floating in a container fully filled with liquid. The cork is tied to the bottom of the container by a rope (Fig. 10.4). The container moves with respect to the ground at constant acceleration $\boldsymbol{a}$. The liquid and the floater are assumed to be at rest with respect to the container. The mass density m' of the cork is smaller than the mass density m of the water, i.e. $m' < m$.

a) Determine the expression of Archimedes' force $\boldsymbol{F}_A$, which is the net force exerted by the fluid on the surface S of the cork,

$$\boldsymbol{F}_A = - \int_S p\, d\boldsymbol{S}$$

b) Determine the angle α between the rope and the vertical axis when the cork is at rest with respect to the container.

Solution:

a) *In the absence of viscous friction, the only external force density exerted on a small volume of liquid is its weight density, i.e.*

$$\boldsymbol{f}^{\text{ext}} = m\,\boldsymbol{g}$$

Since there is no shear, no vorticity and no friction, i.e. $\tau^{fr} = 0$, according to equation (10.81) the divergence of the stress tensor reduces to the opposite of the pressure gradient,

$$\boldsymbol{\nabla} \cdot \boldsymbol{\tau} = - \boldsymbol{\nabla} p$$

Thus, Newton's 2nd law (10.35) can be recast as,

$$- \boldsymbol{\nabla} p = - m\,\boldsymbol{g} + m\,\boldsymbol{a}$$

The integral over the volume of the previous expression can be written as,

$$- \int_V \boldsymbol{\nabla} p\, dV = - m\,(\boldsymbol{g} - \boldsymbol{a}) \int_V dV = - m\,V\,(\boldsymbol{g} - \boldsymbol{a})$$

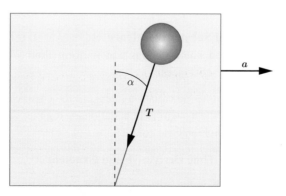

Figure 10.4 An accelerometer is made of a container filled with liquid and a cork tied to the bottom of the container. The container moves with a constant acceleration and the cork leans forward in the direction of the acceleration.

Archimedes' force $\boldsymbol{F}_A$ exerted by the fluid on the surface S of the cork is obtained by applying the divergence theorem,

$$\boldsymbol{F}_A = -\int_S p \, d\boldsymbol{S} = -\int_V \boldsymbol{\nabla} p \, dV = -m \, V (\boldsymbol{g} - \boldsymbol{a})$$

In the particular case where there is no acceleration, i.e. $\boldsymbol{a} = \boldsymbol{0}$, we recover Archimedes' principle,

$$\boldsymbol{F}_A = -m \, V \boldsymbol{g} \qquad \text{(Archimedes' principle)}$$

b) *The cork of mass M' is subjected to three external forces, its weight $M'\boldsymbol{g}$, Archimedes' force $\boldsymbol{F}_A$ and the tension $\boldsymbol{T}$ in the rope. Thus, Newton's 2nd applied on the cork reads,*

$$M'\boldsymbol{g} + \boldsymbol{F}_A + \boldsymbol{T} = M'\boldsymbol{a}$$

The mass of the homogeneous cork can be written as $M' = m' V$ where m' is its mass density. Thus, the tension is given by,

$$\boldsymbol{T} = -m' V (\boldsymbol{g} - \boldsymbol{a}) - \boldsymbol{F}_A = (m - m') V (\boldsymbol{g} - \boldsymbol{a})$$

These vectorial quantities are expressed as,

$$\boldsymbol{T} = T_x \hat{\boldsymbol{x}} + T_y \hat{\boldsymbol{y}} \qquad \boldsymbol{a} = a\hat{\boldsymbol{x}} \qquad \boldsymbol{g} = -g\hat{\boldsymbol{y}}$$

where $\hat{\boldsymbol{x}}$ is the unit vector along the horizontal axis defined positively towards the right (Fig. 10.4) and $\hat{\boldsymbol{y}}$ is the unit vector along the vertical axis defined positively upwards. The angle α is the angle between the rope and the vertical axis, and the tension is collinear to the rope. Thus,

$$\tan\alpha = \frac{T_x}{T_y} = \frac{a}{g} \qquad \text{thus} \qquad \alpha = \arctan\left(\frac{a}{g}\right)$$

which means that the cork leans forwards if $a > 0$ and backwards if $a < 0$.

Exercises

10.1   Chemical Substance Balance

Consider a homogeneous and uniform fluid which consists of different reactive chemical substances.

1. Determine the variation rate $\dot{n}_A$ of the chemical substance density A.
2. When the system is in a stationary regime, determine the condition imposed on the stoichiometric coefficients ν_{aA}.

10.2   Pressure Time Derivative and Gradient

1. Determine the expression of the pressure time derivative.
2. Determine the expression of the pressure gradient.

10.3 Oil and Water Container

A container with a long spout is filled with oil and water in such a way that the spout is filled to its top (Fig. 10.5). The water and the oil surfaces are in contact with the atmosphere. The height h of the water and oil are the same in the container. Determine the expression of the tilt angle α of the spout in terms of the mass densities m_w and m_o of water and oil.

10.4 Floating Tub Stopper

A spherical weight of radius R is blocking a horizontal circular hole at the bottom of a container filled with a liquid (Fig. 10.6). The liquid of mass density m is at a height H above the hole ($H > R/2$) and the lowest point of the sphere is at a height h below it ($h < R/2$). The pressure above the liquid and below the sphere is the atmospheric pressure p_0. Determine Archimedes' force $\boldsymbol{F}_A$ exerted by the liquid on the spherical plug,

$$\boldsymbol{F}_A = - \int_S p\, d\boldsymbol{S}$$

10.5 Temperature Profile of the Earth's Atmosphere

Model the temperature profile $T(z)$ of the Earth's atmosphere as a function of height z. Ignore winds, clouds and the many effects due to the presence of moisture in the air and treat the air as an ideal gas. Assume that the Earth's atmosphere reaches an equilibrium mostly by matter transfer. Suppose that when an air mass moves up or down, it does so adiabatically because the air thermal conductivity is small.

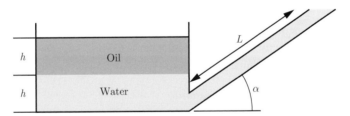

Figure 10.5   A container with equal heights of oil and water, with water filling the spout to its end.

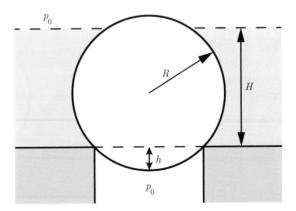

Figure 10.6   A spherical floater used as tub stopper.

a) Show that,

$$T(z) = T_0 - \frac{g}{c_p^*} z$$

where c_p^* is the specific heat at constant pressure per unit mass.

b) Deduce from the temperature profile $T(z)$, the pressure profile $p(z)$ and the mass density profile $m(z)$ of the Earth atmosphere,

$$p(z) = p_0 \left(\frac{T(z)}{T_0} \right)^{c+1} \quad \text{and} \quad m(z) = m_0 \left(\frac{T(z)}{T_0} \right)^c$$

where c is defined in equation (5.60).

10.6 **Stratospheric Balloon** Model the rise of a balloon of mass M, which rises from the ground into the stratosphere [149]. At ground level, the balloon has a volume V_0, which is much smaller than the volume V_{max} it has when it is fully inflated. Nonetheless, the volume V_0 is sufficient to lift the payload. The ballom is filled with helium, which is considered an ideal gas. Use Archimedes' principle given in exercise 10.4.5 and the model of Earth's atmosphere established in exercise 10.5.

a) Find the maximum height z_{max} reached by the balloon.

b) Show that Archimedes' force $\mathbf{F}_A$ exerted on the balloon is uniform as long as the balloon is not fully inflated.

10.7 **Velocity Field Inside a Pipe**

A fluid flows through a pipe which is shaped in such a way that the velocity field depends linearly on the position x along the pipe (Fig. 10.7). At the inlet ($x = 0$), the velocity is v_0. At the outlet ($x = L$), it is $3 v_0$. Determine the acceleration $a(x)$ of a small volume of fluid.

Numerical Application:

$v_0 = 3$ m/s, $L = 0.3$ m.

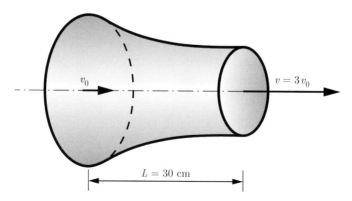

v_0 $v = 3 v_0$ $L = 30$ cm

Figure 10.7 A pipe imposes a certain velocity field $v(x, t)$ to the fluid flowing through it.

10.8 Divergence of a Velocity Field

Establish the continuity equation (10.34) for the mass density by working out the change in mass inside an infinitesimal cubic box centred at a position written in Cartesian coordinates as (x, y, z). The box has square faces orthogonal to the Cartesian axes and the dimensions of the edges of the box are dx, dy and dz. The velocity field is $\mathbf{v}(x, y, z)$.

11

Thermodynamics of Irreversible Processes

Lars Onsager, 1903–1976

L. Onsager attended the Norwegian Institute of Technology in Trondheim and graduated with a degree in chemical engineering in 1925. Following a visit to the Swiss Federal Institute of Technology in Zurich and learning about P. Debye's model of electrolytic solutions, although he disputed it, he was hired there in 1926 to further develop his ideas on electrochemistry. In 1928, having moved to the US and received the news that he was not accepted at Johns Hopkins University based on his poor teaching of general chemistry, he transferred to Brown University where he taught advanced statistical mechanics and published the reciprocity relations that are now named after him. Four years later, he became a professor at Yale University and in 1968 received the Nobel Prize in chemistry.

11.1 Historical Introduction

By the end of the nineteenth century, the physics of transport phenomena consisted of a collection of phenomenological relations. As mentioned in Chapter 3, the term phenomenological means that these relations account well for the observed phenomena, but that there may not be any theoretical framework to justify them. Nowadays, these transport laws are referred to by the name of the scientist who first introduced them.

For example, **Jean Baptiste Joseph, Baron of Fourier** (1768–1830) inferred from experiments a heat equation that relates the time and spatial variations of the temperature. In 1824, he gathered his works into a book entitled: *Analytical theory of heat.* By analogy with the equation of heat which described heat diffusion in solids, **Adolf Eugen Fick** (1829–1901) obtained a diffusion equation for a solute in a solvent. Likewise, **Georg Simon Ohm** (1787–1854) inferred from experiments an equation relating the electric current flowing through an electric conductor and the electric potential difference measured at the ends of the conductor.

Charles Soret (1854–1904), a professor at the University of Geneva whose name has been given to a thermal diffusion effect, studied mixtures of $NaCl$ and KNO_3 in a tube that was heated up at one end and kept cold at the other. The Soret effect can be called **thermophoresis** when the fluid is a suspension. Nowadays, this effect is involved in the production of materials formed by condensation of a metallic substance in the presence of an inert gas. The aggregates thus formed are collected on a cold finger thanks to thermophoresis [69]. **Louis Dufour** (1832–1892), a professor of physics at the Academy of Lausanne, discovered a heat transfer of chemical origin [150, 151, 152]. The Soret and Dufour effects are currently of interest in biophysics [153]. For instance, researchers study how the velocity gradient on the surface of a cell influences its ingestion of nutrients [154, 155]. Others evaluate how a chemical potential variation over the wall of a blood vessel can contribute to the blood shear velocity and thus affect the laminar flow on a prosthesis [156]. The Soret and Dufour effects also appear in heterogenous catalysis [157], and in spintronics [158, 159].

During the nineteenth century, three famous scientists were doing research on the viscosity of liquids and gases. Sir George **Stokes** (1819–1903), Louis **Navier** (1785–1836), and Siméon **Poisson** (1781–1840), were looking for a relationship between internal stresses and the spatial variation of the velocity field in a fluid. Navier and Stokes established an equation of motion for viscous fluids which is still used today, and the science of viscosity is now called **rheology**.

Because these phenomenological laws were discovered during the nineteenth century, many people are under the impression that thermodynamics reached maturity before the beginning of the twentieth century. In fact, it was only in 1940 that Eckart applied thermodynamics to the description of transport phenomena [160, 161]. Thus, this formalism, known as the theory of irreversible processes, came into existence after the dawn of quantum mechanics. Onsager showed that in the neighbourhood of a local thermodynamic equilibrium state, where the generalised forces are small enough, the mathematical structure of the irreversible thermodynamics allows the definition of linear empirical relations generalised by Casimir [162, 163, 164]. The matrix elements relating the generalised current vectors to the generalised force vectors satisfy the symmetry relations called Onsager–Casimir reciprocity relations. Certain authors call the Onsager–Casmir reciprocity relations the linear empirical relations as well as the symmetry relations. We will restrict the definition of this term to the symmetry relations in the remainder of this chapter.

The pioneering work of Eckart remains little known, probably because Eckart did not proceed to a systematic exploitation of his work. On the other hand, Ilya Prigogine started publishing roughly at the same time numerous works on the theory of irreversible processes, notably a very instructive book [165]. Born in Moscow in 1917, his family fled the soviet regime and settled in Belgium, where he was naturalised in 1949. In his career, he was professor in Brussels, Austin Texas and at the University of Chicago. He was awarded the Nobel prize in chemistry in 1977. He later wrote with Isabelle Stengers a remarkable essay on the relations between arts and exact sciences [166]. He conducted research on out-of-equilibrium systems and introduced the notion that dissipative structures can become ordered when a large energy current flows through them [167].

In order to appreciate these ideas, we need to become familiar with the theory of irreversible processes, which is presented in this chapter. We can understand why the description of continuous media is required by considering experiments such as the Joule calorimeter (Fig. 1.1), where oil heats up while stirred, or the temperature increase of balls shaken inside a cylinder. Intuitively, it is clear that friction between solids causes heat. This leads to the question as to how friction phenomena can be accounted for in a liquid. It was not until the twentieth century that friction in a fluid was described using the notion of local entropy production. In this chapter, we will see that this theoretical development can account for the empirical laws describing the transport of matter, electric charge and heat.

11.2 Linear Empirical Relations

The irreversible thermodynamic evolution of a continuous medium consisting of different electrically charged chemical substances is described in equation (10.88) as a sum of dissipative terms. The first two terms on the right-hand side of this equation are of scalar nature and describe the dissipation due to chemical reactions and to the internal friction within the continuous medium. The other terms are of vectorial nature and describe the dissipation due to the transport of chemical substances. Pursuing Onsager's approach, equation (10.88) can be formally written as,

$$\pi_s = \frac{1}{T} \left\{ \sum_i F_i j_i + \sum_\alpha \boldsymbol{F}_\alpha \cdot \boldsymbol{j}_\alpha \right\} \tag{11.1}$$

where F_i and $\boldsymbol{F}_\alpha$ represent scalar and vectorial generalised forces and j_i and $\boldsymbol{j}_\alpha$ represent scalar and vectorial generalised current densities. We use the term 'generalised' in order to account for the fact that the physical dimensions may differ in each term of the sum.

In equation (11.1), there are two types of scalar current and force densities which we index by $i \in \{a, f\}$. The index a refers to the chemical reaction a and the index f refers to viscous friction. The chemical affinity $F_a = \mathcal{A}_a$ and the chemical reaction rate density $j_a = \omega_a$ are associated with the chemical reaction a. The dilatation rate $F_f = \boldsymbol{\nabla} \cdot \boldsymbol{v}$ and the irreversible scalar component of the stress tensor $j_f = \tau^{\text{fr}}$ are associated with the internal friction.

In equation (11.1), there are also two types of vectorial current and force densities which we index by $\alpha \in \{s, A\}$. The index s refers to the entropy density and the index A refers to chemical substance A. The thermal force $\boldsymbol{F}_s = -\boldsymbol{\nabla} T$ and the entropy current density $\boldsymbol{j}_s$ are associated with heat transport. The force $\boldsymbol{F}_A = -\boldsymbol{\nabla}\mu_A - q_A\boldsymbol{\nabla}\varphi$ and the current density $\boldsymbol{j}_A$ are associated with transport of chemical substance A.

The evolution condition of the second law (2.1) requires the positivity (10.23) of the entropy source density, i.e. $\pi_s \geqslant 0$. This is only possible if the entropy source density π_s is, in the neighbourhood of an equilibrium state, composed of a sum of positive definite quadratic forms of the generalised forces. In this neighbourhood, the components of these quadratic forms are independent of the generalised forces. The Curie symmetry principle forbids linear couplings between scalar forces and vectorial forces [168]. Thus, in the neighbourhood of an equilibrium state, the entropy source density π_s is expressed as the sum of a quadratic form of the scalar forces F_i and a quadratic form of the vectorial forces $\boldsymbol{F}_\alpha$,

$$\pi_s = \frac{1}{T}\left(\sum_{i,j} F_i\,(L_{ij}\,F_j) + \sum_{\alpha,\beta} \boldsymbol{F}_\alpha\cdot(\mathsf{L}_{\alpha\beta}\cdot\boldsymbol{F}_\beta)\right) \geqslant 0 \tag{11.2}$$

where the empirical components L_{ij} and $\mathsf{L}_{\alpha\beta}$ are called **Onsager matrix elements**. These are of a different nature: the components L_{ij} are scalars and the components $\mathsf{L}_{\alpha\beta}$ are tensors. These empirical components depend on the state fields s, $\{n_A\}$ and q and their symmetries are given by the Onsager-Casimir reciprocity relations [162, 164],

$$\begin{aligned} L_{ij}\,(s,\{n_A\},q) &= \varepsilon_i\,\varepsilon_j\,L_{ji}\,(s,\{n_A\},q)\\ \mathsf{L}_{\alpha\beta}\,(s,\{n_A\},q) &= \varepsilon_\alpha\,\varepsilon_\beta\,\mathsf{L}_{\beta\alpha}\,(s,\{n_A\},q) \end{aligned} \tag{11.3}$$

where the parameters $\varepsilon_i = \pm 1$, $\varepsilon_j = \pm 1$ and $\varepsilon_\alpha = \varepsilon_\beta = 1$. The parameters ε_i and ε_j are positive if the corresponding generalised scalar forces F_i and F_j are invariant under time reversal, and they are negative in the opposite case. The parameters ε_α and ε_β are positive because the corresponding generalised vectorial forces $\boldsymbol{F}_\alpha$ and $\boldsymbol{F}_\beta$ are invariant under time reversal.

The Onsager–Casimir reciprocity relations (11.3) cannot be established in the framework of a thermodynamic approach. They can be deduced only in a statistical framework because they are a consequence of the reversibility of microscopic dynamics. The Onsager matrices $\{L_{ij}\}$ and $\{\mathsf{L}_{\alpha\beta}\}$ have to be positive in order to ensure the positivity of the entropy source density,

$$\frac{1}{T}\{L_{ij}\} \geqslant 0 \qquad\text{and}\qquad \frac{1}{T}\{\mathsf{L}_{\alpha\beta}\} \geqslant 0 \tag{11.4}$$

In the neighbourhood of a local equilibrium state, where the scalar forces F_i are small enough, the scalar current densities j_i can be expanded to 1st order in terms of the forces F_j. This yields the following scalar linear empirical relations [169],

$$j_i = \sum_j L_{ij}\,F_j \tag{11.5}$$

where the components L_{ij} are independent of the generalised forces F_j. The empirical relations (11.5) describing the irreversibility due to the chemical reactions and to viscous friction are written in terms of the generalised current densities $j_a = \omega_a$ and $j_f = \tau^{\text{fr}}$ and the generalised forces $F_b = \mathcal{A}_b$ and $F_f = \boldsymbol{\nabla}\cdot\boldsymbol{v}$ as,

$$
\begin{cases}
\omega_a = \displaystyle\sum_{b=1}^{n} L_{ab}\, \mathcal{A}_b + L_{af}\, \boldsymbol{\nabla} \cdot \boldsymbol{v} \qquad \forall\ a = 1,\ldots,n \\[2mm]
\tau^{\,\text{fr}} = \displaystyle\sum_{b=1}^{n} L_{fb}\, \mathcal{A}_b + L_{ff}\, \boldsymbol{\nabla} \cdot \boldsymbol{v}
\end{cases}
\tag{11.6}
$$

where the corresponding Onsager matrix (11.4) is positive definite. In terms of the components of the matrix, this condition is expressed as,

$$
\frac{1}{T}
\begin{pmatrix}
L_{11} & \cdots & L_{1n} & L_{1f} \\
\vdots & \ddots & \vdots & \vdots \\
L_{n1} & \cdots & L_{nn} & L_{nf} \\
L_{f1} & \cdots & L_{fn} & L_{ff}
\end{pmatrix}
\geqslant 0
\tag{11.7}
$$

where the indices $a, b = 1, \ldots, n$. In the particular case of a single chemical reaction a, the scalar Onsager–Casimir reciprocity relations (11.3) imply that $L_{fa} = -L_{af}$ and the scalar relations (11.6) reduce to,

$$
\begin{cases}
\omega_a = L_{aa}\, \mathcal{A}_a + L_{af}\, \boldsymbol{\nabla} \cdot \boldsymbol{v} \\[2mm]
\tau^{\,\text{fr}} = -L_{af}\, \mathcal{A}_a + L_{ff}\, \boldsymbol{\nabla} \cdot \boldsymbol{v}
\end{cases}
\tag{11.8}
$$

The condition (11.7) is reduced to,

$$
\frac{1}{T}
\begin{pmatrix}
L_{aa} & L_{af} \\
-L_{af} & L_{ff}
\end{pmatrix}
\geqslant 0
\tag{11.9}
$$

The negative signs in relations (11.8) and (11.9) are due to the fact that the generalised force $F_a = \mathcal{A}_a$ is invariant under time reversal whereas the generalised force $F_f = \boldsymbol{\nabla} \cdot \boldsymbol{v}$ changes its sign according to the transformation law (2.4) of the velocity $\boldsymbol{v}$ under time reversal. This implies that $\varepsilon_a = 1$ and $\varepsilon_f = -1$.

Similarly, in the neighbourhood of a local equilibrium state, the vectorial forces $\boldsymbol{F}_\alpha$ and the vectorial current densities $\boldsymbol{j}_\alpha$ can be expanded to 1st order in terms of the forces $\boldsymbol{F}_\beta$. This yields the following vectorial linear empirical relations [169],

$$
\boldsymbol{j}_\alpha = \sum_{\beta} \mathsf{L}_{\alpha\beta} \cdot \boldsymbol{F}_\beta
\tag{11.10}
$$

where the components $\mathsf{L}_{\alpha\beta}$ are independent of the generalised forces $\boldsymbol{F}_\beta$. The empirical relations (11.10) describe the irreversibility due to heat and matter transport. These two contributions to irreversibility can be written in terms of the generalised current densities $\boldsymbol{j}_s$ and $\boldsymbol{j}_A$ and of the generalised forces $\boldsymbol{F}_s = -\boldsymbol{\nabla} T$ and $\boldsymbol{F}_B = -\boldsymbol{\nabla}\mu_B - q_B \boldsymbol{\nabla}\varphi$ as,

$$
\begin{cases}
\boldsymbol{j}_s = \mathsf{L}_{ss} \cdot (-\boldsymbol{\nabla} T) + \displaystyle\sum_{B=1}^{r} \mathsf{L}_{sB} \cdot \left(-\boldsymbol{\nabla}\mu_B - q_B \boldsymbol{\nabla}\varphi\right) \\[3mm]
\boldsymbol{j}_A = \mathsf{L}_{As} \cdot (-\boldsymbol{\nabla} T) + \displaystyle\sum_{B=1}^{r} \mathsf{L}_{AB} \cdot \left(-\boldsymbol{\nabla}\mu_B - q_B \boldsymbol{\nabla}\varphi\right) \qquad \forall\ A = 1,\ldots,r
\end{cases}
\tag{11.11}
$$

where the corresponding Onsager matrix (11.4) is positive definite. In terms of the components of the matrix, this condition is expressed as,

$$\frac{1}{T}\begin{pmatrix} L_{ss} & L_{s1} & \cdots & L_{sr} \\ L_{1s} & L_{11} & \cdots & L_{1r} \\ \vdots & \vdots & \ddots & \vdots \\ L_{rs} & L_{r1} & \cdots & L_{rr} \end{pmatrix} \geqslant 0 \tag{11.12}$$

where the indices $A, B = 1, \ldots, r$. In the particular case of a single chemical substance A, the vectorial Onsager–Casimir reciprocity relations (11.3) imply that $L_{sA} = L_{As}$, and the vectorial relations (11.11) reduce to,

$$\begin{cases} \boldsymbol{j}_s = L_{ss} \cdot (-\boldsymbol{\nabla} T) + L_{sA} \cdot \left(-\boldsymbol{\nabla} \mu_A - q_A \boldsymbol{\nabla} \varphi \right) \\ \boldsymbol{j}_A = L_{sA} \cdot (-\boldsymbol{\nabla} T) + L_{AA} \cdot \left(-\boldsymbol{\nabla} \mu_A - q_A \boldsymbol{\nabla} \varphi \right) \end{cases} \tag{11.13}$$

and the condition (11.12) is reduced to,

$$\frac{1}{T}\begin{pmatrix} L_{ss} & L_{sA} \\ L_{sA} & L_{AA} \end{pmatrix} \geqslant 0 \tag{11.14}$$

It is possible to deduce from these linear relations the empirical laws of Fourier, Fick and Ohm as well as the Soret and Dufour effects. But before doing so, we examine first the physical meaning of the scalar relations.

11.3 Chemical Reactions and Viscous Friction

The linear empirical relations between scalar quantities describe the irreversibility associated with the chemical reactions between the substances and with the viscous friction in the continuous medium.

11.3.1 Coupled Chemical Reactions

Let us consider different substances A, the amounts of which are linked, due to the chemical reactions a and b. In the absence of dilatation, i.e. $\boldsymbol{\nabla} \cdot \boldsymbol{v} = 0$, the scalar linear relations (11.6) account for a coupling between chemical reactions a and b,

$$\omega_a = \sum_b L_{ab} \left(s, \{n_A\}, q \right) \mathcal{A}_b \tag{11.15}$$

In general, the linear approximation does not describe chemical reactions adequately and the concentration dependence of the coefficients $L_{ab} \left(s, \{n_A\}, q \right)$ has to be taken into account. Let us introduce now an example in which the linear approximation gives a fair description: it is the relaxation of electron spins.

In a quantum mechanical framework, electrons have an intrinsic property called spin. In a metal, we can assume that there are two types of conduction electrons that can be considered as two different substances: a substance U for which the spins are oriented along one spatial direction, and a substance D for which the spins are oriented in the opposite direction.

In this example, there are two possible reactions: a reaction $a : U \rightarrow D$ converting U type electrons into D type electrons and an inverse reaction $b : D \rightarrow U$ converting D type electrons into U type electrons. Then, applying equation (10.26) to a local system at rest, we obtain,

$$\dot{n}_U = \nu_{aU}\,\omega_a + \nu_{bU}\,\omega_b = -\,\omega_a + \omega_b$$
$$\dot{n}_D = \nu_{aD}\,\omega_a + \nu_{bD}\,\omega_b = \omega_a - \omega_b \tag{11.16}$$

where n_U and n_D represent type U and D electron densities. Note that the equations (11.16) imply the conservation of the total electron density $n_U + n_D$, as expected. The chemical affinities, according to equation (10.83), are expressed as,

$$\mathcal{A}_a = \mu_U - \mu_D = -\,\mathcal{A}_b$$

Omitting the explicit dependence on the state fields s, n_A and q, the linear relations (11.15) become,

$$\omega_a = L_{aa}\,\mathcal{A}_a + L_{ab}\,\mathcal{A}_b = (L_{aa} - L_{ab})\,\mathcal{A}_a = (L_{aa} - L_{ab})\,(\mu_U - \mu_D)$$
$$\omega_b = L_{bb}\,\mathcal{A}_b + L_{ba}\,\mathcal{A}_a = (L_{ba} - L_{bb})\,\mathcal{A}_a = (L_{ba} - L_{bb})\,(\mu_U - \mu_D) \tag{11.17}$$

Thus, the empirical equations (11.16) and (11.17) imply [159]:

$$\dot{n}_U = -\,W\,(\mu_U - \mu_D) = -\,\dot{n}_D \tag{11.18}$$

where $W = L_{aa} - L_{ab} - L_{ba} + L_{bb} > 0$ according to condition (11.7).

11.3.2 Viscosity

In the presence of a single chemical reaction at equilibrium, i.e. $\omega_a = 0$, the scalar linear empirical equation (11.6) links the mechanical constraint $\tau^{\,\mathrm{fr}}$ and the dilatation rate $\nabla \cdot \boldsymbol{v}$. This can be written as,

$$\tau^{\,\mathrm{fr}} = \eta\,(s, n_A, q)\,\nabla \cdot \boldsymbol{v} \tag{11.19}$$

where the volume viscosity $\eta\,(s, n_A, q)\,(s, \{n_A\}, q)$ is a scalar coefficient defined as,

$$\eta\,(s, n_A, q) \equiv L_{ff} + \frac{L_{af}^2}{L_{aa}} \tag{11.20}$$

11.4 Transport

The linear empirical relations between vectorial quantities describe the irreversibility associated with transport of substances in a continuous medium. In Chapter 3, we derived two linear empirical relations describing the irreversibility associated with heat and matter transport between simple subsystems. The relation (3.16) is a 'discrete' formulation of Fourier's law for heat transport and equation (3.45) is a 'discrete' formulation of Fick's law for matter transport. In this section, we will establish these laws, and others like Ohm's law, for a continuum.

11.4.1 Fourier's Law

We will use two different approaches in order to clarify the link between the 'discrete' and the 'continuum' formulations. First, we deduce Fourier's law for a continuum using Fourier's law (3.16) for 'discrete' subsystems. Then we establish Fourier's law for a continuum by using the vectorial empirical relations (11.13).

Let us consider an isolated system consisting of two simple subsystems at temperature T^+ and T^-, where $T^+ > T^-$. According to Fourier's 'discrete' law (3.16), the heat transfer between the two subsystems is expressed as,

$$P_Q = \kappa \frac{A}{\ell} \left(T^+ - T^- \right) \tag{11.21}$$

where $\kappa > 0$ is the thermal conductivity, A is the area of the interface between the two subsystems and ℓ is a characteristic length (Fig. 11.1). To obtain a 'continuous' formulation of Fourier's law, we consider that the system is inhomogeneous and that the temperature varies continuously and linearly from the maximum value T^+ on the left to the minimum value T^- on the right. Let ℓ be the length between the two ends of the system and $\hat{x}$ the unit vector oriented from left to right. The temperature gradient is a vector oriented in the direction of the temperature increase that is written as,

$$\boldsymbol{\nabla} T = -\frac{T^+ - T^-}{\ell} \, \hat{x} \tag{11.22}$$

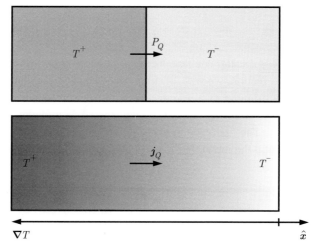

Figure 11.1 The discrete system (top) consists of two simple subsystems at temperature T^+ and T^- and the continuous system (bottom) is subjected to a temperature gradient $\boldsymbol{\nabla} T$ from T^- to T^+.

The heat current density $\boldsymbol{j}_Q$ is a vector oriented in the direction of the temperature decrease and defined as the ratio of the thermal power and the area of the interface,

$$\boldsymbol{j}_Q = \frac{P_Q}{A}\,\hat{\boldsymbol{x}} \tag{11.23}$$

Taking into account definitions (11.22) and (11.23), Fourier's law (11.21) is expressed in a continuous form as,

$$\boldsymbol{j}_Q = -\kappa\,\boldsymbol{\nabla}\,T \tag{11.24}$$

We now use our formalism describing irreversible processes in continuous media. Thus, we consider a solid at rest, i.e. $\boldsymbol{j}_A = \boldsymbol{0}$, consisting of a single chemical substance A. In continuous media, a heat current density is defined as,

$$\boldsymbol{j}_Q = T\boldsymbol{j}_s \tag{11.25}$$

Taking into account definition (11.25), the vectorial relations (11.13) imply that,

$$\boldsymbol{j}_Q = -\boldsymbol{\kappa}\,(s,n_A,q)\cdot\boldsymbol{\nabla}\,T \tag{11.26}$$

where

$$\boldsymbol{\kappa}\,(s,n_A,q) \equiv T\left(\mathsf{L}_{ss} - \mathsf{L}_{sA}\cdot\mathsf{L}_{AA}^{-1}\cdot\mathsf{L}_{sA}\right) \tag{11.27}$$

is the thermal conductivity tensor. The diagonal terms correspond to **Fourier's law** (11.24) [170]. The tensorial relation (11.26) can be inverted,

$$\boldsymbol{\nabla}\,T = -\boldsymbol{\kappa}^{-1}\,(s,n_A,q)\cdot\boldsymbol{j}_Q \tag{11.28}$$

The off-diagonal terms of $\boldsymbol{\kappa}^{-1}$ describe another phenomenon. To illustrate this phenomenon, let us consider for instance a thin slab orthogonal to the z-axis through which a heat current $\boldsymbol{j}_Q$ flows in the x-direction. The tensorial relation (11.28) predicts the existence of a temperature gradient $\boldsymbol{\nabla}\,T$ in the y-direction. In the presence of a magnetic induction field $\boldsymbol{B}$ applied in the z-direction, this phenomenon is known as the **Righi-Leduc effect** [171, 172],

$$\boldsymbol{\nabla}\,T = -\mathcal{R}\,(\boldsymbol{j}_Q \times \boldsymbol{B}) \tag{11.29}$$

where $\mathcal{R}$ is a scalar coefficient.

In this section, we will illustrate each transport law that we derive from our formalism with the example of a homogeneous metal, considered as an isotropic solid containing conduction electrons that can be considered as a chemical substance $A = e$ of electric charge q_e. The thermal conductivity of conduction electrons is measured in the frame of reference of the metal. In the particular case of an isotropic fluid, Fourier's law (11.26) is reduced to,

$$\boldsymbol{j}_Q = -\kappa\,(s,n_e,q)\,\boldsymbol{\nabla}\,T \tag{11.30}$$

where $\boldsymbol{\kappa} = \kappa\,\mathbb{1}$, due to isotropy. Thus, the thermal conductivity coefficient κ is deduced from the expression (11.27),

$$\kappa\,(s,n_e) \equiv T\left(L_{ss} - \frac{L_{se}^2}{L_{ee}}\right) \tag{11.31}$$

11.4.2 Heat Equation

We are now in a position to introduce the notion of **diffusion equations**. In general, a linear relation between a current density of an extensive quantity and the gradient of the conjugate quantity, combined with the continuity equation for the extensive quantity gives rise to a differential equation, which is characterised by having a first derivative with respect to time and a second derivative with respect to spatial variables. Here we show how Fourier's law leads to a diffusion equation for temperature. Later, we will introduce Fick's law for a matter current density and we will find a diffusion equation for substances.

Let us consider an isotropic metal that we consider in the frame of reference where it is at rest, i.e. $\boldsymbol{v} = \boldsymbol{0}$. The conduction electrons are assumed at rest in this frame of reference, i.e. $\boldsymbol{j}_e = \boldsymbol{0}$ where $A = e$. Thus, the current density balance (10.87) yields the equality $\boldsymbol{j}_Q = \boldsymbol{j}_u$. In the absence of any mechanical constraint on the metal, i.e. $\pi_u = 0$, the internal energy continuity equation (10.43) is reduced to,

$$\partial_t u = - \boldsymbol{\nabla} \cdot \boldsymbol{j}_u = - \boldsymbol{\nabla} \cdot \boldsymbol{j}_Q \tag{11.32}$$

Since the conduction electrons are at rest, $\partial_t n_e = 0$. For a constant density, $\partial_t u$ depends only on $\partial_t T$,

$$\partial_t u = c_e \, \partial_t T \tag{11.33}$$

where the specific heat density c_e of the conduction electrons is defined as,

$$c_e \equiv \left. \frac{\partial u}{\partial T} \right|_{n_e} \tag{11.34}$$

In the case where the dilatation of the metal can be neglected, the divergence of Fourier's law (11.30), the continuity equation (11.32) and relation (11.33) yield the **heat equation**,

$$\partial_t T = \lambda \, \boldsymbol{\nabla}^2 \, T \tag{11.35}$$

where the **thermal diffusivity** coefficient λ is defined as,

$$\lambda \equiv \frac{\kappa}{c_e} \tag{11.36}$$

For a homogeneous metal, we used the fact that the thermal diffusivity coefficient λ is independent of position, i.e. $\boldsymbol{\nabla} \lambda = \boldsymbol{0}$.

11.4.3 Heat Diffusion in a Rod

In order to illustrate some of the characteristics of diffusion equations, we find a specific solution for the heat equation (11.35) when we apply it to the case of an infinite rod oriented along the x-axis. We assume that heat diffuses in the rod only. The solution of the heat equation is a temperature profile $T(x, t)$ depending on the position x along the rod and on time t. Along the x-axis, the heat differential equation (11.35) is reduced to,

$$\frac{\partial}{\partial t} T(x, t) = \lambda \frac{\partial^2}{\partial x^2} T(x, t) \tag{11.37}$$

To solve this differential equation, we use the method known as separation of variables. It consists in decomposing the temperature field $T(x, t)$ into a product of fields $A(x)$ and $B(t)$,

$$T(x, t) = A(x) B(t) \tag{11.38}$$

The heat equation (11.37) is then decomposed respectively, in two coupled differential equations,

$$\frac{\lambda}{A(x)} \frac{\partial^2 A(x)}{\partial x^2} = -\lambda k^2$$
$$\frac{1}{B(t)} \frac{\partial B(t)}{\partial t} = -\lambda k^2 \tag{11.39}$$

where λ can be any real positive value and k is called a wave number. The solutions of these differential equations are given by,

$$A(x) = A(0) \, e^{ikx}$$
$$B(t) = B(0) \, e^{-k^2 \lambda t} \tag{11.40}$$

Substituting these solutions (11.40) into expression (11.38) for the temperature field, we find a mathematical solution of the heat equation (11.37) given by

$$T(x, t) = T(0, 0) \, e^{-k^2 \lambda t} \, e^{ikx} \tag{11.41}$$

where $T(0, 0) = A(0) B(0)$ is a constant temperature. This solution depends on the wave number k. Every linear combination of a solution of a differential equation is also a solution. Thus, the integral of the solution (11.41) with respect to the wave number k is also a solution,

$$T(x, t) = C \sqrt{\frac{\lambda}{\pi}} \int_{-\infty}^{\infty} e^{-k^2 \lambda t} e^{ikx} dk \tag{11.42}$$

where the constant C is proportional to $T(0, 0)$. Equation (11.42) can be recast as the integral of a Gaussian,

$$T(x, t) = C \sqrt{\frac{\lambda}{\pi}} e^{-\frac{x^2}{4\lambda t}} \int_{-\infty}^{\infty} e^{-\lambda t \left(k - i \frac{x}{2\lambda t}\right)^2} dk \tag{11.43}$$

The integral of the Gaussian in (11.43) yields $\sqrt{\pi/\lambda t}$, which implies,

$$T(x, t) = \frac{C}{\sqrt{t}} e^{-\frac{x^2}{4\lambda t}} \tag{11.44}$$

This result is illustrated in (Fig. 11.2).

At the initial time at the centre of the rod, i.e. $t = 0$ and $x = 0$, the expression (11.44) of the temperature diverges, i.e. $T(0, 0) = \infty$. Moreover, after an infinite time, the temperature vanishes everywhere, i.e. $T(x, \infty) = 0$. These are the mathematical conditions that correspond to a heat source in contact with the centre of the rod, i.e. at position $x = 0$, during an infinitesimal time interval coinciding with $t = 0$. The heat thus transmitted by the source diffuses throughout the rod, which tends towards a thermal equilibrium state after an infinite time.

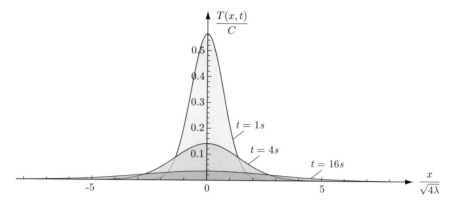

Figure 11.2 The diffusion of a temperature profile $T(x, t)/C$ as a function of the variable $x/\sqrt{4\lambda}$ at different times: $t = \{1\,s, 4\,s, 16\,s\}$.

11.4.4 Fick's Law

As we did for Fourier's law, we deduce Fick's law for a continuum by using Fick's law (3.45) derived for a discrete system. Then, we will establish this law using the empirical relations (11.13).

Let us at first consider an isolated system consisting of two simple subsystems where the chemical potentials of the substance A are μ_A^+ and μ_A^- where $\mu_A^+ > \mu_A^-$. According to Fick's 'discrete' law (3.45), the matter transfer between the two subsystems is expressed as,

$$\dot{N}_A = F_A \frac{A}{\ell}\left(\mu_A^+ - \mu_A^-\right) \tag{11.45}$$

where $F_A > 0$ is Fick's diffusion coefficient of substance A, ℓ is a characteristic length and A is the area of the interface between the subsystems (Fig. 11.3). To obtain a formulation of Fick's law for a continuum, we consider that the system is inhomogeneous and that the chemical potential varies continuously and linearly from the maximum value μ_A^+ on the left to the minimum value μ_A^- on the right. Let ℓ be the length between the two ends of the system and $\hat{x}$ the unit vector oriented from left to right. The chemical potential gradient is a vector oriented in the direction of the chemical potential increase that can be then written as,

$$\nabla \mu_A = -\frac{\mu_A^+ - \mu_A^-}{\ell}\,\hat{x} \tag{11.46}$$

The matter current density $\boldsymbol{j}_A$ is a vector oriented in the direction of the chemical potential decrease and defined as the ratio of the chemical substance variation rate and the area of the interface,

$$\boldsymbol{j}_A = \frac{\dot{N}_A}{A}\,\hat{x} \tag{11.47}$$

Taking into account definitions (11.46) and (11.47), Fick's law (11.45) is expressed for a continuum as,

$$\boldsymbol{j}_A = -F_A \nabla \mu_A \tag{11.48}$$

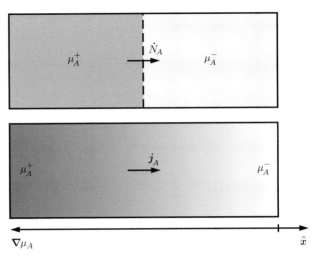

Figure 11.3 The discrete system (top) consists of two simple subsystems in which the chemical potentials are μ_A^+ and μ_A^-. The inhomogeneous system (bottom) is subjected to a chemical potential gradient $\nabla \mu_A$ from μ_A^- to μ_A^+.

Let us now apply the theory of irreversible processes. We consider a single electrically neutral chemical substance A, i.e. $q_A = 0$, dissolved in a non-reactive and electrically neutral fluid in which the temperature is independent of the position, i.e. $\nabla T = \mathbf{0}$. The second vectorial relation (11.13) represents **Fick's law** [173] (11.48),

$$\boldsymbol{j}_A = - \mathsf{F}_A\,(s, n_A) \cdot \nabla \mu_A \tag{11.49}$$

where $\mathsf{F}_A\,(s, n_A) \equiv \mathsf{L}_{AA}\,(s, n_A)$ is the chemical diffusion tensor. In the particular case of an isotropic fluid, Fick's law (11.49) is reduced to,

$$\boldsymbol{j}_A = - F_A\,(s, n_A)\, \nabla \mu_A \tag{11.50}$$

where $\mathsf{F}_A = F_A\,\mathbb{1}$, due to isotropy. Equation (11.50) is known as Fick's first law (11.48).

11.4.5 Diffusion Equation

We obtain a second formulation of Fick's law if we consider the chemical potential μ_A as function of temperature T and densities n_A, i.e. $\mu_A = \mu_A\,(T, n_A)$. If the temperature is independent of position, then $\nabla \mu_A$ depends only on ∇n_A,

$$\nabla \mu_A = \frac{\partial \mu_A}{\partial n_A}\, \nabla n_A$$

Thus, the vectorial relation (11.50) yields Fick's second law,

$$\boldsymbol{j}_A = - D\,(s, \mu_A)\, \nabla n_A \tag{11.51}$$

where the diffusion coefficient $D\,(s, \mu_A)$ is defined as,

$$D\,(s, \mu_A) \equiv F_A\, \frac{\partial \mu_A}{\partial n_A} \tag{11.52}$$

and n_A is the the chemical substance density.

In order to determine the diffusion equation for the chemical substance A in the fluid, we choose the fluid as a frame of reference, i.e. $v = 0$. In the absence of chemical reaction in the fluid, i.e. $\pi_A = 0$, the chemical substance continuity equation (10.26) is reduced to,

$$\partial_t n_A = - \nabla \cdot j_A \tag{11.53}$$

For an isotropic fluid, the **diffusion coefficient** D is independent of position, i.e. $\nabla D = 0$. Thus, the divergence of Fick's second law (11.51) and the continuity equation (11.53) yield the **diffusion equation**,

$$\partial_t n_A = D \nabla^2 n_A \tag{11.54}$$

11.4.6 Dufour Effect

With Fourier's law and Fick's law, we have examples of transport equations that link the current density and the gradient of quantities which are conjugates of one another. We now introduce transport laws known as **cross-effects**, because they link quantities that are not conjugates. Here and in the next section, we identify links between chemical potentials and heat currents or temperature gradients.

Let us consider a single electrically neutral chemical substance A, i.e. $q_A = 0$, dissolved in a non-reactive and electrically neutral fluid in which there is no matter current, i.e. $j_A = 0$. Taking into account the definition of heat current density (11.25), the vectorial relations (11.13) contain the **Dufour effect** [151, 174] describing the heat current induced by a chemical potential gradient $\nabla \mu_A$,

$$j_Q = - \mathsf{D}_A (s, n_A) \cdot \nabla \mu_A \tag{11.55}$$

Here,

$$\mathsf{D}_A (s, n_A) \equiv T \left(\mathsf{L}_{sA} - \mathsf{L}_{ss} \cdot \mathsf{L}_{sA}^{-1} \cdot \mathsf{L}_{AA} \right) \tag{11.56}$$

is a tensor describing this thermochemical effect. In the particular case of an isotropic fluid, the Dufour effect (11.55), is reduced to,

$$j_Q = - D_A (s, n_A) \nabla \mu_A \tag{11.57}$$

where $\mathsf{D}_A = D_A \mathbb{1}$ because of the isotropy of the medium.

11.4.7 Soret Effect

Let us now consider two electrically neutral chemical substances A and B, i.e. $q_A = q_B = 0$, dissolved in a non-reactive and electrically neutral fluid. In the stationary state, that is to say, in the absence of chemical reactions ($\pi_A = \pi_B = 0$) and chemical currents ($j_A = j_B = 0$), the relations (11.11) contain the **Soret effect** [175] describing the gradient of the chemical potential difference between the substances A and B generated by a temperature gradient,

$$\nabla (\Delta\mu) = - \mathsf{S} (s, n_A, n_B) \cdot \nabla T \tag{11.58}$$

where $\Delta\mu = (\mu_A - \mu_B)/2$ is the weighted chemical potential difference and $\mathsf{S} (s, n_A, n_B)$ is a tensor.

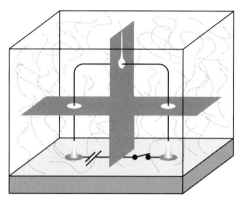

Figure 11.4   In a box filled with smoke, a current flows through a heating wire. Two ribbon-shaped laser beams are made visible by the smoke. The smoke is pulled away from the hot wire because of thermophoresis.

In the particular case of an isotropic fluid, the vectorial relations (11.11), reduce to,

$$\begin{cases} \boldsymbol{j}_s = -L_{ss}\boldsymbol{\nabla}\,T - L_{sA}\boldsymbol{\nabla}\,\mu_A - L_{sB}\boldsymbol{\nabla}\,\mu_B \\ \boldsymbol{j}_A = -L_{As}\boldsymbol{\nabla}\,T - L_{AA}\boldsymbol{\nabla}\,\mu_A - L_{AB}\boldsymbol{\nabla}\,\mu_B \\ \boldsymbol{j}_B = -L_{Bs}\boldsymbol{\nabla}\,T - L_{BA}\boldsymbol{\nabla}\,\mu_A - L_{BB}\boldsymbol{\nabla}\,\mu_B \end{cases} \tag{11.59}$$

The contributions to $\boldsymbol{j}_A$ and $\boldsymbol{j}_B$ proportional to $\boldsymbol{\nabla}\,T$ express the effect known as **thermophoresis**. Thermophoresis is observed easily in a smoke cloud around a hot wire (Fig. 11.4).

Let us now calculate the Soret coefficient in terms of the Onsager coefficients given in equation (11.59). The total current $\boldsymbol{j}_{A+B}$ is defined as the sum of the chemical currents, i.e. $\boldsymbol{j}_{A+B} = \boldsymbol{j}_A + \boldsymbol{j}_B$. Similarly, the diffusion current $\boldsymbol{j}_{A-B}$ of substance A with respect to substance B is defined as the chemical current difference, i.e. $\boldsymbol{j}_{A-B} = \boldsymbol{j}_A - \boldsymbol{j}_B$. The matrix relation (11.59) implies that the total current $\boldsymbol{j}_{A+B}$ and the diffusion current $\boldsymbol{j}_{A-B}$ are given by,

$$\begin{aligned} \boldsymbol{j}_{A+B} &= -\left(L_{As} + L_{Bs}\right)\boldsymbol{\nabla}\,T - \left(L_{AA} + L_{BA}\right)\boldsymbol{\nabla}\,\mu_A - \left(L_{AB} + L_{BB}\right)\boldsymbol{\nabla}\,\mu_B \\ \boldsymbol{j}_{A-B} &= -\left(L_{As} - L_{Bs}\right)\boldsymbol{\nabla}\,T - \left(L_{AA} - L_{BA}\right)\boldsymbol{\nabla}\,\mu_A - \left(L_{AB} - L_{BB}\right)\boldsymbol{\nabla}\,\mu_B \end{aligned} \tag{11.60}$$

The mean chemical potential μ_0 and the deviation $\Delta\mu$ to the mean chemical potential are defined as,

$$\mu_0 \equiv \frac{\mu_A + \mu_B}{2} \qquad \text{and} \qquad \Delta\mu \equiv \frac{\mu_A - \mu_B}{2} \tag{11.61}$$

Using the Onsager relation $L_{AB} = L_{BA}$ and definitions (11.61), the currents (11.60) can be recast as,

$$\begin{aligned} \boldsymbol{j}_{A+B} &= -\left(L_{As} + L_{Bs}\right)\boldsymbol{\nabla}\,T - \left(L_{AA} + L_{BB} + 2L_{AB}\right)\boldsymbol{\nabla}\,\mu_0 \\ &\quad - \left(L_{AA} - L_{BB}\right)\boldsymbol{\nabla}\,(\Delta\mu) \\ \boldsymbol{j}_{A-B} &= -\left(L_{As} - L_{Bs}\right)\boldsymbol{\nabla}\,T - \left(L_{AA} - L_{BB}\right)\boldsymbol{\nabla}\,\mu_0 \\ &\quad - \left(L_{AA} + L_{BB} - 2L_{AB}\right)\boldsymbol{\nabla}\,(\Delta\mu) \end{aligned} \tag{11.62}$$

The first term on the right-hand side of the second equation (11.62) predicts that the relative matter transport $\boldsymbol{j}_{A-B}$ can be induced by a temperature gradient [176].

In the particular case where the system is stationary, the currents vanish, i.e. $\boldsymbol{j}_{A+B} = \boldsymbol{j}_{A-B} = \boldsymbol{0}$. Thus, the equations (11.62) reduce to,

$$\left(\frac{L_{As} + L_{Bs}}{L_{AA} + L_{BB} + 2L_{AB}} - \frac{L_{As} - L_{Bs}}{L_{AA} - L_{BB}} \right) \boldsymbol{\nabla} T =$$
$$\left(\frac{L_{AA} + L_{BB} - 2L_{AB}}{L_{AA} - L_{BB}} - \frac{L_{AA} - L_{BB}}{L_{AA} + L_{BB} + 2L_{AB}} \right) \boldsymbol{\nabla} (\Delta\mu) \tag{11.63}$$

which expresses the scalar equivalent of the Soret effect (11.58),

$$\boldsymbol{\nabla} (\Delta\mu) = - S (s, n_A, n_B) \, \boldsymbol{\nabla} T \tag{11.64}$$

where the thermochemical coefficient $S (s, n_A, n_B)$ of substances A and B is given by,

$$S (s, n_A, n_B) \equiv \frac{1}{2} \frac{L_{AB} (L_{As} - L_{Bs}) + L_{As} L_{BB} - L_{Bs} L_{AA}}{L_{AA} L_{BB} - L_{AB}^2} \, . \tag{11.65}$$

11.4.8 Ohm's Law and Hall Effect

First, we introduce Ohm's law as a consequence of the generalised 'discrete' Fick's law (3.45), applying it to the case of an electrically charged substance. Then, we establish this law using the empirical relations (11.13).

Thus, we consider an isolated system consisting of two simple subsystems where the electrochemical potentials of the electrically charged substance A are $\bar{\mu}_A^+$ and $\bar{\mu}_A^-$ where $\bar{\mu}_A^+ > \bar{\mu}_A^-$. We consider that the chemical potential difference between the two subsystems is negligible compared to the difference in the values of their electrostatic potential. Taking into account definitions $\bar{\mu}_A^+ = \mu_A^+ + q_A \varphi^+$ and $\bar{\mu}_A^- = \mu_A^- + q_A \varphi^-$, in this limit, the electrochemical potential difference is reduced to,

$$\bar{\mu}_A^+ - \bar{\mu}_A^- = q_A \left(\varphi^+ - \varphi^- \right) \tag{11.66}$$

where φ^+ and φ^- are the electrostatic potentials of both substances and q_A is the electric charge of a mole of substance A. The total electric charge of the two subsystems is given by,

$$Q^+ = q_A N_A^+ \qquad \text{and} \qquad Q^- = q_A N_A^- \tag{11.67}$$

Fick's generalised 'discrete' equation applied to an electrically charged substance is obtained by replacing the chemical potentials μ_A^+ and μ_A^- by the electrochemical potentials $\bar{\mu}_A^+$ and $\bar{\mu}_A^-$ in Fick's 'discrete' equation (11.45). Multiplying this generalised equation by the electric charge q_A, taking into account relations (11.66) and (11.67), we obtain the 'discrete' Ohm's law expressing the electric charge transfer between the subsystems $(+)$ and $(-)$ (Fig. 11.5),

$$\dot{Q} = \sigma \frac{A}{\ell} \left(\varphi^+ - \varphi^- \right) \tag{11.68}$$

where $\sigma = q_A^2 F_A > 0$ is the electric conductivity coefficient. To obtain a formulation of Ohm's law for a continuum, we consider that the system is inhomogeneous and that its electrostatic potential varies continuously and linearly, from the maximum value φ^+ on the left to the minimum value φ^- on the right. Let ℓ be the distance between the two ends of the system and $\hat{\boldsymbol{x}}$ the unit vector oriented from left to right. The electrostatic potential

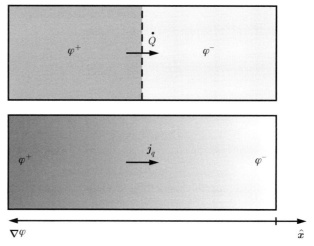

Figure 11.5 A discrete system consists of two simple subsystems of electrostatic potentials φ^+ and φ^- (top). An inhomogeneous system is subjected to a chemical potential gradient $\boldsymbol{\nabla} \varphi$ as the potential goes from φ^- to φ^+ (bottom).

gradient is a vector oriented in the direction of the electrostatic potential increase. It can be written as,

$$\boldsymbol{\nabla} \varphi = -\frac{\varphi^+ - \varphi^-}{\ell} \, \hat{\boldsymbol{x}} \tag{11.69}$$

The electric current density $\boldsymbol{j}_q$ is a vector oriented in the direction of the electrostatic potential decrease. It is defined as the ratio of the electric charge variation rate and the area of the interface,

$$\boldsymbol{j}_q = \frac{\dot{Q}}{A} \, \hat{\boldsymbol{x}} \tag{11.70}$$

Taking into account definitions (11.69) and (11.70), Ohm's law (11.68) for a continuous medium is expressed as,

$$\boldsymbol{j}_q = -\sigma \, \boldsymbol{\nabla} \varphi \tag{11.71}$$

Let us now derive Ohm's law from the thermodynamics of irreversible processes. We consider a single homogeneous chemical substance A, i.e. $\boldsymbol{\nabla} \mu_A = \boldsymbol{0}$, with a position-independent temperature, i.e. $\boldsymbol{\nabla} T = \boldsymbol{0}$. Taking into account definition (10.74) for the conductive electric current density $\boldsymbol{j}_q$, the second vectorial relation (11.13) implies,

$$\boldsymbol{j}_q = -\boldsymbol{\sigma} \, (s, n_A, q) \cdot \boldsymbol{\nabla} \varphi \tag{11.72}$$

where

$$\boldsymbol{\sigma} \, (s, n_A, q) \equiv q_A^2 \, \mathsf{L}_{AA} \tag{11.73}$$

is the electric conductivity tensor. Relation (11.72) can be inverted to express the gradient in terms of the current,

$$\boldsymbol{\nabla} \varphi = -\boldsymbol{\rho} \, (s, n_A, q) \cdot \boldsymbol{j}_q \tag{11.74}$$

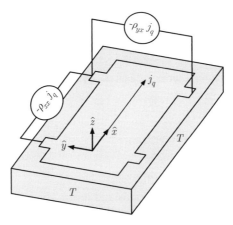

Figure 11.6 A schematic representation of the ohmic voltage and Hall voltage. Typically, these voltages are measured when a magnetic field is applied in the orthogonal direction to the sample plane.

where $\rho\,(s, n_A, q) = \sigma^{-1}\,(s, n_A, q)$ is the electric resistivity tensor. The diagonal elements of the tensor correspond to **Ohm's law** [177]. In reference to the isothermal condition, it could be necessary to specify that it is the **isothermal resistivity**.

The off-diagonal terms of $\rho\,(s, n_A, q)$ describe another phenomenon. To illustrate this phenomenon, let us consider for instance a thin slab orthogonal to the z-axis through which an electric current $\boldsymbol{j}_q$ flows in the x-direction. The tensorial relation (11.74) predicts the existence of a potential gradient $\nabla\,\varphi$ in the y-direction. In the presence of a magnetic induction field $\boldsymbol{B}$ applied in the z-direction, this phenomenon is known as the **Hall effect** [178],

$$\nabla\,\varphi = -\,\mathcal{H}\,\left(\boldsymbol{j}_q \times \boldsymbol{B}\right) \tag{11.75}$$

where $\mathcal{H}$ is a scalar coefficient.

In a metal, the electric conduction is essentially due to the conduction electrons that are considered as a chemical substance $A = e$ of electric charge q_e. For an isotropic metal, Ohm's law (11.72) is reduced to,

$$\boldsymbol{j}_q = -\,\sigma\,(s, n_e, q)\,\nabla\,\varphi \tag{11.76}$$

where $\sigma = \sigma\,\mathbb{1}$ due to isotropy. Thus, the electric conductivity coefficient σ is defined as,

$$\sigma\,(s, n_e, q) \equiv q_e^2\,L_{ee} \tag{11.77}$$

11.4.9 Ettingshausen Effect and Adiabatic Resistivity

Let us now consider a fluid consisting of a single chemical substance A of electric charge q_A, in the absence of heat current density, i.e. $\boldsymbol{j}_Q = \boldsymbol{0}$. Taking into account the definition (10.74) of the conductive electric current density $\boldsymbol{j}_q$, the linear vectorial relations (11.13) imply,

$$\nabla\,T = -\,\mathsf{E}\,(s, n_A, q)\cdot\boldsymbol{j}_q \tag{11.78}$$

where

$$E\left(s, n_A, q\right) \equiv \frac{1}{q_A} \left(L_{sA} - L_{AA} \cdot L_{sA}^{-1} \cdot L_{ss}\right)^{-1} \tag{11.79}$$

The off-diagonal terms of $E\left(s, n_A, q\right)$ imply that an electric current j_q in the x direction can induce a temperature gradient ∇T in the y direction. In the presence of a magnetic induction field B applied in the z-direction, this phenomenon is known as the **Ettingshausen effect** [179],

$$\nabla T = -\mathcal{E}\left(j_q \times B\right) \tag{11.80}$$

where $\mathcal{E}$ is a scalar coefficient.

Under the same conditions and in the absence of chemical effects, i.e. $\nabla \mu_A = 0$, the vectorial relations (11.11) yield a linear relation between the electric current density j_q and the potential gradient $\nabla \varphi$ that gives rise to a second form of Ohm's law,

$$\nabla \varphi = -\rho_{\mathrm{ad}}\left(s, n_A, q\right) \cdot j_q \tag{11.81}$$

where

$$\rho_{\mathrm{ad}}\left(s, n_A, q\right) \equiv \frac{1}{q_A^2} \left(L_{AA} - L_{sA} \cdot L_{ss}^{-1} \cdot L_{sA}\right)^{-1} \tag{11.82}$$

Here, $\rho_{\mathrm{ad}}\left(s, n_A, q\right)$ is called the **adiabatic resistivity** tensor. It is different from the isothermal resistivity tensor (11.74), which is defined for the case when the temperature is independent of position. Thus, resistivity (longitudinal) and Hall resistance do not have the same values under isothermal or adiabatic conditions.

11.4.10 Seebeck Effect and Nernst Effect

In § 11.4.6 and § 11.4.7, we presented cross-effects involving matter transport. In this section, we define two cross-effects that concern heat transport and electrostatic potentials. To do so, we consider a fluid consisting of a single chemical substance A of electric charge q_A, in the absence of matter current, i.e. $j_A = 0$. The chemical potential μ_A is assumed to be independent of position, i.e. $\nabla \mu_A = 0$. The second vectorial relation (11.11) implies,

$$\nabla \varphi = -\varepsilon\left(s, n_A, q\right) \cdot \nabla T \tag{11.83}$$

where

$$\varepsilon\left(s, n_A, q\right) \equiv \frac{1}{q_A} L_{AA}^{-1} \cdot L_{sA} \tag{11.84}$$

The diagonal elements of the tensor correspond to the **Seebeck effect** [180].

The off-diagonal terms of ε describe another phenomenon. In order to illustrate this phenomenon, let us consider a thin slab orthogonal to the z-direction, subjected to a temperature gradient ∇T in the x-direction. The tensorial relation (11.74) predicts the existence of a potential gradient $\nabla \varphi$ in the y-direction. In the presence of a magnetic induction field B applied in the z-direction, this phenomenon is known as the **Nernst effect**,

$$\nabla \varphi = -\mathcal{N}\left(\nabla T \times B\right) \tag{11.85}$$

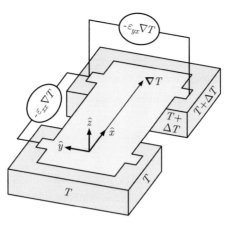

Figure 11.7 A temperature gradient in the x direction induces an electrostatic gradient in the x direction (Seebeck effect) and in the y direction (Nernst effect). Typically, voltage differences are measured at the contacts on the side of the sample and a magnetic field is applied in the direction orthogonal to the sample plane.

where $\mathcal{N}$ is a scalar coefficient. The geometry of the Nernst effect corresponds to that of the Hall effect, where the electric current density $\boldsymbol{j}_q$ is replaced by a temperature gradient $\boldsymbol{\nabla} T$ which is directly linked to a heat current in the same direction, owing to Fourier's law (Fig. 11.7).

In the case of conduction electrons in an isotropic metal, considered as a chemical substance $A = e$ of electric charge $q = e$, the Seebeck effect (11.83) is reduced to,

$$\boldsymbol{\nabla} \varphi = -\varepsilon \left(s, n_e, q\right) \boldsymbol{\nabla} T \tag{11.86}$$

In other words, $\varepsilon = \varepsilon \, \mathbb{1}$ because of the isotropy of the medium. The **Seebeck coefficient** ε is also called the **thermoelectric power**. For conduction electrons in an isotropic metal, equation (11.84) reads,

$$\varepsilon \left(s, n_e, q\right) \equiv \frac{L_{se}}{q_e \, L_{ee}} \tag{11.87}$$

Thermocouples are used as thermometers (§ 11.6.1). The principle of the measurement relies on the Seebeck effect. A typical value for the Seebeck coefficient of a metal is a few tens of microvolts per Kelvin at room temperature.

11.4.11 Joule Effect and Thomson Effect

In this section, we consider heat transport in the presence of a charge current. This will allow us to derive two more transport effects, one of which is very important since it is the Joule effect, commonly known as the heating of a resistor subjected to an electric current. Thus, we consider an isotropic metal containing conduction electrons that are considered as an isotropic chemical substance $A = e$ of electric charge q_e. In this case, the empirical vectorial relations (11.13) reduce to,

$$\begin{cases} \boldsymbol{j}_s = -L_{ss} \, \boldsymbol{\nabla} T - L_{se} \, \boldsymbol{\nabla} \left(\mu_e + q_e \varphi\right) \\ \boldsymbol{j}_e = -L_{se} \, \boldsymbol{\nabla} T - L_{ee} \, \boldsymbol{\nabla} \left(\mu_e + q_e \varphi\right) \end{cases} \tag{11.88}$$

taking into account the fact that $\nabla q_e = 0$. According to relation (11.77), the component L_{ee} can be expressed in terms of the electric conductivity coefficient σ as,

$$L_{ee} = \frac{\sigma}{q_e^2} \tag{11.89}$$

According to relation (11.87) and taking into account equation (11.89), the coefficient L_{se} can be expressed in terms of the electric conductivity coefficient σ and the Seebeck coefficient ε as,

$$L_{se} = q_e \, \varepsilon \, L_{ee} = \frac{\sigma \varepsilon}{q_e} \tag{11.90}$$

According to relation (11.31) and taking into account equations (11.89) and (11.90), the coefficient L_{ss} can be expressed in terms of the electric conductivity σ, the Seebeck coefficient ε and the thermal conductivity κ as,

$$L_{ss} = \frac{\kappa}{T} + \frac{L_{se}^2}{L_{ee}} = \frac{\kappa}{T} + \sigma \varepsilon^2 \tag{11.91}$$

Using equations (11.89), (11.90) and (11.91), the vectorial empirical relations (11.88) can be recast in terms of the coefficients κ, σ and ε characterising the thermodynamics of the conduction electrons in the metal as,

$$\begin{cases} \boldsymbol{j}_s = - \left(\dfrac{\kappa}{T} + \sigma \varepsilon^2 \right) \boldsymbol{\nabla} \, T - \dfrac{\sigma \varepsilon}{q_e} \boldsymbol{\nabla} \, \bar{\mu}_e \\ \boldsymbol{j}_e = - \dfrac{\sigma \varepsilon}{q_e} \boldsymbol{\nabla} \, T - \dfrac{\sigma}{q_e^2} \boldsymbol{\nabla} \, \bar{\mu}_e \end{cases} \tag{11.92}$$

where the electrochemical potential of the electrons is given by,

$$\bar{\mu}_e = \mu_e + q_e \varphi \tag{11.93}$$

Equations (11.92) imply that,

$$\boldsymbol{j}_s = - \frac{\kappa}{T} \boldsymbol{\nabla} \, T + \varepsilon \, q_e \boldsymbol{j}_e \tag{11.94}$$

In the limit where the chemical potential gradient of the electrons $\boldsymbol{\nabla} \, \mu_e$ is negligible with respect to the electrostatic potential gradient $\boldsymbol{\nabla} \, \varphi$, using relations $\boldsymbol{j}_q = q_e \boldsymbol{j}_e$ and $\boldsymbol{j}_Q = T \boldsymbol{j}_s$, the empirical equation (11.94) and the second equation (11.92) are reduced to,

$$\begin{cases} \boldsymbol{j}_Q = - \kappa \, \boldsymbol{\nabla} \, T + T \varepsilon \boldsymbol{j}_q \\ \boldsymbol{j}_q = - \sigma \varepsilon \, \boldsymbol{\nabla} \, T - \sigma \, \boldsymbol{\nabla} \, \varphi \end{cases} \tag{11.95}$$

In the absence of electric current, i.e. $\boldsymbol{j}_q = \boldsymbol{0}$, the first empirical relation (11.95) is reduced to Fourier's law (11.30). The second term on the right-hand side of that relation represents the heat current density generated by the electric current density flowing through the metal. For a uniform temperature gradient, i.e. $\boldsymbol{\nabla} \, T = \boldsymbol{0}$, the second empirical relation (11.95) is reduced to Ohm's law (11.71). The first term on the right-hand side of that equation represents the electric current density generated by the temperature gradient across the metal, which is a combination of Ohm's law (11.71) and the Seebeck effect (11.86).

The Thomson effect, defined below, is observed when the system is in a stationary state. In a stationary state, since there is no 'chemical reaction' between the conduction electrons and the metal, the conduction electron continuity equation (10.26) is reduced to,

$$\nabla \cdot \boldsymbol{j}_e = 0 \tag{11.96}$$

This implies that the electric charge continuity equation is given by,

$$\nabla \cdot \boldsymbol{j}_q = 0 \tag{11.97}$$

since the electric charge q_e is constant.

Taking into account the current density balance (10.87) and the continuity equation (11.96), the power density dissipated by the system in a stationary state $\nabla \cdot \boldsymbol{j}_u$ is given by,

$$\nabla \cdot \boldsymbol{j}_u = \nabla \cdot \boldsymbol{j}_Q + \boldsymbol{j}_e \cdot \nabla \left(\mu_e + q_e \varphi \right) \tag{11.98}$$

Applying the second relation (11.92) and equation (11.95), we find, since $\boldsymbol{j}_q = q_e \boldsymbol{j}_e$, that the dissipated power density (11.98) can be recast as,

$$\nabla \cdot \boldsymbol{j}_u = \nabla \cdot \left(-\kappa \nabla T + T\varepsilon \boldsymbol{j}_q \right) - \boldsymbol{j}_q \cdot \left(\frac{\boldsymbol{j}_q}{\sigma} + \varepsilon \nabla T \right) \tag{11.99}$$

Expanding the terms on the right-hand side of equation (11.99) and using the continuity equation (11.97), equation (11.99) becomes,

$$\nabla \cdot \boldsymbol{j}_u = -\kappa \nabla^2 T + T\boldsymbol{j}_q \cdot \nabla \varepsilon - \frac{\boldsymbol{j}_q^2}{\sigma} \tag{11.100}$$

In a stationary state, the first term on the right-hand side of equation (11.100) vanishes since the heat equation (11.35) implies that this term corresponds to a time variation,

$$-\kappa \nabla^2 T = -c_e \, \partial_t T = 0 \tag{11.101}$$

The thermoelectric power ε is a function of T and n_A. In a metal, the density n_A is independent of position. Therefore, we have that $\nabla \varepsilon$ depends only on ∇T, and we can write,

$$\nabla \varepsilon = \frac{\partial \varepsilon}{\partial T} \nabla T \tag{11.102}$$

Relations (11.101) and (11.102) imply that the dissipated power density (11.100) is reduced to,

$$\nabla \cdot \boldsymbol{j}_u = \tau \boldsymbol{j}_q \cdot \nabla T - \frac{\boldsymbol{j}_q^2}{\sigma} \tag{11.103}$$

where the Thomson coefficient τ is defined as,

$$\tau \equiv T \frac{\partial \varepsilon}{\partial T} \tag{11.104}$$

The first power density term on the right-hand side of equation (11.103) corresponds to the **Thomson effect** [181]. It describes the thermal power density generated by an electric current density flowing along a temperature gradient. The second term corresponds to the **Joule effect** [182]. It describes the thermal power density generated by an electric current density, when the system temperature is independent of position.

11.4.12 Peltier Effect

Joule heating is proportional to the square of the electric current flowing through a metal. It is possible to induce heating or cooling that is proportional to the current when the current flows through the interface between two metals. In order to show this, we consider a junction between two different isotropic metals A and B. A current density $\boldsymbol{j}_q$ flows through this junction. We assume that the temperature T around the junction is independent of position (Fig. 11.8). Then, the empirical relations (11.95) applied to each metal in the absence of a temperature gradient reduce to,

$$\boldsymbol{j}_{QA} = T\boldsymbol{j}_{sA} = T\varepsilon_A\boldsymbol{j}_{qA}$$
$$\boldsymbol{j}_{QB} = T\boldsymbol{j}_{sB} = T\varepsilon_B\boldsymbol{j}_{qB} \tag{11.105}$$

Equations (11.105) illustrate the fact that an entropy current and a heat current are associated with any electric current. The continuity of the electric current density $\boldsymbol{j}_q$ across the junction implies that,

$$\boldsymbol{j}_q = \boldsymbol{j}_{qA} = \boldsymbol{j}_{qB} \tag{11.106}$$

Comparing equations (11.105) and (11.106), we conclude that for metals characterised by different Peltier coefficients, there is a discontinuity of the heat current densities at the junction between the metals, since

$$\varepsilon_A \neq \varepsilon_B \quad \Rightarrow \quad \boldsymbol{j}_{QA} \neq \boldsymbol{j}_{QB} \tag{11.107}$$

Thus, there must be a heat transfer between the system and the exterior. This is called the **Peltier effect** [183].

In view of equations (11.105), (11.106) and (11.108), we find that the heat current density difference $\boldsymbol{j}_{QB} - \boldsymbol{j}_{QA}$ can be written as,

$$\boldsymbol{j}_{QB} - \boldsymbol{j}_{QA} = \pi_{AB}\boldsymbol{j}_q \tag{11.108}$$

where,

$$\pi_{AB} = T\left(\varepsilon_B - \varepsilon_A\right) \tag{11.109}$$

is called the **Peltier coefficient**.

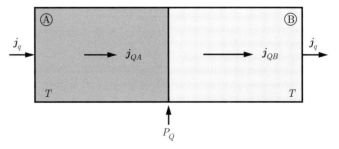

Figure 11.8 Two different metals A and B are kept at temperature T while an electric current density $\boldsymbol{j}_q$ flows through the junction. The difference between the heat current densities $\boldsymbol{j}_{QB}$ and $\boldsymbol{j}_{QA}$ implies that a thermal density P_Q is applied at the junction between the metals.

According to relation (11.108), the junction is a heat source or a heat sink, depending on the sign of the Peltier coefficient π_{AB} and on the orientation of the electric current density $\boldsymbol{j}_q$. Inverting the orientation of the electric current density $\boldsymbol{j}_q$ flowing through the junction, in view of relations (11.108) and (11.109), the orientation of the heat current densities is inverted without modifying the Peltier coefficient π_{AB}. Thus, the thermal power P_Q changes its sign.

Note that the Peltier effect is due to the junction between two materials, while the Seebeck effect is only due to the temperature gradient inside a single material.

11.5 Fluid Dynamics

We conclude this chapter by introducing two very important results concerning incompressible fluids: the Navier–Stokes and the Bernoulli equations.

The Navier–Stokes equation describes a viscous fluid. It is based on Newton's law applied to fluids, our equation of motion (10.35), in which a relation between the stress tensor and the velocity field is applied. We can obtain this relation by expanding on the approach described in this chapter. In the balance equation (10.80), we had obtained a term that was the product of the velocity gradient tensor and the stress tensor. In line with the reasoning presented at the beginning of this chapter, we can establish an Onsager relation between these two tensors and use this description of viscosity effects in the equation of motion (10.35).

The Bernoulli equation is obtained for non-visquous incompressible fluid as a constant of motion deduced from the equation of motion (10.35).

11.5.1 Navier–Stokes Equations

The Navier–Stokes equation describes the motion of a viscous fluid consisting of a single neutral chemical substance A subjected to its own weight and in the presence of shear. In this case, the stress tensor $\boldsymbol{\tau}$ is a generalisation of the expression given in relation (10.81),

$$\boldsymbol{\tau} = \boldsymbol{\tau}^{\,\mathrm{fr}} + \left(\tau^{\,\mathrm{fr}} - p\right)\mathbb{1} \tag{11.110}$$

where the additional term $\boldsymbol{\tau}^{\,\mathrm{fr}}$ is the symmetric trace-free tensor that accounts for the shear of the fluid. We consider that the only external force density $\boldsymbol{f}^{\mathrm{ext}}$ acting on the fluid is its weight density $m\,\boldsymbol{g}$,

$$\boldsymbol{f}^{\mathrm{ext}} = m\,\boldsymbol{g} \tag{11.111}$$

Thus, in view of relations (11.110) and (11.111), Newton's 2nd law (10.35) is recast as,

$$m\dot{\boldsymbol{v}} = m\,\boldsymbol{g} - \boldsymbol{\nabla}p + \boldsymbol{\nabla}\cdot\boldsymbol{\tau}^{\,\mathrm{fr}} + \boldsymbol{\nabla}\tau^{\,\mathrm{fr}} \tag{11.112}$$

By taking into account the shear of the fluid, we introduce an additional term consisting of the contraction of two trace-free stress tensors, i.e. $1/2\left(\boldsymbol{\nabla}\boldsymbol{v} + (\boldsymbol{\nabla}\boldsymbol{v})^T\right) : \boldsymbol{\tau}^{\,\mathrm{fr}}$, in the irreversible evolution equation (10.88). Thus, we can follow the approach presented in § 11.2. The symmetric trace-free stress tensor $\boldsymbol{\tau}^{\text{fr}}$ plays the role of a generalised tensorial force. It is related to the symmetric trace-free velocity gradient $1/2\left(\boldsymbol{\nabla}\boldsymbol{v}+(\boldsymbol{\nabla}\boldsymbol{v})^T\right)$ that plays the role of the generalised tensorial current. For a Newtonian fluid, the tensorial linear empirical relation yields,

$$\boldsymbol{\tau}^{\text{fr}} = \mu\left(s, n_A\right)\left(\boldsymbol{\nabla}\boldsymbol{v}+(\boldsymbol{\nabla}\boldsymbol{v})^T\right) \tag{11.113}$$

where the scalar coefficient $\mu\left(s, n_A\right)$ is the dynamic viscosity of the fluid. Note that the factor $1/2$ appearing in the definition of the symmetric trace-free velocity gradient $1/2\left(\boldsymbol{\nabla}\boldsymbol{v}+(\boldsymbol{\nabla}\boldsymbol{v})^T\right)$ is absorbed in the expression of the dynamic viscosity $\mu\left(s, n_A\right)$. For a neutral chemical substance A, the scalar linear empirical relation (11.19) is reduced to,

$$\tau^{\text{fr}} = \eta\left(s, n_A\right)\boldsymbol{\nabla}\cdot\boldsymbol{v} \tag{11.114}$$

where the scalar coefficient $\eta\left(s, n_A\right)$ is the volume viscosity of the fluid. In view of the linear empirical relations (11.113) and (11.114), Newton's 2nd law (11.112) becomes,

$$m\dot{\boldsymbol{v}} = m\boldsymbol{g} - \boldsymbol{\nabla}p + \boldsymbol{\nabla}\cdot\left(\mu\left(\boldsymbol{\nabla}\boldsymbol{v}+(\boldsymbol{\nabla}\boldsymbol{v})^T\right)\right) + \boldsymbol{\nabla}\left(\eta\,\boldsymbol{\nabla}\cdot\boldsymbol{v}\right) \tag{11.115}$$

We consider that the spatial variations of the dynamic viscosity μ and the volume viscosity η are negligible, i.e. $\boldsymbol{\nabla}\mu = \boldsymbol{0}$ and $\boldsymbol{\nabla}\eta = \boldsymbol{0}$. In this case, equation (11.115) is reduced to the Navier–Stokes equation for a compressible fluid,

$$m\dot{\boldsymbol{v}} = m\boldsymbol{g} - \boldsymbol{\nabla}p + \mu\,\boldsymbol{\nabla}^2\boldsymbol{v} + (\mu+\eta)\,\boldsymbol{\nabla}\left(\boldsymbol{\nabla}\cdot\boldsymbol{v}\right) \tag{11.116}$$

In the particular case where the fluid is incompressible, i.e. $\boldsymbol{\nabla}\cdot\boldsymbol{v} = 0$, the Navier–Stokes equation (11.116) is reduced to,

$$m\dot{\boldsymbol{v}} = m\boldsymbol{g} - \boldsymbol{\nabla}p + \mu\,\boldsymbol{\nabla}^2\boldsymbol{v} \tag{11.117}$$

11.5.2 Bernoulli Theorem

The Bernoulli theorem describes the reversible motion of a fluid consisting of a single neutral chemical substance A subjected to its own weight in the absence of shear and friction, i.e. $\boldsymbol{\tau}^{\text{fr}} = 0$ and $\tau^{\text{fr}} = 0$. In this case, Newton's 2nd law (11.112) reduces to,

$$m\dot{\boldsymbol{v}} = m\boldsymbol{g} - \boldsymbol{\nabla}p \tag{11.118}$$

and thus,

$$m\dot{\boldsymbol{v}}\cdot\boldsymbol{v} = m\boldsymbol{g}\cdot\boldsymbol{v} - \boldsymbol{\nabla}p\cdot\boldsymbol{v} \tag{11.119}$$

For an incompressible fluid, i.e. $\boldsymbol{\nabla}\cdot\boldsymbol{v} = 0$, the continuity equation (10.34) for the mass reduces to,

$$\dot{m} = \frac{dm}{dt} = 0 \tag{11.120}$$

which implies that,

$$m\dot{\boldsymbol{v}}\cdot\boldsymbol{v} = \frac{d}{dt}\left(\frac{1}{2}\,m\,\boldsymbol{v}^2\right) \tag{11.121}$$

and since the gravitational field is constant, i.e. $\dot{g} = 0$,

$$m\,g \cdot v = \frac{d}{dt}\,(m\,g \cdot r) \tag{11.122}$$

Furthermore, since the pressure is a function of the position, i.e. $p = p\,(r)$,

$$\nabla p \cdot v = \frac{dp}{dr} \cdot \frac{dr}{dt} = \frac{dp}{dt} \tag{11.123}$$

Thus, using relations (11.121)–(11.123), equation (11.119) is recast as,

$$\frac{d}{dt}\left(\frac{1}{2}\,m\,v^2 - m\,g \cdot r + p\right) = 0 \tag{11.124}$$

When integrating relation (11.124) over time, we obtain,

$$\frac{1}{2}\,m\,v^2 - m\,g \cdot r + p = \text{const} \tag{11.125}$$

The gravtiational field g and the position r are written in Cartesian coordinates as,

$$g = -g\,\hat{z} \quad \text{and} \quad r = x\,\hat{x} + y\,\hat{y} + z\,\hat{z} \tag{11.126}$$

Finally, using the relation (11.126), equation (11.125) divided by the mass density m yields Bernoulli's theorem,

$$\frac{1}{2}\,v^2 + g\,z + \frac{p}{m} = \text{const} \tag{11.127}$$

11.6 Worked Solutions

11.6.1 Thermocouple

Consider a metal wire A whose ends are in contact with two metal wires B, which are connected to the left 'l' and right 'r' terminals of a voltmeter (Fig. 11.9). One of the junctions between wires A and B is kept at a fixed reference temperature T_1 (melting ice or liquid nitrogen) and the other is at a variable temperature T_2 that we would like to measure. The two terminals of the voltmeter are at the same temperature T, so that the measurement depends only on the temperatures T_1 and T_2 at the ends of the metallic wire A. Consider that the Seebeck coefficients ε_A and ε_B are independent of temperature.

1. Determine the electrochemical potential differences $\bar{\mu}_1 - \bar{\mu}_l$, $\bar{\mu}_2 - \bar{\mu}_1$ and $\bar{\mu}_r - \bar{\mu}_2$.
2. Taking into account the fact that the chemical potential μ_e of the electrons does not depend on temperature T, deduce the electrostatic potential difference $\Delta\varphi = \varphi_r - \varphi_l$ between the poles of the voltmeter.
3. The thermoelectric power ε_{AB} of the thermocouple is defined as the derivative of the electrostatic potential difference $\Delta\varphi$ with respect to the temperature T_2,

$$\varepsilon_{AB} = \frac{\partial\Delta\varphi}{\partial T_2}$$

Express ε_{AB} in terms of the Seebeck coefficients ε_A and ε_B.

A voltmeter detects the voltage drop at the end of a circuit composed of a wire of metal B, a wire of metal A and another wire of metal B. The junctions between wires are at the temperatures T_1 and T_2 as indicated. The terminals of the voltmeter are at the same temperature T.

Solution:

1. *By definition, there is no electron current density in a voltmeter, i.e.* $\boldsymbol{j}_e = \boldsymbol{0}$. *Thus, the second empirical relation (11.92) applied to the metals A and B reads,*

$$\nabla\,\bar{\mu}_A = -\,q_e\,\varepsilon_A\,\nabla\,T_A$$
$$\nabla\,\bar{\mu}_B = -\,q_e\,\varepsilon_B\,\nabla\,T_B$$

The integration of the empirical relations along the metal wires A and B is written as,

$$\bar{\mu}_1 - \bar{\mu}_l = \int d\boldsymbol{r} \cdot \nabla\,\bar{\mu}_B = -\,q_e\,\varepsilon_B \int d\boldsymbol{r} \cdot \nabla\,T_B = -\,q_e\,\varepsilon_B \int_T^{T_1} dT'$$

$$\bar{\mu}_2 - \bar{\mu}_1 = \int d\boldsymbol{r} \cdot \nabla\,\bar{\mu}_A = -\,q_e\,\varepsilon_A \int d\boldsymbol{r} \cdot \nabla\,T_A = -\,q_e\,\varepsilon_A \int_{T_1}^{T_2} dT'$$

$$\bar{\mu}_r - \bar{\mu}_2 = \int d\boldsymbol{r} \cdot \nabla\,\bar{\mu}_B = -\,q_e\,\varepsilon_B \int d\boldsymbol{r} \cdot \nabla\,T_B = -\,q_e\,\varepsilon_B \int_{T_2}^{T} dT'$$

2. *The chemical potential μ_e of the electrons does not depend on temperature. Therefore, it is the same on the terminals 'l' and 'r' of the voltmeter. Thus,*

$$\mu_l = \mu_r$$

Defining the electrochemical potential at the terminals of the voltmeter as,

$$\bar{\mu}_l = \mu_l + q_e\varphi_l$$
$$\bar{\mu}_r = \mu_r + q_e\varphi_r$$

the electrochemical potential difference at the terminals of the voltmeter can be expressed as,

$$\Delta\varphi = \varphi_r - \varphi_l = \frac{1}{q_e}\,(\bar{\mu}_r - \bar{\mu}_l)$$

Applying the values of the electrochemical potential differences in this expression yields

$$\Delta\varphi = -\varepsilon_B\,(T_1 - T) - \varepsilon_B\,(T - T_2) - \varepsilon_A\,(T_2 - T_1)$$
$$= (\varepsilon_B - \varepsilon_A)\,(T_2 - T_1)$$

3. *Differentiating the previous expression with respect to the variable temperature T_2, we obtain an expression for the thermoelectric power,*

$$\varepsilon_{AB} = \frac{\partial\Delta\varphi}{\partial T_2} = \varepsilon_B - \varepsilon_A$$

11.6.2 Seebeck Loop

A magnetised needle fixed on a vertical rod passing through its centre can oscillate in a horizontal plane. Initially, the needle points towards the North. It is mounted in a frame consisting of two different metals A and B, of length ℓ each, forming a loop in a vertical plane (Fig. 11.10). Heating the junction (1) at temperature $T+\Delta T$ and keeping the junction (2) at temperature T causes a deviation of the needle. This is due to the magnetic induction field B generated by the electric current density j_q that circulates in the frame. This is the effect that Seebeck first observed. The thermoelectric materials A and B that form the loop have a length ℓ and a cross-section surface area A, which can be written as,

$$\ell = \int_0^\ell dr \cdot \hat{x} \qquad \text{and} \qquad A = \int_S dS \cdot \hat{x}$$

where $\hat{x}$ is a unit vector oriented anticlockwise along the loop (Fig. 11.10), and the infinitesimal length and surface vectors dr and dS are oriented in the same direction. The temperature difference between the hot and cold ends is given by,

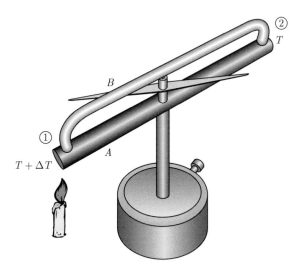

Figure 11.10 In a metallic loop consisting of two metals A and B, a current circulates when junction (1) is heated up to a temperature $T + \Delta T$ while junction (2) is maintained at a temperature T. The magnetic field generated by the current causes a deviation of a compass needle located in the loop.

$$\Delta T = \int_0^\ell d\mathbf{r} \cdot (-\nabla T_A) = \int_0^\ell d\mathbf{r} \cdot \nabla T_B$$

The electric potential differences $\Delta \varphi_A$ and $\Delta \varphi_B$ between the hot and cold ends are given by,

$$\Delta \varphi_A = \int_0^\ell d\mathbf{r} \cdot (-\nabla \varphi_A)$$

$$\Delta \varphi_B = \int_0^\ell d\mathbf{r} \cdot \nabla \varphi_B$$

In a stationary state, the electric charge conservation implies that the electric current densities are the same in each material, i.e. $\boldsymbol{j}_q = \boldsymbol{j}_{q_A} = \boldsymbol{j}_{q_B}$. The electric current I flowing through materials A and B is the integral of the electric current densities $\boldsymbol{j}_{q_A}$ and $\boldsymbol{j}_{q_B}$ over the cross-section area A,

$$I = \int_S \boldsymbol{j}_{q_A} \cdot d\mathbf{S} = \int_S \boldsymbol{j}_{q_B} \cdot d\mathbf{S}$$

1. When the system is in a stationary state, determine the condition imposed on the electric current density $\boldsymbol{j}_q$.
2. Determine the intensity I of the electric current density circulating in the loop, in the limit where the chemical potential gradients $\nabla \mu_A$ and $\nabla \mu_B$ of the electrons in the materials A and B are negligible compared to the gradients of the electrostatic potentials $\nabla (q_e \varphi_A)$ and $\nabla (q_e \varphi_B)$. Express your result in terms of the electrostatic potential differences $\Delta \varphi_A$ and $\Delta \varphi_B$ in materials A and B between junctions (1) and (2).
3. Express the electric current I circulating in the loop in terms of the empirical coefficients σ_A, σ_B, ε_A and ε_B, the length ℓ and the temperature difference ΔT.

Solution:

1. *In a stationary state, the time derivatives vanish, which implies that Fick's equation (11.53) for the conduction electrons is reduced to,*

$$\nabla \cdot \boldsymbol{j}_e = 0$$

Using definition $\boldsymbol{j}_q = q_e \boldsymbol{j}_e$, taking into account the fact that $\nabla q_e = \boldsymbol{0}$, we obtain the stationary condition imposed on the electric current density $\boldsymbol{j}_q$,

$$\nabla \cdot \boldsymbol{j}_q = 0$$

At the junctions between metals A and B, the electric current densities are equal, i.e. $jq_A = jq_B$, because there is no electric charge accumulation.

2. *In the limit where the chemical potential gradients $\nabla \mu_A$ and $\nabla \mu_B$ of the electrons in materials A and B are negligible with respect to the electrostatic potential gradients $\nabla (q_e \varphi_A)$ and $\nabla (q_e \varphi_B)$, the second empirical relation (11.95) is reduced to,*

$$\boldsymbol{j}_{q_A} = \sigma_A \varepsilon_A (-\nabla T_A) + \sigma_A (-\nabla \varphi_A)$$

$$\boldsymbol{j}_{q_B} = -\sigma_B \varepsilon_B \nabla T_B - \sigma_B \nabla \varphi_B$$

The integrals of the electric charge transport equations over the volume are the product of integrals over the cross-section area A times integrals over the length ℓ of the thermoelectric materials,

$$\int_S \boldsymbol{j}_{q_A} \cdot d\boldsymbol{S} \int_0^\ell d\boldsymbol{r} \cdot \hat{\boldsymbol{x}} = \sigma_A \, \varepsilon_A \int_0^\ell d\boldsymbol{r} \cdot (-\boldsymbol{\nabla} T_A) \int_S d\boldsymbol{S} \cdot \hat{\boldsymbol{x}}$$

$$+ \sigma_A \int_0^\ell d\boldsymbol{r} \cdot (-\boldsymbol{\nabla} \varphi_A) \int_S d\boldsymbol{S} \cdot \hat{\boldsymbol{x}}$$

$$\int_S \boldsymbol{j}_{q_B} \cdot d\boldsymbol{S} \int_0^\ell d\boldsymbol{r} \cdot \hat{\boldsymbol{x}} = - \sigma_B \, \varepsilon_B \int_0^\ell d\boldsymbol{r} \cdot \boldsymbol{\nabla} T_B \int_S d\boldsymbol{S} \cdot \hat{\boldsymbol{x}}$$

$$- \sigma_B \int_0^\ell d\boldsymbol{r} \cdot \boldsymbol{\nabla} \varphi_B \int_S d\boldsymbol{S} \cdot \hat{\boldsymbol{x}}$$

The electric charge transport equations integrated over the volume reduce to,

$$I = \sigma_A \, \varepsilon_A \frac{A}{\ell} \Delta T + \sigma_A \, \Delta \varphi_A$$

$$I = - \sigma_B \, \varepsilon_B \frac{A}{\ell} \Delta T - \sigma_B \, \Delta \varphi_B$$

Since the electric current is the same in each thermoelectric material, it can be recast as,

$$I = \frac{A}{2\,\ell} \left((\sigma_A \, \varepsilon_A - \sigma_B \, \varepsilon_B) \, \Delta T + \sigma_A \, \Delta \varphi_A - \sigma_B \, \Delta \varphi_B \right)$$

Thus, if metals A and B are identical, the electric current vanishes, i.e. $I = 0$.

3. *At junctions* (1) *and* (2), *the electrostatic potentials of metals A and B are equal. Thus, the electrostatic potential difference $\Delta \varphi$ between the junctions is the same in both metals,*

$$\Delta \varphi = \Delta \varphi_A = \Delta \varphi_B$$

Thus, the electric current in each material is recast as,

$$I = \frac{A}{\ell} \left(\sigma_A \, \varepsilon_A \, \Delta T + \sigma_A \, \Delta \varphi \right)$$

$$I = - \frac{A}{\ell} \left(\sigma_B \, \varepsilon_B \, \Delta T + \sigma_B \, \Delta \varphi \right)$$

Multiplying the first equation by σ_B and the second by σ_A and then adding them together, we obtain,

$$(\sigma_A + \sigma_B) \, I = \frac{A}{\ell} \, \sigma_A \, \sigma_B \, (\varepsilon_A - \varepsilon_B) \, \Delta T$$

This implies that the intensity of the electric current density j_q can be expressed in terms of the empirical parameters σ_A, σ_B, ε_A, ε_B and the temperature difference ΔT as,

$$I = \frac{A}{\ell} \left(\frac{1}{\sigma_A} + \frac{1}{\sigma_B} \right)^{-1} (\varepsilon_A - \varepsilon_B) \, \Delta T$$

The terms in the first bracket correspond to the equivalent resistivity of the Seebeck loop, taking into account the fact that the metals A and B are connected in series. The terms in the second bracket represent an effective Seebeck coefficient of the loop.

11.6.3 Thermoelectric Junction

Consider a rod consisting of two different metals A and B in thermal contact with one another. Each one has a thickness d and a cross-section surface area A. The metals are characterised by their electric conductivities, σ_A and σ_B, their thermal conductivities κ_A and κ_B, their Seebeck coefficients ε_A and ε_B. These transport properties can all be considered as temperature independent. The end of metal rod A is in contact with a thermal bath at high temperature and the end of metal rod B is in contact with a thermal bath at low temperature. In other words, a temperature difference ΔT is imposed on the whole rod. A constant electric current density $\boldsymbol{j}_q$ is driven through the rod. The electrostatic potential difference between the ends of the rod is $\Delta\varphi$ (Fig. 11.11).

By conservation of the electric charge and because the state of the system is stationary, the electric current density $\boldsymbol{j}_q$ and the heat current density $\boldsymbol{j}_Q$ are conserved at the interface between metals A and B, i.e. $\boldsymbol{j}_q = \boldsymbol{j}_{q_A} = \boldsymbol{j}_{q_B}$ and $\boldsymbol{j}_Q = \boldsymbol{j}_{Q_A} = \boldsymbol{j}_{Q_B}$. The electric current I flowing through materials A and B is the integral of the electric current densities $\boldsymbol{j}_{q_A}$ and $\boldsymbol{j}_{q_B}$ over the cross-section surface area A,

$$I = \int_S \boldsymbol{j}_{q_A} \cdot d\boldsymbol{S} = \int_S \boldsymbol{j}_{q_B} \cdot d\boldsymbol{S}$$

where the infinitesimal surface vector $d\boldsymbol{S}$ is oriented along the electric current density $\boldsymbol{j}_q$. The thermal power P_Q exerted on materials A and B is the integral of the heat current densities $\boldsymbol{j}_{Q_A}$ and $\boldsymbol{j}_{Q_B}$ over the cross-section surface area A,

$$P_Q = \int_S \boldsymbol{j}_{Q_A} \cdot d\boldsymbol{S} = \int_S \boldsymbol{j}_{Q_B} \cdot d\boldsymbol{S}$$

The temperature differences ΔT_A and ΔT_B and the electric potential differences $\Delta\varphi_A$ and $\Delta\varphi_B$ across metals A and B are given by,

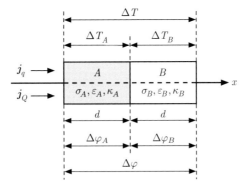

Figure 11.11 A current flows through a rod composed of two different metals, A and B, in electric and thermal contact with one another. The figure indicates the electrostatic potential variations and the temperature variations in each metal. The origin of the x-axis is at the interface between both metals.

$$\Delta T_A = \int_{-d}^{0} d\mathbf{r} \cdot (-\nabla T_A) \qquad \text{and} \qquad \Delta T_B = \int_{0}^{d} d\mathbf{r} \cdot (-\nabla T_B)$$

$$\Delta \varphi_A = \int_{-d}^{0} d\mathbf{r} \cdot (-\nabla \varphi_A) \qquad \text{and} \qquad \Delta \varphi_B = \int_{0}^{d} d\mathbf{r} \cdot (-\nabla \varphi_B)$$

where the infinitesimal length vector $d\mathbf{r}$ is oriented along the electric current density $\mathbf{j}_q$ and the heat current density $\mathbf{j}_Q$. The temperature difference ΔT and the electrostatic potential difference $\Delta \varphi$ across the whole rod satisfy,

$$\Delta T = \Delta T_A + \Delta T_B \qquad \text{and} \qquad \Delta \varphi = \Delta \varphi_A + \Delta \varphi_B$$

The thermoelectric materials A and B have a length d and a cross-section surface area A, which can be written as,

$$d = \int_{-d}^{0} d\mathbf{r} \cdot \hat{\mathbf{x}} = \int_{0}^{d} d\mathbf{r} \cdot \hat{\mathbf{x}} \qquad \text{and} \qquad A = \int_{S} d\mathbf{S} \cdot \hat{\mathbf{x}}$$

where $\hat{\mathbf{x}}$ is a unit vector oriented clockwise along the electric current density $\mathbf{j}_q$ and the heat current density $\mathbf{j}_Q$.

1. Express the electric charge and heat transport equations (11.95) for metals A and B at the interface between the metals in terms of the generalised forces ∇T_A, ∇T_B, $\nabla \varphi_A$, $\nabla \varphi_B$ and the temperature T_{AB} evaluated at the interface between the metals.
2. If the thickness d of the layers is small enough, the gradients can be considered as independent of the position. In this case, integrate the electric charge transport equations between the ends of metals A and B.
3. In the same case, integrate the heat transport equations between the ends of metals A and B.
4. Find expressions for ΔT_A and ΔT_B in terms of I, ΔT and the empirical coefficients.
5. Find expressions for $\Delta \varphi_A$ and $\Delta \varphi_B$ in terms of I, ΔT and the empirical coefficients.
6. Determine an expression for $\Delta \varphi$ in terms of T_{AB}, I, ΔT and the empirical coefficients.

Solution:

1. *The electric charge transport equations for metals A and B are expressed at the interface between the metals as,*

$$\mathbf{j}_{q_A} = -\sigma_A \varepsilon_A \nabla T_A - \sigma_A \nabla \varphi_A$$
$$\mathbf{j}_{q_B} = -\sigma_B \varepsilon_B \nabla T_B - \sigma_B \nabla \varphi_B$$

Likewise, the heat transport equations for metals A and B are expressed at the interface between the metals as,

$$\mathbf{j}_{Q_A} = -\kappa_A \nabla T_A + T_{AB} \varepsilon_A \mathbf{j}_q$$
$$\mathbf{j}_{Q_B} = -\kappa_B \nabla T_B + T_{AB} \varepsilon_B \mathbf{j}_q$$

2. *The integrals of the electric charge transport equations over the volume are the product of integrals over the cross-section area A times integrals over the length d of the metals,*

$$\int_S \boldsymbol{j}_{q_A} \cdot d\boldsymbol{S} \int_{-d}^{0} d\boldsymbol{r} \cdot \hat{\boldsymbol{x}} = \sigma_A \, \varepsilon_A \int_{-d}^{0} d\boldsymbol{r} \cdot (-\boldsymbol{\nabla} T_A) \int_S d\boldsymbol{S} \cdot \hat{\boldsymbol{x}}$$

$$+ \sigma_A \int_{-d}^{0} d\boldsymbol{r} \cdot (-\boldsymbol{\nabla} \varphi_A) \int_S d\boldsymbol{S} \cdot \hat{\boldsymbol{x}}$$

$$\int_S \boldsymbol{j}_{q_B} \cdot d\boldsymbol{S} \int_{0}^{d} d\boldsymbol{r} \cdot \hat{\boldsymbol{x}} = \sigma_B \, \varepsilon_B \int_{0}^{d} d\boldsymbol{r} \cdot (-\boldsymbol{\nabla} T_B) \int_S d\boldsymbol{S} \cdot \hat{\boldsymbol{x}}$$

$$+ \sigma_B \int_{0}^{d} d\boldsymbol{r} \cdot (-\boldsymbol{\nabla} \varphi_B) \int_S d\boldsymbol{S} \cdot \hat{\boldsymbol{x}}$$

Using the integral relations for I, ΔT_A, ΔT_B, $\Delta\varphi_A$ and $\Delta\varphi_B$, the integral of the electric charge transport equation across metals A and B can be written as,

$$I = \frac{A}{d} \left(\sigma_A \, \varepsilon_A \, \Delta T_A + \sigma_A \, \Delta\varphi_A \right) = \frac{A}{d} \left(\sigma_B \, \varepsilon_B \, \Delta T_B + \sigma_B \, \Delta\varphi_B \right)$$

3. *The integrals of the heat transport equations over the volume of metals A and B are the product of integrals over the cross-section area A times integrals over the length d of the metals,*

$$\int_S \boldsymbol{j}_{Q_A} \cdot d\boldsymbol{S} \int_{-d}^{0} d\boldsymbol{r} \cdot \hat{\boldsymbol{x}} = \kappa_A \int_{-d}^{0} d\boldsymbol{r} \cdot (-\boldsymbol{\nabla} T_A) \int_S d\boldsymbol{S} \cdot \hat{\boldsymbol{x}}$$

$$+ T_{AB} \, \varepsilon_A \int_S \boldsymbol{j}_{q_A} \cdot d\boldsymbol{S} \int_{-d}^{0} d\boldsymbol{r} \cdot \hat{\boldsymbol{x}}$$

$$\int_S \boldsymbol{j}_{Q_B} \cdot d\boldsymbol{S} \int_{0}^{d} d\boldsymbol{r} \cdot \hat{\boldsymbol{x}} = \kappa_B \int_{0}^{d} d\boldsymbol{r} \cdot (-\boldsymbol{\nabla} T_B) \int_S d\boldsymbol{S} \cdot \hat{\boldsymbol{x}}$$

$$+ T_{AB} \, \varepsilon_B \int_S \boldsymbol{j}_{q_B} \cdot d\boldsymbol{S} \int_{0}^{d} d\boldsymbol{r} \cdot \hat{\boldsymbol{x}}$$

Using the integral relations for P_Q, I, ΔT_A and ΔT_B, the integral of the heat transport equation across metals A and B can be written as,

$$P_Q = \kappa_A \, \frac{A}{d} \, \Delta T_A + T_{AB} \, \varepsilon_A \, I = \kappa_B \, \frac{A}{d} \, \Delta T_B + T_{AB} \, \varepsilon_B \, I$$

4. *Using the relation $\Delta T_B = \Delta T - \Delta T_A$ in the heat transport equation, we find,*

$$\kappa_A \, \Delta T_A + T_{AB} \, \varepsilon_A \, \frac{d}{A} \, I = \kappa_B \left(\Delta T - \Delta T_A \right) + T_{AB} \, \varepsilon_B \, \frac{d}{A} \, I$$

This can be recast as,

$$\Delta T_A = \frac{\varepsilon_B - \varepsilon_A}{\kappa_A + \kappa_B} \, T_{AB} \, \frac{d}{A} \, I + \frac{\kappa_B}{\kappa_A + \kappa_B} \, \Delta T$$

Using the relation $\Delta T_A = \Delta T - \Delta T_B$ in the heat transport equation, the latter becomes,

$$\Delta T_B = \frac{\varepsilon_A - \varepsilon_B}{\kappa_A + \kappa_B} \, T_{AB} \, \frac{d}{A} \, I + \frac{\kappa_A}{\kappa_A + \kappa_B} \, \Delta T$$

5. *Applying this expression for ΔT_A in the electric charge transport equation, the latter becomes,*

$$I = \sigma_A \frac{\varepsilon_A (\varepsilon_B - \varepsilon_A)}{\kappa_A + \kappa_B} T_{AB} I + \sigma_A \frac{\varepsilon_A \kappa_B}{\kappa_A + \kappa_B} \frac{A}{d} \Delta T + \sigma_A \frac{A}{d} \Delta \varphi_A$$

This can be recast as,

$$\Delta \varphi_A = \left(\frac{1}{\sigma_A} - \frac{\varepsilon_A (\varepsilon_B - \varepsilon_A)}{\kappa_A + \kappa_B} T_{AB} \right) \frac{d}{A} I - \frac{\varepsilon_A \kappa_B}{\kappa_A + \kappa_B} \Delta T$$

Similarly, applying the expression for ΔT_B in the electric charge transport equation, the latter becomes,

$$I = \sigma_B \frac{\varepsilon_B (\varepsilon_A - \varepsilon_B)}{\kappa_A + \kappa_B} T_{AB} I + \sigma_B \frac{\varepsilon_B \kappa_A}{\kappa_A + \kappa_B} \frac{A}{d} \Delta T + \sigma_B \frac{A}{d} \Delta \varphi_B$$

This can be recast as,

$$\Delta \varphi_B = \left(\frac{1}{\sigma_B} - \frac{\varepsilon_B (\varepsilon_A - \varepsilon_B)}{\kappa_A + \kappa_B} T_{AB} \right) \frac{d}{A} I - \frac{\varepsilon_B \kappa_A}{\kappa_A + \kappa_B} \Delta T$$

6. *An expression for the electrostatic potential difference $\Delta \varphi$ at the ends of the rod is obtained by substituting the expressions for $\Delta \varphi_A$ and $\Delta \varphi_B$ in the equation $\Delta \varphi = \Delta \varphi_A + \Delta \varphi_B$,*

$$\Delta \varphi = \left[\left(\frac{1}{\sigma_A} + \frac{1}{\sigma_B} \right) + \frac{(\varepsilon_A - \varepsilon_B)^2}{\kappa_A + \kappa_B} T_{AB} \right] \frac{d}{A} I - \frac{\varepsilon_A \kappa_B + \varepsilon_B \kappa_A}{\kappa_A + \kappa_B} \Delta T$$

The first term in brackets represents Ohm's law. The last term represents the Seebeck effect and the second term in brackets implies the existence of thermal gradients in each metal even in the case where $\Delta T = 0$ [184].

11.6.4 Diffusion Length

Consider a chemical substance consisting of electrically neutral isomeric molecules of types U and D. The chemical reactions a and b transform the isomer of type U into an isomer of type D and vice versa,

$$a : \quad U \to D \qquad and \qquad b : \quad D \to U$$

Assume that the chemical reactions occur in a stationary state and that the fluid is not in expansion, $\nabla \cdot v = 0$.

1. Establish the conditions imposed by the continuity equations of the two isomers when the system is in a stationary state.
2. Find an expression for the chemical reaction rate densities ω_a and ω_b in terms of the chemical potentials μ_U and μ_D.
3. For a temperature that is independent of position, determine the diffusion equation expressed in terms of the chemical potential difference $\mu_U - \mu_D$, considering that the Onsager matrix elements are independent of position.

Solution:

1. *The stoichiometric coefficients associated to the chemical reactions a and b are,*

$$\nu_{aD} = \nu_{bU} = 1 \qquad and \qquad \nu_{aU} = \nu_{bD} = -1$$

In a stationary state, the continuity equations (10.26) for the isomers U and D are given by,

$$\nabla \cdot \boldsymbol{j}_U = \omega_b - \omega_a \qquad and \qquad \nabla \cdot \boldsymbol{j}_D = \omega_a - \omega_b$$

This implies that,

$$\nabla \cdot (\boldsymbol{j}_U + \boldsymbol{j}_D) = 0 \qquad and \qquad \nabla \cdot (\boldsymbol{j}_U - \boldsymbol{j}_D) = 2(\omega_b - \omega_a)$$

2. *Subtracting the empirical relations (11.17), we obtain,*

$$\omega_a - \omega_b = W(\mu_U - \mu_D)$$

where $W = L_{aa} - L_{ab} - L_{ba} + L_{bb}$. This implies that,

$$\nabla \cdot (\boldsymbol{j}_U - \boldsymbol{j}_D) = -2W(\mu_U - \mu_D)$$

3. *For a stationary state, electrically neutral isomers, i.e. $q_U = q_D = 0$ and a uniform temperature, i.e. $\nabla T = \boldsymbol{0}$, the vectorial empirical relations (11.11) are given by,*

$$\boldsymbol{j}_U = -L_{UU} \nabla \mu_U - L_{UD} \nabla \mu_D$$
$$\boldsymbol{j}_D = -L_{DU} \nabla \mu_U - L_{DD} \nabla \mu_D$$

where the Onsager matrix elements L_{UU}, L_{UD}, L_{DU} and L_{DD} are independent of position. The sum and difference of these relations can be recast as,

$$\boldsymbol{j}_U + \boldsymbol{j}_D = -L_{++} \nabla (\mu_U + \mu_D) - L_{+-} \nabla (\mu_U - \mu_D)$$
$$\boldsymbol{j}_U - \boldsymbol{j}_D = -L_{-+} \nabla (\mu_U + \mu_D) - L_{--} \nabla (\mu_U - \mu_D)$$

where the new matrix elements can be expressed in terms of the old ones as,

$$L_{++} = \frac{1}{2} \left(L_{UU} + L_{DD} + L_{UD} + L_{DU} \right)$$

$$L_{+-} = \frac{1}{2} \left(L_{UU} - L_{DD} - L_{UD} + L_{DU} \right)$$

$$L_{-+} = \frac{1}{2} \left(L_{UU} - L_{DD} + L_{UD} - L_{DU} \right)$$

$$L_{--} = \frac{1}{2} \left(L_{UU} + L_{DD} - L_{UD} - L_{DU} \right)$$

In view of the conditions on the divergence of the current densities and the fact that the matrix elements are independent of position, by taking the divergence of the vectorial empirical relations, we obtain,

$$-L_{++} \nabla^2 (\mu_U + \mu_D) - L_{+-} \nabla^2 (\mu_U - \mu_D) = 0$$
$$-L_{-+} \nabla^2 (\mu_U + \mu_D) - L_{--} \nabla^2 (\mu_U - \mu_D) = -2W(\mu_U - \mu_D)$$

The first of these equations can be recast as,

$$\nabla^2 \left(\mu_U + \mu_D \right) = - \frac{L_{+-}}{L_{++}} \, \nabla^2 \left(\mu_U - \mu_D \right)$$

Then, from the second equation, we obtain the stationary diffusion equation

$$\mu_U - \mu_D = \lambda^2 \, \nabla^2 \left(\mu_U - \mu_D \right)$$

where the diffusion length is given by

$$\lambda = \sqrt{ \frac{1}{2W} \left(L_{--} - \frac{L_{+-}^2}{L_{++}} \right) }$$

Here, we have taken into account the Onsager-Casmir reciprocity relations (11.3),

$$L_{DU} = L_{UD} \qquad \Rightarrow \qquad L_{-+} = L_{+-}$$

The matrix elements satisfy the conditions $L_{++} > 0$, $L_{--} > 0$ and $L_{++} L_{--} > L_{+-}^2$.

This problem was inspired by research in spintronics, a new branch of electronics in which researchers seek to make use of the fact that electrons carry a spin. In simple terms, the electron spin can have only two opposite orientations. Thus, this problems applies to spin diffusion, provided we consider the isomers as conduction electrons with either spin up (U) or spin down (D). Then, the diffusion length λ represents the mean distance over which a spin polarised electric current keeps its spin polarisation [185]. The next problem treats another aspect of spintronics.

11.6.5 Giant Magnetoresistance

Consider a bilayer composed of two homogeneous magnetic metallic layers labelled (1) and (2), maintained at a constant temperature T. The sides of this bilayer are covered by two electrodes which are connected to a resistance measurement circuit. A constant electric current is driven through the bilayer. The voltage drop across the bilayer $\Delta\varphi$ is to be determined for each magnetic configuration in order to determine the effective resistance of the bilayer. The two magnetic layers have the same thickness d and their magnetisation vectors are oriented along the vertical axis in either a parallel (p) or an anti-parallel (a) configuration (Fig. 11.12) [186].

In a magnetic metal, the electric conductivity depends on the spin of the conduction electrons with respect to the magnetisation. The spins are oriented either upwards or downwards. Treat the electrons as a chemical substance that has two isomers labelled U and D (for up and down). Both have the electric charge q_e. Assume that there is no transition (or 'chemical reaction') from U to D or from D to U, i.e. we assume that the spin diffusion length λ (c.f. § 11.6.4) is much larger than the thickness d of the sample, $\lambda \gg d$. The electrons with a spin parallel to the magnetisation have a conductivity σ_p, those with a spin anti-parallel to the magnetisation have a conductivity σ_a.

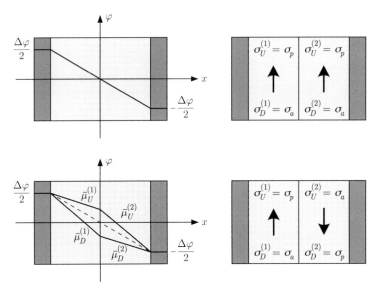

Figure 11.12 A resistance consists of two magnetic metallic layers contacted with two non-magnetic electrodes (dark grey). The magnetisation vectors of the two layers (arrows) are either parallel (top), or anti-parallel (bottom). The conduction electrodes are labelled U or D depending on whether their spin points upwards or downwards. The electrochemical potentials of the electrons U and D in layers 1 are $\mu_U^{(1)}$ and $\mu_D^{(1)}$, and in layers 2 are $\mu_U^{(2)}$ and $\mu_D^{(2)}$, and the electrostatic potential is φ. The conductivity is σ_p when the spin is parallel to the magnetisation, σ_a when it is opposite to the magnetisation.

At constant temperature T, the vectorial empirical relations (11.92) applied to electrons U and D in each layer can be written as,

$$\boldsymbol{j}_U = -\frac{\sigma_U^{(1)}}{q_e^2} \, \nabla \, \bar{\mu}_U \qquad \text{and} \qquad \boldsymbol{j}_D = -\frac{\sigma_D^{(1)}}{q_e^2} \, \nabla \, \bar{\mu}_D$$

$$\boldsymbol{j}_U = -\frac{\sigma_U^{(2)}}{q_e^2} \, \nabla \, \bar{\mu}_U \qquad \text{and} \qquad \boldsymbol{j}_D = -\frac{\sigma_D^{(2)}}{q_e^2} \, \nabla \, \bar{\mu}_D$$

where the electric conductivities $\sigma_U^{(1)}$ and $\sigma_D^{(1)}$ of layers 1 and the electric conductivities $\sigma_U^{(2)}$ and $\sigma_D^{(2)}$ of layers 2 have the values σ_p or σ_a, depending on the relative orientation of the spin with respect to the magnetisation.

1. Determine the variation of the electrochemical potential across each layer for each spin direction, $\Delta\bar{\mu}_U^{(1)}$, $\Delta\bar{\mu}_U^{(2)}$, $\Delta\bar{\mu}_D^{(1)}$, $\Delta\bar{\mu}_D^{(2)}$ in terms of $\Delta\varphi$ and the electrochemical potentials at the interface, $\bar{\mu}_U(0)$ and $\bar{\mu}_D(0)$. In order to keep the expressions simple, choose the zero of the chemical potential scale to be that of electrons in the electrodes.
2. Using the continuity conditions for the electron current densities $\boldsymbol{j}_U$ and $\boldsymbol{j}_D$ at the interface, determine the currents of electrons I_U and I_D in terms of $\sigma_U^{(1)}$, $\sigma_U^{(2)}$, $\sigma_D^{(1)}$ and $\sigma_D^{(2)}$ and $\Delta\varphi$.
3. Express Ohm's law for the bilayer in terms of $\sigma_U^{(1)}$, $\sigma_U^{(2)}$, $\sigma_D^{(1)}$ and $\sigma_D^{(2)}$. Find an expression for the effective electric conductivity σ of the bilayer in terms of the electric conductivities of the U and D electrons in layers (1) and (2).

4. Determine the effective electric conductivity σ_- in terms of σ_p and σ_a when the bilayer is in a parallel magnetic configuration. Likewise, find an expression for the effective electric conductivity σ_+ when the bilayer is in the anti-parallel magnetic configuration.

5. Compare the effective resistivities $\rho_\pm$ of the bilayer in its two magnetic configurations. Use the following notation for the resistivities ρ_p and ρ_a corresponding to the conductivities σ_p and σ_a,

$$\rho_p = \rho_0 - \Delta\rho \qquad \text{and} \qquad \rho_a = \rho_0 + \Delta\rho$$

where $\Delta\rho < \rho_0$.

Solution:

1. *At the electrodes, i.e. at $x = -d$ and $x = d$, the chemical potential of the electrons does not depend on the spin. We can choose the zero of the chemical potential to be at the electrodes,*

$$\mu_U(-d) = \mu_D(-d) = \mu_U(d) = \mu_D(d) = 0$$

At the electrodes, the electrochemical potentials of U and D electrons reduce to the electrostatic potentials of the electrodes (Fig. 11.12),

$$\bar{\mu}_U(-d) = \bar{\mu}_D(-d) = q_e\varphi(-d) = q_e\frac{\Delta\varphi}{2}$$

$$\bar{\mu}_U(d) = \bar{\mu}_D(d) = q_e\varphi(d) = -q_e\frac{\Delta\varphi}{2}$$

The electrochemical potential variations of U and D electrons across layers (1) and (2) are given by,

$$\Delta\bar{\mu}_U^{(1)} = \bar{\mu}_U(-d) - \bar{\mu}_U(0) = q_e\frac{\Delta\varphi}{2} - \bar{\mu}_U(0)$$

$$\Delta\bar{\mu}_U^{(2)} = \bar{\mu}_U(0) - \bar{\mu}_U(d) = q_e\frac{\Delta\varphi}{2} + \bar{\mu}_U(0)$$

$$\Delta\bar{\mu}_D^{(1)} = \bar{\mu}_D(-d) - \bar{\mu}_D(0) = q_e\frac{\Delta\varphi}{2} - \bar{\mu}_D(0)$$

$$\Delta\bar{\mu}_D^{(2)} = \bar{\mu}_D(0) - \bar{\mu}_D(d) = q_e\frac{\Delta\varphi}{2} + \bar{\mu}_D(0)$$

2. *The currents of electrons I_U and I_D are independent of the position between the electrodes because we assume that there is no reaction $U \to D$ or $D \to U$. Thus the currents of electrons I_U and I_D are given by,*

$$I_U = \int_S \boldsymbol{j}_U \cdot d\boldsymbol{S}$$

$$I_D = \int_S \boldsymbol{j}_D \cdot d\boldsymbol{S}$$

where the infinitesimal surface element $d\boldsymbol{S}$ is oriented along the electron current densities $\boldsymbol{j}_U$ and $\boldsymbol{j}_D$. The electrochemical potential differences $\Delta\bar{\mu}_U^{(1)}$, $\Delta\bar{\mu}_U^{(2)}$, $\Delta\bar{\mu}_D^{(1)}$ and $\Delta\bar{\mu}_D^{(2)}$ across each layer can be written as,

$$\Delta\bar{\mu}_U^{(1)} = \int_{-d}^{0} d\boldsymbol{r} \cdot (-\boldsymbol{\nabla}\,\bar{\mu}_U) \qquad and \qquad \Delta\bar{\mu}_U^{(2)} = \int_{0}^{d} d\boldsymbol{r} \cdot (-\boldsymbol{\nabla}\,\bar{\mu}_U)$$

$$\Delta\bar{\mu}_D^{(1)} = \int_{-d}^{0} d\boldsymbol{r} \cdot (-\boldsymbol{\nabla}\,\bar{\mu}_D) \qquad and \qquad \Delta\bar{\mu}_D^{(2)} = \int_{0}^{d} d\boldsymbol{r} \cdot (-\boldsymbol{\nabla}\,\bar{\mu}_D)$$

where the infinitesimal length element $d\boldsymbol{r}$ is oriented along the electron current densities $\boldsymbol{j}_U$ and $\boldsymbol{j}_D$. The thermoelectric materials 1 and 2 have a length d and a cross-section surface area A, which can be written as,

$$d = \int_{-d}^{0} d\boldsymbol{r} \cdot \hat{\boldsymbol{x}} = \int_{0}^{d} d\boldsymbol{r} \cdot \hat{\boldsymbol{x}} \qquad and \qquad A = \int_{S} d\boldsymbol{S} \cdot \hat{\boldsymbol{x}}$$

where the unit vector $d\hat{\boldsymbol{x}}$ is oriented along the electron current densities $\boldsymbol{j}_U$ and $\boldsymbol{j}_D$. The integrals of the electron transport equations over the volume are the product of integrals over the cross-section area A times integrals over the length d of the thermoelectric materials,

$$\int_{S} \boldsymbol{j}_U \cdot d\boldsymbol{S} \int_{-d}^{0} d\boldsymbol{r} \cdot \hat{\boldsymbol{x}} = \frac{\sigma_U^{(1)}}{q_e^2} \int_{-d}^{0} d\boldsymbol{r} \cdot (-\boldsymbol{\nabla}\,\bar{\mu}_U) \int_{S} d\boldsymbol{S} \cdot \hat{\boldsymbol{x}}$$

$$\int_{S} \boldsymbol{j}_U \cdot d\boldsymbol{S} \int_{0}^{d} d\boldsymbol{r} \cdot \hat{\boldsymbol{x}} = \frac{\sigma_U^{(2)}}{q_e^2} \int_{0}^{d} d\boldsymbol{r} \cdot (-\boldsymbol{\nabla}\,\bar{\mu}_U) \int_{S} d\boldsymbol{S} \cdot \hat{\boldsymbol{x}}$$

$$\int_{S} \boldsymbol{j}_D \cdot d\boldsymbol{S} \int_{-d}^{0} d\boldsymbol{r} \cdot \hat{\boldsymbol{x}} = \frac{\sigma_D^{(1)}}{q_e^2} \int_{-d}^{0} d\boldsymbol{r} \cdot (-\boldsymbol{\nabla}\,\bar{\mu}_D) \int_{S} d\boldsymbol{S} \cdot \hat{\boldsymbol{x}}$$

$$\int_{S} \boldsymbol{j}_D \cdot d\boldsymbol{S} \int_{0}^{d} d\boldsymbol{r} \cdot \hat{\boldsymbol{x}} = \frac{\sigma_D^{(2)}}{q_e^2} \int_{0}^{d} d\boldsymbol{r} \cdot (-\boldsymbol{\nabla}\,\bar{\mu}_D) \int_{S} d\boldsymbol{S} \cdot \hat{\boldsymbol{x}}$$

The electron transport equations integrated over the volume reduce to,

$$I_U = \frac{\sigma_U^{(1)}}{q_e^2} \frac{A}{d} \Delta\bar{\mu}_U^{(1)} = \frac{\sigma_U^{(2)}}{q_e^2} \frac{A}{d} \Delta\bar{\mu}_U^{(2)}$$

$$I_D = \frac{\sigma_D^{(1)}}{q_e^2} \frac{A}{d} \Delta\bar{\mu}_D^{(1)} = \frac{\sigma_D^{(2)}}{q_e^2} \frac{A}{d} \Delta\bar{\mu}_D^{(2)}$$

Using the expressions of $\Delta\bar{\mu}_U^{(1)}$, $\Delta\bar{\mu}_U^{(2)}$, $\Delta\bar{\mu}_D^{(1)}$ and $\Delta\bar{\mu}_D^{(2)}$ established in section 1, the previous conditions become,

$$\sigma_U^{(1)} \left(q_e \frac{\Delta\varphi}{2} - \bar{\mu}_U(0) \right) = \sigma_U^{(2)} \left(q_e \frac{\Delta\varphi}{2} + \bar{\mu}_U(0) \right)$$

$$\sigma_D^{(1)} \left(q_e \frac{\Delta\varphi}{2} - \bar{\mu}_D(0) \right) = \sigma_D^{(2)} \left(q_e \frac{\Delta\varphi}{2} + \bar{\mu}_D(0) \right)$$

This implies that,

$$\bar{\mu}_U(0) = \frac{\sigma_U^{(1)} - \sigma_U^{(2)}}{\sigma_U^{(1)} + \sigma_U^{(2)}} q_e \frac{\Delta\varphi}{2}$$

$$\bar{\mu}_D(0) = \frac{\sigma_D^{(1)} - \sigma_D^{(2)}}{\sigma_D^{(1)} + \sigma_D^{(2)}} q_e \frac{\Delta\varphi}{2}$$

Applying the previous relations to the expressions of $\Delta\bar{\mu}_U^{(1)}$, $\Delta\bar{\mu}_U^{(2)}$, $\Delta\bar{\mu}_D^{(1)}$ and $\Delta\bar{\mu}_D^{(2)}$ obtained in section 1, we find,

$$\Delta\bar{\mu}_U^{(1)} = \left(1 - \frac{\sigma_U^{(1)} - \sigma_U^{(2)}}{\sigma_U^{(1)} + \sigma_U^{(2)}}\right) q_e \frac{\Delta\varphi}{2} = \frac{\sigma_U^{(2)}}{\sigma_U^{(1)} + \sigma_U^{(2)}} q_e \Delta\varphi$$

$$\Delta\bar{\mu}_U^{(2)} = \left(1 + \frac{\sigma_U^{(1)} - \sigma_U^{(2)}}{\sigma_U^{(1)} + \sigma_U^{(2)}}\right) q_e \frac{\Delta\varphi}{2} = \frac{\sigma_U^{(1)}}{\sigma_U^{(1)} + \sigma_U^{(2)}} q_e \Delta\varphi$$

$$\Delta\bar{\mu}_D^{(1)} = \left(1 - \frac{\sigma_D^{(1)} - \sigma_D^{(2)}}{\sigma_D^{(1)} + \sigma_D^{(2)}}\right) q_e \frac{\Delta\varphi}{2} = \frac{\sigma_D^{(2)}}{\sigma_D^{(1)} + \sigma_D^{(2)}} q_e \Delta\varphi$$

$$\Delta\bar{\mu}_D^{(2)} = \left(1 + \frac{\sigma_D^{(1)} - \sigma_D^{(2)}}{\sigma_D^{(1)} + \sigma_D^{(2)}}\right) q_e \frac{\Delta\varphi}{2} = \frac{\sigma_D^{(1)}}{\sigma_D^{(1)} + \sigma_D^{(2)}} q_e \Delta\varphi$$

Taking into account the previous expressions, the currents of electrons I_U and I_D can be written as,

$$I_U = \frac{\sigma_U^{(1)} \sigma_U^{(2)}}{\sigma_U^{(1)} + \sigma_U^{(2)}} \frac{A}{d} \frac{\Delta\varphi}{q_e} = \left(\frac{1}{\sigma_U^{(1)}} + \frac{1}{\sigma_U^{(2)}}\right)^{-1} \frac{A}{d} \frac{\Delta\varphi}{q_e}$$

$$I_D = \frac{\sigma_D^{(1)} \sigma_D^{(2)}}{\sigma_D^{(1)} + \sigma_D^{(2)}} \frac{A}{d} \frac{\Delta\varphi}{q_e} = \left(\frac{1}{\sigma_D^{(1)}} + \frac{1}{\sigma_D^{(2)}}\right)^{-1} \frac{A}{d} \frac{\Delta\varphi}{q_e}$$

3. *Ohm's law (11.76) applied to the bilayer reads,*

$$\boldsymbol{j}_q = -\sigma \boldsymbol{\nabla} \varphi$$

The electrostatic potential difference $\Delta\varphi$ is given by,

$$\Delta\varphi = \int_{-d}^{d} d\boldsymbol{r} \cdot (-\boldsymbol{\nabla} \varphi)$$

The integrals of Ohm's law over the volume is the product of the integral over the cross-section area A times the integral over the length $2d$ of the thermoelectric materials,

$$\int_S \boldsymbol{j}_q \cdot d\boldsymbol{S} \int_{-d}^{d} d\boldsymbol{r} \cdot \hat{\boldsymbol{x}} = \sigma \int_{-d}^{d} d\boldsymbol{r} \cdot (-\boldsymbol{\nabla} \varphi) \int_S d\boldsymbol{S} \cdot \hat{\boldsymbol{x}}$$

Ohm's law integrated over the volume reduce to,

$$I = \sigma \frac{A}{2d} \Delta\varphi$$

The total electric current I flowing through the metallic material can be written in terms of the currents of electrons I_U and I_D as,

$$I = q_e \left(I_U + I_D\right)$$

Taking into account the expressions for I_U and I_D obtained in section 2, the electric current I is expressed as,

$$I = 2 \left(\left(\frac{1}{\sigma_U^{(1)}} + \frac{1}{\sigma_U^{(2)}} \right)^{-1} + \left(\frac{1}{\sigma_D^{(1)}} + \frac{1}{\sigma_D^{(2)}} \right)^{-1} \right) \frac{A}{2d} \Delta\varphi$$

Comparing these two expressions of Ohm's law, we deduce an effective electric conductivity σ which can be written in terms of the electric conductivities $\sigma_U^{(1)}$, $\sigma_U^{(2)}$, $\sigma_D^{(1)}$ and $\sigma_D^{(2)}$ as,

$$\sigma = 2 \left(\left(\frac{1}{\sigma_U^{(1)}} + \frac{1}{\sigma_U^{(2)}} \right)^{-1} + \left(\frac{1}{\sigma_D^{(1)}} + \frac{1}{\sigma_D^{(2)}} \right)^{-1} \right)$$

4. In the parallel magnetic configuration, the electric conductivities of each layer are defined as,

$$\sigma_U^{(1)} = \sigma_U^{(2)} = \sigma_p \qquad and \qquad \sigma_D^{(1)} = \sigma_D^{(2)} = \sigma_a$$

Comparing the two expressions of Ohm's law, we can write the electric conductivity σ_- between the two electrodes as,

$$\sigma_- = \sigma_p + \sigma_a$$

In the anti-parallel magnetic configuration, with an upward magnetisation in the first layer and a downward magnetisation in the second layer, the electric conductivities of each layer are given by,

$$\sigma_U^{(1)} = \sigma_D^{(2)} = \sigma_p \qquad and \qquad \sigma_U^{(2)} = \sigma_D^{(1)} = \sigma_a$$

Comparing the two expressions of Ohm's law, we can write the electric conductivity σ_+ between the two electrodes as,

$$\sigma_+ = 4 \left(\frac{1}{\sigma_p} + \frac{1}{\sigma_a} \right)^{-1}$$

5. The resistivities are defined as the inverse of the conductivities (11.74),

$$\rho_- = \frac{1}{\sigma_-} \qquad and \qquad \rho_p = \frac{1}{\sigma_p}$$

$$\rho_+ = \frac{1}{\sigma_+} \qquad and \qquad \rho_a = \frac{1}{\sigma_a}$$

The resistivities ρ_- and ρ_+ are written in terms of the resistivities ρ_0 and $\Delta\rho$ as,

$$\rho_- = \left(\frac{1}{\rho_p} + \frac{1}{\rho_a} \right)^{-1} = \frac{\rho_p \, \rho_a}{\rho_p + \rho_a} = \frac{(\rho_0 - \Delta\rho)(\rho_0 + \Delta\rho)}{\rho_0 - \Delta\rho + \rho_0 + \Delta\rho} = \frac{\rho_0}{2} - \frac{(\Delta\rho)^2}{2\rho_0}$$

$$\rho_+ = \frac{\rho_p + \rho_a}{4} = \frac{\rho_0 - \Delta\rho + \rho_0 + \Delta\rho}{4} = \frac{\rho_0}{2}$$

Comparing these two expressions, we find the following inequality for the resistivity across the layers,

$$\rho_+ > \rho_-$$

Thus, if the bilayer is in the anti-parallel magnetic configuration, its resistivity is larger than if it is in the parallel magnetic configuration.

For certain magnetic metals, the relative change of resistivity between the two magnetic configurations can become very large. This effect is called giant magnetoresistance. It was discovered in 1988 and gave rise to a Nobel Prize in physics, awarded to **Albert Fert** and **Peter Grünberg**.

Exercises

11.1   Heat Diffusion Equation

Show that the temperature profile (11.44),

$$T(x,t) = \frac{C}{\sqrt{t}} \exp\left(-\frac{x^2}{4\lambda t}\right)$$

where T is the temperature and x the spatial coordinate, is a solution of the heat diffusion equation (11.37).

11.2   Thermal Dephasing

A long copper rod of thermal diffusivity λ is heated at one end with a flame passing underneath it periodically, while the other end is located so far away that it remains at room temperature T_0. Consider the rod as a one-dimensional system with a periodic temperature variation of amplitude ΔT at $x = 0$, given by,

$$T(0,t) = T_0 + \Delta T \cos(\omega t)$$

where x is the spatial coordinate along the wire. Once the rod has reached a regime at which every point of the rod has a periodic temperature variation, show that the temperature profile is given by,

$$T(x,t) = T_0 + \Delta T \exp\left(-\frac{x}{d}\right) \cos\left(\omega t - \frac{x}{d}\right) \qquad \text{where} \qquad d = \sqrt{\frac{2\lambda}{\omega}}$$

The temperature oscillation at position x is dephased by an angle $-x/d$ with respect to the oscillation at position $x = 0$. The amplitude of oscillation is attenuated by a factor $\exp(-x/d)$.

11.3   Heat Equation with Heat Source

The heat diffusion equation was established in § 11.4.2, in the absence of any source term due to a conductive electric current density, i.e. $\boldsymbol{j_q} = q_e \boldsymbol{j_e} = \boldsymbol{0}$. Show that for an electric conductor in the presence of a conductive electric current density $\boldsymbol{j_q}$, the heat equation becomes,

$$\partial_t T = \lambda \boldsymbol{\nabla}^2 T - \frac{\tau}{c} \boldsymbol{j_q} \cdot \boldsymbol{\nabla} T + \frac{\boldsymbol{j_q^2}}{\sigma c}$$

where λ is the thermal diffusivity, σ is the electric conductivity, τ is the Thomson coefficient of the electric conductor and c is the specific heat density of the conduction electrons.

11.4   Joule Heating in a Wire

Estimate the temperature profile of a wire of length L and radius r when an electric current I is driven through it, from the left end to the right end, causing it to heat up. The wire has an electric conductivity σ and a thermal conductivity κ. The heat propagates down the wire to its ends, i.e. no heat is dissipated at the surface of the wire. The Thomson heating is negligible compared to the Joule heating. The left end and right end are kept at a constant temperature T_0. Determine the temperature profile $T(x)$ along the wire when it has reached a stationary state.

11.5   Thomson Heating in a Wire

Estimate the temperature profile of a wire of length L and radius r when an electric current I is driven through it, from the left end to the right end, causing it to heat up. The wire has an electric conductivity σ and a Thomson coefficient τ. The heat is entirely carried by the wire to its ends. The heat dissipated at the surface of the wire and the Joule heating are assumed negligible. The left end is kept at a constant temperature T_0. Determine the temperature profile $T(x)$ along the wire when it has reached a stationary state. Also give an expression for the temperature at the right end in terms of the Thomson coefficient τ and the electric resistance R of the wire.

11.6   Heat Exchanger

A heat exchanger is made up of two identical pipes separated by an impermeable diathermal wall of section area A, thickness h and thermal conductivity κ. In both pipes, a liquid flows at uniform velocities $\boldsymbol{v}_1 = v_1 \hat{\boldsymbol{x}}$ and $\boldsymbol{v}_2 = -v_2 \hat{\boldsymbol{x}}$, with $v_1 > 0$ and $v_2 > 0$, where $\hat{\boldsymbol{x}}$ is the unit vector that is parallel to the liquid flow in pipe 1. The temperature T_1 of the liquid in pipe 1 is larger than the temperature T_2 of the liquid in pipe 2, i.e. $T_1 > T_2$. Thus, there is a heat current density $\boldsymbol{j}_Q = j_Q \hat{\boldsymbol{y}}$, with $j_Q > 0$ going across the wall separating the pipes, where $\hat{\boldsymbol{y}}$ is the unit vector orthogonal to the wall and oriented positively from pipe 1 to pipe 2. There is no liquid current density across the wall, i.e. $\boldsymbol{j}_C = \boldsymbol{0}$. Heat conductivity is considered negligible in the direction of the flow and yet large enough to ensure a homogeneous temperature across any section of both pipes. Consider that the heat exchanger has reached a stationary state.

a) Show that the temperature profiles in the fluids are given by the differential equations,

$$\partial_x T_1 = -\frac{\kappa}{h\ell\, c_1\, v_1}\,(T_1 - T_2)$$
$$\partial_x T_2 = \frac{\kappa}{h\ell\, c_2\, v_2}\,(T_1 - T_2)$$

where c_1 and c_2 are the specific heat densities of liquids 1 and 2, and κ is the thermal conductivity of the diathermal wall and ℓ is a characteristic length for the thermal transfer.

b) Show that the convective heat current density $j = c_1 v_1 T_1 + c_2 v_2 T_2$ is homogeneous.

c) Determine the temperature difference $\Delta T(x) = T_1(x) - T_2(x)$.

d) Determine the temperature profiles $T_1(x)$ and $T_2(x)$.

e) Show that on a distance that is short enough, i.e. $x/d \ll 1$,

$$T_1(x) = \frac{j + c_2 v_2 \Delta T(0)}{c_1 v_1 + c_2 v_2} - \frac{\kappa \Delta T(0)}{h \ell c_1 v_1} x$$

$$T_2(x) = \frac{j - c_1 v_1 \Delta T(0)}{c_1 v_1 + c_2 v_2} + \frac{\kappa \Delta T(0)}{h \ell c_2 v_2} x$$

11.7   Harman Method

A rod is connected at both ends to electrodes. The electric wires that connect the rod to each electrode are strong enough to carry an electric current flowing through the rod and yet thin enough for the heat transfer to be negligible. The contact resistance and the heat radiated from the rod are negligible. In these experimental conditions, an adiabatic measurement of the resistivity of the material can be performed. As Harman suggested in his seminal paper [187], experimental conditions can be found such that the Joule and Thomson heating have negligible effects. Use the empirical linear equations (11.92) to show that the adiabatic resistivity thus measured is given by,

$$\rho_{\text{ad}} = \rho \left(1 + \frac{\varepsilon^2}{\kappa \rho} T \right)$$

where $\rho = 1/\sigma$ is the isothermal resistivity, κ is the thermal conductivity and ε is the Seebeck coefficient of the rod material.

11.8   Peltier Generator

A Peltier generator is made up of two thermoelectric elements connected in series (Fig. 11.13). One side of the Peltier generator is maintained at temperature T^+ and the other temperature T^-. The electric current I generated by the Peltier generator flows through the thermoelectric materials labelled 1 and 2. The plate which is heated up to temperature T^+ connects electrically these two materials but is not electrically available to the user. Its electric potential is V^+. The other ends of the thermoelectric materials are on the cold side, at a temperature T^-. They are connected to the electric leads of the device. A load resistance R_0 is connected to these leads. The voltage V designates the electric potential difference between the leads.

Analyse the operation of this generator using the electric charge and heat transport equations,

$$\boldsymbol{j}_{q_1} = -\sigma_1 \varepsilon_1 \boldsymbol{\nabla} T_1 - \sigma_1 \boldsymbol{\nabla} \varphi_1 \quad \text{and} \quad \boldsymbol{j}_{Q_1} = -\kappa_1 \boldsymbol{\nabla} T_1 + T_1 \varepsilon_1 \boldsymbol{j}_{q_1}$$

$$\boldsymbol{j}_{q_2} = -\sigma_2 \varepsilon_2 \boldsymbol{\nabla} T_2 - \sigma_2 \boldsymbol{\nabla} \varphi_2 \quad \text{and} \quad \boldsymbol{j}_{Q_2} = -\kappa_2 \boldsymbol{\nabla} T_2 + T_2 \varepsilon_2 \boldsymbol{j}_{q_2}$$

The thermoelectric materials 1 and 2 have a length d and a cross-section surface area A, which can be written as,

$$d = \int_0^d d\boldsymbol{r} \cdot \hat{\boldsymbol{x}} \qquad A = \int_S d\boldsymbol{S} \cdot \hat{\boldsymbol{x}}$$

where $\hat{\boldsymbol{x}}$ is a unit vector oriented clockwise along the electric current density $\boldsymbol{j}_q$, and the infinitesimal length and surface vectors $d\boldsymbol{r}$ and $d\boldsymbol{S}$ are oriented in the same direction. The temperature difference between the hot and cold ends is given by,

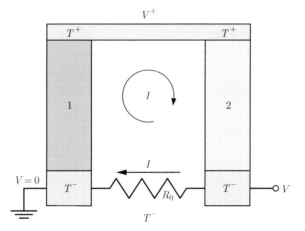

Figure 11.13 A Peltier generator has a load represented by the resistance R_0 connected to its leads. V is the voltage between the leads. The electric bridge at V^+ is not accessible to the user. The regions marked 1 and 2 represent the two thermoelectric materials. The regions marked T^+ and T^- are the hot and cold sides of the device.

$$\Delta T = T^+ - T^- = \int_0^d d\boldsymbol{r} \cdot \boldsymbol{\nabla} T_1 = \int_0^d d\boldsymbol{r} \cdot (-\boldsymbol{\nabla} T_2)$$

Likewise, the electric potential differences $\Delta \varphi_1$ and $\Delta \varphi_2$ between the hot and cold ends are written as,

$$\Delta \varphi_1 = V^+ = \int_0^d d\boldsymbol{r} \cdot \boldsymbol{\nabla} \varphi_1$$

$$\Delta \varphi_2 = V^+ - V = \int_0^d d\boldsymbol{r} \cdot (-\boldsymbol{\nabla} \varphi_2)$$

The electric charge conservation implies that the electric current densities are the same in each material, i.e. $\boldsymbol{j}_{q_1} = \boldsymbol{j}_{q_2}$. The electric current I flowing through materials 1 and 2 is the integral of the electric current densities $\boldsymbol{j}_{q_1}$ and $\boldsymbol{j}_{q_2}$ over the cross-section area A,

$$I = \int_S \boldsymbol{j}_{q_1} \cdot d\boldsymbol{S} = \int_S \boldsymbol{j}_{q_2} \cdot d\boldsymbol{S}$$

According to the relation (10.104), the thermal powers P_{Q_1} and P_{Q_2} are the integrals of the heat current densities $\boldsymbol{j}_{Q_1}$ and $\boldsymbol{j}_{Q_2}$ flowing through materials 1 and 2 over the cross-section area A,

$$P_{Q_1} = \int_S (-\boldsymbol{j}_{Q_1}) \cdot d\boldsymbol{S} \qquad P_{Q_2} = \int_S \boldsymbol{j}_{Q_2} \cdot d\boldsymbol{S}$$

Determine:

a) the thermal power P'_Q applied on the hot side of the device when no electric current flows through the device.

b) the effective electric resistance R of the two thermoelectric materials when the temperatures are equal, i.e. $T^+ = T^-$, and no electric current flows through the resistance R_0, i.e. when $R_0 = \infty$. Instead, an electric current flows through the thermoelectric materials.

c) the electric current I as a function of the temperature difference ΔT.
d) the thermodynamic efficiency of the generator defined as,

$$\eta = \frac{R_0 \, I^2}{P_Q}$$

where here, P_Q is the thermal power at the hot side when the electric current is flowing through the device. Show that the optimum load resistance is given by

$$\frac{R_0}{R} = \sqrt{1 + \zeta}$$

where ζ is a dimensionless parameter given by [188],

$$\zeta = \frac{T^+ \, (\varepsilon_1 - \varepsilon_2)^2}{(\kappa_1 + \kappa_2) \left(\dfrac{1}{\sigma_1} + \dfrac{1}{\sigma_2} \right)}$$

11.9 ZT Coefficient of a Thermoelectric Material

The transport properties of a thermoelectric material of cross-section area A and length L are defined by the transport equations,

$$\boldsymbol{j}_q = -\sigma \epsilon \, \boldsymbol{\nabla} T - \sigma \boldsymbol{\nabla} \varphi \qquad \text{and} \qquad \boldsymbol{j}_Q = -\kappa \, \boldsymbol{\nabla} T + T \varepsilon \boldsymbol{j}_q$$

in conformity with relations (11.92), where $\boldsymbol{\nabla} \mu_e = \boldsymbol{0}$, and (11.95). The efficiency η of the thermoelectric material is defined as,

$$\eta = -\frac{P_q}{P_Q}$$

where P_Q is the heat entering at the hot end and P_q is the electric power defined as,

$$P_q = \int_V \boldsymbol{j}_q \cdot (-\boldsymbol{\nabla} \varphi) \, dV$$

Write the efficiency η in terms of the ratio [189],

$$r = \frac{I}{\kappa} \frac{L}{A} \frac{1}{\Delta T}$$

where I is the electric current flowing through the thermoelectric material. In the limit where the thermoelectric effect is much smaller than the thermal power, i.e. $r\varepsilon \ll 1/T^+$, show that the optimal efficiency η is given by,

$$\eta = \left(1 - \frac{T^-}{T^+} \right) \frac{\sigma \, \varepsilon^2}{4 \, \kappa} \, T^+$$

The coefficient $(\sigma \varepsilon^2 / \kappa) \, T^+$ is called the 'ZT coefficient' of the thermoelectric material. The term in brackets is the Carnot efficiency.

11.10 Transverse Transport Effects

A transport equation such as Ohm's law (11.74),

$$\nabla \varphi = - \rho \cdot j_q$$

relates two vectors, which are the conductive electric current density j_q and electric potential gradient $\nabla \varphi$, through a linear application, which is the electric resistivity ρ. Mathematically, a vector is a rank-1 tensor and a linear application between two vectors is a rank-2 tensor.

a) Show that the electric resistivity ρ can be decomposed into the sum of a symmetric part ρ^s and an antisymmetric part ρ^a.

b) Show that the antisymmetric part ρ^a has a contribution to the transport that can be written as,

$$\nabla^a \varphi = - \rho^a \left(\hat{u} \times j_q \right)$$

where $\nabla^a \varphi$ is the antisymmetric part of the electric potential gradient and $\hat{u}$ is a unit axial vector.

The decomposition and the expression for the antisymmetric part of the electric potential gradient is a general result that applies for any empirical linear relation between a current density vector and a generalised force vector.

11.11 Hall Effect

An isotropic conductor is in the presence of a magnetic induction field B. The electric resistivity rank-2 tensor is a function of the magnetic induction field B and Ohm's law is written as,

$$\nabla \varphi = - \rho (B) \cdot j_q$$

The reversibility of the dynamics at the microscopic scale implies that the transpose of the electric resistivity tensor is obtained by reversing the orientation of the magnetic induction field B [190]. Thus,

$$\rho^T (B) = \rho (- B)$$

This result cannot be established in a thermodynamic framework but requires the use of a statistical physics. In a linear electromagnetic framework, when the magnetic induction field B is applied orthogonally to the conductive electric current density j_q, show that Ohm's law can be written as,

$$\nabla \varphi = - \rho \cdot j_q - \mathcal{H} j_q \times B$$

where the first term is Ohm's law (11.74) in the absence of a magnetic induction field B and the second term is the Hall effect (11.75) in a direction that is orthogonal to the magnetic induction field B and to the conductive electric current density. Use the result established in § 11.10.

11.12 Heat Transport and Crystal Symmetry

A crystal consists of a honeycomb lattice. It is invariant under a rotation of angle $\theta = \pi/6$ in the horizontal plane around the vertical axis. This means that the physical properties of the crystal are the same after such a rotation. Show that the symmetric thermal conductivity tensor κ is written in components as,

$$\kappa = \begin{pmatrix} \kappa_\perp & 0 & 0 \\ 0 & \kappa_\perp & 0 \\ 0 & 0 & \kappa_\| \end{pmatrix}$$

where $\kappa_\|$ is the thermal conductivity along the vertical rotation axis and $\kappa_\perp$ is the thermal conductivity in the horizontal plane of rotation.

11.13 Planar Ettingshausen Effect

In this chapter, several examples of a current density in one direction inducing the gradient of an intensive quantity in another direction were shown. These effects are referred to by the name of their discoverers: Righi-Leduc (11.29), Hall (11.75), Nernst (11.85), Ettingshausen (11.80). The latter refers to a temperature gradient induced by an orthogonal electric charge current density. It was pointed out recently that this effect can occur in a crystal, which consists of two types of electric charge carriers and presents a strong crystalline anisotropy in the plane where the heat and electric charge transport take place. No magnetic induction field needs to be applied orthogonally in order to observe this effect [191].

The material has two types of electric charge carriers, electrons (e) and holes (h). Assume that no 'chemical reaction' takes place between them. The thermoelectric properties are isotropic, i.e. the same in all directions. Therefore, the Seebeck tensors for the electrons and holes are given by,

$$\varepsilon_e = \begin{pmatrix} \varepsilon_e & 0 \\ 0 & \varepsilon_e \end{pmatrix} \qquad \text{and} \qquad \varepsilon_h = \begin{pmatrix} \varepsilon_h & 0 \\ 0 & \varepsilon_h \end{pmatrix}$$

However, the conductivities differ greatly in two orthogonal directions. Therefore, the conductivity tensors are given by,

$$\sigma_e = \begin{pmatrix} \sigma_{e,aa} & 0 \\ 0 & \sigma_{e,bb} \end{pmatrix} \qquad \text{and} \qquad \sigma_h = \begin{pmatrix} \sigma_{h,aa} & 0 \\ 0 & \sigma_{h,bb} \end{pmatrix}$$

where a and b label the a-axis and the b-axis that are orthogonal crystalline axes.

Consider an electric charge transport along the x-axis at an angle θ from the a-axis and show that this electric current density j_q induces a heat current density j_Q along the y-axis. This is the planar Ettingshausen effect. It can be understood by establishing the following facts:

a) Show that the Seebeck tensor for this crystal is given by [176],

$$\varepsilon = (\sigma_e + \sigma_h)^{-1} \cdot (\sigma_e \cdot \varepsilon_e + \sigma_h \cdot \varepsilon_h)$$

b) Show that the Seebeck tensor for this crystal is diagonal and written as,

$$\varepsilon = \begin{pmatrix} \varepsilon_{aa} & 0 \\ 0 & \varepsilon_{bb} \end{pmatrix}$$

where the diagonal component ε_{aa} is different from ε_{bb} in general. The matrix is given here for a vector basis along the crystalline a-axis and b-axis.

c) Write the components of the Seebeck tensor with respect to the coordinate basis (x, y),

$$\varepsilon = \begin{pmatrix} \varepsilon_{xx} & \varepsilon_{xy} \\ \varepsilon_{yx} & \varepsilon_{yy} \end{pmatrix}$$

in terms of the diagonal components ε_{aa} and ε_{bb} of the Seebeck tensor with respect to the coordinate basis (a, b).

d) The heat current density $\boldsymbol{j}_Q$ is related to the electric charge current density $\boldsymbol{j}_q$ by,

$$\boldsymbol{j}_Q = \boldsymbol{\Pi} \cdot \boldsymbol{j}_q$$

which is a local version of the Peltier effect (11.108). The Peltier tensor is related to the Seebeck tensor by,

$$\boldsymbol{\Pi} = T \varepsilon$$

In particular, for an electric charge current density $\boldsymbol{j}_q = j_{q,x} \hat{\boldsymbol{x}}$, where $\hat{\boldsymbol{x}}$ is a unit vector along the x-axis, show that the component $j_{Q,y}$ along the y-axis of the heat current density $\boldsymbol{j}_Q = j_{Q,x} \hat{\boldsymbol{x}} + j_{Q,y} \hat{\boldsymbol{y}}$, where $\hat{\boldsymbol{y}}$ is a unit vector along the y-axis, is given by,

$$j_{Q,y} = \frac{1}{2} T \left(\varepsilon_{aa} - \varepsilon_{bb} \right) \sin \left(2\, \theta \right) j_{q,x}$$

Thus, the planar Ettingshausen effect is maximal for an angle $\theta = \pi/4$.

11.14 Turing Patterns

A biological medium consists of two substances 1 and 2 of densities n_1 and n_2. This medium is generating both substances by processes characterised by the matter source densities $\pi_1 (n_1, n_2)$ and $\pi_2 (n_1, n_2)$. The substances 1 and 2 can diffuse inside this medium. The matter current densities $\boldsymbol{j}_1$ and $\boldsymbol{j}_2$ follow Fick's law (11.51),

$$\boldsymbol{j}_1 = -D_1 \boldsymbol{\nabla} n_1 \qquad \text{and} \qquad \boldsymbol{j}_2 = -D_2 \boldsymbol{\nabla} n_2$$

where $D_1 > 0$ and $D_2 > 0$ are the homogeneous diffusion constants of substances 1 and 2. The medium has a fixed volume, which means that its expansion rate vanishes, i.e. $\boldsymbol{\nabla} \cdot \boldsymbol{v} = 0$. Thus, the matter continuity equations for substances 1 and 2 are given by,

$$\dot{n}_1 + \boldsymbol{\nabla} \cdot \boldsymbol{j}_1 = \pi_1 (n_1, n_2) \qquad \text{and} \qquad \dot{n}_2 + \boldsymbol{\nabla} \cdot \boldsymbol{j}_2 = \pi_2 (n_1, n_2)$$

At equilibrium, the system is assumed to be homogeneous and characterised by the densities n_{01} and n_{02} of substances 1 and 2. In the neighbourhood of the equilibrium, the matter source densities $\pi_1 (n_1, n_2)$ and $\pi_2 (n_1, n_2)$ are given to first-order in terms of the density perturbations $\Delta n_1 = n_1 - n_{01}$ and $\Delta n_2 = n_2 - n_{02}$ by,

$$\pi_1 (n_1, n_2) = \Omega_{11}\, \Delta n_1 + \Omega_{12}\, \Delta n_2$$
$$\pi_2 (n_1, n_2) = \Omega_{21}\, \Delta n_1 + \Omega_{22}\, \Delta n_2$$

where the coefficients $\Omega_{11}, \Omega_{12}, \Omega_{21}, \Omega_{22}$ are given by,

$$\Omega_{11} = \frac{\partial \pi_1}{\partial n_1} \qquad \Omega_{12} = \frac{\partial \pi_1}{\partial n_2} \qquad \Omega_{21} = \frac{\partial \pi_2}{\partial n_1} \qquad \Omega_{22} = \frac{\partial \pi_2}{\partial n_2}$$

Assume that the processes to generate substances 1 and 2 are the two chemical reactions $1 \xrightarrow{a} 2$ and $2 \xrightarrow{b} 1$ described by the stoichiometric coefficients $\nu_{a1} = -1$, $\nu_{a2} = 1$, $\nu_{b1} = 1$, $\nu_{b2} = -1$ and the reaction rate densities ω_a and ω_b. Assume that the temperature T and the chemical potentials μ_1 and μ_2 are homogeneous, i.e. $\nabla T = \mathbf{0}$ and $\nabla \mu_1 = \nabla \mu_2 = \mathbf{0}$. Analyse the evolution of the density perturbations Δn_1 and Δn_2 by using the following instructions:

a) Express the coefficients Ω_{11}, Ω_{12}, Ω_{21}, Ω_{22} in terms of the total density $n = n_1 + n_2$, the density perturbations Δn_1 and Δn_2, the temperature T and a scalar $W \geq 0$, which is a linear combination of Onsager matrix elements L_{aa}, L_{ab}, L_{ba} and L_{bb}. Begin by using the second law, i.e. $\pi_s \geq 0$, and the relation (8.68) for a mixture of ideal gas.
b) Determine the coupled time evolution equations for the density perturbations Δn_1 and Δn_2.
c) Show that under the condition imposed in $a)$ the relation,

$$\begin{pmatrix} \Delta n_1 (t) \\ \Delta n_2 (t) \end{pmatrix} = e^{\lambda t} \, \cos (\mathbf{k} \cdot \mathbf{r} + \varphi) \begin{pmatrix} \Delta n_1 (0) \\ \Delta n_2 (0) \end{pmatrix}$$

is a solution of the coupled time evolution equations with $\lambda < 0$.

11.15 Ultramicroelectrodes

In electrochemistry, the observed current is generally determined by ion diffusion in the electrolyte. It was found that diffusion-limited currents can be avoided by using very small electrodes, known as '*ultramicroelectrodes*' [192, 193, 194]. In order to capture how conductive current densities (also called diffusion current densities) vary with the size of the electrode, consider a spherical electrode and a conductive matter current density with spherical symmetry, $\mathbf{j}_A = j_{Ar} \hat{\mathbf{r}} \equiv j_r \hat{\mathbf{r}}$. Show that when the system reaches a stationary state, the conductive matter current density is non-zero. The analysis of the transient behaviour would show that the stationary state is reached faster when the electrode is smaller. [195] In spherical coordinates (r, θ, ϕ), taking into account the spherical symmetry of the conductive matter current density, i.e. $\partial/\partial\theta = 0$ and $\partial/\partial\phi = 0$, the matter diffusion equation (11.54) for a solute of concentration $c (r, t)$ reads,

$$\frac{\partial c (r, t)}{\partial t} = D \left(\frac{\partial^2 c (r, t)}{\partial r^2} + \frac{2}{r} \frac{\partial c (r, t)}{\partial r} \right)$$

The boundary conditions are,

$$c (r > r_0, t = 0) = c^* \qquad \text{and} \qquad \lim_{r \to \infty} c (r, t) = c^*$$

where c^* is the concentration very far away from the electrode and r_0 is the radius of the electrode. According to relation (11.51), the conductive matter current density scalar j_r that characterises this electrode is,

$$j_r (r_0, t) = -D \left. \frac{\partial c (r, t)}{\partial r} \right|_{r=r_0}$$

Establish the following results:

a) The diffusion equation recast in terms of the function $w(r, t) = r c(r, t)$ has the structure of a diffusion equation where the spherical coordinate r plays an analogous role to a Cartesian coordinate.

b) The diffusion equation,

$$\frac{\partial w(r, t)}{\partial t} = D \frac{\partial^2 w(r, t)}{\partial r^2}$$

admits the solution,

$$w(r, t) = B \int_{\nu_0}^{\nu} \exp\left(-\nu'^2\right) d\nu' \qquad \text{where} \qquad \nu = \frac{r}{2\sqrt{Dt}}$$

First, write $w(r, t) = f(\eta)$ where the variable η is a dimensionless function of r and t given by,

$$\eta(r, t) = \frac{r^2}{Dt}$$

c) In the limit where the radius of the electrode is negligible, i.e. $r = 0$, the scalar conductive matter current density is given by,

$$j_r(0, t) = \frac{B}{8\sqrt{D} t^{3/2}}$$

d) After a transient behaviour, the scalar conductive matter current density reaches a stationary value,

$$j_r(r_0, \infty) = -\frac{D c^*}{r_0}$$

11.16   Effusivity

Two long blocks, made up of different homogeneous materials, are at temperatures T_1 and T_2 when they are brought into contact with one another. The interface quickly reaches a temperature T_0 given by,

$$T_0 = \frac{E_1 T_1 + E_2 T_2}{E_1 + E_2}$$

where $E_1 = \sqrt{\kappa_1 c_1} > 0$ and $E_2 = \sqrt{\kappa_2 c_2} > 0$ are called the **effusivities** of materials 1 and 2, where κ_1 and κ_2 are the thermal conductivities and c_1 and c_2 the specific heat per unit volume of both materials. If material 1 is very hot, but it has a low thermal conductivity κ_1 and specific heat c_1, and to the contrary material 2 has large thermal conductivity κ_2 and specific heat c_2, then the temperature of the interface T_0 is almost T_2, i.e. material 2 does not 'feel the heat' of material 1. Establish this result by using the following instructions:

a) Consider an x-axis normal to the interface with $x = 0$ at the interface, $x < 0$ in material 1 and $x > 0$ in material 2. Let $T_1(x, t)$ and $T_2(x, t)$ be the solutions of the heat diffusion equation (11.35) in materials 1 and 2. Determine the boundary conditions on $T_1(x, t)$ and $T_2(x, t)$ at the interface.

b) Using an approach that is analogous to the one presented in § 11.4.3, show that the general solutions for the temperature profiles $T_1(x, t)$ and $T_2(x, t)$ are given by,

$$T_1(x, t) = C_1 + D_1 \operatorname{erf}\left(\frac{x}{2\sqrt{\lambda_1 t}}\right) \qquad \text{where} \qquad x \leq 0$$

$$T_2(x, t) = C_2 + D_2 \operatorname{erf}\left(\frac{x}{2\sqrt{\lambda_2 t}}\right) \qquad \text{where} \qquad x \geq 0$$

where $\operatorname{erf}(\nu)$ is the error function defined as,

$$\operatorname{erf}(\nu) = \frac{2}{\sqrt{\pi}} \int_0^\nu \exp\left(-s^2\right) ds$$

and C_1, C_2, D_1 and D_2 are constant coefficients.

c) Use the boundary conditions to determine the coefficients in terms of the temperatures T_0, T_1 and T_2. Show that the temperature T_0 is given by the effusivity relation just after the two blocks have reached a common temperature at the interface.

Part IV

EXERCISES AND SOLUTIONS

Thermodynamic System and First Law

The exercises given in the last section of Chapter 1 are presented here with their solutions.

1.1 State Function: Mathematics

Consider the function $f(x, y) = y \exp(ax) + xy + bx \ln y$ where a and b are constants.

a) Calculate $\dfrac{\partial f(x, y)}{\partial x}$, $\dfrac{\partial f(x, y)}{\partial y}$ and $df(x, y)$

b) Calculate $\dfrac{\partial^2 f(x, y)}{\partial x\, \partial y}$

Solution:

The partial derivatives and differential of the function $f(x, y) = y \exp(ax) + xy + bx \ln y$ are given by,

a) $\dfrac{\partial f(x, y)}{\partial x} = ay \exp(ax) + y + b \ln y$

   $\dfrac{\partial f(x, y)}{\partial y} = \exp(ax) + x + \dfrac{bx}{y}$

   $df(x, y) = (ay \exp(ax) + y + b \ln y)\, dx + \left(\exp(ax) + x + \dfrac{bx}{y} \right) dy$

b) $\dfrac{\partial^2 f(x, y)}{\partial x\, \partial y} = a \exp(ax) + 1 + \dfrac{b}{y}$

1.2 State Function: Ideal Gas

An ideal gas is characterised by the relation $pV = NRT$ where p is the pressure of the gas, V is the volume, T is the temperature, N is the number of moles of gas and R is a constant.

a) Calculate the differential $dp(T, V)$

b) Calculate $\dfrac{\partial}{\partial T} \left(\dfrac{\partial p(T, V)}{\partial V} \right)$ and $\dfrac{\partial}{\partial V} \left(\dfrac{\partial p(T, V)}{\partial T} \right)$

Solution:

The differential and partial derivatives of the pressure $p\,(T, V)$ of an ideal gas are given by,

a) $dp\,(T, V) = \dfrac{NR}{V}\,dT - \dfrac{NRT}{V^2}\,dV.$

b) $\dfrac{\partial}{\partial T}\left(\dfrac{\partial p\,(T, V)}{\partial V}\right) = \dfrac{\partial}{\partial V}\left(\dfrac{\partial p\,(T, V)}{\partial T}\right) = -\dfrac{NR}{V^2}.$

1.3 State Function: Rubber Cord

A rubber cord of length L, which is a known state function $L\,(T, F)$ of the temperature T of the cord and of the forces of magnitude F applied at each end to stretch it. Two physical properties of the cord are :

a) the Young modulus, defined as $E = \dfrac{L}{A}\left(\dfrac{\partial L}{\partial F}\right)^{-1}$, where A is the cord cross section area.

b) the thermal expansion coefficient $\alpha = \dfrac{1}{L}\dfrac{\partial L}{\partial T}.$

Determine how much the length of the cord varies if its temperature changes by ΔT and at the same time the force F changes by ΔF. Assume that $\Delta T \ll T$ and $\Delta F \ll F$. Express ΔL in terms of E and α.

Solution:

Applying definition (1.7) to the differential of the length $L\,(T, F)$ of the rubber cord, the change of length of the rubber cord is given by,

$$\Delta L = \frac{\partial L}{\partial T}\,\Delta T + \frac{\partial L}{\partial F}\,\Delta F$$

which can be recast as,

$$\Delta L = L\left(\frac{1}{L}\frac{\partial L}{\partial T}\right)\Delta T + \frac{L}{A}\left(\frac{L}{A}\left(\frac{\partial L}{\partial F}\right)^{-1}\right)^{-1}\Delta F$$

Using the two physical properties of the cord, we obtain an expression for the change of length of the rubber cord,

$$\Delta L = L\alpha\,\Delta T + \frac{L}{AE}\,\Delta F$$

1.4 State Function: Volume

A liquid is filling a container that has the form of a cone of angle α around a vertical axis (Fig. 1.1). The liquid enters the cone from the apex through a hole of diameter d at a

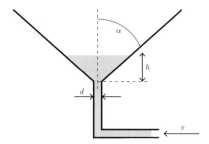

Figure 1.1   A liquid enters into a funnel with a laminar flow of velocity v in a tube of diameter d. The funnel is a cone of opening angle α. The cone axis is vertical.

velocity $v(t) = kt$ where k is a constant. When the surface of the liquid is at height $h(t)$, the volume is $V(t) = \dfrac{1}{3}\pi \tan^2 \alpha\, h^3(t)$. Initially, at time $t = 0$, the height $h(0) = 0$. Find an expression for the rate of change of volume $\dot{V}(t)$ and determine $h(t)$.

Solution:

The volume rate of change of the liquid is obtained by taking the time derivative of the volume $V(t) = \dfrac{1}{3}\pi \tan^2 \alpha\, h^3(t)$,

$$\dot{V}(t) = \pi \tan^2 \alpha\, h^2(t)\, \dot{h}(t)$$

where the angle α is a constant. The volume rate of change of the liquid inflow is expressed as,

$$\dot{V}(t) = \pi \left(\frac{d}{2}\right)^2 v(t) = \pi \frac{kd^2}{4} t$$

Equating the two previous equations yields,

$$\tan^2 \alpha\, h^2(t)\, \dot{h}(t) = \frac{kd^2}{4} t$$

which can be recast as,

$$h^2(t)\, dh(t) = \frac{kd^2}{4\tan^2 \alpha}\, t\, dt$$

When integrating, we obtain,

$$\int_{h(0)}^{h(t)} h'^2(t')\, dh'(t') = \frac{kd^2}{4\tan^2 \alpha} \int_0^t t'\, dt'$$

The result of this integration is,

$$\frac{1}{3} h^3(t) = \frac{kd^2}{4\tan^2 \alpha} \frac{1}{2} t^2$$

Thus, the height of liquid inside the cone is,

$$h(t) = \left(\frac{3\,kd^2}{8\tan^2 \alpha}\right)^{\frac{1}{3}} t^{\frac{2}{3}}$$

1.5 Cyclic Rule for the Ideal Gas

An ideal gas is characterised by the relation $pV = NRT$ as in § 1.2 where the pressure $p(T, V)$ is a function of T and V, the temperature $T(p, V)$ is a function of p and V and the volume (T, p) is a function of T and p. Calculate,

$$\frac{\partial p(T, V)}{\partial T} \frac{\partial T(p, V)}{\partial V} \frac{\partial V(T, p)}{\partial p}$$

Solution:

The partial derivative of the pressure, the temperature and the volume of an ideal gas that satisfies the relation $pV = NRT$ are given by,

$$\frac{\partial p(T, V)}{\partial T} = \frac{\partial}{\partial T}\left(\frac{NRT}{V}\right) = \frac{NR}{V}$$

$$\frac{\partial T(p, V)}{\partial V} = \frac{\partial}{\partial V}\left(\frac{pV}{NR}\right) = \frac{p}{NR}$$

$$\frac{\partial V(T, p)}{\partial p} = \frac{\partial}{\partial p}\left(\frac{NRT}{p}\right) = -\frac{NR}{p^2} = -\frac{V}{p}$$

Thus,

$$\frac{\partial p(T, V)}{\partial T} \frac{\partial T(p, V)}{\partial V} \frac{\partial V(T, p)}{\partial p} = -1$$

This result can be generalised, as shown in § 4.7.2.

1.6 Evolution of Salt Concentration

A basin contains $N_s(t)$ moles of salt dissolved in $N_w(t)$ moles of water. The basin receives fresh water at a constant rate Ω_w^{in}. This water is assumed to be thoroughly mixed in the basin so that the salt concentration can be considered homogeneous. The salty water comes out of the basin at a constant rate $\Omega_{sw}^{\text{out}} = \Omega_s^{\text{out}}(t) + \Omega_w^{\text{out}}(t)$, where $\Omega_s^{\text{out}}(t)$ and $\Omega_w^{\text{out}}(t)$ are the salt and water outflow rates. Determine the salt concentration,

$$c_s(t) = \frac{N_s(t)}{N_s(t) + N_w(t)}$$

as a function of time t for the given initial conditions $N_s(0)$ and $N_w(0)$.

Solution:

The time derivative of the amount of salt in the basin is equal to the salt outflow rate and the time derivative of the amount of water is the sum of the water inflow and outflow rates,

$$\dot{N}_s(t) = \Omega_s^{\text{out}}(t)$$
$$\dot{N}_w(t) = \Omega_w^{\text{in}} + \Omega_w^{\text{out}}(t)$$

where Ω_w^{in} is the fresh water inflow rate (positive), and Ω_s^{out} and Ω_w^{out} are the salt and water outflow rates (negative). Since a matter flow is an extensive quantity, the constant salty water outflow rate Ω_{sw}^{out} is the sum of the salt outflow rate $\Omega_s^{\text{out}}(t)$ and the water outflow rate $\Omega_w^{\text{out}}(t)$,

$$\Omega_{sw}^{\text{out}} = \Omega_s^{\text{out}}(t) + \Omega_w^{\text{out}}(t)$$

Water and salt are assumed to be thoroughly mixed in the basin so that the salt concentration can be considered homogeneous. Thus, the salt outflow rate $\Omega_s^{\text{out}}(t)$ is equal to the product of its concentration $c_s(t)$ in the basin and of the salty water outflow rate Ω_{sw}^{out},

$$\Omega_s^{\text{out}}(t) = c_s(t)\,\Omega_{sw}^{\text{out}}$$

Applying this equation for $\Omega_s^{\text{out}}(t)$ in the balance equation for the salt in the basin, using the definition of the molar concentration,

$$c_s(t) = \frac{N_s(t)}{N_s(t) + N_w(t)}$$

and dividing by $N_s(t)$, we obtain,

$$\frac{\dot{N}_s(t)}{N_s(t)} = \frac{\Omega_{sw}^{\text{out}}}{N_s(t) + N_w(t)}$$

Adding up the first two balance equations, we obtain the balance equation for the salty water in the basin,

$$\dot{N}_s(t) + \dot{N}_w(t) = \Omega_w^{\text{in}} + \Omega_{sw}^{\text{out}}$$

Since the term on the right hand side is constant, we integrate this equation with respect to time from $t = 0$ onwards,

$$N_s(t) + N_w(t) = \left(\Omega_w^{\text{in}} + \Omega_{sw}^{\text{out}}\right) t + N_s(0) + N_w(0)$$

Applying this into the equation for $\dot{N}_s(t)/N_s(t)$, we obtain,

$$\frac{\dot{N}_s(t)}{N_s(t)} = \frac{\Omega_{sw}^{\text{out}}}{\left(\Omega_w^{\text{in}} + \Omega_{sw}^{\text{out}}\right) t + N_s(0) + N_w(0)}$$

The time integral of this equation is given by,

$$\ln\left(\frac{N_s(t)}{N_s(0)}\right) = \frac{\Omega_{sw}^{\text{out}}}{\Omega_w^{\text{in}} + \Omega_{sw}^{\text{out}}} \ln\left(\frac{\left(\Omega_w^{\text{in}} + \Omega_{sw}^{\text{out}}\right) t + N_s(0) + N_w(0)}{N_s(0) + N_w(0)}\right)$$

The exponential of this integrated expression yields,

$$N_s(t) = N_s(0) \left(1 + \frac{\left(\Omega_w^{\text{in}} + \Omega_{sw}^{\text{out}}\right) t}{N_s(0) + N_w(0)}\right)^{\frac{\Omega_{sw}^{\text{out}}}{\Omega_w^{\text{in}} + \Omega_{sw}^{\text{out}}}}$$

Applying the equations for $N_s(t)$ and $N_s(t) + N_w(t)$ in the expression for the molar salt concentration $c_s(t)$, we obtain,

$$c_s(t) = \frac{N_s(0)}{\left(\Omega_w^{in} + \Omega_{sw}^{out}\right)t + N_s(0) + N_w(0)} \left(1 + \frac{\left(\Omega_w^{in} + \Omega_{sw}^{out}\right)t}{N_s(0) + N_w(0)}\right)^{\frac{\Omega_{sw}^{out}}{\Omega_w^{in} + \Omega_{sw}^{out}}}$$

This can be recast as,

$$c_s(t) = \frac{N_s(0)}{N_s(0) + N_w(0)} \left(1 + \frac{\left(\Omega_w^{in} + \Omega_{sw}^{out}\right)t}{N_s(0) + N_w(0)}\right)^{-\frac{1}{1 + \Omega_{sw}^{out}/\Omega_w^{in}}}$$

1.7 Capilarity: Contact Angle

Capilarity effects are taken into account by considering that the energy of the system contains contributions that are proportional to the surface area of the interfaces between the different parts of the system. For a drop of wetting liquid on a horizontal surface (Fig. 1.2), where the drop is assumed to have a spherical shape, the internal energy is expressed as $U(h, R) = \left(\gamma_{s\ell} - \gamma_{sg}\right)\pi a^2 + \gamma_{\ell g} A$ where $a = R\sin\theta = \sqrt{2Rh - h^2}$ is the radius and $A = 2\pi Rh$ is the surface area of the spherical cap of height h at the intersection of the sphere of radius R and the solid substrate. The parameters $\gamma_{s\ell}, \gamma_{sg}, \gamma_{\ell g}$ characterise the substances and are independent of the drop shape. Show that the contact angle θ is given by,

$$\left(\gamma_{s\ell} - \gamma_{sg}\right) + \gamma_{\ell g}\cos\theta = 0$$

by minimising the internal energy $U(h, R)$ under the condition that the volume $V(h, R) = \frac{\pi}{3}h^2(3R - h) = V_0$ is constant.

Solution:

In order to minimise the internal energy $U(h, R)$, use the Lagrange multipliers method to impose the condition that the volume of the drop is fixed, i.e. $V(h, R) = V_0$. The function $F(h, R, \lambda)$ to be minimised is,

$$\begin{aligned} F(h, R, \lambda) &= U(h, R) - \lambda\left(V(h, R) - V_0\right) \\ &= \left(\gamma_{s\ell} - \gamma_{sg}\right)\pi\left(2Rh - h^2\right) + \gamma_{\ell g} 2\pi Rh \\ &\quad - \lambda\left(\frac{\pi}{3}h^2(3R - h) - V_0\right) \end{aligned}$$

where λ is the Lagrange multiplier. According to this method, the partial derivative of the function $F(h, R, \lambda)$ with respect to h has to vanish,

$$\frac{\partial F}{\partial h} = \left(\gamma_{s\ell} - \gamma_{sg}\right)2\pi(R - h) - \gamma_{\ell g}2\pi R + \lambda\pi\left(2Rh - h^2\right) = 0$$

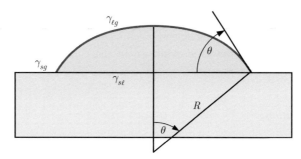

Figure 1.2 A drop of liquid on a horizontal substrate is assumed to have a spherical shape. The angle θ is called the contact angle. Aurface tension is defined for each of the three interfaces: solid–liquid ($\gamma_{s\ell}$), solid–gas (γ_{sg}) and liquid–gas ($\gamma_{\ell g}$).

which yields an expression for the Lagrange multiplier,

$$\lambda = \left(\frac{2}{2Rh - h^2}\right)(R - h)\left(\gamma_{s\ell} - \gamma_{sg}\right) + \left(\frac{2}{2Rh - h^2}\right)R\gamma_{\ell g}$$

The partial derivative of the function $F(h, R, \lambda)$ with respect to R has to vanish as well,

$$\frac{\partial F}{\partial R} = \left(\gamma_{s\ell} - \gamma_{sg}\right)2\pi h + \gamma_{\ell g}2\pi h - \lambda\pi h^2 = 0$$

which yields another expression for the Lagrange multiplier,

$$\lambda = \frac{2}{h}\left(\gamma_{s\ell} - \gamma_{sg}\right) + \frac{2}{h}\gamma_{\ell g}$$

Equating the two expressions for the Lagrange multiplier λ yields,

$$(R - h)\left(\gamma_{s\ell} - \gamma_{sg}\right) + R\gamma_{\ell g} = (2R - h)\left(\gamma_{s\ell} - \gamma_{sg}\right) + (2R - h)\gamma_{\ell g}$$

which reduces to,

$$\left(\gamma_{s\ell} - \gamma_{sg}\right) + \left(\frac{R - h}{R}\right)\gamma_{\ell g} = 0$$

By graphical inspection (Fig. 1.2),

$$\cos\theta = \frac{R - h}{R}$$

Thus, we find that,

$$\left(\gamma_{s\ell} - \gamma_{sg}\right) + \gamma_{\ell g}\cos\theta = 0$$

1.8 Energy: Thermodynamics Versus Mechanics

A weight of mass M is hanging from a rope. The force $\boldsymbol{F}$ applied to the rope is such that the weight is lowered vertically at a velocity $\boldsymbol{v}$, which may vary with time.

a) Determine the expression for the time evolution of the mechanical energy E', which is the sum of the kinetic and potential energies.

b) Determine the time evolution of the energy E of the system according to the first law (1.11).

Solution:

a) From the standpoint of mechanics, the projection of Newton's law of motion for the weight, $\boldsymbol{F} + M\boldsymbol{g} = M\boldsymbol{a}$, along a coordinate axis Oz pointing downwards yields,

$$-F + Mg = M\ddot{z}$$

The time evolution of the mechanical energy E' is obtained by multiplying this result by $\dot{z}$,

$$\frac{d}{dt}\left(\frac{1}{2}M\dot{z}^2 - Mgz\right) = -F\dot{z}$$

Since the mechanical energy E' is the sum of the kinetic and potential energies,

$$E' = \frac{1}{2}M\dot{z}^2 - Mgz$$

the previous result can be recast as,

$$\dot{E}' = -F\dot{z}$$

b) From the standpoint of thermodynamics, the energy of the system E is expressed as,

$$E = \frac{1}{2}M\dot{z}^2 + U$$

where U is the internal energy of the system. Since the system consists of the mass M only, its weight is an external force. Thus, the gravitational potential energy is not included in the energy E of the system. Since the internal energy U is a function of the state variables of the system only, it is independent of the height z in the Earth's gravitational field. Since there is no heat transfer between the weight and the environment, the thermal power vanishes, i.e. $P_Q = 0$. Furthermore, the weight is assumed rigid, which implies that the mechanical deformation power vanishes as well, i.e. $P_W = 0$. The external power is due to the weight $M\boldsymbol{g}$ and the force $\boldsymbol{F}$ that can change the kinetic energy of the system,

$$P^{\text{ext}} = \boldsymbol{F} \cdot \boldsymbol{v} + M\boldsymbol{g} \cdot \boldsymbol{v} = -F\dot{z} + Mg\dot{z}$$

The first law is expressed as, $\dot{E} = P^{\text{ext}}$, which implies that,

$$\dot{E} = (-F + Mg)\dot{z}$$

Since the internal energy of this particular system is constant, i.e. $\dot{U} = 0$, the previous result reduces to,

$$\frac{d}{dt}\left(\frac{1}{2}M\dot{z}^2\right) = (-F + Mg)\dot{z}$$

1.9 Damped Harmonic Oscillator

A one-dimensional harmonic oscillator of mass M and spring constant k is subjected to a friction force $\boldsymbol{F}_f(t) = -\lambda \boldsymbol{v}(t)$ where $\boldsymbol{v}(t)$ is the velocity of the point mass and $\lambda > 0$. Using a coordinate axis Ox where the origin O corresponds to the position of the point mass when the harmonic oscillator is at rest, the equation of motion reads,

$$\ddot{x} + 2\gamma\dot{x} + \omega_0^2 x = 0$$

where $\omega_0^2 = k/M$ and $\gamma = \lambda/(2M)$. In the weak damping regime, where $\gamma \ll \omega_0$, the position can be expressed as

$$x(t) = Ce^{-\gamma t}\cos(\omega_0 t + \phi)$$

where C and ϕ are integration constants.

a) Express the mechanical energy $E(t)$ in terms of the coefficients k, C and γ.
b) Calculate the power $P(t)$ dissipated due to the friction force $\boldsymbol{F}_f(t)$ during one oscillation period.

Solution:

a) The mechanical energy of the damped harmonic oscillator of a point mass M and spring constant k subjected to a damping coefficient γ, where $\gamma \ll \omega_0$ and $\omega_0^2 = k/M$, is the sum of its kinetic energy $T(t)$ and the elastic potential energy $V_e(t)$ of the spring,

$$E(t) = T(t) + V_e(t) = \frac{1}{2}Mv^2(t) + \frac{1}{2}kr^2(t)$$

where $v^2(t) = \dot{x}^2(t)$ is the velocity squared and $r^2(t) = x^2(t)$ is the displacement squared along the axis Ox. The rest position of the point mass is at the origin. Thus,

$$E(t) = \frac{1}{2}M\dot{x}^2(t) + \frac{1}{2}kx^2(t)$$

Since the position coordinate $x(t)$ reads,

$$x(t) = Ce^{-\gamma t}\cos(\omega_0 t + \phi)$$

we have,

$$x^2(t) = C^2 e^{-2\gamma t}\cos^2(\omega_0 t + \phi)$$

$$\dot{x}^2(t) = \omega_0^2 C^2 e^{-2\gamma t}\sin^2(\omega_0 t + \phi) = \frac{k}{M}C^2 e^{-2\gamma t}\sin^2(\omega_0 t + \phi)$$

Hence, the mechanical energy is recast as,

$$E(t) = \frac{1}{2}kC^2 e^{-2\gamma t}\left(\sin^2(\omega_0 t + \phi) + \cos^2(\omega_0 t + \phi)\right) = \frac{1}{2}kC^2 e^{-2\gamma t}$$

b) The power dissipated by the friction force $\boldsymbol{F}_f(t)$ is equal to the time derivative of the mechanical energy,

$$P(t) = \dot{E}(t) = \frac{d}{dt}\left(\frac{1}{2}kC^2 e^{-2\gamma t}\right) = -\gamma kC^2 e^{-2\gamma t}$$

Entropy and Second Law

The exercises given in the last section of Chapter 2 are presented here with their solutions.

2.1 Entropy as a State Function

Determine which of the following functions may represent the entropy of a system of positive temperature. In these expressions, E_0 and V_0 are constants representing an energy and a volume, respectively.

a)

$$S(U, V, N) = NR \ln \left(1 + \frac{U}{NE_0} \right) + \frac{RU}{E_0} \ln \left(1 + \frac{NE_0}{U} \right)$$

b)

$$S(U, V, N) = \frac{RU}{E_0} \exp \left(-\frac{UV}{NE_0} \right)$$

c)

$$S(U, V, N) = \frac{NRU}{\left(V^3 / V_0^3 \right) E_0}$$

d)

$$S(U, V, N) = R \frac{U^{3/5} V^{2/5}}{E_0^{3/5} V_0^{2/5}}$$

Solution:

Since the entropy $S(U, V, N)$ is an extensive state function and the state variables U, V and N are also extensive, they have to satisfy the identity $S(\lambda U, \lambda V, \lambda N) = \lambda S(U, V, N)$ where λ is a positive integer. Expressions $b)$ and $c)$ do not satisfy this identity. Moreover, since the partial derivative of the entropy with respect to the internal energy in expression $b)$ is negative, this expression also needs to be rejected because it has a negative temperature. Thus, only expressions $a)$ and $d)$ may represent the entropy of a system of positive temperature.

2.2 Work as a Process-Dependent Quantity

Three processes are performed on a gas from a state given by (p_1, V_1) to a state given by (p_2, V_2) :

a) an isochoric process followed by an isobaric process,
b) an isobaric process followed by an isochoric process,
c) a process where $p\,V$ remains constant.

Compute for the three processes the work performed on the gas from the initial to the final state. These processes are assumed to be reversible. Determine the analytical results first, then give numerical values in joules.

Numerical Application:

$p_1 = p_0 = 1$ bar, $V_1 = 3\,V_0, p_2 = 3\,p_0, V_2 = V_0 = 1\,\text{l}.$

Solution:

There is no work performed on the gas during an isochoric process, only during the isobaric process or during the process where $p\,V$ remains constant.

a) The work performed on the gas by an isochoric process followed by an isobaric process is given by,

$$W = -\int_{V_1}^{V_2} p\,dV = -p_2 \int_{V_1}^{V_2} dV = -3\,p_0 \int_{3\,V_0}^{V_0} dV$$
$$= -3\,p_0\,(3\,V_0 - V_0) = 6\,p_0\,V_0 = 600\text{ J}$$

b) The work performed on the gas by an isobaric process followed by an isochoric process is given by,

$$W = -\int_{V_1}^{V_2} p\,dV = -p_1 \int_{V_1}^{V_2} dV = -p_0 \int_{3\,V_0}^{V_0} dV$$
$$= -p_0\,(3\,V_0 - V_0) = 2\,p_0\,V_0 = 200\text{ J}$$

c) The work performed on the gas by a process where $p\,V$ remains constant, i.e. $p\,V = p_1\,V_1 = \text{const}$, is given by,

$$W = -\int_{V_1}^{V_2} p\,dV = -p_1\,V_1 \int_{V_1}^{V_2} \frac{dV}{V} = -3\,p_0\,V_0 \int_{3\,V_0}^{V_0} \frac{dV}{V}$$
$$= -3\,p_0\,V_0 \ln\left(\frac{V_0}{3\,V_0}\right) = 3\,p_0\,V_0 \ln 3 = 330\text{ J}$$

2.3 Bicycle Pump

Air is compressed inside the inner tube of a bike using a manual bicycle pump. The handle of the pump is brought down from an initial position x_2 to a final position x_1 where $x_1 < x_2$ and the norm of the force is assumed to be given by,

$$F(x) = F_{\max} \frac{x_2 - x}{x_2 - x_1}$$

The process is assumed to be reversible and the cylinder of the pump has a cross section A. Determine in terms of the atmospheric pressure p_0,

a) the work W_h performed by the hand on the handle of the pump,
b) the pressure $p(x)$,
c) the work W_{12} performed on the system according to relation (2.42).

Numerical Application:

$F_{\max} = 10$ N, $x_1 = 20$ cm, $x_2 = 40$ cm, $A = 20$ cm^2 and $p_0 = 10^5$ Pa.

Solution:

a) The work performed by the hand on the handle of the pump is,

$$W_h = - \int_{V_2}^{V_1} p(V) \, dV = - \int_{x_2}^{x_1} F(x) \, dx$$

$$= - \frac{F_{\max}}{x_2 - x_1} \int_{x_2}^{x_1} (x_2 - x) \, dx$$

$$= - \left(\frac{F_{\max}}{x_2 - x_1} \right) \left(x_2 (x_1 - x_2) - \frac{1}{2} (x_1^2 - x_2^2) \right)$$

$$= \frac{1}{2} F_{\max} (x_2 - x_1) = 1 \text{ J}$$

b) The net pressure $p(x)$ exerted on the air inside the inner tube of the bike when the handle of the pump is at position x is the sum of the atmospheric pressure p_0 and the force $F(x)$ exerted by the piston on the air divided by the surface A,

$$p(x) = p_0 + \frac{F(x)}{A} = p_0 + \frac{F_{\max}}{A} \frac{x_2 - x}{x_2 - x_1}$$

$$= \left(10^5 + 5 \cdot 10^3 (2 - 5x) \right) \text{ N m}^{-2}$$

c) The work performed on the system by the atmosphere and by the person pushing the handle of the pump is,

$$W_{12} = - \int_{x_2}^{x_1} p_0 A \, dx - \int_{x_2}^{x_1} p(x) A \, dx = p_0 A (x_2 - x_1) + W_h = 41 \text{ J}$$

2.4 Rubbing Hands

Rubbing hands together is a dissipative process that we would like to model and quantify.

a) Determine the mechanical power P_W dissipated by friction during this process, in terms of the friction force F^{fr} and the the mean relative velocity v of a hand with respect to the other.

b) At room temperature T, determine the entropy production rate Π_S of this process.

Numerical Application:

$\|F^{\text{fr}}\| = 1\,\text{N}$, $\|v\| = 0.1\,\text{m/s}$ and $T = 25°\text{C}$

Solution:

a) With respect to the centre of mass frame of reference of the hands, each hand is moving with a velocity of norm $\|v\|/2$. The friction force F^{fr} is opposed to the relative motion of the hands, i.e. $F^{\text{fr}} \cdot v < 0$. The mechanical power is written,

$$P_W = \left(-F^{\text{fr}}\right) \cdot \left(\frac{v}{2}\right) + \left(F^{\text{fr}}\right) \cdot \left(-\frac{v}{2}\right) = -F^{\text{fr}} \cdot v = 0.1\,\text{W}$$

b) It can be presumed here that no heat is transferred to the environment, so that the evolution equation (2.18) for S implies $\dot{S} = \Pi_S$. The evolution equation (1.29) for the internal energy implies $\dot{U} = P_W$. As the internal energy U is a function of the entropy S only, $\dot{U} = T\dot{S}$. Consequently, the entropy production rate is given by,

$$\Pi_S = \frac{P_W}{T} = -\frac{F^{\text{fr}} \cdot v}{T} = 3.36 \cdot 10^{-4}\,\text{W/K}$$

2.5 Heating by Stirring

In an experiment similar to the Joule experiment (Fig. 1.1), an electric motor is used instead of a weight to stir the liquid. The thermal power P_Q, assumed to result from the friction, is known. The coefficient c_M, which represents the heat per unit mass and temperature, is known and it is assumed to be independent of temperature.

a) Deduce the temperature rise ΔT after stirring for a time Δt.
b) Find an expression for the entropy variation ΔS during this process, which started at temperature T_0.

Numerical Application:

$M = 200\,\text{g}$, $P_Q = 19\,\text{W}$, $c_M = 3\,\text{J}\,\text{g}^{-1}\text{K}^{-1}$, $\Delta t = 120\,\text{s}$, $T_0 = 300\,\text{K}$.

Solution:

a) According to the first law, the increase of internal energy ΔU during a time interval Δt is given by,

$$\Delta U = P_Q \Delta t$$

The increase of internal energy ΔU for a temperature increase ΔT of the liquid yields,

$$\Delta U = M c_M \Delta T$$

Thus, by comparing these two equations, the temperature increase ΔT is found to be,

$$\Delta T = \frac{P_Q \Delta t}{M c_M} = 3.8\,K.$$

b) The first law (2.22) is written as,

$$\dot{U} = T\dot{S} = P_Q \qquad \text{thus} \qquad \dot{S} = \frac{P_Q}{T}$$

According to *a*) the temperature is given by,

$$T = T_0 + \frac{P_Q}{M c_M} t$$

which implies that,

$$dS = \frac{P_Q}{T} dt = \frac{P_Q \, dt}{T_0 + \frac{P_Q}{M c_M} t} = M c_M \left(\frac{\frac{P_Q}{M c_M T_0} dt}{1 + \frac{P_Q}{M c_M T_0} t} \right)$$

When integrating this equation over time, we obtain,

$$S(t) = \int_{S_0}^{S(t)} dS = \int_0^t M c_M \left(\frac{\frac{P_Q}{M c_M T_0} dt'}{1 + \frac{P_Q}{M c_M T_0} t'} \right)$$

$$= S_0 + M c_M \ln \left(1 + \frac{P_Q t}{M c_M T_0} \right)$$

Thus, the entropy increase during the stirring process is,

$$\Delta S = S(\Delta t) - S_0 = M c_M \ln \left(1 + \frac{P_Q \Delta t}{M c_M T_0} \right)$$

2.6 Swiss Clock

A Swiss watchmaker states in a flyer the mechanical power P_W dissipated by a specific clock (Fig. 2.1). The work provided to the clock is due to the temperature fluctuations ΔT around room temperature T that let the clock run during a time t. We consider that the atmospheric pressure p_{ext} is equal to the pressure of the gas p, i.e. $p_{ext} = p$. The pressure

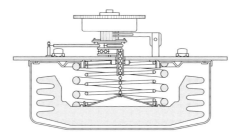

Figure 2.1 A clock receives its energy from the gas capsule (shaded area). The gas expands and contracts under the effects of the temperature fluctuations in the room.

p and the volume V of the gas are related by the ideal gas law $pV = NRT$ where R is the ideal gas constant. Consider that the gas in the capsule is always at equilibrium with the air outside of the capsule (pressure and temperature are the same inside and outside). From the watchmaker's information, estimate the volume V of the gas capsule used to run this clock.

Numerical Application:

$P_W = 0.25 \cdot 10^{-6}$ W, $T = 25°C$, $\Delta T = 1°C$, $p_{ext} = 10^5$ Pa and $t = 1$ day.

Solution:
According to relation (2.41), the infinitesimal work performed on the gas in the volume V inside the capsule, at constant atmospheric pressure p, is given by,

$$\delta W = -p\, dV$$

The ideal gas law implies that,

$$dV = \frac{NR}{p} dT$$

Thus, the infinitesimal work performed on the gas by the environment is given by,

$$\delta W = -NR\, dT$$

The initial state i corresponds to the "hotter" equilibrium state just after a fluctuation and the final state f corresponds to the "colder" equilibrium state just before the next fluctuation. The work W_{if} performed on the gas between the initial state i, characterised by the thermodynamic quantities ($V_i = V + \Delta V$, $T_i = T + \Delta T$) and the final state f, characterised by the thermodynamic quantities ($V_f = V$, $T_f = T$), is obtained when integrating the previous equation,

$$W_{if} = -\int_{T_i}^{T_f} NR\, dT = NR\Delta T$$

Moreover, the ideal gas law evaluated in the final state implies that,

$$W_{if} = \frac{pV}{T}\Delta T$$

Thus,

$$V = \frac{W_{if}T}{p\,\Delta T}$$

The work W_{if} must be equal to the product $P_W t$ of the mechanical power P_W required to run the clock and the time t during which it runs, i.e. $W_{if} = P_W t$. Thus, the volume V of the capsule is given by,

$$V = \frac{P_W t T}{p\,\Delta T} = 128 \text{ cm}^3$$

2.7 Reversible and Irreversible Gas Expansion

A mole of gas undergoes an expansion through two different processes. The gas satisfies the equation of state $p V = NR T$ where R is a constant, N the number of moles, p the pressure, T the temperature and V the volume of the gas. The initial and final temperatures are T_0. The walls of the gas container are diathermal. However, if a process takes place extremely fast, the walls can be considered as adiabatic. The initial pressure of the gas is p_1, the final pressure is p_2. Express the work performed on the gas in terms of p_1, p_2 and T_0 for the following processes:

a) a reversible isothermal process,
b) an extremely fast pressure variation, such that the external pressure on the gas is p_2 during the expansion, then an isochoric process during which the temperature again reaches the equilibrium temperature T_0,

Solution:

To compute the work performed on the gas undergoing a process from the initial state (V_1, p_1, T_0) to the final state (V_2, p_2, T_0), we use the state equation,

$$p_1 V_1 = p_2 V_2 = NR T_0 \qquad \text{thus} \qquad \frac{V_2}{V_1} = \frac{p_1}{p_2}$$

a) The work performed on the gas during a reversible expansion from the initial state (V_1, p_1, T_0) to the final state (V_2, p_2, T_0) is obtained by calculating the integral expression (2.42),

$$W_{12} = - \int_{V_1}^{V_2} p \, dV = - NR T_0 \int_{V_1}^{V_2} \frac{dV}{V} = - NR T_0 \ln \left(\frac{V_2}{V_1} \right)$$

$$= - NR T_0 \ln \left(\frac{p_1}{p_2} \right) = NR T_0 \ln \left(\frac{p_2}{p_1} \right)$$

b) There is no work performed on the gas during an isochoric process since the volume is constant. However, to determine the work performed by the environment at pressure $p^{\text{ext}} = p_2$ on the gas during an irreversible isobaric process from the initial state (V_1, p_1, T_0) to the final state (V_2, p_2, T_0), we need to integrate the general expression (2.36) for the mechanical power P_W over time,

$$W_{12} = \int_{t_1}^{t_2} P_W \, dt = - \int_{V_1}^{V_2} p^{\text{ext}} \, dV = - \int_{V_1}^{V_2} p_2 \, dV = - p_2 \int_{V_1}^{V_2} dV$$

$$= - p_2 V_2 + p_2 V_1 = \frac{p_2}{p_1} (p_1 V_1) - p_2 V_2 = NR T_0 \left(\frac{p_2}{p_1} - 1 \right)$$

Thermodynamics of Subsystems

The exercises given in the last section of Chapter 3 are presented here with their solutions.

3.1 Thermalisation of Two Separate Gases

An isolated system consisting of two closed subsystems A and B is separated by a diathermal wall. Initially, they are held at temperatures T_A^i and T_B^i. Subsystem A contains N_A moles of gas. The internal energy of the gas is given by $U_A = c_A N_A R T_A$, where T_A is the temperature of the gas, R is a positive constant and c_A is a dimensionless coefficient. Likewise, there are N_B moles of gas in subsystem B and the internal energy of the gas is given by $U_B = c_B N_B R T_B$.

a) Determine the change of the internal energy U_A due to the thermalisation process.
b) Compare the initial temperature T_B^i and the final temperature T_f of the system if the size of subsystem B is much larger than that of subsystem A.

Solution:

According to § 3.2, the equilibrium under the circumstances specified in this exercise is characterised by equal temperatures for both subsystems. Since the whole system is isolated, the total energy U is conserved which means that the initial value U_i is equal to the final value U_f. Thus, we have,

$$U_i = c_A N_A R\, T_A^i + c_B N_B R\, T_B^i = c_A N_A R\, T_f + c_B N_B R\, T_f = U_f$$

which implies that,

$$T_f = \frac{c_A N_A T_A^i + c_B N_B T_B^i}{c_A N_A + c_B N_B}$$

a) The internal energy variation is given by,

$$\Delta U_A = c_A N_A R \left(T_f - T_A^i \right) = c_A N_A R \left(\frac{c_A N_A T_A^i + c_B N_B T_B^i}{c_A N_A + c_B N_B} - T_A^i \right)$$

which is recast as,

$$\Delta U_A = R \left(\frac{1}{c_A N_A} + \frac{1}{c_B N_B} \right)^{-1} \left(T_B^i - T_A^i \right)$$

b) The final temperature of the system T_f can be recast as,

$$T_f = \frac{1}{1 + \dfrac{c_A}{c_B} \dfrac{N_A}{N_B}} \left(T_B^i + \frac{c_A}{c_B} \frac{N_A}{N_B} T_A^i \right)$$

If subsystem B is much larger than subsystem A, it will contain much more gas, i.e. $N_A \ll N_B$. In this limit, the final temperature of the system T_f is the initial temperature T_B^i of the subsystem B,

$$T_f = T_B^i$$

Thus, the temperature T_B^i remains constant during the thermalisation process. In other words, the large system does not change its temperature when put in contact with the small subsystem A. We will formally introduce the notion of thermal bath in § 4.5.1.

3.2 Thermalisation of Two Separate Substances

The entropy of a particular substance is given in terms of its internal energy U and number of moles N as,[1]

$$S(U, V, N) = NR \ln \left(1 + \frac{U}{NE_0} \right) + \frac{RU}{E_0} \ln \left(1 + \frac{NE_0}{U} \right)$$

where R and E_0 are positive constants. A system consists of two subsystems containing such a substance, with N_A moles in subsystem A and N_B moles of it in subsystem B. When the subsystems are set in thermal contact, their initial temperatures are T_A^i and T_B^i. Determine the final temperature T_f of the system.

Solution:

As shown in § 3.2, the equilibrium is characterised by equal temperatures for both subsystems. Thus, we have to find an expression for the temperature of this substance. We defined the temperature as the partial derivative of the internal energy U with respect to the entropy S. When taking the inverse of that relation, according to a well-known mathematical rule, which can be found in § 4.7.2, we obtain,

$$\frac{1}{T} = \frac{\partial S}{\partial U} = \frac{R}{E_0} \ln \left(1 + \frac{NE_0}{U} \right)$$

When inverting this relation, we find an expression for the internal energy,

$$U = \frac{NE_0}{\exp \left(\dfrac{E_0}{RT} \right) - 1}$$

[1] G. Carrington, *Basic Thermodynamics*, Oxford Science Publications, Oxford University Press, New York (1994).

Then, as in exercise 3.1, we apply the internal energy conservation to the total system and we find,

$$\frac{E_0}{R\,T_f} = \ln\left(1 + \left(\frac{\dfrac{N_A}{N_A + N_B}}{\exp\left(\dfrac{E_0}{RT_A^i}\right) - 1} + \frac{\dfrac{N_A}{N_A + N_B}}{\exp\left(\dfrac{E_0}{RT_B^i}\right) - 1}\right)^{-1}\right)$$

3.3 Diffusion of a Gas through a Permeable Wall

Analyse the time evolution of a gas consisting of one substance that diffuses through a permeable wall. Thus, consider an isolated system containing N moles of gas, consisting of two subsystems of equal volumes separated by a fixed and permeable wall, with $N_1(t)$ moles of gas in subsystem 1 and $N_2(t)$ in subsystem 2. In order to be able to find the time evolution, model the chemical potentials in each subsystem by assuming that they are proportional to the amount of matter. In order to simplify the expressions in the solution, write,

$$\mu_1(N_1) = \frac{\ell}{FA}\frac{N_1}{2\tau}$$
$$\mu_2(N_2) = \frac{\ell}{FA}\frac{N_2}{2\tau}$$

where $\tau > 0$ will be identified as a specific diffusion time, $F > 0$ is the Fick diffusion coefficient and $\ell > 0$ is a specific length. Initially, there are N_0 moles in the subsystem 1, i.e. $N_1(0) = N_0$, and $N - N_0$ moles in the subsystem 2, i.e. $N_2(0) = N - N_0$. Determine the evolution of the number of moles $N_1(t)$ and $N_2(t)$. Find the number of moles in each subsystem when equilibrium is reached.

Solution:

The amount of gas in the system is equal to the sum of the amounts of gas in the subsystems,

$$N = N_1(t) + N_2(t)$$

According to the irreversible diffusion equation (3.45), the variation rate $\dot{N}_1$ of the number of moles of gas in the subsystem 1 is given by

$$\dot{N}_1 = F\frac{A}{\ell}\left(\mu_2 - \mu_1\right) = \frac{1}{2\tau}\left(N_2(t) - N_1(t)\right) = -\frac{1}{\tau}\left(N_1(t) - \frac{N}{2}\right)$$

The integral of this evolution equation is written as,

$$\int_{N_0}^{N_1(t)} \frac{dN_1'}{N_1' - \frac{N}{2}} = -\frac{1}{\tau}\int_0^t dt'$$

and the result of this integration is given by,

$$\ln \left(\frac{N_1(t) - \frac{N}{2}}{N_0 - \frac{N}{2}} \right) = -\frac{t}{\tau}$$

which implies that,

$$N_1(t) = \frac{N}{2} + \left(N_0 - \frac{N}{2} \right) \exp \left(-\frac{t}{\tau} \right)$$

Given that $N_2(t) = N - N_1(t)$, we have,

$$N_2(t) = \frac{N}{2} + \left(\frac{N}{2} - N_0 \right) \exp \left(-\frac{t}{\tau} \right)$$

Thus, at equilibrium, when $t \to \infty$,

$$N_1(\infty) = N_2(\infty) = \frac{N}{2}$$

which means that there is the same amount of gas in each subsystem. Thus, at equilibrium, the system is homogeneous.

3.4 Mechanical Damping by Heat Flow

An isolated system of volume V_0 consists of two subsystems, labelled 1 and 2, separated by an impermeable and moving diathermal wall of mass M and of negligible volume. Both subsystems contain a gas. The presure p, the volume V, the number of moles N and the temperature T of the gas are related by the equation $pV = NRT$ where R is a positive constant (see § 5.6). The internal energy of this gas is given by $U = cNRT$ where c is a dimensionless coefficient (see § 5.7). Initially, both subsystems are at the temperature T_i. Subsystem 1 is in a state characterised by a volume V_{1i} and a pressure p_{1i}. Likewise, subsystem 2 is characterised by a pressure p_{2i} and a volume V_{2i}. Determine,

a) The number of moles N_1 and N_2 in subsystems 1 and 2.
b) The final temperature T_f when the system has reached equilibrium.
c) The final volumes V_{1f} and V_{2f} of the subsystems when the system has reached equilibrium.
d) The final pressure p_f when the system has reached equilibrium.
e) Determine the entropy variation between the initial state and the final equilibrium state and show that, for the particular case where $N_1 = N_2 = N$, the result implies an increase in entropy.
f) As in § 3.3, assume that the wall is able to transfer heat fast enough that the temperature T stays the same on both sides of the wall, which means that the heat transfer is reversible. Take into account the kinetic energy of the wall, neglect any heat stored inside the wall. Show that the wall comes to its equilibrium position with a velocity v that decays exponentially with a time constant τ inversely proportional to the thermo-hydraulic resistance R_{th}.

Solution:

a) Since the wall is impermeable, the numbers of moles N_1 and N_2 are constant. Thus, in the initial state,

$$N_1 = \frac{p_{1i}V_{1i}}{R\,T_i} \quad \text{and} \quad N_2 = \frac{p_{2i}V_{2i}}{R\,T_i}$$

b) Since N_1 and N_2 are constant, the conservation of total internal energy implies that the temperature is constant, i.e. $T_f = T_i$,

$$c\,N_1R\,T_i + c\,N_2R\,T_i = c\,N_1R\,T_f + c\,N_2R\,T_f$$

c) Owing to the equation of state for each subsystem in the final state, we have,

$$p_f V_{1f} = N_1 R\,T_f \quad \text{and} \quad p_f V_{2f} = N_2 R\,T_f$$

Similarly, the equation of state for the system in the final state is written as,

$$p_f V_0 = (N_1 + N_2)\,R\,T_f$$

By identifying the final pressure p_f in the relations above, we obtain the final volumes of the subsystems,

$$V_{1f} = \frac{N_1}{N_1 + N_2}\,V_0 \quad \text{and} \quad V_{2f} = \frac{N_2}{N_1 + N_2}\,V_0$$

d) The final pressure p_f is given by,

$$p_f = \frac{(N_1 + N_2)\,R\,T_f}{V_0}$$

It is the pressure of a gas of $N_1 + N_2$ moles at temperature T_f occupying the volume V_0 of the whole system.

e) Although the process is irreversible, as explained in § 3.3, we can make use of the fact that entropy is a state function and compute how it would change if the system underwent a reversible process. Thus, we consider that subsystem 1 and 2 undergo an isothermal process at the temperature $T_i = T_f = T$, which implies that their internal energy is constant, i.e. $dU_1 = c\,N_1R\,dT = 0$ and $dU_2 = c\,N_2R\,dT = 0$. Hence, we can write

$$dU_1 = -p_1 dV_1 + T dS_1 = -\frac{N_1 R T}{V_1}\,dV_1 + T dS_1 = 0$$

$$dU_2 = -p_2 dV_2 + T dS_2 = -\frac{N_2 R T}{V_2}\,dV_2 + T dS_2 = 0$$

The integration of these relations between the initial state i and the final state f yields,

$$\Delta S_{1\,if} = N_1 R \ln\left(\frac{V_{1f}}{V_{1i}}\right) \quad \text{and} \quad \Delta S_{2\,if} = N_2 R \ln\left(\frac{V_{2f}}{V_{2i}}\right)$$

In the particular case where $N_1 = N_2 = N$, $V_{1f} = V_{2f} = V_0/2$, we have $V_{2i} = V_0 - V_{1i}$. Thus, the entropy variation is given by,

$$\Delta S_{if} = \Delta S_{1\,if} + \Delta S_{2\,if} = NR \ln \left(\frac{V_{1f}}{V_{1i}} \right) + NR \ln \left(\frac{V_{2f}}{V_{2i}} \right)$$

$$= -NR \ln \left(\frac{V_{1i}}{V_{1f}} \frac{V_{2i}}{V_{2f}} \right) = -NR \ln \left(4 \frac{V_{1i}}{V_0} \left(1 - \frac{V_{1i}}{V_0} \right) \right)$$

where $0 < V_{1i}/V_0 < 1$. The argument of the logarithm is unity when $V_{1i}/V_0 = 1/2$ and then the entropy variation vanishes, i.e. $\Delta S_{if} = 0$. This means that in this particular case nothing happens as the initial state is already the equilibrium state. For all other values of V_{1i}/V_0, the argument of the logarithm is found between 0 and 1, hence the entropy variation is positive.

f) The internal energy of the system is the sum of the internal energies of the subsystem and the kinetic energy of the wall,

$$U = U_1 + U_2 + \frac{1}{2} M v^2$$

Since the system is isolated the total internal energy is constant. Thus,

$$\dot{U} = \dot{U}_1 + \dot{U}_2 + M v \dot{v} = 0$$

Given that both subsystems are at temperature T, the time derivative of the internal energy of the subsystems is given by,

$$\dot{U}_1 = T \dot{S}_1 - p_2 \dot{V}_1 \qquad \text{and} \qquad \dot{U}_2 = T \dot{S}_2 - p_1 \dot{V}_2$$

Hence, we have,

$$\dot{U} = T \left(\dot{S}_1 + \dot{S}_2 \right) - p_2 \dot{V}_1 - p_1 \dot{V}_2 + M v \dot{v} = 0$$

Furthermore, since the heat transfer between the two subsystems is reversible, the total entropy of the system is constant. Thus,

$$\dot{S} = \dot{S}_1 + \dot{S}_2 = 0$$

This implies that,

$$M \dot{v} = \frac{1}{v} \left(p_2 \dot{V}_1 + p_1 \dot{V}_2 \right)$$

which is Newton's law of motion for the wall. Since the volume of the system is constant, $\dot{V}_2 = - \dot{V}_1$. Furthermore, according to the law (3.31), Newton's law of motion can be recast as,

$$M \dot{v} = \frac{1}{v} \left(p_2 - p_1 \right) \dot{V}_1 = - \frac{R_{th}}{v} \dot{V}_1^2$$

Assuming that the velocity of the wall v is positive when the volume of the subsystem 1 increasases, we have $\dot{V}_1 = A v$ where A is the area of the wall. Thus, the deceleration of the wall is given by,

$$\dot{v} = - \frac{R_{th} A^2}{M} v$$

The integration of this relation from the initial velocity v_i to the velocity $v(t)$ yields,

$$v(t) = v_i \exp\left(-\frac{t}{\tau}\right)$$

where $\tau = M/\left(R_{\text{th}} A^2\right)$ is the mechanical damping time constant.

3.5 Entropy Production by Thermalisation

In exercise 3.6.2 devoted to the thermalisation of two blocks, show for the particular case where $N_1 = N_2 = N$ that the entropy variation,

$$\Delta S = 3N_1 R \ln\left(\frac{T_f}{T_1}\right) + 3N_2 R \ln\left(\frac{T_f}{T_2}\right)$$

is strictly positive.

Solution:

In the particular case where $N_1 = N_2 = N$, the final temperature is given by,

$$T_f = \frac{1}{2}(T_1 + T_2)$$

which implies that the entropy variation is recast as,

$$\Delta S = 3NR \ln\left(\frac{T_f^2}{T_1 T_2}\right) = 3NR \ln\left(\frac{(T_1 + T_2)^2}{4 T_1 T_2}\right)$$

Since the initial temperatures T_1 and T_2 are different, taking into account the identity,

$$(T_1 - T_2)^2 = T_1^2 + T_2^2 - 2 T_1 T_2 > 0 \qquad \text{implying} \qquad T_1^2 + T_2^2 > 2 T_1 T_2$$

the entropy variation is strictly positive,

$$\Delta S > 3NR \ln\left(\frac{4 T_1 T_2}{4 T_1 T_2}\right) = 0$$

as it should for an irreversible process.

3.6 Entropy Production by Heat Transfer

An isolated system consists of two subsystems labelled 1 and 2 analysed in exercise 3.6.1. Using the second law (2.2), show that in a stationary state when $T_1 > T_2$ the entropy production rate Π_S of the whole system is positive when heat flows across these two subsystems despite the fact that, according to equation (2.23), $\Pi_{S_1} = \Pi_{S_2} = 0$.

Solution:

Recall that the entropy production rate of the system cannot be simply obtained by adding the entropy production rates in each subsystem (see equation (3.14)). The system is in a stationary state, which implies that the time derivative of the entropy vanishes, i.e. $\dot{S} = 0$. Thus, according to the second law (2.2), the entropy production rate is equal to the opposite of the entropy exchange rate, i.e.

$$\dot{S} = I_S + \Pi_S = 0 \qquad \text{and thus} \qquad \Pi_S = -I_S$$

The entropy exchange rate I_S is the sum of the entropy inflow rate $P_Q^{(01)}/T_1$ from the environment into subsystem 1 and the entropy outflow rate $-P_Q^{(20)}/T_2$ from subsystem 2 to the environment, where the negative sign characterises a loss of entropy. Thus, we have,

$$\Pi_S = \frac{P_Q^{(20)}}{T_2} - \frac{P_Q^{(01)}}{T_1}$$

Since $T_1 > T_2$, heat flows from subsystem 1 to subsystem 2, thus $P_Q^{(12)} > 0$. As established in problem 3.6.1 for a stationary heat flow, $P_Q^{(01)} = P_Q^{(12)} = P_Q^{(20)}$, thus

$$\Pi_S = \left(\frac{1}{T_2} - \frac{1}{T_1} \right) P_Q^{(12)} > 0$$

3.7 Thermalisation by Radiation

An isolated system consists of two blocks made of the same substance (Fig. 3.1). The internal energies of blocks 1 and 2 are $U_1 = C_1 T_1$ and $U_2 = C_2 T_2$ where C_1 and C_2 are two positive constants. Two sides of the blocks face each other exactly. The area of each side is A and they are separated by a fixed air gap. We neglect the heat conductivity of the air in the gap. The radiative thermal power that block i exerts on block j, where $i,j = 1,2$, is given by,

$$P_Q^{(ij)} = \sigma A \left(T_i(t)^4 - T_j(t)^4 \right)$$

where σ is a constant coefficient.

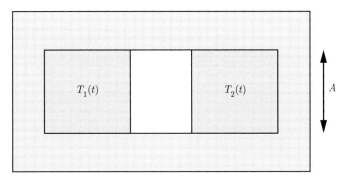

Figure 3.1   Two blocks of the same material face each other, separated by air. Thermal conduction through the air and convection are neglected. The blocks come to a thermal equilibrium due to heat exchange by radiation.

a) Determine the final temperature T_f of the system when it reaches equilibrium.
b) Derive the time evolution equation for $T_1(t)$ and $T_2(t)$.
c) Consider the particular case where $C_1 = C_2 = C$ and the limit of small temperature variations, i.e. $T_1(t) = T_f + \Delta T_1(t)$ and $T_2(t) = T_f + \Delta T_2(t)$ with $\Delta T_1(t) \ll T_f$ and $\Delta T_2(t) \ll T_f$ at all times. Show that the temperature difference $\Delta T(t) = \Delta T_1(t) - \Delta T_2(t)$ is exponentially decreasing.

Solution:

a) Applying the internal energy conservation law to the whole system, we obtain the final temperature at equilibrium,

$$T_f = \frac{C_1 T_1 + C_2 T_2}{C_1 + C_2}$$

b) The first law applied to each subsystem reads,

$$\dot{U}_1 = C_1 \dot{T}_1 = P_Q^{(21)} = \sigma A \left(T_2^4 - T_1^4 \right)$$
$$\dot{U}_2 = C_2 \dot{T}_2 = P_Q^{(12)} = \sigma A \left(T_1^4 - T_2^4 \right)$$

Thus, the temperature time evolution equations are given by,

$$\dot{T}_1 = \frac{\sigma A}{C_1} \left(T_2^4 - T_1^4 \right) \qquad \text{and} \qquad \dot{T}_2 = \frac{\sigma A}{C_2} \left(T_1^4 - T_2^4 \right)$$

c) In case where $C_1 = C_2 = C$, the time derivatives of the temperature variations yield,

$$\Delta \dot{T}_1 = \frac{\sigma A}{C} \left((T_f + \Delta T_2)^4 - (T_f + \Delta T_1)^4 \right)$$
$$\Delta \dot{T}_2 = \frac{\sigma A}{C} \left((T_f + \Delta T_1)^4 - (T_f + \Delta T_2)^4 \right)$$

In the limit where $\Delta T_1 \ll T_f$ and $\Delta T_2 \ll T_f$, these results reduce to,

$$\Delta \dot{T}_1 = \frac{4\,\sigma A}{C} T_f^3 \left(\Delta T_2 - \Delta T_1 \right)$$
$$\Delta \dot{T}_2 = \frac{4\,\sigma A}{C} T_f^3 \left(\Delta T_1 - \Delta T_2 \right)$$

Subtracting these two equations yields,

$$\Delta \dot{T} = \frac{8\,\sigma A}{C} T_f^3 \, \Delta T$$

where $\Delta T = \Delta T_1 - \Delta T_2$. After integration, we obtain,

$$\Delta T(t) = \Delta T(0) \, \exp\left(-\frac{t}{\tau} \right)$$

where $\Delta T(0) = T_1(0) - T_2(0)$ and the damping time constant τ is,

$$\tau = \frac{C}{8\,\sigma A\, T_f^3}$$

Thermodynamic Potentials

The exercises given in the last section of Chapter 4 are presented here with their solutions.

4.1 Adiabatic Compression

A gas is characterised by the enthalpy $H(S, p) = C_p T$, where C_p is a constant (called heat capacity and defined in § 5.2), and by $pV = NRT$, where p is its pressure, V its volume, T its temperature and N the number of moles of gas. An adiabatic reversible compression brings the pressure from p_1 to p_2 where $p_2 > p_1$. The initial temperature is T_1. Determine the temperature T_2 at the end of the compression.

Solution:

For a reversible adiabatic process, we have $dS = 0$. Then, the enthalpy differential is given by,

$$dH = C_p \, dT = T \, dS + V \, dp = V \, dp$$

Since $pV = NRT$, it can be recast as,

$$\frac{dT}{T} = \frac{NR}{C_p} \frac{dp}{p}$$

The integration of this relation from the initial state (T_1, p_1) to the final state (T_2, p_2) yields,

$$\ln \left(\frac{T_2}{T_1} \right) = \frac{NR}{C_p} \ln \left(\frac{p_2}{p_1} \right)$$

The exponentiation of this equation yields the temperature at the end of the compression,

$$T_2 = T_1 \left(\frac{p_2}{p_1} \right)^{\frac{NR}{C_p}}$$

4.2 Irreversible Heat Transfer

A cylinder closed by a piston contains N moles of a diatomic gas characterised by $U = (5/2) NRT$ and by $pV = NRT$, as in exercise 4.1. The gas has a temperature T when it is brought in contact with a heat reservoir at temperature T_{ext}, causing an irreversible process to occur. The pressure p of the gas is equal to the constant pressure p_{ext} of the environment at all times, i.e. $p = p_{ext} = $ const. Determine the amount of heat exchanged.

Numerical Application:

$N = 0.5 \, \text{mol}$, $T = 450 \, \text{K}$ and $T_{ext} = 300 \, \text{K}$.

Solution:

The enthalpy of the gas is given by,

$$H = U + p V = \frac{5}{2} NR T + NR T = \frac{7}{2} NR T$$

According to the relation (4.61), the heat exchanged yields,

$$Q_{if} = \Delta H_{if} = \frac{7}{2} NR (T_f - T_i) = 2.18kJ.$$

4.3 Internal Energy as Function of T and V

Establish the expression of the differential of the internal energy $dU\big(S(T,V),V\big)$ as a function of the temperature T and the volume V. In the particular case of a gas that satisfies the relation $p V = NR T$, show that $dU\big(S(T,V),V\big)$ is proportional to dT.

Solution:

According to the mathematical definition (4.76), the differential $dU\big(S(T,V),V\big)$ is expressed as,

$$dU\big(S(T,V),V\big) = \left(\frac{\partial U\big(S(T,V),V\big)}{\partial S(T,V)} \frac{\partial S(T,V)}{\partial T} \right) dT$$

$$+ \left(\frac{\partial U\big(S(T,V),V\big)}{\partial S(S,V)} \frac{\partial S(T,V)}{\partial V} + \frac{\partial U\big(S(T,V),V\big)}{\partial V} \right) dV$$

Using the definitions (2.9), (2.10), (4.77) and the Maxwell relation (4.71), we obtain,

$$dU\big(S(T,V),V\big) = \frac{\partial U}{\partial T}\bigg|_V dT + \left(T \frac{\partial p(T,V)}{\partial T} - p(T,V) \right) dV$$

In the particular case of a gas that satisfies the relation $p V = NR T$, the terms inside the brackets cancel each other out and the differential reduces to,

$$dU\big(S(T,V),V\big) = \frac{\partial U}{\partial T}\bigg|_V dT$$

which is indeed proportional to dT.

4.4 Grand Potential

The **grand potential** $\Phi\left(T, V, \{\mu_A\}\right)$, also known as the **Landau free energy**, is a thermodynamical potential obtained by performing Legendre transformations of the internal energy $U\left(S, V, \{N_A\}\right)$. Use Legendre transformations to express the thermodynamical potential $\Phi\left(T, V, \{\mu_A\}\right)$ in terms of the thermodynamical potential F. Also determine the differential $d\Phi\left(T, V, \{\mu_A\}\right)$.

Solution:

To obtain the grand potential $\Phi\left(T, V, \{\mu_A\}\right)$, we perform Legendre transformations on the internal energy $U\left(S, V, \{N_A\}\right)$ with respect to the entropy S and the number of moles N_A of every substance A,

$$\Phi = U - \frac{\partial U}{\partial S} S - \sum_A \frac{\partial U}{\partial N_A} N_A = U - TS - \sum_A \mu_A N_A$$

$$= F - \sum_A \mu_A N_A = -p V$$

Differentiating the grand potential $\Phi\left(T, V, \{\mu_A\}\right)$ yields,

$$d\Phi = dU - T dS - S dT - \sum_A \mu_A dN_A - \sum_A N_A d\mu_A$$

$$= -S dT - p dV - \sum_A N_A d\mu_A$$

4.5 Massieu Functions

Two of the **Massieu functions** are functions of the following state variables:

1. $J\left(\dfrac{1}{T}, V\right)$

2. $Y\left(\dfrac{1}{T}, \dfrac{p}{T}\right)$

The Massieu functions are obtained by performing Legendre transformations of the state function entropy $S\left(U, V\right)$ with respect to the state variables U and V. Use Legendre transformations to express the Massieu functions $J\left(\dfrac{1}{T}, V\right)$ and $Y\left(\dfrac{1}{T}, \dfrac{p}{T}\right)$ in terms of the thermodynamical potentials F and G. Determine also the differentials $dJ\left(\dfrac{1}{T}, V\right)$ and $dY\left(\dfrac{1}{T}, \dfrac{p}{T}\right)$.

Solution:

The entropy $S(U, V)$ as a state function reads,

$$S = \frac{1}{T} U + \frac{p}{T} V$$

and its differential is written as,

$$dS = \frac{1}{T} dU + \frac{p}{T} dV$$

Thus,

$$\frac{\partial S}{\partial U} = \frac{1}{T} \quad \text{and} \quad \frac{\partial S}{\partial V} = \frac{p}{T}$$

To obtain the Massieu function $J\left(\frac{1}{T}, V\right)$, we perform a Legendre transformation on the entropy $S(U, V)$ with respect to the internal energy U,

$$J = S - \frac{\partial S}{\partial U} U = S - \frac{U}{T} = -\frac{F}{T}$$

Likewise, to obtain the Massieu function $Y\left(\frac{1}{T}, \frac{p}{T}\right)$, also called the **Planck function**, we perform two Legendre transformations on the entropy $S(U, V)$ with respect to the internal energy U and the volume V,

$$Y = S - \frac{\partial S}{\partial U} U - \frac{\partial S}{\partial V} V = S - \frac{U}{T} - \frac{pV}{T} = -\frac{G}{T}$$

Differentiating the Massieu function $J\left(\frac{1}{T}, V\right)$ yields,

$$dJ = dS - \frac{1}{T} dU - U d\left(\frac{1}{T}\right) = -U d\left(\frac{1}{T}\right) + \frac{p}{T} dV$$

Similarly, differentiating the Massieu function $Y\left(\frac{1}{T}, \frac{p}{T}\right)$ yields,

$$dY = dS - \frac{1}{T} dU - U d\left(\frac{1}{T}\right) - \frac{p}{T} dV - V d\left(\frac{p}{T}\right) = -U d\left(\frac{1}{T}\right) - V d\left(\frac{p}{T}\right)$$

4.6 Gibbs–Helmoltz Equations

a) Show that

$$U(S, V) = -T^2 \frac{\partial}{\partial T}\left(\frac{F(T, V)}{T}\right)$$

where $T \equiv T(S, V)$ is to be understood as a function of S and V.

b) Show that

$$H(S,p) = -T^2 \frac{\partial}{\partial T}\left(\frac{G(T,p)}{T}\right)$$

where $T \equiv T(S,p)$ is to be understood as a function of S and p.

Solution:

a) The internal energy U is related to the free energy F and expressed in terms of the state variables S and V as,

$$U(S,V) = F\Big(T(S,V),V\Big) + T(S,V)\,S$$

Using the definition (4.26) and the chain rule, it can be recast as,

$$U(S,V) = F\Big(T(S,V),V\Big) - T(S,V)\,\frac{\partial F\Big(T(S,V),V\Big)}{\partial T}$$

$$= -T(S,V)^2\,\frac{\partial}{\partial T}\left(\frac{F\Big(T(S,V),V\Big)}{T}\right)$$

b) Likewise, the enthalpy H is related to the Gibbs free energy G and expressed in terms of the state variables S and p as,

$$H(S,p) = G\Big(T(S,p),p\Big) + T(S,p)\,S$$

Using the definition (4.40) and the chain rule, it can be recast as,

$$H(S,p) = G\Big(T(S,p),p\Big) - T(S,p)\,\frac{\partial G\Big(T(S,p),p\Big)}{\partial T}$$

$$= -T(S,p)^2\,\frac{\partial}{\partial T}\left(\frac{G\Big(T(S,p),p\Big)}{T}\right)$$

4.7 Pressure in a Soap Bubble

A soap bubble is a system consisting of two subsystems. Subsystem (f) is the thin film and subsystem (g) is the gas enclosed inside the film. The surrounding air is a thermal bath. The equilibrium is characterised by the minimum of the free energy F of the system. The differential of the free energy dF reads,

$$dF = -\,(S_g + S_f)\,dT + 2\,\gamma\,dA - (p - p_0)\,dV$$

where A is the surface area of the soap film and V the volume of the bubble. The parameter γ is called the ***surface tension***. It characterises the interactions at the interface between the liquid and the air. Since the soap film has two such interfaces, there is a factor 2 in front of the parameter γ. The surface tension γ is an intensive variable that plays an analogous role for a surfacic system as the pressure p for a volumic system. However, the force due to pressure of a gas is exerted outwards whereas the force due to the surface tension is exerted inwards. This is the reason why the signs of the corresponding two terms in dF differ. The term $p - p_0$ is the pressure difference between the pressure p inside the bubble and the atmospheric pressure p_0. Consider the bubble to be a sphere of radius r and show that,

$$p - p_0 = \frac{4\gamma}{r}$$

Solution:

Since the surrounding air is a thermal bath, the temperature is constant, i.e. $dT = 0$. For a spherical bubble, the area differential is given by,

$$dA = 4\pi (r + dr)^2 - 4\pi r^2 = 4\pi \left(2 r dr + dr^2\right) \approx 8\pi r dr$$

where we neglect the second-order term in dr^2. The volume differential is given by,

$$dV = \frac{4\pi}{3} (r + dr)^3 - \frac{4\pi}{3} r^3 = \frac{4\pi}{3} \left(3 r^2 dr + 3 r dr^2 + dr^3\right) \approx 4\pi r^2 dr$$

where we neglect the second-order term in dr^2 and the third-order term in dr^3. At equilibrium, the free energy F is minimum. Thus,

$$dF = 16\pi \gamma r dr - 4\pi (p - p_0) r^2 dr = 0$$

which implies that the pressure difference is given by,

$$p - p_0 = \frac{4\gamma}{r}$$

4.8 Pressure in a Droplet

Determine the hydrostatic pressure p inside a droplet, as a function of its radius r (Fig. 4.1). Assume that the drop (d) forms at the end of a short thin tube mounted at the end of vertical cylindrer containing the liquid (l). When a drop forms at the end of the tube, the change in the container height is negligible. If the height of the liquid above the tip of the tube is h, then the hydrostatic pressure is $p = p_0 + \rho g h$, where ρ is the volumetric mass density of the liquid, and g characterises the gravitation at the surface of the Earth. For this liquid, the differential of the free energy reads,

$$dF = - (S_l + S_d)\, dT + \gamma\, dA - (p - p_0)\, dV$$

Show that

$$p - p_0 = \frac{2\gamma}{r} = \rho g h$$

Solution:

The pressure difference is obtained by performing the same calculation as for the soap bubble, using the surface tension γ instead of 2γ.

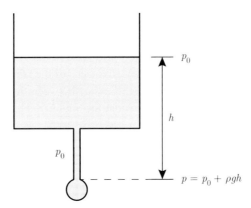

Figure 4.1 Principle of a setup that could be used to estimate the influence of surface tension on the pressure inside a liquid drop. The container is wide enough, so that when a drop is forming, the change in height of the liquid is negligible. The system is in a thermal bath at constant temperature T.

4.9 Isothermal Heat of Surface Expansion

A system consists of a thin film of surface area A, of internal energy $U(S, A)$, where

$$dU = T\, dS + \gamma\, dA$$

Hence, the surface tension is given by

$$\gamma(S, A) = \frac{\partial U(S, A)}{\partial A}$$

Express the heat Q_{if} to provide to the film for a variation $\Delta A_{if} = A_f - A_i$ of the surface of the film through an isothermal process at temperature T, that brings the film from an initial state i to a final state f, in terms of $\gamma(T, A)$ and its partial derivatives.

Solution:

Perform a Legendre transformation on the internal energy $U(S, A)$ with respect to the entropy to define the free energy and derive its differential,

$$dF(T, A) = -S(T, A)\, dT + \gamma(T, A)\, dA$$

where

$$\gamma(T, A) = \frac{\partial F(T, A)}{\partial A} \qquad \text{and} \qquad S(T, A) = -\frac{\partial F(T, A)}{\partial T}$$

For an isothermal process, we can compute the heat Q_{if} as,

$$Q_{if} = T\, \Delta S_{if} = T\frac{\partial S(T, A)}{\partial A}\, \Delta A_{if}$$

The Schwarz theorem applied to free energy $F(T, A)$ yields,

$$\frac{\partial}{\partial A}\left(\frac{\partial F}{\partial T}\right) = \frac{\partial}{\partial T}\left(\frac{\partial F}{\partial A}\right)$$

which leads to the Maxwell relation,

$$\frac{\partial S\,(T,A)}{\partial A} = -\,\frac{\partial \gamma\,(T,A)}{\partial T}$$

Hence, the heat is given by,

$$Q_{if} = -\,T\frac{\partial \gamma\,(T,A)}{\partial T}\,\Delta A_{if}$$

4.10 Thermomechanical Properties of an Elastic Rod

An state of an elastic rod is described by the state variables entropy S and length L. The differential of the internal energy $U\,(S,L)$ of the rod is written as,

$$dU = \frac{\partial U\,(S,L)}{\partial S}\,dS + \frac{\partial U\,(S,L)}{\partial L}\,dL = T\,(S,L)\,dS + f\,(S,L)\,dL$$

Note that $f(S,L)$ has the units of a force. The longitudinal stress τ on the rod is $\tau = \dfrac{f}{A}$, where A is the cross-section of the rod. We neglect any change of A due to f. The physical properties of the rod material are given by the linear thermal expansion coefficient at constant stress,

$$\alpha = \frac{1}{L}\frac{\partial L\,(T,f)}{\partial T},$$

and the isothermal Young modulus,

$$E = \frac{L}{A}\frac{\partial f(T,L)}{\partial L}.$$

Make use of these two physical properties of the material to answer the following questions :

a) Compute the partial derivative of the rod's stress τ in the rod changes with respect to its temperature when its length is fixed. Consider that the cross-section A is independent of the temperature.

b) Determine the heat transfer during an isothermal variation of the length of the rod ΔL_{if} from an initial state i to a final state f in terms of α and E.

c) Compute the partial derivative of the rod's length L with respect to its temperature T.

Solution:

a) Applying the cyclic rule (4.81) to the force $F\,(T,L)$ we obtain,

$$\frac{\partial f}{\partial T}\frac{\partial T}{\partial L}\frac{\partial L}{\partial f} = -1$$

and thus

$$\frac{\partial f}{\partial T} = -\,\frac{\partial L}{\partial T}\frac{\partial f}{\partial L} = -\,\alpha A E$$

Since the cross section A is independent of the temperature, the stress in the rod varies with temperature as,

$$\frac{\partial \tau}{\partial T} = - \alpha E$$

b) At constant temperature T, the infinitesimal heat transfer is written as,

$$\delta Q = T\, dS\,(T,L) = T\frac{\partial S}{\partial L}\,dL$$

Thus, after integration, we obtain the heat transfer for an isothermal process from an intial state i to a final state f,

$$Q_{if} = T\frac{\partial S}{\partial L}\,\Delta L_{if}$$

The differential of the free energy is,

$$dF = - S\,(T,L)\,dT + f\,(S,L)\,dL$$

The Schwarz theorem applied to free energy $F\,(T,L)$ yields,

$$\frac{\partial}{\partial L}\left(\frac{\partial F}{\partial T}\right) = \frac{\partial}{\partial T}\left(\frac{\partial F}{\partial L}\right)$$

which leads to the Maxwell relation,

$$- \frac{\partial S\,(T,A)}{\partial L} = \frac{\partial f\,(S,L)}{\partial T}$$

Using the Maxwell relation and the cyclic rule (4.81), the heat transfer can be recast as,

$$Q_{if} = - T\frac{\partial f}{\partial T}\,\Delta L_{if} = T\frac{\partial L}{\partial T}\frac{\partial f}{\partial L}\,\Delta L_{if} = \alpha\,T\,A\,E\,\Delta L_{if}$$

c) For an abiabatic process, we need to determine the derivative of the length $L\,(S,T)$ with respect to temperature when the entropy is kept constant. Using the cyclic rule (4.81),

$$\frac{\partial L}{\partial T} = - \frac{\partial T}{\partial S}\frac{\partial S}{\partial L}$$

When identifying the expressions of the heat transfer Q_{if} obtained above, we find,

$$\frac{\partial S}{\partial L} = \alpha\,A\,E$$

Thus,

$$\frac{\partial L}{\partial T} = - \frac{\alpha\,A\,E\,T}{c_L} \qquad \text{where} \qquad c_L = T\frac{\partial S\,(T,L)}{\partial T}$$

4.11 Chemical Power

An open system consists of a fluid of a single substance kept between two pistons sliding inside a cylinder with adiabatic walls. Matter enters and exits the cylinder in two specific

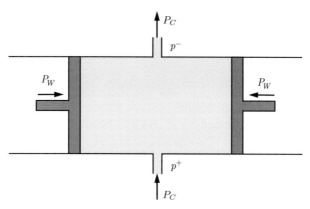

Figure 4.2 Two pistons slide in a cylinder that contains a fluid that enters and exits the system. The pressure at the entrance is p^+ and the pressure at the exit is p^-. The mechanical power generated by the pistons on the system is P_W and the chemical power generated by the matter flows is P_C.

locations. These two matter flows generate a chemical power P_C. The pressure at the entrance is p^+ and the pressure at the exit is p^-. The pistons exert a mechanical power P_W on the fluid. Since the walls are adiabatic there is a heat transfer through convection but no through conduction, i.e. $P_Q = 0$ (Fig. 4.2). For this open system, show that the chemical power P_C generated by the matter flow can be written as,

$$P_C = h^+ \dot{N}^+ - h^- \dot{N}^-$$

where $\dot{N}^+$ and $\dot{N}^-$ are the rates of substance entering and exiting the system and h^+ and h^- are the molar enthalpies entering and exiting the system.

Solution:

The rates of variation of substance and of volume satisfy the following conservation laws,

$$\dot{N} = \dot{N}^+ - \dot{N}^- \qquad \text{and} \qquad \dot{V} = \dot{V}^+ - \dot{V}^-$$

where N is the number of moles of substance in the system, V is the volume of the system, $\dot{V}^+$ and $\dot{V}^-$ are the volumes entering and exiting the system per unit of time. The molar internal energies u^+ and u^- and the molar volumes v^+ and v^- entering and exiting the system per unit of time are defined as,

$$u^+ = \frac{U^+}{N^+} \qquad u^- = \frac{U^-}{N^-} \qquad v^+ = \frac{V^+}{N^+} \qquad \text{and} \qquad v^- = \frac{V^-}{N^-}$$

where U^+ and U^- are the internal energies entering and exiting the system. The time derivative of the internal energy of the system is the difference between the rate of variation of the internal energy due to matter entering the system, i.e. $u^+ \dot{N}^+$, and to matter exiting the system, i.e. $- u^- \dot{N}^-$. Thus,

$$\dot{U} = u^+ \dot{N}^+ - u^- \dot{N}^-$$

According to relation (2.16), the mechanical power exerted by the environment on the gas is the difference between the mechanical exerted at the entrance, i.e. $-p^+ \dot{V}^+$, and the mechanical power exerted at the exit, i.e. $p^- \dot{V}^-$. Hence,

$$P_W = -p^+ \dot{V}^+ + p^- \dot{V}^-$$

Since there is no conductive heat transfer, the thermal power vanishes, i.e. $P_Q = 0$. According to the first law (1.28), the chemical power then is written as,

$$P_C = \dot{U} - P_W = u^+ \dot{N}^+ - u^- \dot{N}^- + p^+ \dot{V}^+ - p^- \dot{V}^-$$

Thus, the chemical power exerted by the matter flow consists of two convective contributions : the convective heat transfers $u^+ \dot{N}^+ - u^- \dot{N}^-$ and the mechanical actions $p^+ \dot{V}^+ - p^- \dot{V}^-$. Taking into account the definition of the entering and exiting volumes per unit of time,

$$\dot{V}^+ = v^+ \dot{N}^+ \qquad \text{and} \qquad \dot{V}^- = v^- \dot{N}^-$$

and the entering and exiting molar enthalpies,

$$h^+ = u^+ + p^+ v^+ \qquad \text{and} \qquad h^- = u^- + p^- v^-$$

the chemical power reduces then to,

$$P_C = h^+ \dot{N}^+ - h^- \dot{N}^-$$

Calorimetry

The exercises given in the last section of Chapter 5 are presented here with their solutions.

5.1 Heat Transfer as a Function of *V* and *p*

The infinitesimal heat transfer δQ is expressed as a function of the state variables T and V in equation (5.4). It was done as a function of the state variables T and p in equation (5.17). Express the infinitesimal heat transfer δQ as a function of V and p.

Solution:

The infinitesimal heat transfer δQ can be expressed as a function of V and p as,

$$\delta Q = T(V,p)\, dS(V,p) = L_V(V,p)\, dV + L_p(V,p)\, dp$$

where

$$L_V(V,p) = T(V,p)\,\frac{\partial S(V,p)}{\partial V} \qquad \text{and} \qquad L_p(V,p) = T(V,p)\,\frac{\partial S(V,p)}{\partial p}$$

5.2 Bicycle Pump

A bicycle pump takes a volume ΔV of air at atmospheric pressure p_0 and constant temperature T_0 and compresses it so that it enters a tire that has a volume V_0. The air inside the tire is initially at atmospheric pressure p_0 and can be considered as an ideal gas. Determine the number of times n the user has to pump air into the tire to reach a pressure p_f. Assume that the pump is designed so that the air in the tire is always at temperature T_0.

Numerical Application:

$V_0 = 50\,\mathrm{l}$, $\Delta V = 1.2\,\mathrm{l}$ and $p_f = 2.5\,p_0$.

Solution:

The initial and final number of moles of air inside the tire of volume V_0 at temperature T_0 are given by,

$$N_0 = \frac{p_0\, V_0}{R\, T_0}$$

and

$$N_f = \frac{p_f V_0}{R T_0} \quad \text{thus} \quad \frac{N_f}{N_0} = \frac{p_f}{p_0}$$

The additional number of moles of air pumped into the tire each time are,

$$\Delta N = \frac{p_0 \Delta V}{R T_0} \quad \text{and} \quad N_f = N_0 + n \Delta N$$

Thus,

$$\frac{N_f}{N_0} = 1 + n \frac{\Delta N}{N_0} = 1 + n \frac{p_0 \Delta V}{N_0 R T_0} = 1 + n \frac{\Delta V}{V_0} = \frac{p_f}{p_0}$$

which implies that,

$$n = \left(\frac{p_f}{p_0} - 1 \right) \frac{V_0}{\Delta V} = 62.5$$

This means that the air has to be pumped 63 times in order to reach a final pressure p_f that is at least $2.5\, p_0$.

5.3 Heat Transfer at Constant Pressure

A gas container is thermally isolated except for a small hole that insures that the pressure inside the container is equal to the atmospheric pressure p_0. Initially, the container holds N_i moles of gas at a temperature T_i. The molar specific heat of the gas at constant pressure is c_p. The gas is heated up to a temperature T_f by a resistive coil in the cylinder. As the gas temperature rises, some of the gas is released through the small hole. Assume that for the gas remaining in the cylinder, the process is reversible and neglect the specific heat of the heater. Determine:

a) the volume V_0 of the container.
b) the number of moles ΔN leaving the container in this process.
c) the heat transfer Q_{if} to accomplish this process.

Numerical Application:

$p_0 = 10^5$ Pa, $N_0 = 10$ moles, $T_0 = 273$ K, $c_p = 29.1$ J K^{-1} mol^{-1}, $T_f = 293$ K.

Solution:

a) The volume of the container is,

$$V_0 = \frac{N_i R T_i}{p_0} = 227\,l$$

b) The final number of moles is,

$$N_f = \frac{p_0 V_0}{R T_f} = N_i \frac{T_i}{T_f}$$

Thus, the number of moles leaving the container in this process is,

$$\Delta N = N_i - N_f = N_i \left(1 - \frac{T_i}{T_f} \right) = 0.68 \text{ mol}$$

c) Using the result established in b) the heat transfer yields,

$$Q_{if} = \int_i^f N c_p \, dT = c_p N_i T_i \int_{T_i}^{T_f} \frac{dT}{T} = c_p N_i T_i \ln \left(\frac{T_f}{T_i} \right) = 5.6 \text{ kJ}$$

5.4 Specific Heat of a Metal

A metallic block of mass M is brought to a temperature T_0 and plunged into a calorimeter filled with a mass M' of water. The system consisting of the metallic block and the water container is considered as isolated. In this process, the water temperature rises from T_i to T_f, the equilibrium temperature. The specific heat of the water per unit mass is $c_{M'}^*$. Determine the specific heat per unit mass of the metal c_M^* in the temperature range used in this experiment. Consider that the water container is made of a material with a negligible specific heat.

Numerical Application:

$M = 0.5$ kg, $M' = 1$ kg, $T_0 = 120°C$, $T_i = 16°C$, $T_f = 20°C$ and $c_{M'}^* = 4,187$ J kg^{-1} K^{-1}.

Solution:

Since the system consisting of the metallic block and the water is isolated, the variation of internal energy vanishes,

$$\Delta U_{if} = M c_M^* (T_f - T_0) + M' c_{M'}^* (T_f - T_i) = 0$$

which yields the specific heat of the metal,

$$c_M^* = c_{M'}^* \frac{M'}{M} \frac{T_f - T_i}{T_0 - T_f} = 335 \text{ J kg}^{-1} \text{ K}^{-1}$$

5.5 Work in Adiabatic Compression

An ideal gas undergoes a reversible adiabatic compression from an initial volume V_i and initial pressure p_i to a final pressure p_f. Determine the work W_{if} performed on the gas during this process.

Numerical Application:

$V_i = 1$ l, $p_i = 5 \cdot 10^5$ Pa, $p_f = 2p_i$, $c = 5/2$ (see definition (5.62)).

Solution:

For an adiabatic compression, the work performed on the gas is expressed as,

$$W_{if} = \Delta U_{if} = c N_0 R (T_f - T_i) = c \left(p_f V_f - p_i V_i \right)$$

where

$$p_f V_f^\gamma = p_i V_i^\gamma \qquad \text{thus} \qquad V_f = \left(\frac{p_i}{p_f} \right)^{\frac{1}{\gamma}} V_i = \left(\frac{p_i}{p_f} \right)^{\frac{c}{c+1}} V_i$$

Thus,

$$W_{if} = c V_i \left(p_f \left(\frac{p_i}{p_f} \right)^{\frac{c}{c+1}} - p_i \right) = 28.5 \, \text{J}$$

5.6 Slopes of Isothermal and Adiabatic Processes

For an ideal gas, show that at any point on a Clapeyron (p, V) diagram, the absolute value of the slope is greater for an adiabatic process (A) than an isothermal process (I).

Solution:

An isothermal process (I) in a Clapeyron diagram is characterised by,

$$p V \equiv C_I = \text{const} \qquad \text{where} \qquad C_I = NR T$$

Thus, for an isothermal process,

$$p = \frac{C_I}{V} \qquad \text{and} \qquad \frac{dp}{dV} = -\frac{C_I}{V^2} = -\frac{NR T}{V} \frac{1}{V} = -\frac{p}{V}$$

An adiabatic process (A) in a Clapeyron diagram is characterised by,

$$p V^\gamma \equiv C_A = \text{const} \qquad \text{where} \qquad C_A = p^{1-\gamma} (NR T)^\gamma$$

Thus, for an adiabatic process,

$$p = \frac{C_A}{V^\gamma} \qquad \text{and} \qquad \frac{dp}{dV} = -\frac{\gamma C_A}{V^{\gamma+1}} = -\gamma \left(\frac{NR T}{p V} \right)^\gamma \frac{p}{V} = -\gamma \frac{p}{V}$$

The slopes of both processes are negative in the Clapeyron diagram. Since $\gamma > 1$, the absolute value of the slope of the adiabatic process is greater than that of the isothermal process.

5.7 Adsorption Heating of Nanoparticles

The process whereby molecules bind to a metallic surface is called adsorption. Here, molecules are adsorbed on Pt nanoparticles. The specific heat of an average Pt nanoparticle is C_V. The heat transferred to an average Pt nanoparticle during the adsorption of molecules is Q_{if}. Determine the temperature increase $\Delta T_{if} = T_f - T_i$ of an average Pt nanoparticle, assuming that the system consisting of the nanoparticles and the gas is isolated.

Numerical Application:

$C_V = 1.4 \cdot 10^{-18}$ J K^{-1}, $Q_{if} = 6.5 \cdot 10^{-16}$ J.

Solution:

During the adsorption of molecules, the temperature increase of an average Pt nanoparticle is,

$$\Delta T_{if} = \frac{Q_{if}}{C_V} = 460 \text{ K}$$

5.8 Thermal Response Coefficients

The thermal response of a homogeneous system subjected to an infinitesimal heat transfer δQ is characterised by coefficients defined in equations (5.4) and (5.17) when either the state variables (T, V) or (T, p) are used.

a) Find a relation between the latent heat of expansion $L_V (T, V)$ and the latent heat of compression $L_p (T, p)$.

b) Express the latent heat of compression $L_p (T, p)$ in terms of the specific heat at constant volume $C_V (T, V)$ and the specific heat at constant pressure $C_p (T, p)$.

Solution:

a) The thermal response is written in terms of the temperature T and the volume V as,

$$\delta Q = C_V (T, V) \, dT + L_V (T, V) \, dV$$

The thermal response is written in terms of the temperature T and the pressure p as,

$$\delta Q = C_p (T, p) \, dT + L_p (T, p) \, dp$$

which can be recast in terms of the temperature T and the volume V as,

$$\delta Q = C_p (T, p) \, dT + L_p (T, p) \left(\frac{\partial p (T, V)}{\partial T} \, dT + \frac{\partial p (T, V)}{\partial V} \, dV \right)$$
$$= \left(C_p (T, p) + L_p (T, p) \frac{\partial p (T, V)}{\partial T} \right) dT + L_p (T, p) \frac{\partial p (T, V)}{\partial V} \, dV$$

The identification of the terms multiplying the volume differential dV in the two expressions for the thermal response δQ written in terms of the temperature T and the volume V yields the relation,

$$L_V (T, V) = L_p (T, p) \frac{\partial p (T, V)}{\partial V}$$

b) The identification of the terms multiplying the temperature differential dT in the two expressions for the thermal response δQ written in terms of the temperature T and the volume V yields the relation,

$$C_V(T, V) = C_p(T, p) + L_p(T, p) \frac{\partial p(T, V)}{\partial T}$$

Using relation (4.80) for the inverse of a partial derivative, this result can be recast as,

$$L_p(T, p) = \Big(C_V(T, V) - C_p(T, p) \Big) \frac{\partial T(p, V)}{\partial p}$$

6 Phase Transitions

The exercises given in the last section of Chapter 6 are presented here with their solutions.

6.1 Melting Ice

A mixture of ice and water is heated up in such a way that the ice melts. The ice melts at a rate r and the molar latent heat of ice melting is $\ell_{s\ell}$.

a) Determine the thermal power P_Q transferred to the ice.
b) Determine the entropy rate of change $\dot{S}$.

Numerical Application:

$\Omega = 2.0 \cdot 10^{-2}$ mol s^{-1}, $\ell_{s\ell} = 6.0 \cdot 10^3$ J mol^{-1}.

Solution:

a) The thermal power transferred to the ice is,

$$P_Q = \ell_{s\ell}\,\Omega = 120 \text{ W}$$

b) The entropy rate of change is,

$$\dot{S} = \frac{P_Q}{T_m} = \frac{\ell_{s\ell}\,\Omega}{T_m} = 0.44 \text{ W K}^{-1}$$

where $T_m = 273$ K is the ice melting temperature.

6.2 Cooling Water with Ice Cubes

Water is cooled with ice cubes (Fig. 6.10). The water and the ice cubes are considered as an isolated system. Initially, the ice cubes are at melting temperature T_0 and the water at temperature T_i. The total initial mass of ice is M' and the initial mass of water is M. The latent heat of melting of ice per unit mass is $\ell_{s\ell}^*$ and the specific heat per unit mass of water is c_V^*.

a) Determine the final temperature T_m of the water.
b) Determine the final temperature T_m of the water if melted ice (i.e. water) had been added at melting temperature T_0 instead of ice.

Numerical Application:

$M = 0.45$ kg, $M' = 0.05$ kg, $T_i = 20°C$, $T_0 = 0°C$, $\ell_{s\ell}^* = 3.33 \cdot 10^6$ J kg^{-1} and $c_V^* = 4.19 \cdot 10^3$ J kg^{-1} K^{-1}.

Solution:

Heat is transferred from the water to melt the ice and increases its temperature while the temperature of the water decreases.

a) Since the system is isolated the total variation of internal energy vanishes,

$$\Delta U_{if} = M' \ell_{sf}^* + M' c_V^* (T_f - T_0) + M c_V^* (T_f - T_i) = 0$$

Thus, the final temperature T_f is,

$$T_f = \frac{M T_i + M' T_0}{M + M'} - \frac{M' \ell_{sf}^*}{(M + M') c_V^*} = 283 \text{ K} = 10°C$$

b) If the melted ice had been added at melting temperature T_0 instead of ice, there would be no more melting. Thus, the final temperature T_f would be,

$$T_f = \frac{M T_i + M' T_0}{M + M'} = 291 \text{ K} = 18°C$$

6.3 Wire through Ice without Cutting

A steel wire is wrapped over a block of ice with two heavy weights attached to the ends of the wire. The wire passes through a block of ice without cutting the block in two. The ice melts under the wire and the water freezes again above the wire. The wire is considered a rigid rod of negligible mass laying on the ice block with an area of contact A. The two weights of mass M each are hanging at both ends of the wire (Fig. 6.1). The entire system is at atmospheric pressure p_0 and the ice is held at temperature $T_m - \Delta T$ where T_m is the melting temperature at atmospheric pressure. The latent heat of melting of ice is $\ell_{s\ell}$, the molar volume of water v_ℓ and the molar volume of ice is v_s. Determine the minimal mass M of each weight for this experiment to succeed, i.e. for the wire to pass through the ice block.

Solution:

The process of ice melting due to the pressure exerted by the weights is represented by a vertical line on the diagram $p(T)$ (Fig 6.2). The pressure variation Δp between the atmospheric pressure p_0 and the pressure $p_0 + \Delta p$ at ice melting is expressed on the (p, T) diagram as,

$$\Delta p = \int_{p_0}^{p_0 + \Delta p} dp = \int_{T_m}^{T_m - \Delta T} \frac{dp}{dT} dT$$

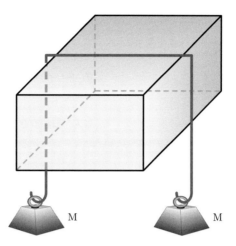

Figure 6.1 A steel wire wrapped over a block of ice with two heavy weights hanging on both sides passes through the ice cutting the block in two.

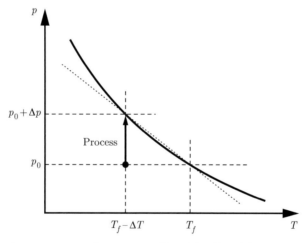

Figure 6.2 (T, s) diagram where the curve represents the coexistence of the solid and liquid phases. The ice melting process is due to a pressure variation from point $(T_m - \Delta T, p_0)$ to point $(T_m - \Delta T, p_0 + \Delta p)$.

Using the Clausius–Clapeyron relation (6.50) where the ice latent heat of melting $\ell_{s\ell}$ is considered as constant, the pressure variation Δp is expressed as, i.e.

$$\Delta p = -\frac{\ell_{s\ell}}{v_s - v_\ell} \int_{T_m}^{T_m - \Delta T} \frac{dT}{T} = \frac{\ell_{s\ell}}{v_s - v_\ell} \ln\left(\frac{T_m}{T_m - \Delta T}\right)$$

The pressure variation Δp that allows ice to melt is equal to the pressure exerted by the minimal weight of the two masses on the area of contact A between the wire and the ice block,

$$\Delta p = \frac{2 M g}{A}$$

Equating both expressions for Δp, we obtain the minimal value for the mass M of each weight,

$$M = \frac{A\,\ell_{s\ell}}{2\,g\,(v_s - v_\ell)}\,\ln\left(\frac{T_m}{T_m - \Delta T}\right)$$

6.4 Dupré's Law

A liquid is at equilibrium with its vapour. The vapour is assumed to be an ideal gas. The liquid has a molar latent heat of vaporisation $\ell_{\ell g}$ that depends on temperature, with $\ell_{\ell g} = A - B\,T$, where A and B are constants. Apply the Clausius–Clapeyron relation (6.50) and consider that the molar volume of the liquid phase is negligible compared to the vapour phase, i.e. $v_\ell \ll v_g$. Use the ideal gas law (5.47) for the vapour phase. Show that at equilibrium at a temperature T, the vapour pressure p depends on temperature according to Dupré's law,

$$\ln\left(\frac{p}{p_0}\right) = \frac{A}{R}\left(\frac{1}{T_0} - \frac{1}{T}\right) - \frac{B}{R}\ln\left(\frac{T}{T_0}\right)$$

where p_0 is the vapour pressure at T_0.

Solution:

Neglecting the molar volume of the liquid phase compared to the vapour phase and using the ideal gas law, i.e. $v_g = R\,T/p$, the Clausius–Clapeyron relation (6.50) can be written as,

$$\frac{dp}{dT} = \frac{\ell_{\ell g}}{T\,v_g} = \frac{A - B\,T}{T\,v_g} = \frac{p\,(A - B\,T)}{R\,T^2}$$

which can be recast as,

$$\frac{dp}{p} = \frac{A}{R}\frac{dT}{T^2} - \frac{B}{R}\frac{dT}{T}$$

The integration of the relation in the initial state (p_0, T_0) to the final state (p, T) yields Dupré's law,

$$\ln\left(\frac{p}{p_0}\right) = \frac{A}{R}\left(\frac{1}{T_0} - \frac{1}{T}\right) - \frac{B}{R}\ln\left(\frac{T}{T_0}\right)$$

6.5 Hydropneumatic Accumulator

A container contains a substance in gaseous and liquid phases at room temperature (Fig. 6.3). The container is closed by a piston of surface area A, held back by a spring of elastic constant k. We neglect the mass of the piston. For simplicity, we neglect the volume of the liquid compared to that of the gas. The atmospheric pressure is p_0 and assumed independent of temperature.

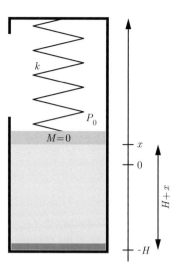

Figure 6.3 A cylinder containing a substance in liquid and gas phases is closed by a spring-loaded piston. The zero of the coordinate x is at the rest position of the spring. The mass of the piston is neglected.

a) Determine the temperature derivative of the gas pressure $\dfrac{dp}{dT}$ when there is no liquid phase present in the container.

b) Determine the temperature derivative of the gas pressure $\dfrac{dp}{dT}$ when liquid is present.

Solution:

a) The mechanical equilibrium condition requires the force exerted by the gas pressure to be equal and opposite to the elastic force exerted by the spring,

$$(p - p_0)\, A = kx \qquad \text{thus} \qquad p = p_0 + \frac{kx}{A}$$

The temperature derivative of the pressure reads,

$$\frac{dp}{dT} = \frac{k}{A}\frac{dx}{dT} = \frac{k}{A^2}\frac{dV}{dT} \qquad \text{thus} \qquad \frac{dV}{dT} = \frac{A^2}{k}\frac{dp}{dT}$$

since $dV = A\, dx$. The temperature derivative of the ideal gas equation of state $pV = NRT$ implies that,

$$\frac{dp}{dT} = \frac{NR}{V} - \frac{NRT}{V^2}\frac{dV}{dT} = \frac{p}{T} - \frac{A^2}{k}\frac{p}{V}\frac{dp}{dT}$$

which can be recast as,

$$\frac{dp}{dT} = \left(1 + \frac{p}{V}\frac{A^2}{k}\right)^{-1}\frac{p}{T} = \left(1 + \frac{p_0 + \dfrac{kx}{A}}{\dfrac{kH}{A} + \dfrac{kx}{A}}\right)^{-1}\frac{p}{T} \approx \frac{p}{T}$$

since $V = A\,(H + x)$. Note that the terms in brackets are nearly equal to one since $kx/A \ll kH/A$ and $p_0 \ll kH/A$ because the pressure kH/A needed to compress the gas entirely (such that its volume vanishes) is much larger than the pressure exerted by the spring kx/A or by the atmospheric pressure p_0.

b) Using the Clausius–Clapeyron relation (6.50), deducing the entropy of vaporisation $\Delta s_{\ell g} \equiv s_g - s_\ell$ from relation (6.44) and using the ideal gas equation of state $v_g = R T/p$, where $v_g \gg v_\ell$, we can write,

$$\frac{dp}{dT} = \frac{s_g - s_\ell}{v_g - v_\ell} = \frac{\Delta s_{\ell g}}{R}\frac{p}{T}$$

Note that according to Trouton's rule[2], at standard pressure the entropy of vaporisation $\Delta s_{\ell g}$ of most liquids is about 85 J mol^{-1} K^{-1}. Thus, we can consider that $\Delta s_{\ell g}/R \gg 1$. This means that the presence of the liquid enhances the ratio $\dfrac{dp}{dT}$ considerably compared to what it would be for a gas. In other words, the pressure change for a given change in temperature is much larger in the presence of the liquid phase.

6.6 Positivity of Thermal Response Coefficients

To establish the positivity of the specific heat at constant pressure C_p compressibility coefficient at constant temperature κ_T (see relations (6.31)), follow the steps given below:[3]

a) Show that the Mayer relation (5.42) can be recast as,

$$C_p = C_V + \frac{\alpha^2}{\kappa_T} V T$$

where α is the thermal coefficient of expansion,

$$\alpha = \frac{1}{V}\frac{\partial V(T,p)}{\partial T} \qquad and \qquad \kappa_T = -\frac{1}{V}\frac{\partial V(T,p)}{\partial p}$$

b) Show that

$$\frac{\partial^2 F(T,V)}{\partial V^2} = \frac{\dfrac{\partial^2 U}{\partial S^2}\dfrac{\partial^2 U}{\partial V^2} - \left(\dfrac{\partial^2 U}{\partial S \partial V}\right)^2}{\dfrac{\partial^2 U}{\partial S^2}}$$

c) Conclude from these two results that $\kappa_T \geq 0$ and $C_p \geq 0$.

Solution:

a) The Mayer relation yields,

$$C_p = C_V + T\frac{\partial p}{\partial T}\frac{\partial V}{\partial T}$$

---

[2] F. Trouton, *On Molecular Latent Heat*, Philosophical Magazine, **18**, 54–57 (1884).
[3] H. B. Callen, *Thermodynamics and an Introduction to Thermostatistics*, Wiley, New York, 2nd edition (1985), §8.2.2.

According to the cyclic rule,

$$\frac{\partial p}{\partial T}\frac{\partial T}{\partial V}\frac{\partial V}{\partial p} = -1 \qquad \text{thus} \qquad \frac{\partial p}{\partial T} = -\frac{\partial V}{\partial T}\frac{\partial p}{\partial V} = \frac{\alpha}{\kappa_T}$$

Moreover,

$$\frac{\partial V}{\partial T} = \alpha V$$

Thus, the Mayer relation can be recast as,

$$C_p = C_V + \frac{\alpha^2}{\kappa_T} V T$$

b) Using the mathematical definition (4.77),

$$\frac{\partial^2 F(T,V)}{\partial V^2} = \frac{d}{dV}\left(\frac{dF\left(T(S,V),V\right)}{dV}\right) = -\frac{dp\left(T(S,V),V\right)}{dV}$$

$$= -\frac{\partial p}{\partial T}\frac{\partial T}{\partial V} - \frac{\partial p}{\partial V} = -\frac{\partial p}{\partial T}\frac{\partial}{\partial V}\left(\frac{\partial U}{\partial S}\right) + \frac{\partial}{\partial V}\left(\frac{\partial U}{\partial V}\right)$$

Thus,

$$\frac{\partial^2 F(T,V)}{\partial V^2} = \frac{\partial^2 U}{\partial V^2} - \frac{\partial p}{\partial T}\frac{\partial^2 U}{\partial S\,\partial V}$$

According to the cyclic rule,

$$\frac{\partial p}{\partial T}\frac{\partial T}{\partial S}\frac{\partial S}{\partial p} = -1 \qquad \text{thus} \qquad \frac{\partial p}{\partial T} = -\frac{\partial p}{\partial S}\frac{\partial S}{\partial T}$$

which is then recast as,

$$\frac{\partial p}{\partial T} = -\frac{\dfrac{\partial p}{\partial S}}{\dfrac{\partial T}{\partial S}} = \frac{\dfrac{\partial}{\partial S}\left(\dfrac{\partial U}{\partial V}\right)}{\dfrac{\partial}{\partial S}\left(\dfrac{\partial U}{\partial S}\right)} = \frac{\dfrac{\partial^2 U}{\partial S\,\partial V}}{\dfrac{\partial^2 U}{\partial S^2}}$$

Hence,

$$\frac{\partial^2 F(T,V)}{\partial V^2} = \frac{\dfrac{\partial^2 U}{\partial S^2}\dfrac{\partial^2 U}{\partial V^2} - \left(\dfrac{\partial^2 U}{\partial S\,\partial V}\right)^2}{\dfrac{\partial^2 U}{\partial S^2}}$$

c) According to relation (6.32), the compressibility coefficient κ_T is expressed as,

$$\kappa_T = V\left(\frac{\partial^2 F}{\partial V^2}\right)^{-1}$$

According to relation (6.22), the numerator in the expression for $\partial^2 F(T,V)/\partial V^2$ is positive and according to relation (6.14), the denominator is also positive. Thus, since the volume V is positive, the compressibility coefficient κ_T is positive, i.e. $\kappa_T \geq 0$. Furthermore, since the temperature is positive, the Mayer relation requires that $C_p \geq C_V$. Since C_V is positive according to relation (6.29), this in turn implies that C_p is positive, i.e. $C_p \geq 0$.

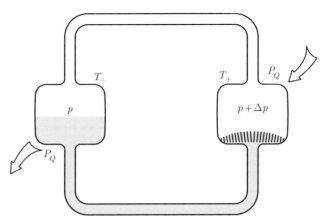

Figure 6.4 Principle of a heat pipe : At the hot side, the liquid passes through a wick and vaporises at pressure $p + \Delta p$. At the cold side, the vapour condenses at pressure p.

6.7 Heat Pipe

Heat pipes are devices used to transfer heat over a certain distance. A typical heat pipe looks like a metal rod, but modern versions, that are used for example to cool the hottest part of a phone, have a flat geometry. The principe of heat pipes is also considered in aerospace research.[4] Here, we will examine a simple model to understand the principle of a heat pipe (Fig. 6.4). The pressure difference Δp is modelled in a linear approximation, with $\Delta p = R_p \, \Omega$, where Ω is the rate of substance flowing down the pipe. The system is considered in a stationary state, so that the heat transfer P_Q is the same (in absolute value) on both sides. The heat of evaporation $\ell_{\ell g}$ is given and assumed independent of temperature. The temperature difference $\Delta T = T_+ - T_-$ is assumed small in order to simplify the calculations. Neglect the molar volume of the liquid v_ℓ compared to that of the vapour v_g and treat it as an ideal gas. Express the heat transfer P_Q as a function of the temperature difference ΔT.

Solution:

We note that there is coexistence of the liquid and vapour phases on both ends. Therefore, if there is coexistence of phases at temperature T_- and pressure p, in the limit where $v_\ell \ll v_g$ and $v_g = R\,T/p$ the Clausius–Clapeyron relation (6.50) implies that,

$$\frac{\Delta p}{\Delta T} = \frac{\ell_{\ell g}}{T \, v_g} = \frac{\ell_{\ell g} \, p}{R T^2}$$

At the hot side, the thermal transfer is $P_Q = \ell_{\ell g} \, \Omega$ where $\Omega = \Delta p / R_p$. Thus, we have,

$$P_Q = \frac{\ell_{\ell g}}{R_p} \Delta p = \frac{\ell_{\ell g}^2 \, p}{R_p \, R \, T^2} \Delta T$$

[4] P. R. Mashaei, M. Shahryari, S. Madani, *Analytical Study of Multiple Evaporator Heat Pipe with Nanofluid; a Smart Material for Satellite Equipment Cooling Application*, Aerosp. Sci. Technol. **59**, 112–121 (2016).

6.8 Vapour Pressure of Liquid Droplets

Consider a cloud of droplets and assume that they all have the same diameter r. According to the Laplace formula (exercise 4.8), the pressure $p(r)$ inside the droplets of radius r is related to the vapour pressure $p_0(r)$ by,

$$p(r) = p_0(r) + \frac{2\gamma}{r}$$

where γ is the surface tension. We note p_∞ the vapour pressure for an infinite radius. At temperature T, show that,

$$p_0(r) = p_\infty + \frac{2\gamma}{r} \frac{p_\infty v_\ell}{RT}$$

where v_ℓ is the molar volume of liquid, in the limit where $p_\infty v_\ell \ll RT$ since the molar volume of liquid is much smaller than the molar volume of gas.

Solution:

The equilibrium between the liquid and the gas is determined by the condition,

$$\mu_\ell\big(T, p(r)\big) = \mu_g\big(T, p_0(r)\big) \qquad \text{thus} \qquad \mu_\ell(T, p_\infty) = \mu_g(T, p_\infty)$$

We assume that the effect is small and do a first-order series expansion of the terms on both sides of the first equation around $\mu_\ell(T, p_\infty)$ and $\mu_g(T, p_\infty)$,

$$\mu_\ell(T, p_\infty) + \frac{\partial \mu_\ell}{\partial p}\big(p(r) - p_\infty\big) = \mu_g(T, p_\infty) + \frac{\partial \mu_g}{\partial p_0}\big(p_0(r) - p_\infty\big)$$

thus taking into account the second equation and the Laplace formula,

$$\frac{\partial \mu_\ell}{\partial p}\left(p_0(r) + \frac{2\gamma}{r} - p_\infty\right) = \frac{\partial \mu_g}{\partial p_0}\big(p_0(r) - p_\infty\big)$$

The Schwarz theorem applied to the Gibbs free energy $G(T, p, N)$ yields,

$$\frac{\partial}{\partial p}\left(\frac{\partial G}{\partial N}\right) = \frac{\partial}{\partial N}\left(\frac{\partial G}{\partial p}\right)$$

which gives the following Maxwell relations for the liquid and the gas,

$$\frac{\partial \mu_\ell}{\partial p} = \frac{\partial V_\ell}{\partial N_\ell} = v_\ell \qquad \text{and} \qquad \frac{\partial \mu_g}{\partial p_0} = \frac{\partial V_g}{\partial N_g} = v_g = \frac{RT}{p_\infty}$$

The last partial derivative is evaluated at $p_0 = p_\infty$. Therefore,

$$v_\ell\left(p_0(r) + \frac{2\gamma}{r} - p_\infty\right) = \frac{RT}{p_\infty}\big(p_0(r) - p_\infty\big)$$

which implies that the vapour pressure is given by,

$$p_0(r) = p_\infty + \frac{2\gamma}{r}\left(\frac{\dfrac{p_\infty v_\ell}{RT}}{1 - \dfrac{p_\infty v_\ell}{RT}}\right)$$

In the limit where $p_\infty v_\ell \ll R T$, the vapour pressure becomes,

$$p_0(r) = p_\infty + \frac{2\gamma}{r} \frac{p_\infty v_\ell}{R T}$$

6.9 Melting Point of Nanoparticles

The surface tension modifies the melting point of particles. The effect is important effect when the diameter is in the nanometer range. A differential equation has to be written for $T_m(r)$, the melting temperature of particles of radius r. In order to perform this thermodynamical analysis, assume that the pressure p_s inside the particles is defined.[5] At atmospheric pressure p_0 and for infinitely large particles, the melting temperature is noted T_∞. The surface tension is γ_s for a solid particle and γ_l for a liquid one. According to exercise 4.8, the Laplace pressure $p_s(r)$ for a solid nanoparticle and the Laplace pressure $p_\ell(r)$ for a liquid nanoparticle are given by,

$$p_s(r) = \frac{2\gamma_s}{r} \qquad \text{and} \qquad p_\ell(r) = \frac{2\gamma_\ell}{r}$$

Determine the temperature difference $T_\infty - T_m(r)$ in terms of the latent heat of melting $\ell_{s\ell} = T_\infty(s_\ell - s_s)$ and the molar volumes v_s and v_ℓ that are both assumed to be independent of the radius r. Therefore, perform a series expansion in terms of the radius r on the chemical equilibrium condition. This result is known as the **Gibbs–Thomson equation**. For some materials, a lowering of the melting temperature can be expected, i.e. $T_m(r) < T_\infty$. This effect has been observed on individual nanoparticles by electron microscopy.[6] It is used to sinter ceramics at low temperatures.[7]

Solution:

The chemical equilibrium between an infinitely large solid particle and the liquid is given by,

$$\mu_s(T_\infty, p_0) = \mu_\ell(T_\infty, p_0)$$

For a given radius r, this chemical equilibrium relation becomes,

$$\mu_s\left(T_m(r), p_s(r)\right) = \mu_\ell\left(T_m(r), p_\ell(r)\right)$$

Thus, for a radius $r + dr$, this relation becomes,

$$\mu_s\left(T_m(r+dr), p_s(r+dr)\right) = \mu_\ell\left(T_m(r+dr), p_\ell(r+dr)\right)$$

[5] J.-P. Borel, A. Chatelain, *Surface Stress and Surface Tension: Equilibrium and Pressure in Small Particles*, Surf. Sci. **156**, 572–579 (1985).

[6] Ph. Buffat, J.-P. Borel, *Size Effect on the Melting Temperature of Gold Particles*, Phys. Rev. A **13** (6), 2287 (1976).

[7] R. W. Siegel, *Cluster-Assembled Nanophase Materials*, A. Rev. Mater. Sci. **21**, 559 (1991).

When expanding the pressure and the melting temperature to first-order in terms of the radius r, this result is recast as,

$$\mu_s \left(T_m(r) + \frac{dT_m}{dr} dr, \ p_s(r) + \frac{dp_s}{dr} dr \right)$$

$$= \mu_\ell \left(T_m(r) + \frac{dT_m}{dr} dr, \ p_\ell(r) + \frac{dp_\ell}{dr} dr \right)$$

Furthermore, when expanding the chemical potential to first-order in terms of the melting temperature and the pressure, we obtain,

$$\mu_s \left(T_m(r), p_s(r) \right) + \frac{\partial \mu_s}{\partial T_m} \frac{dT_m}{dr} dr + \frac{\partial \mu_s}{\partial p_s} \frac{dp_s}{dr} dr$$

$$= \mu_\ell \left(T_m(r), p_\ell(r) \right) + \frac{\partial \mu_\ell}{\partial T_m} \frac{dT_m}{dr} dr + \frac{\partial \mu_\ell}{\partial p_\ell} \frac{dp_\ell}{dr} dr$$

The expression for the Laplace pressure implies that,

$$\frac{dp_s}{dr} = -\frac{2\gamma_s}{r^2} \quad \text{and} \quad \frac{dp_\ell}{dr} = -\frac{2\gamma_\ell}{r^2}$$

Thus, the chemical equilibrium condition is reduced to,

$$\left(\frac{\partial \mu_s}{\partial T_m} - \frac{\partial \mu_\ell}{\partial T_m} \right) dT_m = 2 \left(\gamma_s \frac{\partial \mu_s}{\partial p_s} - \gamma_\ell \frac{\partial \mu_\ell}{\partial p_\ell} \right) \frac{dr}{r^2}$$

The Gibbs free energy differential reads,

$$dG = -S\,dT + V\,dp + \mu\,dN$$

which implies that,

$$\frac{\partial \mu}{\partial T} = \frac{\partial}{\partial T} \left(\frac{\partial G}{\partial N} \right) = \frac{\partial}{\partial N} \left(\frac{\partial G}{\partial T} \right) = -\frac{\partial S}{\partial N} = -s$$

$$\frac{\partial \mu}{\partial p} = \frac{\partial}{\partial p} \left(\frac{\partial G}{\partial N} \right) = \frac{\partial}{\partial N} \left(\frac{\partial G}{\partial p} \right) = \frac{\partial V}{\partial N} = v$$

The differential equation becomes,

$$(s_\ell - s_s)\,dT_m = 2\,(\gamma_s v_s - \gamma_\ell v_\ell)\,\frac{dr}{r^2}$$

The integration of this equation from temperature T_∞ and radius $r = \infty$ to temperature $T_m(r)$ and radius r yields,

$$(s_\ell - s_s) \left(T_m(r) - T_\infty \right) = -\frac{2}{r} (\gamma_s v_s - \gamma_\ell v_\ell)$$

Since $\ell_{s\ell} = T_\infty (s_\ell - s_s)$,

$$T_\infty - T_m(r) = \frac{2\,T_\infty}{\ell_{s\ell}\,r} (\gamma_s v_s - \gamma_\ell v_\ell)$$

Thus, if $\gamma_\ell v_\ell < \gamma_s v_s$, then $T_m(r) < T_\infty$, which is the case for certain metals.

6.10 Work on a van der Waals Gas

A mole of oxygen, considered as a van der Waals gas, undergoes a reversible isothermal expansion at fixed temperature T_0 from an initial volume V_i to a final volume V_f. Determine the work W_{if} performed on the van der Waals gas in terms of the parameters a, and b.

Numerical Application:

$T_0 = 273$ K, $V_i = 22.3 \cdot 10^{-3}$ m^3, $V_f = 3V_i$, $p_0 = 1.013 \cdot 10^5$ Pa, $a = 0.14$ Pa m^6 and $b = 3.2\ 10^{-6}$ m^3.

Solution:

For one mole of oxygen at temperature T_0, the van der Waals equation of state (6.64) reads,

$$p = \frac{R T_0}{V - b} - \frac{a}{V^2}$$

Thus, the work is expressed as,

$$W_{if} = - \int_{V_i}^{V_f} p \, dV = - R T_0 \int_{V_i}^{V_f} \frac{dV}{V - b} - a \int_{V_i}^{V_f} \frac{dV}{V^2}$$

$$= - R T_0 \ln \left(\frac{V_f - b}{V_i - b} \right) - a \left(\frac{1}{V_f} - \frac{1}{V_i} \right) = - 2.49 \cdot 10^3 \text{ J}$$

6.11 Inversion Temperature of the Joule–Thomson Process

A van der Waals gas is going through a Joule-Thomson process that keeps the enthalpy H constant (problem 4.8.3). A van der Waals gas in characterised by the following equations of state,

$$p = \frac{NR T}{V - Nb} - \frac{N^2 a}{V^2} \qquad \text{and} \qquad U = c NR T - \frac{N^2 a}{V}$$

and the amount of gas is constant, i.e. $N = $ const. Use the condition $dH = 0$ in order to obtain an expression for the derivative $\dfrac{dT}{dV}$. Determine the temperature T_0 at which this derivative changes sign.

Solution:

The enthalpy for a van der Waals gas reads,

$$H = U + p V = (c + 1) NR T - \frac{2 N^2 a}{V} + \frac{N^2 b R T}{V - Nb}$$

and its differential is given by,

$$dH = (c + 1) NR \, dT + 2 N^2 a \frac{dV}{V^2} + \frac{N^2 b R}{V - Nb} \, dT - \frac{N^2 b R T}{(V - Nb)^2} \, dV = 0$$

which implies that,

$$\frac{dT}{dV} = \frac{Nb}{(c+1)(V-Nb)^2 + Nb(V-Nb)} \left(T - \frac{2a}{bR} \frac{(V-Nb)^2}{V^2} \right)$$

The derivative changes sign at the temperature at which dT/dV vanishes, which takes place at temperature,

$$T_0 = \frac{2a}{bR} \frac{(V-Nb)^2}{V^2}$$

6.12 Lever Rule

A phase diagram is drawn for a mixture of two substances at a fixed pressure p with a liquid phase and a gaseous phase (Fig. 6.5). The substances are labelled 1 and 2 and the diagram is shown as a function of the concentration c_1 of substance 1. There is a range of temperature for which there is coexistence of two phases. Answer the following questions, treating the concentrations c_1^A and c_1^B as given values.

a) Apply the Gibbs phase rule (6.62) to find the number of degrees of freedom when two phases coexist at a fixed pressure p.
b) We distill a substance 1 with an initial concentration c_1^A by heating the liquid up to the temperature T_C. Determine the final concentration of substance 1 after distillation.
c) A mixture with a concentration c_1^C of substance 1 is put in a container. The mixture is brought to a temperature T_C while the pressure remains at p. Establish that,

$$N_\ell \left(c_1^C - c_1^A \right) = N_g \left(c_1^B - c_1^C \right)$$

where N_ℓ the amount of mixture in the liquid phase, and N_g that in the gas phase. This is known as the **lever rule**.

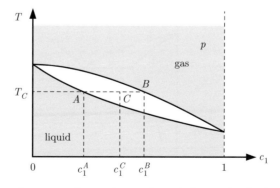

Figure 6.5   Phase diagram of a binary mixture presenting two phases and an exclusion zone (see § 6.4).

Solution:

a) According to Gibbs phase rule (6.62) $r = 2$ and $m = 2$ imply that $f = 2$. If the pressure p is fixed then there is only one degree of freedom. If we choose c_1 as the variable, then the temperature where the two phases coexist can be read on the phase diagram for each value of c_1.

b) When the liquid mixture reaches the temperature T_C, a vaporisation takes place. After condensation, the concentration of substance 1 is c_1^B.

c) The concentration c_1^A, c_1^B and c_1^C are defined as,

$$c_1^A = \frac{N_{\ell 1}}{N_\ell} \quad \text{and} \quad c_1^B = \frac{N_{g1}}{N_g} \quad \text{and} \quad c_1^C = \frac{N_{\ell 1} + N_{g1}}{N_\ell + N_g}$$

where $N_\ell = N_{\ell 1} + N_{\ell 2}$ and $N_g = N_{g1} + N_{g2}$ are the amounts of substances 1 and 2 in the liquid phase ℓ and the gaseous phase g. Thus,

$$(N_\ell + N_g)\, c_1^C = N_\ell\, c_1^A + N_g\, c_1^B$$

which implies that

$$N_\ell \left(c_1^C - c_1^A \right) = N_g \left(c_1^B - c_1^C \right)$$

This result would be obtained if we had two 'weights' N_ℓ and N_g hanging from the ends of a mechanical lever of length AB at equilibrium around its axis at C.

6.13 Eutectic

A phase diagram is drawn for a mixture of two substances at a fixed pressure p with one liquid phase and two solid phases (Fig 6.6). The substances are labelled 1 and 2 and the diagram is shown as a function of the concentration c_1 of substance 1. This diagram presents what is called a **eutectic** point. At the eutectic concentration c_1^E, the freezing

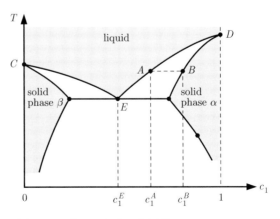

Figure 6.6   Phase diagram of a binary mixture presenting a eutectic point at E.

temperature is the lowest. In particular, it is lower than the freezing temperatures of the pure substances (points C and D). At the eutectic, the liquid freezes into a mixture of two solid phases, the α and β phases.

a) Consider a liquid at concentration c_1^A. As the temperature is lowered, the point A is reached. Describe qualitatively what happens then.
b) Describe what happens if a liquid of composition c_1^E is cooled.

Solution:

a) At point A, the solid phase α starts to precipitate. Notice that the concentration c_1^B of solid thus formed is different from the concentration c_1^A of liquid.
b) At point E, the entire solution becomes a solid alloy. The concentrations of the solid phases α and β in the alloy are given by the lever rule with respect to point E.

Heat Engines

The exercises given in the last section of Chapter 7 are presented here with their solutions.

7.1 Refrigerator

A thermoelectric refrigerator becomes cold by expelling heat into the environment at a temperature T^+. The power supplied to the device is P_W and the thermal power corresponding to the rejected heat is P_Q. Determine the lowest temperature T^- that the system can reach if it had an optimal efficiency.

Numerical Application:

$P_W = 100$ W, $P_Q = 350$ W and $T^+ = 25°$ C.

Solution:

The lowest temperature that the system can reach is obtained when the refrigerator operates a Carnot refrigeration cycle. Then, the cooling coefficient of performance (7.42) of a Carnot cycle is given by (7.49),

$$\varepsilon_C^- = \frac{P_Q}{P_W} = \frac{T^-}{T^+ - T^-}$$

Thus, the lowest temperature T^- is given by,

$$T^- = \frac{P_Q}{P_Q + P_W} T^+ = -41° \text{ C}$$

7.2 Power Plant Cooled by a River

A power plant operates between a hot reservoir consisting of a combustion chamber or a nuclear reactor and a cold reservoir consisting of the water of a river. It is modelled as a thermal machine operating between the hot reservoir at temperature T^+ and the cold reservoir at temperature T^-. Analyse this power plant by using the following instructions :

a) Determine the maximum efficiency η_C of this power plant and the thermal power P_{Q+} describing the heat exchange with the combustion chamber.

b) Assume that its real efficiency is $\eta = k\eta_C$ and find the thermal power P_{Q^-} describing the heat exchange with the river.

c) Determine the temperature difference ΔT of the water flowing at a rate $\dot{V}$ down the river. The water has a density m and a specific heat at constant pressure per unit of mass c_p^*.

Numerical Application:

$P_W = -750\,\text{MW}$, $T^+ = 300°\,\text{C}$, $T^- = 19°\,\text{C}$, $k = 60\,\%$, $\dot{V} = 200\,\text{m}^3/\text{s}$, $m = 1,000\,\text{kg/m}^3$ and $c_p^* = 4,181\,\text{J/kg K}$.

Solution:

We mainly examined thermodynamic cycles composed of distinct processes in chapter 7. Here, we consider a power plant as a heat engine in a continuous regime, so it is more natural to speak in terms of its power output P_W, the thermal power P_{Q^+} at the hot reservoir and the thermal power P_{Q^-} at the cold reservoir.

a) The maximal efficiency is the Carnot efficiency (7.46). When this efficiency is written in terms of powers, the definition (7.38) reads,

$$\eta_C = -\frac{P_W}{P_{Q^+}} = 1 - \frac{T^-}{T^+} = 49\,\%$$

Thus, the thermal power P_{Q^+} describing the heat exchange with the heat source is given by,

$$P_{Q^+} = -P_W\,\frac{T^+}{T^+ - T^-} = 1.53\,\text{GW}$$

b) Since the nuclear power plant is in a steady state, the first law (1.29) is written as,

$$\dot{U} = P_W + P_{Q^+} + P_{Q^-} = 0$$

which implies that,

$$P_{Q^-} = -\left(P_W + P_{Q^+}\right)$$

The efficiency (7.38) reads,

$$\eta = k\eta_C = -\frac{P_W}{P_{Q^+}} \qquad \text{thus} \qquad P_{Q^+} = -\frac{P_W}{k\eta_C}$$

Hence,

$$P_{Q^-} = P_W\left(\frac{1 - k\eta_C}{k\eta_C}\right) = -1.80\,\text{GW}$$

c) Since the mass of water M is the product of the water density m and volume V, the mass flux is given by,

$$\dot{M} = m\,\dot{V}$$

The thermal power is given by,

$$P_{Q^-} = \dot{M}c_p^*\,\Delta T = m\,\dot{V}c_p^*\,\Delta T$$

which implies that the temperature difference is given by,

$$\Delta T = \frac{P_{Q^-}}{m \, \dot{V} \, c_p^*} = -2°C$$

7.3 Braking Cycle

A system is made up of a vertical cylinder which is sealed at the top and closed by a piston at the bottom. A valve A controls the intake of gas at the top and an exhaust valve B (also at the top) is held back by a spring that exerts a constant pressure p_2 on the valve. The system goes through the following processes:

- $0 \longrightarrow 1$: the piston is at the top of the cylinder; valve A opens up and the piston is lowered into it so that some of the gas at atmospheric pressure $p_0 = p_1$ is added to the cylinder. The gas is at room temperature T_1. Valve B is closed. The maximum volume occupied by the incoming gas is V_1.
- $1 \longrightarrow 2$: Valve A is now closed and the piston moves upward, fast enough so that the process can be considered adiabatic. Valve B remains closed as long as the pressure during the rise of the piston is lower than p_2. As the piston continues in its rise, the gas reaches pressure $p_2 = 10p_1$, at a temperature T_2 in a volume V_2. Assume a reversible adiabatic process for which equations (5.90) and (5.83) apply.
- $2 \longrightarrow 3$: As the piston keeps moving up, valve B opens up, the pressure is $p_3 = p_2$ and the gas is released in the environment while valve A still remains closed until the piston reaches the top, where $V_3 = V_0 = 0$.
- $3 \longrightarrow 0$: Valve B closes and valve A opens up. The system is ready to start over again.

Analyse this cycle by using the following instructions :

a) Draw the (p, V) diagram for the three processes that the system is undergoing.
b) Determine the temperature T_2 and the volume V_2.
c) Find the work W performed per cycle.

Numerical Application:

$V_0 = V_3 = 0, p_0 = p_1 = 10^5$ Pa, $V_1 = 0.25$ l, $T_1 = 27°$ C and $\gamma = 1.4$.

Solution:

a) The (p, V) diagram consists of an isobaric expansion $0 \longrightarrow 1$, an adiabatic compression $1 \longrightarrow 2$, an isobaric contraction $2 \longrightarrow 3$, and an isochoric decompression $3 \longrightarrow 0$ (Fig. 7.1).
b) For the adiabatic compression, the adiabatic condition (5.83) is written as,

$$T_1^{\gamma} \, p_1^{1-\gamma} = T_2^{\gamma} \, p_2^{1-\gamma}$$

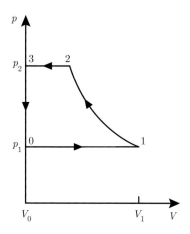

Figure 7.1 (p, V) diagram of the braking cycle (exercise § 7.3).

which implies that,

$$T_2 = T_1 \left(\frac{p_1}{p_2}\right)^{\frac{1-\gamma}{\gamma}} = T_1 \left(\frac{1}{10}\right)^{\frac{1-\gamma}{\gamma}} = 579\,\mathrm{K}$$

The adiabatic condition (5.90) is written as,

$$p_1 V_1^\gamma = p_2 V_2^\gamma$$

which implies that,

$$V_2 = V_1 \left(\frac{p_1}{p_2}\right)^{\frac{1}{\gamma}} = V_1 \left(\frac{1}{10}\right)^{\frac{1}{\gamma}} = 0.0481$$

c) The work performed on the gas over the entire cycle is the sum of the works performed during the four processes,

$$W = W_{01} + W_{12} + W_{23} + W_{30}$$

The work performed during the isobaric processes is,

$$W_{01} + W_{23} = -p_1 \int_0^{V_1} dV - p_2 \int_{V_2}^0 dV = -p_1 V_1 + p_2 V_2$$

There is no work performed during the isochoric process,

$$W_{30} = 0$$

The work performed during the adiabatic process is,

$$W_{12} = -\int_{V_1}^{V_2} p\,dV = -p_1 V_1^\gamma \int_{V_1}^{V_2} \frac{dV}{V^\gamma} = \frac{p_1 V_1^\gamma}{\gamma - 1} \left(V_2^{1-\gamma} - V_1^{1-\gamma}\right)$$

$$= \frac{1}{\gamma - 1} \left(p_2 V_2 - p_1 V_1\right)$$

Thus,

$$W = \frac{\gamma}{\gamma - 1} \left(p_2 V_2 - p_1 V_1 \right) = \frac{\gamma}{\gamma - 1} p_1 \left(10 V_2 - V_1 \right) = 80.5 \, \text{kJ}$$

Since the work W is positive, the system can act as a brake for whatever mechanism drives the motion of the piston.

7.4 Lenoir Cycle

The Lenoir cycle is a model for the operation of a combustion engine patented by Jean Joseph Etienne Lenoir in 1860 (Figs. 7.2 and 7.3). This idealised cycle is defined by three reversible processes :

- $1 \longrightarrow 2$ isochoric compression
- $2 \longrightarrow 3$ adiabatic expansion
- $3 \longrightarrow 1$ isobaric contraction

Assume that the cycle is performed on an ideal gas characterised by the coefficient c found in relation (5.62). The following values of some state variables of the gas are assumed to be known : the pressure p_1, volumes V_1 and V_3, temperature T_1 and the number of moles of gas N. Analyse this cycle by using the following instructions :

a) Draw the (p, V) and (T, S) diagrams of the cycle.
b) Determine the entropy variation ΔS_{12} of the gas during the isochoric process $1 \longrightarrow 2$.
c) Express the temperature T_2 in terms of the heat exchanged Q_{12} during the isochoric process $1 \longrightarrow 2$.
d) Determine the pressure p_2 in terms of the pressure p_1, the volume V_1 and the heat exchanged Q_{12}.
e) Determine the pressure p_3 in terms of the pressure p_2 and volumes V_2 and V_3.
f) Determine the work W_{23} performed during the adiabatic process $2 \longrightarrow 3$ and the heat Q_{23} exchanged during this process.
g) Find the work W_{31} performed during the isobaric process $3 \longrightarrow 1$ and the heat Q_{31} exchanged during this process.
h) Find the efficiency of the cycle η_L defined in conformity with relation (7.38) as,

$$\eta_L = - \frac{W_{23} + W_{31}}{Q_{12}}$$

Express the efficiency η_L in terms of the temperatures T_1, T_2 and T_3.

Solution:

a) For the isochoric process, $V = V_1 = V_2 = \text{const}$ (Fig. 7.2). For the adiabatic process according to relation (5.90), $p(V) = \text{const}/V^\gamma$ where $\gamma > 1$ and $\text{const} = p_1 V_1^\gamma = p_2 V_2^\gamma$, which is a monotonously decreasing convex function of V. For the isobaric process, $p(V) = p_3 = p_1 = \text{const}$.

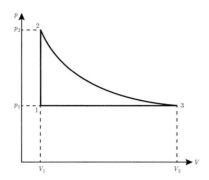

Figure 7.2 (p, V) diagram of a Lenoir cycle.

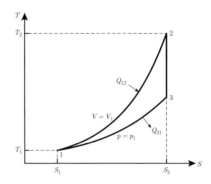

Figure 7.3 (T, S) diagram of a Lenoir cycle.

For the isochoric process, according to relation (7.20), $T(S) = T_1 \exp\left((S - S_1)/c\,NR\right)$, which is a monotonously increasing function of S (Fig. 7.3). For the adiabatic process, $S = S_1 = S_2 = \text{const}$. For the isobaric process, according to relation (7.23), $T(S) = T_1 \exp\left((S - S_1)/(c + 1)\,NR\right)$, which is a monotonously increasing function of S.

b) According to relation (7.20), the work performed during the isochoric process vanishes,

$$\Delta S_{12} = c\,NR\,\ln\left(\frac{T_2}{T_1}\right) = (c + 1)\,NR\,\ln\left(\frac{V_1}{V_3}\right)$$

c) According to relation (7.19), the heat exchanged during the isochoric process is written as,

$$Q_{12} = \Delta U_{12} = \int_{U_1}^{U_2} dU = c\,NR\int_{T_1}^{T_2} dT = c\,NR\,(T_2 - T_1)$$

Thus,

$$T_2 = T_1 + \frac{Q_{12}}{c\,NR}$$

d) Using the ideal gas equation of state (5.47), the pressure p_2 is given by,

$$p_2 = \frac{NR\,T_2}{V_2} = \frac{NR}{V_1}\left(T_1 + \frac{Q_{12}}{c\,NR}\right) = p_1 + \frac{Q_{12}}{c\,V_1}$$

e) According to relation (5.90) for an adiabatic process,

$$p_3 = p_2 \left(\frac{V_2}{V_3} \right)^{\gamma}$$

f) According to relations (7.14) and (7.13), the work performed during the adiabatic process is given by,

$$W_{23} = \Delta U_{23} = c\,NR \int_{T_2}^{T_3} dT = c\,NR\,(T_3 - T_2)$$

and there is no heat exchanged,

$$Q_{23} = \int_2^3 T dS = 0$$

g) According to relation (7.21) and (7.22) characterising the isobaric process, the work performed can be written as,

$$W_{31} = - \int_3^1 p\,dV = - p_1 \int_{V_3}^{V_1} dV = - p_1\,(V_1 - V_3) = NR\,(T_3 - T_1)$$

and the heat exchanged is given by,

$$Q_{31} = \Delta H_{31} = \int_{H_3}^{H_1} dH = (c+1)\,NR \int_{T_3}^{T_1} dT = (c+1)\,NR\,(T_1 - T_3)$$

This heat transfer is not taking place during an isothermal process. Thus, it cannot be described by a heat transfer to a thermal reservoir.

h) Applying the results above to the definition of η_L and using the coefficient $\gamma = (c+1)/c$, we can write,

$$\eta_L = - \frac{W_{23} + W_{31}}{Q_{12}} = \frac{c\,(T_2 - T_3) + (T_1 - T_3)}{c\,(T_2 - T_1)}$$

$$= \frac{T_1 + c\,T_2 - (c+1)\,T_3}{c\,(T_2 - T_1)} = \frac{(\gamma - 1)\,T_1 + T_2 - \gamma\,T_3}{T_2 - T_1} = 1 - \gamma\,\frac{T_3 - T_1}{T_2 - T_1}$$

7.5 Otto Cycle

The Otto cycle is a model for a spark ignition engine and represents the mode of operation of most non-diesel car engines. It consists of four processes when the system is closed, and of two additional isobaric processes when the system is open, corresponding to air intake and exhaust. Thus, we have,

- $0 \longrightarrow 1$ isobaric air intake
- $1 \longrightarrow 2$ adiabatic compression
- $2 \longrightarrow 3$ isochoric heating
- $3 \longrightarrow 4$ adiabatic expansion
- $4 \longrightarrow 1$ isochoric cooling
- $1 \longrightarrow 0$ isobaric gas exhaust

Assume that the adiabatic processes are reversible and that the gas is an ideal gas characterised by the coefficient c found in relation (5.62) and coefficient $\gamma = (c+1)/c$. The following values of state variables are assumed to be known : the pressure p_1, the volumes $V_1 = V_4$ and $V_2 = V_3$, the temperature T_3, and the number of moles N of air at the intake. Analyse this cycle by using the following instructions :

a) Draw the (p, V) and (T, S) diagrams of the cycle. On the (p, V) diagram, show also the intake and exhaust processes.

b) Describe what the engine does in each of the processes.

c) Explain why an exchange of air with the exterior is needed.

d) On the (p, V) and (T, S) diagrams determine the relation between the area enclosed in the cycles and the work W and the heat Q per cycle.

e) Determine all the state variables at points 1, 2, 3 and 4 of the cycle, i.e. find p_2, p_3, p_4, T_2 and T_4.

f) Compute the work W performed per cycle and the heat Q exchanged during a cycle.

g) Determine the efficiency of the Otto cycle,

$$\eta_O = -\frac{W}{Q^+}$$

where $Q^+ = Q_{23}$.

Solution:

a) For the isobaric processes, $p(V) = p_1 = p_2 = $ const (Fig. 7.4). For the adiabatic processes, according to relation (5.90), $p(V) = \text{const}/V^\gamma$ where $\gamma > 1$ and const $= p_1 V_1^\gamma = p_2 V_2^\gamma$ or const $= p_3 V_3^\gamma = p_4 V_4^\gamma$, which are monotonously decreasing convex functions of V. For the isochoric processes, $V = V_2 = V_3 = $ const or $V = V_4 = V_1 = $ const.

For the adiabatic processes, $S = S_1 = S_2 = $ const and $S = S_3 = S_4 = $ const (Fig. 7.5). For the isochoric processes, according to relation (7.20), $T(S) = T_1 \exp((S - S_1)/c\,NR)$ and $T(S) = T_2 \exp((S - S_2)/c\,NR)$, which are monotonously increasing functions of S.

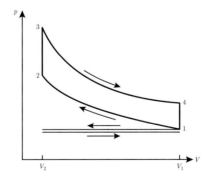

(p, V) diagram of an Otto cycle.

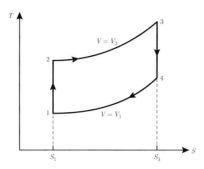

Figure 7.5 (T, S) diagram of an Otto cycle.

b) During the isobaric air intake $0 \longrightarrow 1$, a mass of air is brought in the cylinder at constant atmospheric pressure p_1 as the piston moves and the volume inside the cylinder increases from V_2 to V_1. During the adiabatic compression $1 \longrightarrow 2$, the air inside the cylinder is compressed adiabatically by the piston from an initial volume V_1 to a final volume V_2. During the isochoric heating $2 \longrightarrow 3$, the fuel-air mixture is lit up. During the adiabatic expansion $3 \longrightarrow 4$, the gas is expanded adiabatically from an initial volume V_3 to a final volume V_4, moving the piston back to its initial position. At this point, the gas occupies a volume V_4. During the isochoric cooling $4 \longrightarrow 1$, heat is transferred to the environment until the pressure comes back to the atmospheric pressure p_1. Finally, during the isobaric gas exhaust $4 \longrightarrow 0$, the gas is removed from the cylinder at constant atmospheric pressure p_1 as the piston moves and the volume inside the cylinder decreases from V_1 to V_2.

c) An engine that runs according to the Otto cycle is a combustion engine. This means that oxygen is essential for the chemical combustion reaction to take place. After each ignition, fresh air needs to enter the cylinder in order to run a chemical combustion reaction again.

d) The area enclosed in the cycle on the (p, V) diagram is written as,

$$\oint p\, dV = \int_{V_1}^{V_2} p\, dV + \int_{V_3}^{V_4} p\, dV = -W_{12} - W_{34} = -W$$

since $W_{23} = W_{41} = 0$. Thus, the area enclosed in the cycle on the (p, V) diagram represents the opposite of the work performed per cycle W.

The area enclosed in the cycle on the (T, S) diagram is written as,

$$\oint T\, dS = \int_{S_2}^{S_3} T\, dS + \int_{S_4}^{S_1} T\, dS = Q_{23} + Q_{41} = Q$$

since $Q_{12} = Q_{34} = 0$. Thus, the area enclosed in the cycle on the (T, S) diagram represents the heat Q exchanged per cycle. Since the internal energy U is a state function, we must have $Q = -W$, as stated in relation (7.6).

e) Using relation (5.90) and the ideal gas equation of state (5.47), the pressures are given by,

$$p_2 = p_1 \left(\frac{V_1}{V_2}\right)^\gamma \qquad p_3 = \frac{N R T_3}{V_2} \qquad p_4 = \frac{N R T_3}{V_1} \left(\frac{V_2}{V_1}\right)^{\gamma-1}$$

and the temperatures by,

$$T_1 = \frac{p_1 V_1}{NR} \qquad T_2 = \frac{p_1 V_1}{NR} \left(\frac{V_1}{V_2}\right)^{\gamma-1} \qquad T_4 = T_3 \left(\frac{V_2}{V_1}\right)^{\gamma-1}$$

f) According to relation (7.14), the work performed during the adiabatic compression and expansion is written as,

$$W_{12} = \Delta U_{12} = c\,NR \int_{T_1}^{T_2} dT = c\,NR\,(T_2 - T_1)$$

$$W_{34} = \Delta U_{34} = c\,NR \int_{T_3}^{T_4} dT = c\,NR\,(T_4 - T_3)$$

The work performed per cycle is given by,

$$W = W_{12} + W_{34} = c\,NR\,(T_4 - T_3 + T_2 - T_1)$$

According to relation (7.19), the heat exchanged during the isochoric heating and cooling is written as,

$$Q_{23} = \Delta U_{23} = \int_{U_2}^{U_3} dU = c\,NR \int_{T_2}^{T_3} dT = c\,NR\,(T_3 - T_2)$$

$$Q_{41} = \Delta U_{41} = \int_{U_4}^{U_1} dU = c\,NR \int_{T_4}^{T_1} dT = c\,NR\,(T_1 - T_4)$$

The heat exchanged per cycle is given by,

$$Q = Q_{23} + Q_{41} = c\,NR\,(T_3 - T_2 + T_1 - T_4)$$

g) Using the definition of the efficiency (7.38) we obtain,

$$\eta_O = -\frac{W}{Q^+} = -\frac{W}{Q_{23}} = -\frac{c\,(T_4 - T_3 + T_2 - T_1)}{c\,(T_3 - T_2)} = 1 - \frac{T_4 - T_1}{T_3 - T_2}$$

7.6 Atkinson Cycle

James Atkinson was a British engineer who designed several combustion engines. The thermodynamic cycle bearing his name is a modification of the Otto cycle intended to improve its efficiency. The trade-off in achieving higher efficiency is a decrease in the work performed per cycle. The idealised Atkinson cycle consists of the following reversible processes:

- 1 ⟶ 2: adiabatic compression
- 2 ⟶ 3: isochoric heating
- 3 ⟶ 4: isobaric heating
- 4 ⟶ 5: adiabatic expansion
- 5 ⟶ 6: isochoric cooling
- 6 ⟶ 1: isobaric cooling

Assume that the adiabatic processes are reversible and that the cycle is operated on an ideal gas characterised by,

$$pV = NRT \qquad\qquad U = cNRT \qquad\qquad \gamma = \frac{c+1}{c}$$

The following physical quantities that characterise the cycle are assumed to be known : volumes V_1, V_2 and V_6, pressures p_1 and p_3, temperature T_5, and the number of moles N of gas. Analyse this cycle by using the following instructions :

a) Draw the (p, V) diagram of the Atkinson cycle.
b) Determine the pressures p_2, p_4, p_5, p_6, the volumes V_3, V_4, V_5 and temperatures T_1, T_2, T_3, T_4, T_6, in terms of the known physical quantities.
c) Find the works W_{12}, W_{23}, W_{34}, W_{45}, W_{56}, W_{61} and the work W performed per cycle.
d) Find the heat transfers Q_{12}, Q_{23}, Q_{34}, Q_{45}, Q_{56}, Q_{61} and the heat $Q^+ = Q_{23} + Q_{34}$ provided to the gas.
e) Determine the efficiency of the Atkinson cycle,

$$\eta_A = -\frac{W}{Q^+}$$

Solution:

a) For the adiabatic process, according to relation (5.90), $p(V) = \text{const}/V^\gamma$ where $\gamma > 1$ and $\text{const} = p_1 V_1^\gamma = p_2 V_2^\gamma$ or $\text{const} = p_4 V_4^\gamma = p_5 V_5^\gamma$, which are monotonously decreasing convex functions of V (Fig. 7.6). For the isochoric processes, $V = V_2 = V_3 = \text{const}$ or $V = V_5 = V_6 = \text{const}$. For the isobaric processes, $p(V) = p_3 = p_4 = \text{const}$ or $p(V) = p_6 = p_1 = \text{const}$.
b) Using relation (5.90) and the ideal gas equation of state (5.47), the pressures are given by,

$$p_2 = p_1 \left(\frac{V_1}{V_2}\right)^\gamma \qquad\qquad p_4 = p_3 \qquad\qquad p_5 = \frac{NRT_5}{V_6} \qquad\qquad p_6 = p_1$$

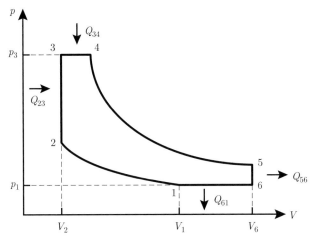

Figure 7.6 (p, V) diagram of the Atkinson cycle.

the volumes by,

$$V_3 = V_2 \qquad V_4 = \left(\frac{NRT_5}{p_3}\right)^{\frac{1}{\gamma}} V_6^{\frac{\gamma-1}{\gamma}} \qquad V_5 = V_6$$

The temperatures are written as,

$$T_1 = \frac{p_1 V_1}{NR} \qquad T_2 = \frac{p_1 V_1}{NR}\left(\frac{V_1}{V_2}\right)^{\gamma-1} \qquad T_3 = \frac{p_3 V_3}{NR}$$

$$T_4 = \left(\frac{p_3 V_6}{NR}\right)^{\frac{\gamma-1}{\gamma}} T_5^{\frac{1}{\gamma}} \qquad T_6 = \frac{p_1 V_6}{NR}$$

c) According to relation (7.14), the work performed during the adiabatic compression and expansion is written as,

$$W_{12} = \Delta U_{12} = c\,NR \int_{T_1}^{T_2} dT = c\,NR\,(T_2 - T_1)$$

$$W_{45} = \Delta U_{45} = c\,NR \int_{T_4}^{T_5} dT = c\,NR\,(T_5 - T_4)$$

According to relation (7.18), there is no work performed during the isochoric heating and cooling,

$$W_{23} = W_{56} = 0$$

According to relation (7.21), the work performed during the isobaric processes can be written as,

$$W_{34} = -\int_3^4 p\,dV = -p_3 \int_{V_3}^{V_4} dV = -p_3\,(V_4 - V_3) = NR\,(T_4 - T_3)$$

$$W_{61} = -\int_6^1 p\,dV = -p_1 \int_{V_6}^{V_1} dV = -p_1\,(V_1 - V_6) = NR\,(T_1 - T_6)$$

The work performed per cycle reads,

$$W = W_{12} + W_{34} + W_{45} + W_{61}$$
$$= c\,NR\,(T_2 - T_1 + T_5 - T_4) + NR\,(T_4 - T_3 + T_1 - T_6)$$

d) According to relation (7.13), there is no heat exchanged during the adiabatic compression and expansion,

$$Q_{12} = Q_{45} = 0$$

According to relation (7.19), the heat exchanged during the isochoric heating and cooling is given by,

$$Q_{23} = \Delta U_{23} = \int_{U_2}^{U_3} dU = c\,NR \int_{T_2}^{T_3} dT = c\,NR\,(T_3 - T_2)$$

$$Q_{56} = \Delta U_{56} = \int_{U_5}^{U_6} dU = c\,NR \int_{T_5}^{T_6} dT = c\,NR\,(T_6 - T_5)$$

According to relation (7.22), the heat exchanged during the isobaric processes can be written as,

$$Q_{34} = \Delta H_{34} = \int_{H_3}^{H_4} dH = (c+1)\,NR \int_{T_3}^{T_4} dT = (c+1)\,NR\,(T_4 - T_3)$$

$$Q_{61} = \Delta H_{61} = \int_{H_6}^{H_1} dH = (c+1)\,NR \int_{T_6}^{T_1} dT = (c+1)\,NR\,(T_1 - T_6)$$

The heat provided from the hot reservoir reads,

$$Q^+ = Q_{23} + Q_{34} = c\,NR\,(T_3 - T_2) + (c+1)\,NR\,(T_4 - T_3)$$

e) Using the definition of the efficiency (7.38), we obtain,

$$\begin{aligned}
\eta_A &= -\frac{W}{Q^+} = -\frac{c\,(T_2 - T_1 + T_5 - T_4) + (T_4 - T_3 + T_1 - T_6)}{c\,(T_3 - T_2) + (c+1)\,(T_4 - T_3)} \\
&= \frac{(T_1 - T_2 + T_4 - T_5) + (\gamma - 1)\,(T_3 - T_4 + T_6 - T_1)}{(T_3 - T_2) + \gamma\,(T_4 - T_3)}
\end{aligned}$$

7.7 Refrigeration Cycle

An ideal gas characterised by the coefficient c found in relation (5.62) and the coefficient $\gamma = (c+1)/c$ undergoes a refrigeration cycle consisting of four reversible processes (Fig. 7.7):

- $1 \longrightarrow 2$: adiabatic compression
- $2 \longrightarrow 3$: isobaric compression
- $3 \longrightarrow 4$: isochoric cooling
- $4 \longrightarrow 1$: isobaric expansion

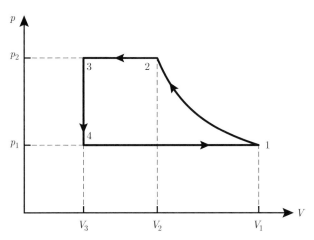

Figure 7.7 (p, V) diagram of a refrigeration cycle

Analyse this cycle by using the following instructions :

a) Determine the volume V_2 in terms of the volumes V_1 and V_3 and the pressures p_1 and p_2.
b) Find the entropy variation ΔS_{23} during the isobaric compression.
c) Determine the heat exchanged Q_{23} during the isobaric compression.
d) Assume now that instead of an ideal gas a fluid is used, which is entirely in a gaseous state at point 2 and completely in a liquid state at point 3. The isobaric compression $2 \longrightarrow 3$ is then a phase transition occurring at temperature T and characterised by the molar latent heat of vaporisation $\ell_{\ell g}$. Determine the entropy variation ΔS_{23} during the phase transition in terms of the number of moles N of fluid, the volume V_2, the pressure p_2 and the molar latent heat of vaporisation $\ell_{\ell g}$, assuming that $p V = NRT$ in the gas phase.

Solution:

a) Using the adiabatic condition (5.90), the volume V_2 can be written as,

$$V_2 = V_1 \left(\frac{p_1}{p_2} \right)^{\frac{1}{\gamma}}$$

b) According to relation (7.23), there entropy variation during the isobaric compression is given by,

$$\Delta S_{23} = \int_{S_2}^{S_3} dS = (c + 1) NR \int_{T_2}^{T_3} \frac{dT}{T} = (c + 1) NR \ln \left(\frac{T_3}{T_2} \right)$$

c) According to relation (7.22), the heat exchanged during the isobaric compression is written as,

$$Q_{23} = \Delta H_{23} = \int_{H_2}^{H_3} dH = (c + 1) NR \int_{T_2}^{T_3} dT = (c + 1) NR (T_3 - T_2)$$

d) According to relation (2.43) for an isothermal process like the phase transition at temperature T,

$$Q_{23} = T \int_{S_2}^{S_3} dS = T (S_3 - S_2) = T \Delta S_{23}$$

According to relations (6.43), (6.44) and the ideal gas equation of state (5.47),

$$\Delta S_{23} = \frac{Q_{23}}{T} = - \frac{Q_{\ell g}}{T} = - \frac{N \ell_{\ell g}}{T} = - \frac{N^2 R \ell_{\ell g}}{p_2 V_2}$$

7.8 Rankine Cycle

An ideal gas characterised by the coefficient c found in relation (5.62) and the coefficient $\gamma = (c + 1) / c$ undergoes a Ranking engine cycle consisting of four reversible processes:

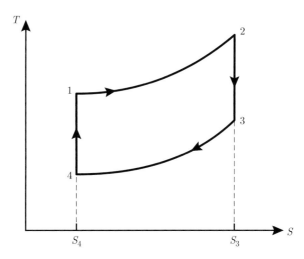

Figure 7.8 (T, S) diagram of a Rankine cycle operated on an ideal gas

- $1 \longrightarrow 2$: isobaric expansion
- $2 \longrightarrow 3$: adiabatic expansion
- $3 \longrightarrow 4$: isobaric compression
- $4 \longrightarrow 1$: adiabatic compression

Thus, the cycle in a (T, S) diagram as shown in Fig. 7.8.
  Analyse this cycle by using the following instructions:

a) Draw the (p, V) diagram of a Rankine cycle for an ideal gas.
b) Determine the works performed W_{12}, W_{23}, W_{34} and W_{41} and the work performed per cycle W in terms of the enthalpies H_1, H_2, H_3 and H_4.
c) Find the heat provided by the hot reservoir $Q^+ = Q_{12}$ in terms of the enthalpies H_1, H_2, H_3 and H_4.
d) Determine the efficiency of the Rankine cycle for an ideal fluid defined as,

$$\eta_R = -\frac{W}{Q^+}$$

Solution:

a) For the isobaric processes, $p\,(V) = p_1 = p_2 = \text{const}$ or $p\,(V) = p_3 = p_4 = \text{const}$ (Fig. 7.9). For the adiabatic processes, according to relation (5.90), $p\,(V) = \text{const}/V^\gamma$ where $\gamma > 1$ and $\text{const} = p_1\,V_1^\gamma = p_4\,V_4^\gamma$ or $\text{const} = p_2\,V_2^\gamma = p_3\,V_3^\gamma$, which are monotonously decreasing convex function of V.
b) According to relation (7.21), the works performed during the isobaric expansion and compression are given by,

$$W_{12} = -\int_1^2 p\,dV = -p\int_{V_1}^{V_2} dV = -p\,(V_2 - V_1) = -NR\,(T_2 - T_1)$$

$$W_{34} = -\int_3^4 p\,dV = -p\int_{V_3}^{V_4} dV = -p\,(V_4 - V_3) = -NR\,(T_4 - T_3)$$

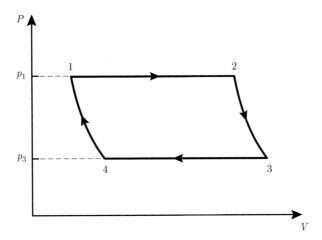

(p, V) diagram of a Rankine cycle

According to relation (5.65), these works can be recast in terms of the enthalpies as,

$$W_{12} = \frac{1}{c+1} (H_1 - H_2) = \frac{\gamma - 1}{\gamma} (H_1 - H_2)$$

$$W_{34} = \frac{1}{c+1} (H_3 - H_4) = \frac{\gamma - 1}{\gamma} (H_3 - H_4)$$

According to relation (7.14), the works performed during the adiabatic expansion and compression are given by,

$$W_{23} = \Delta U_{23} = c\,NR \int_{T_2}^{T_3} dT = c\,NR\,(T_3 - T_2)$$

$$W_{41} = \Delta U_{41} = c\,NR \int_{T_2}^{T_3} dT = c\,NR\,(T_1 - T_4)$$

According to relation (5.65), these works can be recast in terms of the enthalpies as,

$$W_{23} = \frac{c}{c+1} (H_3 - H_2) = \frac{1}{\gamma} (H_3 - H_2)$$

$$W_{41} = \frac{c}{c+1} (H_1 - H_4) = \frac{1}{\gamma} (H_1 - H_4)$$

The work performed per cycle is given by,

$$
\begin{aligned}
W &= W_{12} + W_{23} + W_{34} + W_{41} \\
&= \frac{\gamma - 1}{\gamma} (H_1 - H_2 + H_3 - H_4) + \frac{1}{\gamma} (H_3 - H_2 + H_1 - H_4) \\
&= H_1 - H_2 + H_3 - H_4
\end{aligned}
$$

c) According to relation (7.22), the heat exchanged during the isobaric expansion is given by,

$$Q^+ = Q_{12} = \Delta H_{12} = H_2 - H_1$$

d) Using definition (7.38) of the efficiency, we obtain,

$$\eta_R = -\frac{W}{Q^+} = -\frac{W}{Q_{12}} = -\frac{H_1 - H_2 + H_3 - H_4}{H_2 - H_1} = 1 - \frac{H_3 - H_4}{H_2 - H_1}$$

7.9 Rankine Cycle for a Biphasic Fluid

An engine consists of a boiler, a condenser, a turbine and a water pump (Fig. 7.10). This engine is operating a Rankine cycle on a biphasic fluid (Fig. 7.11). The cycle consists of five processes:

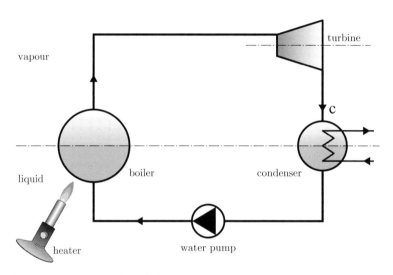

Figure 7.10   Diagram of the Rankine engine for a biphasic fluid.

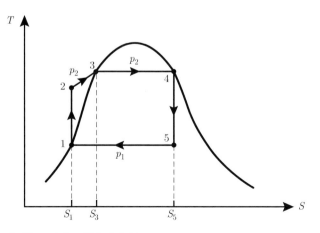

Figure 7.11 (T, S) diagram of the Rankine cycle for a biphasic fluid.

- $1 \longrightarrow 2$: The fluid coming out of the turbine is completely condensed (1). The liquid goes then through an isentropic compression from an initial pressure p_1 to a final pressure p_2.
- $2 \longrightarrow 3$: The liquid is heated up at constant pressure p_2 by the boiler. It undergoes an isobaric heating until it reaches the vaporisation temperature (3).
- $3 \longrightarrow 4$: The liquid is vaporised at constant pressure p_2. It goes through a phase transition until saturation is reached (4).
- $4 \longrightarrow 5$: The fluid undergoes an isentropic expansion from an initial pressure p_2 to a final pressure p_1.
- $5 \longrightarrow 1$: The fluid is condensed at constant pressure p_1. It goes through a phase transition until full condensation is reached (1).

Analyse this cycle by using the following instructions:

a) Determine the heat provided by the boiler $Q^+ = Q_{23} + Q_{34}$, the heat released at the condenser $Q^- = Q_{51}$ in terms of the enthalpies per unit mass h_1^*, h_2^*, h_4^* and h_5^* and the M of fluid undergoing this cycle (Fig. 7.11).
b) Find the work performed by the pump W_{12} and the work performed on the turbine W_{45} in terms of the enthalpies per unit mass h_1^*, h_2^* and h_5^* and the mass M by using the results obtained for the open system presented in § 4.11 and assuming that the mechanical power is due to the chemical power P_C of the fluid flowing through the pump and the turbine, i.e. $P_W = P_C$.
c) Determine the efficiency of the Rankine cycle for a biphasic fluid defined as,

$$\eta_R = -\frac{W}{Q^+}$$

Solution:

a) The isobaric heating and the vaporisation both occur at constant pressure p_2. According to relation (4.61), the heat provided by the boiler is written as,

$$Q^+ = Q_{23} + Q_{34} = \Delta H_{23} + \Delta H_{34} = \Delta H_{24} = M\left(h_4^* - h_2^*\right)$$

The condensation occurs at constant pressure p_1. According to relation (4.61), the heat provided to the condenser is given by,

$$Q^- = Q_{51} = \Delta H_{51} = M\left(h_1^* - h_5^*\right)$$

b) In view of the expression obtained for the chemical power P_C exerted on an open system due to matter flows in § 4.11 and assuming that the mechanical power P_W is due to the chemical power P_C of the fluid flowing through the pump or the turbine, we have,

$$P_W = P_C = \dot{M}\left(h_f^* - h_i^*\right)$$

where $\dot{M}$ is the mass flux, h_i^* and h_f^* are the constant initial and final enthalpies per unit mass. When integrating this result over time, we find the work performed during the process i $\longrightarrow$ f,

$$W_{if} = M\left(h_f^* - h_i^*\right)$$

Thus, the work W_{12} performed by the pump on the fluid is,

$$W_{12} = M \left(h_2^* - h_1^* \right)$$

and the work W_{45} performed by the fluid on the turbine is,

$$W_{45} = M \left(h_5^* - h_4^* \right)$$

c) Using the definition of the efficiency (7.38), we obtain,

$$\eta_R = -\frac{W}{Q^+} = -\frac{W_{12} + W_{45}}{Q^+} = -\frac{h_2^* - h_1^* + h_5^* - h_4^*}{h_4^* - h_2^*} = 1 - \frac{h_5^* - h_1^*}{h_4^* - h_2^*}$$

Chemistry and Electrochemistry

The exercises given in the last section of Chapter 8 are presented here with their solutions.

8.1 Oxidation of Ammonia

The chemical reaction of ammonia oxidation reads,

$$4\,NH_3 + 5\,O_2 \rightarrow 4\,NO + 6\,H_2O$$

Consider that initially this reaction is taking place with $N_{NH_3}(0)$ moles of NH_3 and $N_{O_2}(0)$ moles of O_2. Find the amount of NH_3, O_2, NO and H_2O at the end of the reaction.

Numerical Application:

$N_{NH_3}(0) = 2\,mol$, $N_{O_2}(0) = 2\,mol$, $N_{NO}(0) = 0\,mol$ and $N_{H_2O}(0) = 0\,mol$.

Solution:

The stoichiometric coefficients of the chemical reaction,

$$4\,NH_3 + 5\,O_2 \rightarrow 4\,NO + 6\,H_2O$$

are $\nu_{NH_3} = -4$, $\nu_{O_2} = -5$, $\nu_{NO} = 4$ and $\nu_{H_2O} = 6$. According to relation (8.6), the time evolution of the number of moles of a substance A is given by,

$$N_A(t) = N_A(0) + \nu_A\,\xi(t)$$

where $\xi(t)$ is the extent of the chemical reaction. The molecular oxygen O_2 will run out first. The reaction will stop at time t_f given by,

$$N_{O_2}(t_f) = N_{O_2}(0) + \nu_{O_2}\,\xi(t_f) = 0$$

which implies that the final extent of the reaction yields,

$$\xi(t_f) = -\frac{N_{O_2}(0)}{\nu_{O_2}} = \frac{2}{5}\,mol$$

At the end of the reaction, the amount of NH_3 is,

$$N_{NH_3}(t_f) = N_{NH_3}(0) + \nu_{NH_3}\,\xi(t_f) = 0.4\,mol$$

the amount of NO is,

$$N_{NO}(t_f) = N_{NO}(0) + \nu_{NO}\,\xi(t_f) = 1.6\,mol$$

and the amount of H_2O is,

$$N_{H_2O}(t_f) = N_{H_2O}(0) + \nu_{H_2O}\, \xi(t_f) = 2.4\,\text{mol}$$

8.2 Acetylene Lamp

Acetylene (C_2H_2) can be produced through a chemical reaction between water (H_2O) and calcium carbide (CaC_2):

$$CaC_2\,(s) + 2\,H_2O\,(l) \;\rightarrow\; C_2H_2\,(g) + Ca(OH)_2\,(s)$$

where (s) and (l) indicate whether the substance is solid or liquid. A cave explorer considers using an acetylene torch, known to consume this gas at a volume rate of $\dot{V}$ (at standard conditions of temperature and pressure). As the expedition is due to last a time t, find the amount of calcium carbide that the explorer would need if he chose this type of light source. Determine the amount of water used by this torch during this time.

Numerical application:

$T = 0\,^\circ C$, $p = 10^5$ atm, $\dot{V} = 10$ l/h and $t = 8$ h.

Solution:

Acetylene can be considered as an ideal gas. Thus, at standard conditions of pressure and temperature, the number of moles of C_2H_2 needed for this exploration is,

$$N_{C_2H_2} = \frac{p\,V}{R\,T} = \frac{p\,\dot{V}t}{R\,T} = 3.52\,\text{mol}$$

Since the stoichiometric coefficients of acetylene and calcium carbide are equal and opposite, i.e. $\nu_{C_2H_2} = -\nu_{CaC_2} = 1$, it means that $N_{CaC_2} = N_{C_2H_2} = 3.52$ mol of calcium carbide are consumed and produce 3.52 mol of acetylene. The mass m_{CaC_2} of calcium carbide is the product of the number of moles $N_{C_2H_2}$ and the molar mass $M_{CaC_2} = 64\,g$,

$$m_{CaC_2} = N_{CaC_2} M_{CaC_2} = N_{C_2H_2} M_{CaC_2} = 225\,\text{g}$$

Since the stoichiometric coefficient of water is twice the opposite stoichiometric coefficient of acetylene, i.e. $\nu_{H_2O} = -2\,\nu_{C_2H_2} = 2$, it means that $N_{H_2O} = 2N_{C_2H_2} = 7.05$ mol of water are consumed in order to produce 3.52 mol of acetylene. The mass m_{H_2O} of water is the product of the number of moles N_{H_2O} and the molar mass $M_{H_2O} = 18\,g$,

$$m_{H_2O} = N_{H_2O} M_{H_2O} = 2\,N_{C_2H_2} M_{H_2O} = 127\,\text{g}$$

8.3 Coupled Chemical Reactions

The oxidation of methane can take place according to either one of the following reactions:

$$CH_4 + 2\,O_2 \quad \xrightarrow{\;1\;} \quad CO_2 + 2\,H_2O$$
$$2\,CH_4 + 3\,O_2 \quad \xrightarrow{\;2\;} \quad 2\,CO + 4\,H_2O$$

When the reactions stop at time t_f because all the methane is burned, the total mass of the products (CO_2, CO, H_2O) is

$$m(t_f) = m_{CO_2}(t_f) + m_{CO}(t_f) + m_{H_2O}(t_f)$$

Determine the initial mass of methane $m_{CH_4}(0)$ in terms of the total mass of the products $m(t_f)$ and the mass of water $m_{H_2O}(t_f)$.

Numerical Application:

$m(t_f) = 24.8\,g$ and $m_{H_2O}(t_f) = 12.6\,g$.

Solution:

The stoichiometric coefficients of the chemical reaction,

$$CH_4 + 2\,O_2 \ \overset{1}{\longrightarrow} \ CO_2 + 2\,H_2O$$
$$2\,CH_4 + 3\,O_2 \ \overset{2}{\longrightarrow} \ 2\,CO + 4\,H_2O$$

are $\nu_{1,CH_4} = -1$, $\nu_{1,O_2} = -2$, $\nu_{1,CO_2} = 1$, $\nu_{1,H_2O} = 2$, $\nu_{2,CH_4} = -2$, $\nu_{1,O_2} = -3$, $\nu_{2,CO} = 2$ and $\nu_{2,H_2O} = 4$. According to relation (8.11), the time evolution of the number of moles of a substance A taking part in the coupled reactions 1 and 2 is given by,

$$N_A(t) = N_A(0) + \nu_{1,A}\,\xi_1(t) + \nu_{2,A}\,\xi_2(t)$$

The reactions stop at time t_f when all the methane is burnt. Thus, according to relation (8.11), we write,

$$N_{CH_4}(t_f) = N_{CH_4}(0) + \nu_{1,CH_4}\,\xi_1(t_f) + \nu_{2,CH_4}\,\xi_2(t_f) = 0$$

which implies that the initial number of moles of methane yields,

$$N_{CH_4}(0) = \xi_1(t_f) + 2\,\xi_2(t_f)$$

Initially, there is no water, i.e. $N_{H_2O}(0) = 0$. Thus, according to relation (8.11), we write,

$$N_{H_2O}(t_f) = N_{H_2O}(0) + \nu_{1,N_{H_2O}}\,\xi_1(t_f) + \nu_{2,N_{H_2O}}\,\xi_2(t_f)$$

Since $m_{H_2O}(t_f) = 12.6\,g$ and $M_{H_2O} = 18\,g$, we obtain the following identity,

$$N_{H_2O}(t_f) = \frac{m_{H_2O}(t_f)}{M_{H_2O}} = 2\,\xi_1(t_f) + 4\,\xi_2(t_f) = 0.7\,mol$$

Initially, there is no carbon dioxide and monoxide, i.e. $N_{CO_2}(0) = 0$ and $N_{CO}(0) = 0$. Thus, according to relation (8.6), the time evolution of the carbon dioxide and monoxide are given by,

$$N_{CO_2}(t_f) = N_{CO_2}(0) + \nu_{1,CO_2}\,\xi_1(t_f) = \xi_1(t_f)$$
$$N_{CO}(t_f) = N_{CO}(0) + \nu_{2,CO}\,\xi_2(t_f) = 2\,\xi_2(t_f)$$

The total mass $m(t_f)$ of products of their number of moles times their molar masses,

$$m(t_f) = N_{CO_2}(t_f)\,M_{CO_2} + N_{CO}(t_f)\,M_{CO} + N_{H_2O}(t_f)\,M_{H_2O} = 24.8\,g$$

which implies that,

$$\frac{M_{CO_2}}{m(t_f)}\,N_{CO_2}(t_f) + \frac{M_{CO}}{m(t_f)}\,N_{CO}(t_f) + \frac{M_{H_2O}}{m(t_f)}\,N_{H_2O}(t_f) = 1\,mol$$

and can be recast as,

$$\left(\frac{M_{CO_2}}{m\,(t_f)} + 2\,\frac{M_{H_2O}}{m\,(t_f)} \right) \xi_1\,(t_f) + \left(2\,\frac{M_{CO}}{m\,(t_f)} + 4\,\frac{M_{H_2O}}{m\,(t_f)} \right) \xi_2\,(t_f) = 1\,\text{mol}$$

Since $m\,(t_f) = 24.8\,\text{g}$, $M_{H_2O} = 18\,\text{g}$, $M_{CO} = 28\,\text{g}$ and $M_{CO_2} = 48\,\text{g}$, we obtain the following identity,

$$3.39\,\xi_1\,(t_f) + 5.26\,\xi_2\,(t_f) = 1\,\text{mol}$$

Solving the system of equations,

$$2\,\xi_1\,(t_f) + 4\,\xi_2\,(t_f) = 0.7\,\text{mol}$$
$$3.39\,\xi_1\,(t_f) + 5.26\,\xi_2\,(t_f) = 1\,\text{mol}$$

we find that,

$$\xi_1\,(t_f) = 0.10\,\text{mol} \qquad \text{and} \qquad \xi_2\,(t_f) = 0.12\,\text{mol}$$

Since $M_{CH_4} = 16\,\text{g}$, the initial mass of methane $m_{CH_4}\,(t_f)$ burned in this reaction is,

$$m_{CH_4}\,(t_f) = N_{CH_4}\,(t_f)\,M_{CH_4} = \left(\xi_1\,(t_f) + 2\,\xi_2\,(t_f) \right) M_{CH_4} = 5.4\,\text{g}$$

8.4 Variance

The variance v is the number of degrees of freedom of a system consisting of r substances in m phases taking part in n chemical reactions. The variance v is obtained by subtracting n constraints from the number of degrees of freedom f determined by the Gibbs phase rule (6.62),

$$v = f - n = r - m - n + 2$$

The pressure p and the temperature T are not fixed. Otherwise, there are additional constraints to fix p and T. Apply this concept to the following situation.

a) Determine the variance v of methane cracking described by the chemical reaction:

$$CH_4\,(g) \leftrightarrows C\,(g) + 2\,H_2\,(g)$$

b) A system consists of three phases and there is one chemical reaction between the substances. The variance is known to be two. Determine the number r of substances in the system.

c) A system is at a fixed temperature and consists of three phases. Its variance is known to be two and there are two chemical reactions between the substances. Determine the number r of substances in the system.

Solution:

a) There are three substances (CH_4, C and H_2), i.e. $r = 3$. There is one gaseous phase, i.e. $m = 1$. There is one chemical reaction, i.e. $n = 1$. Thus, the variance v is,

$$v = r - m - n + 2 = 3$$

b) There are three phases, i.e. $m = 3$. There is one chemical reaction, i.e. $n = 1$. The variance is known to be two, i.e. $v = 2$. Thus, the number r of substances is,

$$r = v + m + n - 2 = 4$$

c) There are three phases, i.e. $m = 3$. There are two chemical reactions, i.e. $n = 2$. The variance is known to be two, i.e. $v = 2$. However, there is one additional constraint due to the fact that the temperature is fixed. In this particular case, there is one fewer degree of freedom. Thus, the variance is given by,

$$v = r - m - n + 1$$

which implies that the number r of substances is,

$$r = v + m + n - 1 = 6$$

8.5 Enthalpy of Formation

a) There are two isomers of butane: butane (C_4H_{10}) and isobutane (methylpropane) (iso-C_4H_{10}). Determine the standard enthalpy of isomerisation Δh° of butane to isobutane in terms of the enthalpies of formation of the two isomers, $h_{C_4H_{10}}$ and $h_{iso-C_4H_{10}}$.

b) The lunar module 'Eagle' of the Apollo mission was propelled using the energy released by the reaction:

$$H_2NN(CH_3)_2(l) + 2\,N_2O_4(l) \rightarrow 3\,N_2(g) + 2\,CO_2(g) + 4\,H_2O(g)$$

Determine the molar enthalpy Δh° of this exothermic reaction in terms of the enthalpies of formation of the reactants, $h_{H_2NN(CH_3)_2(l)}$, $h_{N_2O_4(l)}$ and of the products $h_{N_2(g)}$ $h_{CO_2(g)}$, h_{H_2O}.

c) The combustion of acetylene (C_2H_2) is described by the chemical reaction:

$$C_2H_2(g) + \frac{5}{2}O_2(g) \rightarrow 2\,CO_2(g) + H_2O(l)$$

Determine the enthalpy of formation $h_{C_2H_2}$ of acetylene (C_2H_2) in terms of the molar enthalpies $h_{O_2(g)}$, $h_{CO_2(g)}$, $h_{H_2O(g)}$, the molar enthalpy of the reaction Δh° and the vaporisation molar enthalpy of water h_{vap}.

Numerical Application:

a) $h_{C_4H_{10}} = -2,877\,kJ/mol$, $h_{i-C_4H_{10}} = -2,869\,kJ/mol$,

b) $h_{H_2O(g)} = -242\,kJ/mol$, $h_{CO_2(g)} = -394\,kJ/mol$,
$h_{N_2(g)} = 0\,kJ/mol$, $h_{N_2O_4(l)} = 10\,kJ/mol$, $h_{H_2NN(CH_3)_2(l)} = 52\,kJ/mol$
c) $\Delta h^\circ = -1,300\,kJ/mol$, $h_{vap} = 44\,kJ/mol$, $h_{O_2(g)} = 0\,kJ/mol$,

Solution:

a) According to Hess' law (8.53), the standard enthalpy of isomerisation Δh° of butane to isobutane is the difference between the enthalpy of formation of isobutane and the enthalpy of formation of butane,

$$\Delta h^\circ = h_{\text{iso-}C_4H_{10}} - h_{C_4H_{10}} = 8\,\text{kJ/mol}$$

which implies that the isomerisation is endothermic since $\Delta h^\circ > 0$.

b) The molar enthalpy released Δh° by this exothermic reaction is obtained by applying Hess' law (8.53),

$$\Delta h^\circ = 3\,h_{N_2(g)} + 2\,h_{CO_2(g)} + 4\,h_{H_2O(g)} - h_{H_2NN(CH_3)_2(l)} - 2\,h_{N_2O_4(l)}$$
$$= -1{,}828\,\text{kJ/mol}$$

c) The molar enthalpy released Δh° by the combustion of acetylene (C_2H_2) is obtained by applying Hess' law (8.53),

$$\Delta h^\circ = 2\,h_{CO_2(g)} + h_{H_2O(l)} - h_{C_2H_2} - \frac{5}{2}\,h_{O_2}$$

where the molar enthalpy of gaseous water $h_{H_2O(g)}$ is equal to the sum of the molar enthalpy of liquid water $h_{H_2O(l)}$ and the vaporisation molar enthalpy of water h_{vap},

$$h_{H_2O(l)} = h_{H_2O(g)} + h_{\text{vap}}$$

Thus, the enthalpy of formation $h_{C_2H_2}$ of acetylene (C_2H_2) is given by,

$$h_{C_2H_2} = 2\,h_{CO_2(g)} + h_{H_2O(g)} - \frac{5}{2}\,h_{O_2} - \Delta h^\circ - h_{\text{vap}} = 226\,\text{kJ/mol}$$

8.6 Work and Heat of a Chemical Reaction

Steel wool is placed inside a cylinder filled with molecular oxygen O_2, considered as an ideal gas. A piston ensures a constant pressure of the gas. The steel wool reacts with the molecular oxygen to form iron rust Fe_2O_3,

$$2\,Fe + \frac{3}{2}\,O_2 \;\rightarrow\; Fe_2O_3$$

The reaction is slow, so that the gas remains at ambient temperature T_0. Determine the heat Q_{if}, the work W_{if} and the internal energy variation ΔU_{if} in terms of the enthalpy of reaction ΔH_{if} for a reaction involving two moles of iron.

Numerical Application:

$\Delta H_{if} = -830\,\text{kJ}$, $T_0 = 25^\circ\text{C}$.

Solution:

Since the system is coupled to a work reservoir at constant pressure, according to the relation (4.61) the heat transfer Q_{if} is equal to the enthalpy of reaction,

$$Q_{if} = \Delta H_{if} = -830\,\text{kJ}$$

According to relation (2.28), the work W_{if} performed on the ideal gas at constant pressure p_0 is given by,

$$W_{if} = - \int_{V_i}^{V_f} p_0 \, dV = - p_0 \int_{V_i}^{V_i - \Delta V} dV = p_0 \, \Delta V$$

When two moles of iron are consumed $3/2$ moles of molecular oxygen O_2 are oxidised, i.e. $\Delta N = 3/2$. The change in volume of the molecular oxygen ΔV at constant pressure p_0 is expressed as,

$$p_0 \, \Delta V = \Delta N R \, T_0$$

Thus, the work W_{if} is recast as,

$$W_{if} = p_0 \, \Delta V = \Delta N R \, T_0 = 3,715 \, \text{kJ}$$

According to the first law (1.44), the internal energy variation ΔU_{if} is given by,

$$\Delta U_{if} = W_{if} + Q_{if} = 2,885 \, \text{kJ}$$

8.7 Mass Action Law: Esterification

The Fischer esterification reaction is given by,

$$R\text{-}(C{=}O)\text{-}OH + R\text{-}OH \; \rightleftarrows \; R\text{-}(C{=}O)\text{-}OR + H_2O$$

Determine the equilibrium constant K of this reaction in terms of the concentrations of the reactants $c_{R\text{-}(C{=}O)\text{-}OH}$, $c_{R\text{-}OH}$ and of the products $c_{R-(C{=}O)-OR}$ and c_{H_2O} at equilibrium.

Numerical Application:

$c_{R\text{-}(C{=}O)\text{-}OH} = 1/3$, $c_{R\text{-}OH} = 1/3$, $c_{R-(C{=}O)-OR} = 2/3$, $c_{H_2O} = 2/3$.

Solution:

In this reaction, the stoichiometric coefficients of the products are equal to 1 and the stoichiometric coefficients of the reactants are equal to -1. By applying the mass action law (8.80), we determine the equilibrium constant,

$$K = \frac{c_{R-(C{=}O)-OR} \, c_{H_2O}}{c_{R\text{-}(C{=}O)\text{-}OH} \, c_{R\text{-}OH}} = 4$$

8.8 Mass Action Law: Carbon Monoxide

In a reactor of volume V_0, initially empty, solid carbon is introduced in an excess amount together with $N_{CO_2(g)}(0)$ moles of carbon dioxide. The reactor is brought to temperature T_0 and the system reaches a chemical equilibrium,

$$CO_2(g) + C(s) \; \rightleftarrows \; 2 \, CO(g)$$

At equilibrium, which occurs at time $t = t_f$, the density of the gases relative to air is δ. Determine:

a) the pressure $p(t_f)$ in the reactor.
b) the equilibrium constant K.
c) the variance v as defined in exercise (8.4).

Numerical Application:

$V_0 = 1\,1$, $T_0 = 1,000°C$, $N_{CO_2(g)}(0) = 0.1$ mol, $\delta = 1.24$, $M_{air} = 29$ g.

Solution:

a) The stoichiometric coefficients of the reaction,

$$CO_2(g) + C(s) \rightleftarrows 2\,CO(g)$$

are $\nu_{CO_2(g)} = -1$, $\nu_{C(s)} = -1$ and $\nu_{CO(g)} = 2$. According to relation (8.6), the time evolution of the number of moles of $CO_2(g)$ and $CO(g)$ are given by,

$$N_{CO_2(g)}(t) = N_{CO_2(g)}(0) - \xi(t)$$
$$N_{CO(g)}(t) = 2\,\xi(t)$$

since $N_{CO(g)}(0) = 0$. The total number of moles of air is equal to the total number of moles of gas,

$$N_{air}(t) = N_{gas}(t) = N_{CO_2(g)}(t) + N_{CO(g)}(t) = N_{CO_2(g)}(0) + \xi(t)$$

The density δ of the gases relative to air at chemical equilibrium occurring at $t = t_f$ is,

$$\delta = \frac{N_{CO_2(g)}(t_f)\,M_{CO_2} + N_{CO(g)}(t_f)\,M_{CO}}{N_{gas}(t_f)\,M_{air}}$$

$$= \frac{\left(N_{CO_2(g)}(0) - \xi(t_f)\right)M_{CO_2} + 2\,\xi(t_f)\,M_{CO}}{\left(N_{CO_2(g)}(0) + \xi(t_f)\right)M_{air}}$$

Thus, the extent of the reaction is,

$$\xi(t_f) = \frac{N_{CO_2(g)}(0)\left(M_{CO_2} - M_{air}\,\delta\right)}{M_{air}\,\delta + M_{CO_2} - 2\,M_{CO}} = 0.043\,\text{mol}$$

since $M_{CO} = 28$ g and $M_{CO_2} = 48$ g. The gas pressure $p(t_f)$ inside the reactor is given by the ideal gas law,

$$p(t_f) = \frac{N_{gas}(t_f)\,R\,T_0}{V_0} = \frac{\left(N_{CO_2(g)}(0) + \xi(t_f)\right)R\,T_0}{V_0} = 1.51 \cdot 10^6\,\text{Pa}$$

b) The concentrations of gases at equilibrium are,

$$c_{CO} = \frac{N_{CO(g)}(t_f)}{N_{gas}(t_f)} = \frac{2\,\xi(t_f)}{N_{CO_2(g)}(0) + \xi(t_f)}$$

$$c_{CO_2} = \frac{N_{CO_2(g)}(t_f)}{N_{gas}(t_f)} = \frac{N_{CO_2(g)}(0) - \xi(t_f)}{N_{CO_2(g)}(0) + \xi(t_f)}$$

By applying the mass action law (8.80), we determine that the equilibrium constant is,

$$K = \frac{(c_{CO})^2}{c_{CO_2}} = \frac{4\,\xi^2\,(t_f)}{\left(N_{CO_2(g)}\,(0)\right)^2 - \xi^2\,(t_f)} = 0.91$$

c) There are three substances, $CO(g)$, $CO_2(g)$, $C(S)$, i.e. $r = 3$, two phases (g), (s), i.e. $m = 2$, and one chemical reaction, i.e. $n = 1$. Since the temperature T_0 is fixed, there is an additional constraint. The expression for the variance is,

$$v = r - m - n + 1 = 1$$

8.9 Entropy of Mixing

A gas container of fixed volume V is divided into two compartments by an impermeable fixed wall. One compartment contains ideal gas 1, the other ideal gas 2. Both sides are at pressure p and temperature T. When the wall is removed, the system reaches equilibrium. During this process, going from an initial state i to a final state f, the system is held at constant temperature T. There is no chemical interaction between the two gases. Therefore, the mixture is an ideal gas also.

a) Determine the internal energy variation ΔU_{if} during this process.
b) Show that the total entropy variation ΔS_{if} is given by (Fig. 8.1),

$$\Delta S_{if} = -\,(N_1 + N_2)\,R\sum_{A=1}^{2} c_A \ln (c_A)$$

Solution:

a) The internal energy remains constant during this process, because the temperature of the two ideal gases is constant and the internal energy of an ideal gas is proportional to

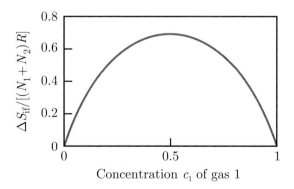

Figure 8.1   Entropy of mixing, as a function of the concentration c_1 of gas 1.

its temperature. Note that the mixture of two non-interacting ideal gases is also an ideal gas. Thus,

$$\Delta U_{if} = U_f - U_i = 0$$

c) According to the Euler equation (4.7), we express the initial internal energy $U_i(S_1, S_2, N_1, N_2)$ and the final internal energy $U_f(S, N_1, N_2)$ in terms of their state variables,

$$U_i(S_1, S_2, N_1, N_2) = T(S_1 + S_2) + \sum_{A=1}^{2} \mu_A(T,p) N_A$$

$$U_f(S, N_1, N_2) = TS + \sum_{A=1}^{2} \mu_A(T,p,c_A) N_A$$

Since the internal energy remains constant,

$$\Delta U_{if} = T(S - S_1 - S_2) + \sum_{A=1}^{2} \Big(\mu_A(T,p,c_A) - \mu_A(T,p) \Big) N_A = 0$$

Using the expression for the entropy variation,

$$\Delta S_{if} = S_f - S_i = S - S_1 - S_2$$

and the chemical potential of an ideal gas mixture (8.68),

$$\mu_A(T,p,c_A) = \mu_A(T,p) + RT \ln(c_A)$$

the previous relation can be recast as,

$$T\Delta S_{if} + RT \sum_{A=1}^{2} \ln(c_A) N_A = 0$$

Using expression (8.35) for the concentration $c_A = N_A / (N_1 + N_2)$, the entropy of mixing ΔS_{if} is found to be,

$$\Delta S_{if} = - (N_1 + N_2) R \sum_{A=1}^{2} c_A \ln(c_A)$$

where $\Delta S_{if} \geq 0$ since $0 \leq c_A \leq 1$. This expression for the entropy variation is analogous to the *von Neuman* entropy.

8.10 Raoult's Law

A container at a pressure p and temperature T contains two substances 1 and 2, present both in liquid and gas phases. Estimate the partial pressure p_A of substance A in the gas phase ($A = 1, 2$) as a function of the concentrations $c_1^{(\ell)}$ and $c_2^{(\ell)}$ of substances 1 and 2 in the liquid phase. Raoult's law relates the partial pressure p_A of substance A to the saturation pressure p_A°,

$$p_A = p_A^\circ \, c_A^{(\ell)}$$

where the saturation pressure p_A° is the pressure that the pure substance A would have in the gas phase in equilibrium with the liquid phase at temperature T. Establish Raoult's law by assuming that the liquid and gas mixtures can be treated as ideal mixtures (§ 8.5.2) and by considering that molar volumes in the liquid phase are negligible compared to molar volumes in the gaseous phase.

Solution:

The chemical potentials of substance A in the liquid and gas phases depend on the concentration $c_A^{(g)}$ in the gas phase and $c_A^{(\ell)}$ in the liquid phase according to the ideal mixture mixing law (8.68),

$$\mu_A^{(\ell)}\left(T,p,c_A^{(\ell)}\right) = \mu_A^{(\ell)}(T,p) + RT\ln\left(c_A^{(\ell)}\right)$$

$$\mu_A^{(g)}\left(T,p,c_A^{(g)}\right) = \mu_A^{(g)}(T,p) + RT\ln\left(c_A^{(g)}\right)$$

Here, $\mu_A^{(\ell)}(T,p)$ and $\mu_A^{(g)}(T,p)$ are the chemical potential of the pure substance in the liquid and gas phases. When a concentration appears in the argument, then the substance is part of a mixture. Since the problem refers to the saturation pressure p_A°, we want to introduce it in the relations above. For the gaseous phase, we simply apply relation (8.58) and write,

$$\mu_A^{(g)}\left(T,p,c_A^{(g)}\right) = \mu_A^{(g)}(T,p_A^\circ) + RT\ln\left(\frac{p}{p_A^\circ}\right) + RT\ln\left(c_A^{(g)}\right)$$

For the liquid phase, we turn to relation (8.85), that was established for incompressible liquids, and write,

$$\mu_A^{(\ell)}\left(T,p,c_A^{(\ell)}\right) = \mu_A^{(\ell)}(T,p_A^\circ) + (p - p_A^\circ)\,v_A^{(\ell)} + RT\ln\left(c_A^{(\ell)}\right)$$

The equilibrium condition for substance A in the mixture reads,

$$\mu_A^{(g)}\left(T,p,c_A^{(g)}\right) = \mu_A^{(\ell)}\left(T,p,c_A^{(\ell)}\right)$$

The saturation pressure p_A° is defined by the equilibrium between the liquid and the gas phases of the pure substance, which is characterised by,

$$\mu_A^{(g)}(T,p_A^\circ) = \mu_A^{(\ell)}(T,p_A^\circ)$$

Therefore, the equality of the chemical potentials of the substance A in the gas and liquid phases yields,

$$RT\ln\left(\frac{p}{p_A^\circ}\right) + RT\ln\left(c_A^{(g)}\right) = (p - p_A^\circ)\,v_A^{(\ell)} + RT\ln\left(c_A^{(\ell)}\right)$$

which can be recast as,

$$RT\ln\left(\frac{p\,c_A^{(g)}}{p_A^\circ\,c_A^{(\ell)}}\right) = (p - p_A^\circ)\,v_A^{(\ell)}$$

According to relation (8.67), the partial pressure of substance A in the gaseous phase is $p_A = p\, c_A^{(g)}$, and according to relation (8.89),

$$(p - p_A^\circ)\, v_A^{(g)} = RT$$

which implies that,

$$\ln\left(\frac{p_A}{p_A^\circ\, c_A^{(\ell)}}\right) = \frac{v_A^{(\ell)}}{v_A^{(g)}}$$

Since the molar volume in the liquid phase is negligible compared to the molar volumes in the gas phase, i.e. $v_A^{(\ell)} \ll v_A^{(g)}$,

$$\ln\left(\frac{p_A}{p_A^\circ\, c_A^{(\ell)}}\right) \simeq 0$$

Thus, we recover Raoult's law,

$$p_A = p_A^\circ\, c_A^{(\ell)}$$

8.11 Boiling Temperature of Salt Water

Consider a mixture of water and salt with a low salt concentration. Use the ideal mixture law (8.68) to evaluate the chemical potential of water in the salt solution. Recall that according to relation (8.51), for any substance A in any given phase, $\mu_A(T) = h_A - T s_A$. Assume that near the boiling temperature T_0 of pure water, the molar enthalpy h_A and the molar entropy s_A of the liquid and vapour phases do not depend on temperature. Determine the boiling temperature variation $T - T_0$ as a function of salt concentration c_A.

Solution:

Since the salt concentration is c_A, the water concentration is $1 - c_A$, where $c_A \ll 1$. Since the mixture is assumed to be ideal, the chemical potential of water is written as,

$$\mu_A^{(\ell)}(T, 1 - c_A) = \mu_A^{(\ell)}(T) + RT \ln(c_A) \simeq \mu_A^{(\ell)}(T) - RT c_A$$

When water vapour and salt water are at equilibrium, the chemical potentials of water in the liquid and gaseous phases are equal. Thus, we have for pure water at boiling temperature T_0,

$$\mu_A^{(\ell)}(T_0) = \mu_A^{(g)}(T_0)$$

When the salt water solution is at boiling temperature T, the same equilibrium condition reads,

$$\mu_A^{(\ell)}(T, 1 - c_A) = \mu_A^{(g)}(T)$$

The difference between these two conditions reads,

$$\mu_A^{(\ell)}(T, 1 - c_A) - \mu_A^{(\ell)}(T_0) = \mu_A^{(g)}(T) - \mu_A^{(g)}(T_0)$$

Using the ideal mixture relation, it can be recast as,

$$\mu_A^{(\ell)}(T) - \mu_A^{(\ell)}(T_0) - RTc_A = \mu_A^{(g)}(T) - \mu_A^{(g)}(T_0)$$

Now, we express the chemical potentials in terms of the molar enthalpies and the molar entropies,

$$h_A^{(\ell)} - Ts_A^{(\ell)} - h_A^{(\ell)} + T_0 s_A^{(\ell)} - RTc_A = h_A^{(g)} - Ts_A^{(g)} - h_A^{(g)} - T_0 s_A^{(g)}$$

which reduces to,

$$(T - T_0) s_A^{(\ell)} + RTc_A = (T - T_0) s_A^{(g)}$$

Thus, the boiling temperature variation in the presence of salt is,

$$T - T_0 = \frac{RTc_A}{s_A^{(g)} - s_A^{(\ell)}}$$

8.12 Battery Potential

Apply the general definition of the battery potential,

$$\Delta\varphi = -\frac{1}{zF_F} \sum_A \nu_{aA}\,\mu_A$$

to the Daniell cell (§ 8.7.4) and show that it yields relation (8.108). Show that the battery potential can be written as,

$$\Delta\varphi = \frac{\mathcal{A}_a}{zF_F} = -\frac{\Delta_a G}{zF_F} = -\frac{\Delta_a H - T\Delta_a S}{zF_F}$$

where

$$\Delta_a H = \sum_A \nu_{aA}\,h_A \qquad \text{and} \qquad \Delta_a S = \sum_A \nu_{aA}\,s_A$$

Solution:

Apart from the electrons which are produced at the anode and consumed at the cathode inside the Daniell cell, there are four other substances in this electrochemical reactions, i.e. Cu^{2+}, Cu, Zn^{2+} and Zn. Thus, the battery potential is given by,

$$\Delta\varphi = -\frac{1}{zF_F} \sum_A \nu_{aA}\,\mu_A$$

$$= -\frac{1}{zF_F} \left(\nu_{+Cu^{2+}}\,\mu_{Cu^{2+}} + \nu_{+Cu}\,\mu_{Cu} + \nu_{-Zn^{2+}}\,\mu_{Zn^{2+}} + \nu_{-Zn}\,\mu_{Zn} \right)$$

where the $+$ sign labels the reduction occurring at the cathode and the $-$ sign labels the oxidation occurring at the anode. According to the redox equations at equilibrium (8.96) and (8.95), the stoichiometric coefficients $\nu_{+Cu^{2+}} = -1, \nu_{+Cu} = 1, \nu_{-Zn^{2+}} = 1, \nu_{-Zn} = -1$, and the electrovalence is $z = 2$. Thus, the battery potential reduces to relation (8.108),

$$\Delta\varphi = \frac{1}{2F_F} \left((\mu_{Cu^{2+}} - \mu_{Cu}) - (\mu_{Zn^{2+}} - \mu_{Zn}) \right)$$

According to relation (8.18),

$$\Delta\varphi = -\frac{1}{zF_F} \sum_A \nu_{aA}\, \mu_A = \frac{\mathcal{A}_a}{zF_F}$$

In view of relation (8.16),

$$\Delta\varphi = -\frac{1}{zF_F} \sum_A \nu_{aA}\, \mu_A = -\frac{\Delta_a G}{zF_F}$$

According to relation (8.51),

$$\Delta\varphi = -\frac{1}{zF_F} \sum_A \nu_{aA}\, \mu_A = -\sum_A \nu_{aA}\, (h_A - T s_A) = -\frac{\Delta_a H - T\Delta_a S}{zF_F}$$

8.13 Thermogalvanic Cell

Consider an electrochemical cell made of two half-cells that are identical except that they are maintained at different temperatures. This is called a thermogalvanic cell. Determine the thermogalvanic coefficient,

$$\alpha = \frac{\partial \Delta\varphi}{\partial T} = \frac{\Delta_a S}{zF_F}$$

using the definition of the battery potential introduced in exercise (8.12).

Solution:

According to the definition of the battery potential introduced in exercise (8.12), the thermogalvanic coefficient is given by,

$$\alpha = \frac{\partial \Delta\varphi}{\partial T} = -\frac{1}{zF_f} \frac{\partial}{\partial T} \left(\Delta_a H - T\Delta_a S \right) = \frac{\Delta_a S}{zF_F}$$

Note that a typical value for the Seebeck coefficient of a metal (equation 11.86) is much smaller than a typical value of the thermogalvanic coefficient. Thermogalavanic cells have been considered to convert heat to energy.[8]

8.14 Gas Osmosis

An isolated system consists of two rigid subsystems of volumes V_1 and V_2 separated by a rigid and porous membrane. Helium (He) can diffuse through the membrane, but oxygen (O_2) cannot. We label the gases as A for helium and B for oxygen. The whole system is in thermal equilibrium at all times. Each gas can be considered an ideal gas, i.e. they satisfy

[8] S. W. Lee, Y. Yang, H.-W. Lee, H. Ghasemi, D. Kraemer, G. Chen, Y. Cui, *An Electrochemical System for Efficiently Harvesting Low-Grade Heat Energy*, Nat. Commun. **5**, 3942 (2014).

Initial

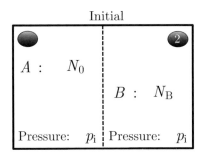

Final

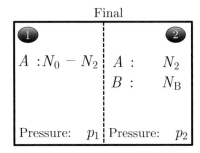

Figure 8.2 A system is divided into two by an osmotic membrane which lets substance A diffuse through it, but not substance B.

the equations of state (5.46) and (5.47), namely, $pV = NRT$ and $U = cNRT$. The gas mixture obeys the ideal mixture relation (8.68), that is,

$$\mu_A (T, p, c_A) = \mu_A (T, p) + RT \ln (c_A)$$
$$\mu_B (T, p, c_B) = \mu_B (T, p) + RT \ln (c_B)$$

where $\mu_A (T, p)$ and $\mu_B (T, p)$ are the chemical potentials of substances A and B when they are pure, c_A and c_B are the concentrations of A and B. Initially, the system contains N_0 moles of helium in subsystem 1, and N_B moles of oxygen in subsystem 2 (Fig. 8.2). The numbers of moles N_0 and N_B are chosen so that the initial pressure p_i is the same in both subsystems. At all times, each subsystem is assumed to be homogeneous. Designate by N_1 and N_2 the number of moles of helium in subsystems 1 and 2, respectively.

a) At equilibrium, show that $\mu_A (T, p_1) = \mu_A (T, p_2, c_A)$.
b) Deduce from the previous result a relation between the pressures p_1 and p_2 when the two sub-systems reach equilibrium. Express c_A, p_1 and p_2 in terms of N_2. Determine p_1 and p_2 in terms of the initial pressure p_i under the condition of equal volume, i.e. $V_1 = V_2 = V_0$.

Solution:

a) Applying Gibbs' relation (2.12) to the helium in each subsystem yields,

$$\dot{U}_1 = T\dot{S}_1 + \mu_1 \dot{N}_1 \qquad \text{and} \qquad \dot{U}_2 = T\dot{S}_2 + \mu_2 \dot{N}_2$$

which implies that,

$$\dot{S} = \dot{S}_1 + \dot{S}_2 = \frac{1}{T} (\dot{U}_1 + \dot{U}_2) - \frac{\mu_1}{T} \dot{N}_1 - \frac{\mu_2}{T} \dot{N}_2$$

Since the system is isolated $\dot{U} = 0$, which implies that $\dot{U}_1 = -\dot{U}_2$. The helium conservation law implies that $\dot{N}_1 = -\dot{N}_2$. Thus,

$$\dot{S} = -\frac{\mu_1 - \mu_2}{T} \dot{N}_1$$

and equivalently,

$$\frac{\partial S}{\partial N_1} = -\frac{\mu_1 - \mu_2}{T}$$

According to the second law, the total entropy S of the system reaches a maximum at equilibrium. Thus, at equilibrium,

$$\frac{\partial S}{\partial N_1} = 0 \quad \text{(equilibrium)}$$

which implies that $\mu_1 \equiv \mu_A(T, p_1)$ is equal to $\mu_2 \equiv \mu_A(T, p_2, c_A)$,

$$\mu_A(T, p_1) = \mu_A(T, p_2, c_A) \quad \text{(equilibrium)}$$

b) Using the ideal mixture relation (8.68),

$$\mu_A(T, p_2, c_A) = \mu_A(T, p_2) + RT \ln(c_A)$$

which implies that at chemical equilibrium,

$$\mu_A(T, p_1) = \mu_A(T, p_2) + RT \ln(c_A)$$

Moreover, according to relation (8.58),

$$\mu_A(T, p_1) = \mu_A(T, p_2) + RT \ln\left(\frac{p_1}{p_2}\right)$$

Comparing the two previous relations and using the definition of the concentration c_A, we obtain,

$$c_A = \frac{p_1}{p_2} \quad \text{and} \quad c_A = \frac{N_2}{N_2 + N_0}$$

The ideal gas equation of state implies that,

$$p_1 = \frac{(N_0 - N_2)\, RT}{V_0} \quad \text{and} \quad p_2 = \frac{(N_0 + N_2)\, RT}{V_0}$$

Using the last four equations, we obtain,

$$N_2 = \frac{N_0}{2}$$

Thus,

$$p_1 = \frac{N_0 RT}{2\, V_0} \quad \text{and} \quad p_2 = \frac{3\, N_0 RT}{2\, V_0}$$

Moreover, since,

$$p_i = \frac{N_0 RT}{V_0}$$

we have that,

$$p_1 = \frac{1}{2} p_i \quad \text{and} \quad p_2 = \frac{3}{2} p_i$$

8.15 Osmosis Power Plant

At sea level, water from the outlet of a river is diverted to a power plant that operates on the principle of osmosis. A turbine is installed in the pipe that brings the river water to an osmotic membrane separating the clear water from the salt water of the sea. The sea water at the location of the membrane is assumed to have a constant low salt concentration c, i.e. $c \ll 1$. The pure water pressure in the river and in the sea is p_0. Because of osmosis, water is driven from the river through the turbine and then across the osmotic membrane into the sea. Just after the turbine and before the membrane the pressure is $p_1 = p_0 - \Delta p$. Calculate the mechanical power of the water flowing through the turbine,

$$P_W = \Delta p \, \dot{N} v$$

where v is the molar volume of water and $\dot{N}$ are the number of moles per unit time flowing through the osmotic membrane. The hydrodynamics of the turbine is such that we can assume,

$$\Delta p = \frac{R_H \dot{N}}{v}$$

so that $P_W = R_H \dot{N}^2$, i.e. it is similar to the form of the Joule power for electrical heating. Use the ideal mixture relation (8.68) to determine the mechanical power P_W. Since the salt concentration c is low enough, we can assume at ambient temperature that $\Delta \mu \gg RTc$. Show that the mechanical power is given by,

$$P_W = \Delta \mu \, \dot{N}$$

where $\Delta \mu$ is the chemical potential drop between the river and the sea water.

Solution:

The water chemical potential difference between the two sides of the membrane is given by,

$$\Delta \mu = \mu (p_0, 1 - c) - \mu (p_1)$$

Here, we use the notation of § 8.6. In particular, when a chemical potential does not depend on a concentration, it means that we are referring to the pure substance. According to relation (8.85), we have for pure water,

$$\mu (p_1) = \mu (p_0) + v (p_1 - p_0) = \mu (p_0) - v \Delta p$$

For the salt water, assuming $c \ll 1$, we use relation (8.84) and write,

$$\mu (p_0, 1 - c) = \mu (p_0) - RTc$$

Hence, since $\Delta \mu \gg RTc$, we find that,

$$\Delta \mu = v \Delta p - RTc \simeq v \Delta p$$

Thus, using the previous relation and the assumption $v \Delta p = R_H \dot{N}$,

$$\Delta \mu = R_H \dot{N}$$

Therefore, the mechanical power is given by,

$$P_W = R_H \dot{N}^2 = \Delta \mu \, \dot{N}$$

This expression is analogous to the electrical power, expressed as the product of a current (i.e. $\dot{N}$) and potential difference (i.e. $\Delta \mu$).

9 Matter and Electromagnetic Fields

The exercises given in the last section of Chapter 9 are presented here with their solutions.

9.1 Vapour Pressure of a Paramagnetic Liquid

At constant temperature T, show that the vapour pressure $p(T, \boldsymbol{B})$ of a paramagnetic liquid has the following dependence on the magnetic induction field $\boldsymbol{B}$,

$$p(T, \boldsymbol{B}) - p(T, \boldsymbol{0}) = -p(T, \boldsymbol{0}) \, \frac{c \, \boldsymbol{B}^2}{2 \, \mu_0 \, R \, T^2}$$

in the small field limit, i.e. $c \, \boldsymbol{B} \ll \mu_0 \, R \, T^2$. The liquid phase satisfies the Curie equation of state (9.106),

$$\boldsymbol{m}_\ell = \frac{c \, \boldsymbol{B}}{\mu_0 \, T}$$

where $\boldsymbol{m}_\ell$ is the molar magnetisation. The molar volume v_ℓ of the liquid phase can be neglected compared to the molar volume v_g of the gaseous phase, i.e. $v_\ell \ll v_g$.

Solution:

Since the gaseous phase is non-magnetic, i.e. $\boldsymbol{m}_g = \boldsymbol{0}$, and the temperature is fixed, i.e. $dT = 0$, the Gibbs-Duhem equations (9.15) for the liquid and gaseous phases read,

$$d\mu_\ell = v_\ell \, dp - \boldsymbol{m}_\ell \cdot d\boldsymbol{B} \qquad \text{and} \qquad d\mu_g = v_g \, dp$$

The vapour pressure is defined at chemical equilibrium between the liquid and gaseous phases,

$$d\mu_\ell = d\mu_g$$

which is expressed in the limit where $v_\ell \ll v_g$ as,

$$v_g \, dp + \boldsymbol{m}_\ell \cdot d\boldsymbol{B} = 0$$

According to the ideal gas equation of state (5.47),

$$v_g = \frac{R \, T}{p}$$

Using the Curie equation of state the chemical equilibrium equation implies that,

$$\frac{dp}{p} = - \frac{c\,\boldsymbol{B}}{\mu_0\,R\,T^2} \cdot d\boldsymbol{B}$$

When integrating from a zero initial magnetic induction field $\boldsymbol{B}$, we obtain

$$\ln\left(\frac{p\,(T,\boldsymbol{B})}{p\,(T,\boldsymbol{0})}\right) = - \frac{c\,\boldsymbol{B}^2}{2\,\mu_0\,R\,T^2}$$

Thus,

$$\frac{p\,(T,\boldsymbol{B})}{p\,(T,\boldsymbol{0})} = \exp\left(- \frac{c\,\boldsymbol{B}^2}{2\,\mu_0\,R\,T^2}\right)$$

In the small field limit, i.e. $c\,\boldsymbol{B} \ll \mu_0\,R\,T^2$, the previous result reduces to,

$$\frac{p\,(T,\boldsymbol{B})}{p\,(T,\boldsymbol{0})} = 1 - \frac{c\,\boldsymbol{B}^2}{2\,\mu_0\,R\,T^2}$$

Hence,

$$p\,(T,\boldsymbol{B}) - p\,(T,\boldsymbol{0}) = -p\,(T,\boldsymbol{0})\,\frac{c\,\boldsymbol{B}^2}{2\,\mu_0\,R\,T^2}$$

The dependence of vapour pressure on the magnetic induction field was observed for liquid oxygen.[9] The contribution of nuclear spins to the magnetic field dependence of the vapour pressure of He_3 was also observed and analysed.[10]

9.2 Magnetic-Field Induced Adsorption or Desorption

We consider a substance that can be either in a gaseous phase inside a rigid container or in an absorbed phase on the surface of a substrate inside it. We analyse the equilibrium reached by the substance in the gaseous and adsorbed phases. The entire system is at a fixed temperature T. We assume that when the substance is adsorbed, it acquires a magnetisation $\boldsymbol{M}$. The reason whether and why this might happen has been the subject of much research.[11] The gas has no magnetisation. Use relation (8.58) for the pressure dependence of the chemical potential of the gaseous phase and assume that the chemical potential of the adsorbed phase is independent of pressure. Find the dependence on magnetic induction field $\boldsymbol{B}$ of the pressure p.

Solution:

Let the gaseous phase be labelled 1 and the adsorbed phase labelled 2. The chemical equilibrium is characterised by the equivalence of the chemical potentials,

$$\mu_1\,(p) = \mu_2\,(\boldsymbol{B})$$.

[9] K. Nishigaki, M. Takeda, *The Effect of Magnetic Fields on the Vapor Pressure of Liquid Oxygen*, Physica B 194–196 (1994)

[10] J. Kopp, *The Vapour Pressure of Helium-3 in High Magnetic Fields*, Cryogenics 271–273 (1967).

[11] S. Ozeki, J. Miyamoto, S. Ono, C. Wakai, T. Watanabe, *Water-Solid Interactions under Steady Magnetic Fields : Magnetic Field-Induced Adsorption and Desorption of Water*, J. Phys. Chem. **100**, 4205–4212 (1996).

since the gas is assumed to be non-magnetic and the chemical potential of the substance in the adsorbed phase is independent of pressure. To first-order, this equilibrium condition can be recast as,

$$\mu_1\left(p_0\right) + \frac{\partial \mu_1}{\partial p}\left(p - p_0\right) = \mu_2\left(\boldsymbol{0}\right) + \frac{\partial \mu_2}{\partial \boldsymbol{B}} \cdot \boldsymbol{B}$$

According to relation (8.56),

$$\frac{\partial \mu_1}{\partial p} = \frac{\partial}{\partial N}\left(\frac{NRT}{p}\right) = \frac{RT}{p}$$

According to relation (9.29), the differential of the magnetic free enthalpy density of the adsorbed phase yields,

$$dg_m\left(T, N_2, \boldsymbol{B}\right) = -s_\ell\, dT + \mu_2\, dN_2 - \boldsymbol{M} \cdot d\boldsymbol{B}$$

Applying the Schwarz theorem (4.67) to the magnetic free enthalpy density $g_m\left(T, N_2, \boldsymbol{B}\right)$ we obtain,

$$\frac{\partial}{\partial \boldsymbol{B}}\left(\frac{\partial g_m}{\partial N_2}\right) = \frac{\partial}{\partial N_2}\left(\frac{\partial g_m}{\partial \boldsymbol{B}}\right)$$

which yields the Maxwell relation,

$$\frac{\partial \mu_2}{\partial \boldsymbol{B}} = -\frac{\partial \boldsymbol{M}}{\partial N_2} \equiv -\boldsymbol{m}_2$$

where $\boldsymbol{m}_2$ is the molar magnetisation of the adsorbed substance. This implies that the chemical equilibrium relation can be recast as,

$$\mu_1\left(p_0\right) + \frac{RT}{p}\left(p - p_0\right) = \mu_2\left(\boldsymbol{0}\right) - \boldsymbol{m}_2 \cdot \boldsymbol{B}$$

Applying the chemical equilibrium condition when the magnetic induction field is zero, which is,

$$\mu_1\left(p_0\right) = \mu_2\left(\boldsymbol{0}\right)$$

we find the magnetic induction field $\boldsymbol{B}$ dependence of the pressure p,

$$\frac{p - p_0}{p} = -\frac{\boldsymbol{m}_2 \cdot \boldsymbol{B}}{RT}$$

9.3 Magnetic Battery

A U-shaped tube contains a solution of paramagnetic salt, such as $CoSO_4$. One side of the U-shaped tube, considered as a subsystem 1, is in a magnetic induction field $\boldsymbol{B}$, the other side, considered as a subsystem 2 is at $\boldsymbol{B} = \boldsymbol{0}$. The initial concentration of paramagnetic Co^{++} salt is c_0 and the ideal mixture relation applies. Find the electric potential difference $\Delta\varphi$ in terms of the magnetic induction field $\boldsymbol{B}$ and the molar magnetisation $\boldsymbol{m}_1$ of electrodes made of the same material dipped in the the salt solution. Determine also the concentration ratio of paramagnetic Co^{++} salt in both subsystems at equilibrium in terms of the magnetic induction field $\boldsymbol{B}$.

Solution:

When the subsystems reach equilibrium, their chemical potentials are equal,

$$\mu_1 (c_1, \boldsymbol{B}) = \mu_2 (c_2, \boldsymbol{0})$$

where c_1 and c_2 are the concentrations of paramagnetic Co^{++} salt in both subsystems at equilibrium. To first-order, this equilibrium condition can be recast as,

$$\mu_1 (c_1, \boldsymbol{0}) + \frac{\partial \mu_1}{\partial \boldsymbol{B}} \cdot \boldsymbol{B} = \mu_2 (c_2, \boldsymbol{0})$$

According to relation (9.29), the differential of the magnetic free enthalpy density of subsystem 1 yields,

$$dg_m (T, N_1, \boldsymbol{B}) = - s_1 \, dT + \mu_1 \, dN_1 - \boldsymbol{M} \cdot d\boldsymbol{B}$$

Applying the Schwarz theorem (4.67) to the magnetic free enthalpy density $g_m (T, N_1, \boldsymbol{B})$ we obtain,

$$\frac{\partial}{\partial \boldsymbol{B}} \left(\frac{\partial g_m}{\partial N_1} \right) = \frac{\partial}{\partial N_1} \left(\frac{\partial g_m}{\partial \boldsymbol{B}} \right)$$

which yields the Maxwell relation,

$$\frac{\partial \mu_1}{\partial \boldsymbol{B}} = - \frac{\partial \boldsymbol{M}}{\partial N_1} \equiv - \boldsymbol{m}_1$$

where $\boldsymbol{m}_1$ is the molar magnetisation of the paramagnetic Co^{++} salt in subsystem 1. Thus, the equilibrium condition is recast as,

$$\mu_1 (c_1, \boldsymbol{0}) - \mu_2 (c_2, \boldsymbol{0}) = \boldsymbol{m}_1 \cdot \boldsymbol{B}$$

According to relation (8.105), the Nernst potential $\Delta \varphi$, which is the concentration-dependent potential difference between the two electrodes, yields,

$$\Delta \varphi = \mu_1 (c_1, \boldsymbol{0}) - \mu_2 (c_2, \boldsymbol{0}) = \frac{R T}{z F_F} \ln \left(\frac{c_1}{c_2} \right) = \boldsymbol{m}_1 \cdot \boldsymbol{B}$$

Thus, the concentration ratio c_1/c_2 in terms of the magnetic induction field $\boldsymbol{B}$ is given by,

$$\frac{c_1}{c_2} = \exp \left(\frac{z F_F}{R T} (\boldsymbol{m}_1 \cdot \boldsymbol{B}) \right)$$

9.4 Electrocapilarity

Electrowetting is a means of driving microfluidic motion by modifying the surface tension.[12] Here, we determine the effect of an electrostatic potential difference $\Delta \varphi$ applied

---

[12] T. M. Squires, S. R. Quake, *Microfluidics: Fluid Physics at the Nanoliter Scale*, Rev. Mod. Phys. **77**, 977 (2005).

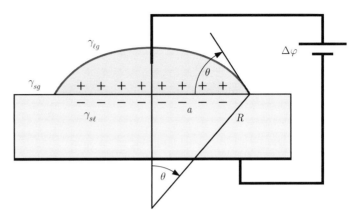

Figure 9.1   Contact angle θ for a drop for an applied electrostatic potential difference $\Delta\varphi$.

between the solid and liquid phases on the surface tension. Since the liquid is now electrically charged a double layer forms at the interface between the solid and the liquid phases. The capacitance per unit area c of this double layer is supposed to be known. Determine the electrically charged surface tension $\bar{\gamma}_{s\ell}$ in terms of the neutral surface tension $\gamma_{s\ell}$.

Solution:

According to § 1.7, the neutral surface tension $\gamma_{s\ell}$ is the internal energy per unit area of the interface between the solid and liquid phases. In this exercise, the electrostatic potential difference $\Delta\varphi$ is the state variable. Therefore, the thermodynamic potential is the electrostatic enthalpy density. In view of relation (9.69), at constant electrostatic potential difference $\Delta\varphi$, the electrically charged surface tension $\bar{\gamma}_{s\ell}$ is the sum of the neutral surface tension $\gamma_{s\ell}$ and the electrostatic enthalpy per unit surface $-1/2\,c\,\Delta\varphi^2$ where c is the capacitance per unit surface. Thus,

$$\bar{\gamma}_{s\ell} = \gamma_{s\ell} - \frac{1}{2}\,c\,\Delta\varphi^2$$

which means that the presence of an applied electrostatic potential difference $\Delta\varphi$ lowers the surface tension at the interface between the solid and liquid phases.

9.5 Magnetic Clausius–Clapeyron Equation

Some magnetic materials undergo a first-order phase transition when a magnetic induction field $\boldsymbol{B}$ is applied. These materials are of interest because some of them have a strong magnetocaloric effect, which could be used for refrigeration applications.[13] In these

---

[13] A. M. Tishin, Y. I. Spichkin, *Recent Progress in Magnetocaloric Effect : Mechanisms and Potential Applications*, Int. J. Refrigeration **37**, 223–229 (2014).

materials, at any given temperature T, there is a critical magnetic induction field $\boldsymbol{B}$ in which the two phases coexist. The two phases, labelled 1 and 2, are characterised by their molar magnetisation $\boldsymbol{m}_1$ and $\boldsymbol{m}_2$ and their molar entropies per unit volume s_1 and s_2. Show that, the temperature derivative of the magnetic induction field $\boldsymbol{B}$ satisfies the magnetic Clausius-Clapeyron equation,

$$\frac{d\boldsymbol{B}}{dT} = -\left(s_1 - s_2\right)\left(\boldsymbol{m}_1 - \boldsymbol{m}_2\right)^{-1}$$

which is analogous to the Clausius-Clapeyron equation (6.49).

Solution:

According to the Gibbs-Duhem relation (9.15) for a system with a single substance, the differentials of the chemical potentials are given by,

$$d\mu_1 = -s_1\, dT - \boldsymbol{m}_1 \cdot d\boldsymbol{B}$$
$$d\mu_2 = -s_2\, dT - \boldsymbol{m}_2 \cdot d\boldsymbol{B}$$

The coexistence between these two phases is characterised by,

$$d\mu_1 = d\mu_2$$

which implies that,

$$\left(s_1 - s_2\right) dT + \left(\boldsymbol{m}_1 - \boldsymbol{m}_2\right) \cdot d\boldsymbol{B} = 0$$

Thus, the magnetic Clausius-Clapeyron equation is found to be,

$$\frac{d\boldsymbol{B}}{dT} = -\left(s_1 - s_2\right)\left(\boldsymbol{m}_1 - \boldsymbol{m}_2\right)^{-1}$$

9.6 Magnetocaloric Effect

The magnetocaloric effect is defined by the temperature variation observed in a material under adiabatic conditions when the applied magnetic induction field $\boldsymbol{B}$ varies. The effect becomes particularly large when the material undergoes a phase transition due to the magnetic induction field variation. An application of this effect was considered for space exploration.[14] The applied magnetic induction field may be produced by permanent magnets.[15] Show that the infinitesimal temperature variation dT during an adiabatic process characterised by an infinitesimal magnetic induction field variation $d\boldsymbol{B}$ at constant pressure, is given by,

$$dT = -\frac{T}{c_B}\frac{\partial \boldsymbol{M}}{\partial T} \cdot d\boldsymbol{B}$$

where c_B is the specific heat per unit volume at constant magnetic induction field $\boldsymbol{B}$.

[14] C. Hagmann, D. J. Benfod, P. L. Richards, *Paramagnetic Salt Pill Design for Magnetic Refrigerators Used in Space Applications*, Cryogencis, **34** (3) 213–219 (1994).
[15] X. Bohigas, E. Molins, A. Roig, J. Tejada, X. X. Zhang, *Room-Temperature Magnetic Refrigerator Using Permanent Magnets*, IEEE Trans. Mag. **36** (3), 538–544 (2000).

Solution:

A reversible adiabatic process is charaterised by a constant entropy. Thus,

$$ds\left(T, \boldsymbol{B}\right) = \frac{\partial s\left(T, \boldsymbol{B}\right)}{\partial T}\, dT + \frac{\partial s\left(T, \boldsymbol{B}\right)}{\partial \boldsymbol{B}} \cdot d\boldsymbol{B} = 0$$

According to relation (9.122), the specific heat per unit volume at constant magnetic induction field $\boldsymbol{B}$ is defined as,

$$c_{\boldsymbol{B}} = T\frac{\partial s\left(T, \boldsymbol{B}\right)}{\partial T}$$

which implies that,

$$dT = -\frac{T}{c_{\boldsymbol{B}}}\frac{\partial s}{\partial \boldsymbol{B}} \cdot d\boldsymbol{B}$$

According to relation (9.29), the differential of the magnetic free enthalpy density yields,

$$dg_m\left(T, \boldsymbol{B}\right) = -s\, dT - \boldsymbol{M} \cdot d\boldsymbol{B}$$

Applying the Schwarz theorem (4.67) to the magnetic free enthalpy density $g_m\left(T, \boldsymbol{B}\right)$ we obtain,

$$\frac{\partial}{\partial \boldsymbol{B}}\left(\frac{\partial g_m}{\partial T}\right) = \frac{\partial}{\partial T}\left(\frac{\partial g_m}{\partial \boldsymbol{B}}\right)$$

which yields the Maxwell relation,

$$\frac{\partial s}{\partial \boldsymbol{B}} = \frac{\partial \boldsymbol{M}}{\partial T}$$

Thus, the temperature differential can be recast as,

$$dT = -\frac{T}{c_{\boldsymbol{B}}}\frac{\partial \boldsymbol{M}}{\partial T} \cdot d\boldsymbol{B}$$

9.7 Kelvin Probe

Two parallel plates are made of two metals 1 and 2 inside a vacuum chamber (Fig. 9.2). The chemical potentials of electrons in metals 1 and 2 are μ_1 and μ_2. When contact is established between the two metals, conduction electrons flow until equilibrium is reached. As a result of this electron flow, there is a charge Q and $-Q$ at the surface of the electrodes. These electric charges produce an electric field $\boldsymbol{E}$ between the plates. According to relation (9.67) the corresponding electrostatic potential difference $\Delta\varphi$ is related to the charge Q by,

$$Q = C\,\Delta\varphi$$

where C is the capacitance of the capacitor. The plates have a surface area A and they are separated by a distance d. It can be shown that the capacitance C is given by,

$$C = \frac{\varepsilon_0\, A}{d}$$

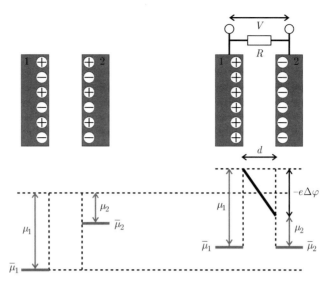

Figure 9.2   A parallel plate capacitor is made of metals 1 and 2 which have electrochemical potentials $\bar{\mu}_1$ and $\bar{\mu}_2$ and chemical potentials μ_1 and μ_2. On the left, the plates are not connected and the electrochemical potentials are not equal. On the right, when conduction electrons of charge e (sign included) are allowed to flow, the electrochemical potentials are equal. If the spacing d varies over time, electric charges flow through the resistance R and a voltage V can be measured.

where ε_0 is the electric permittivity of vacuum. For practical applications, the flow of electric charges is too difficult to detect. It is easier to make the separation distance d oscillate in time, which causes C to vary, hence Q oscillates in time, which implies that an electric current of intensity I flows through the resistance R. Determine the voltage V detected between the plates as a function of $\dot{d}$, μ_1 and μ_2.

Solution:

On the left, the electrochemical potentials $\bar{\mu}_1$ and $\bar{\mu}_2$ are different since the plates are not in contact (Fig. 9.2). On the right, the electrochemical potentials are equal since the plates are not in contact and electrochemical equilibrium is reached (§ 3.4). According to relation (9.97), the electrochemical potentials $\bar{\mu}_1$ and $\bar{\mu}_2$ are expressed in terms of the chemical potentials μ_1 and μ_2, the electrostatic potentials φ_1 and φ_2 and the electric charge $q = e$,

$$\bar{\mu}_1 = \mu_1 + e\,\varphi_1 \qquad \text{and} \qquad \bar{\mu}_2 = \mu_2 + e\,\varphi_2$$

where e is the electric charge of an electron (sign included). At chemical equilibrium,

$$\bar{\mu}_1 = \bar{\mu}_2$$

Thus, the electrostatic potential difference is given by (Fig. 9.2),

$$\Delta\varphi = \varphi_1 - \varphi_2 = \frac{\mu_2 - \mu_1}{e}$$

The intensity I of the electric current is defined as the time derivative of the electric charge Q. This is only due to the oscillation of the separation distance d in time. Thus,

$$I = \frac{dQ}{dt} = \frac{d}{dt}\left(C\,\Delta\varphi\right) = \frac{d}{dt}\left(\frac{\varepsilon_0 A}{d}\right)\Delta\varphi = -\frac{\varepsilon_0 A}{d}\frac{\dot{d}}{d}\Delta\varphi = -C\frac{\dot{d}}{d}\Delta\varphi$$

According to Ohm's law (§ 11.4.8), the detected voltage V is the product of the resistance R and the intensity I of the electric current,

$$V = RI = -RC\frac{\dot{d}}{d}\Delta\varphi = -\frac{RC}{e}\frac{\dot{d}}{d}(\mu_2 - \mu_1)$$

If the physical characteristics of the Kelvin probe are known, namely the resistance R, the capacitance C, the electric charge e, the distance d and its time derivative $\dot{d}$, then the voltage V provides a measure of the chemical potential μ_1 of the conduction electrons in metal 1 with respect to the chemical potential μ_2 of the conduction electrons in metal 2. This setup was analysed recently because of its relevance to electrocatalysis.[16]

9.8 Electromechanical Circuit

In the electromechanical system presented here (Fig. 9.3), a metallic plate 1 is fixed mechanically and a metallic plate 2 vibrates under the effect of a spring of elastic constant k, of length z and of natural length z_0. The distance between the plates is $\ell - z$ and their surface area is A. The capacitance of the parallel plate capacitor is,

$$C = \frac{\varepsilon_0 A}{\ell - z}$$

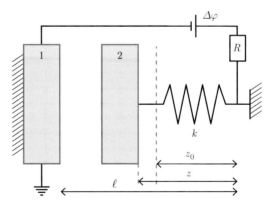

Figure 9.3 A parallel plate capacitor is made of metals 1 and 2. The metallic plate 1 is fixed mechanically. The metallic plate 2 vibrates under the effect of a spring of elastic constant k. When it vibrates, conduction electrons flow through the resistance R and the spring. The electrostatic potential difference $\Delta\varphi$ is maintained constant by the power supply.

[16] P. Peljo, J. A. Manzares, H. H. Girault, *Contact Potentials, Fermi Level Equilibriation, and Surface Charging*, Langmuir **32**, 5765–5775, (2016).

The electrostatic potential difference $\Delta\varphi$ between the plates is maintained constant by a power supply, which generates a flow of conduction electrons through the circuit. The chemical potential of the conduction electrons on plate 1 and 2 are μ_1 and μ_2. At equilibrium, we assume that the deviation of the plate $\Delta z = z - z_0$ is small enough, i.e. $\Delta z \ll z_0$. Determine this deviation Δz to first-order.

Solution:

This system consists of the parallel plate capacitor of capacitance and the spring. Since the electrostatic potential difference $\Delta\varphi$ is held constant, the thermodynamic potential is the electrostatic enthalpy H_e. The total electrostatic enthalpy of the system is the sum of the electrostatic enthalpy (9.69), the elastic energy, and the chemical energy of the conductions electrons on both plates,

$$H_e\left(\Delta\varphi, z\right) = -\frac{1}{2}C\left(\Delta\varphi\right)^2 + \frac{1}{2}k\left(z - z_0\right)^2 + \mu_1 N_1 + \mu_2 N_2$$

where N_1 and N_2 are the number of moles of conduction electrons on plates 1 and 2. The electric charges on plates 1 and 2 are $-Q$ and Q. Thus,

$$Q = C\Delta\varphi = e N_1 = -e N_2$$

where e is the electric charge of an electron (sign included). This implies that,

$$N_1 = -N_2 = \frac{C\Delta\varphi}{e} = \frac{\varepsilon_0 A}{e}\frac{\Delta\varphi}{\ell - z}$$

Thus, the total electrostatic enthalpy is recast as,

$$H_e\left(\Delta\varphi, z\right) = -\frac{1}{2}\frac{\varepsilon_0 A}{\ell - z}\left(\Delta\varphi\right)^2 + \frac{1}{2}k\left(z - z_0\right)^2 + \left(\mu_1 - \mu_2\right)\frac{\varepsilon_0 A}{e}\frac{\Delta\varphi}{\ell - z}$$

Since the potential difference $\Delta\varphi$ is maintained constant by the power supply, at equilibrium, i.e. $z = z_0 + \Delta z$, the partial derivative of the electrostatic potential energy with respect to z vanishes,

$$\left.\frac{\partial H_e}{\partial z}\right|_{z=z_0+\Delta z} = 0$$

which is written as,

$$-\frac{1}{2}\frac{\varepsilon_0 A}{\left(\ell - z_0 - \Delta z\right)^2}\left(\Delta\varphi\right)^2 + k\Delta z + \left(\mu_1 - \mu_2\right)\frac{\varepsilon_0 A}{e}\frac{\Delta\varphi}{\left(\ell - z_0 - \Delta z\right)^2} = 0$$

and recast as,

$$\Delta z\left(\ell - z_0 - \Delta z\right)^2 = \frac{\varepsilon_0 A}{k}\left(\frac{1}{2}\left(\Delta\varphi\right)^2 - \frac{\mu_1 - \mu_2}{e}\Delta\varphi\right)$$

Thus, to first-order in Δz, the deviation of the plate is given by,

$$\Delta z = \frac{\varepsilon_0 A}{k\left(\ell - z_0\right)^2}\left(\frac{1}{2}\left(\Delta\varphi\right)^2 - \frac{\mu_1 - \mu_2}{e}\Delta\varphi\right)$$

Thus, we have derived an expression for the deviation of the moving plate that depends on the electrostatic potential difference between the plates and on the chemical potential difference of the conduction electrons on them. We have assumed that the chemical potential does not depend on the concentration of conduction electrons on the plates. This assumption is valid for bulk metals. When the electrode is very small, the concentration of conduction electrons on the plate needs to be taken into account.[17]

---

[17] C. Chen, V. V. Deshpande, M. Koshino, S. Lee, A. Gondarenko, A. H. MacDonald, Ph. Kim, J. Hone, *Modulation of Mechanical Resonance by Chemical Potential Oscillation in Graphene*, Nat. Phys. **7**, 242–245 (2015).

Thermodynamics of Continuous Media

The exercises given in the last section of Chapter 10 are presented here with their solutions.

10.1 Chemical Substance Balance

Consider a homogeneous and uniform fluid which consists of different reactive chemical substances.

1. Determine the variation rate $\dot{n}_A$ of the chemical substance density A.
2. When the system is in a stationary regime, determine the condition imposed on the stoichiometric coefficients ν_{aA}.

Solution:

1. For a uniform system, the divergence of the velocity $\boldsymbol{v}$ and the divergence of the chemical current density $\boldsymbol{j}_A$ vanish,

$$\nabla \cdot \boldsymbol{v} = 0 \qquad \text{and} \qquad \nabla \cdot \boldsymbol{j}_A = 0$$

Thus, the continuity equation (10.26) for the chemical substance A reduces to,

$$\dot{n}_A = \sum_{a=1}^{n} \omega_a \nu_{aA}$$

2. For a stationary regime,

$$\dot{n}_A = 0$$

which implies that the stoichiometric coefficients satisfy the condition,

$$\sum_{a=1}^{n} \omega_a \nu_{aA} = 0$$

10.2 Pressure Time Derivative and Gradient

1. Determine the expression of the pressure time derivative.
2. Determine the expression of the pressure gradient.

Solution:

1. Taking into account equations (10.86) and (10.71), the internal energy density u is given by,

$$u = Ts - p + \sum_{A=1}^{r} \mu_A n_A + q\varphi$$

The internal energy density time derivative u is expressed as,

$$\dot{u} = s\dot{T} + T\dot{s} - \dot{p} + \sum_{A=1}^{r} (n_A \dot{\mu}_A + \mu_A \dot{n}_A) + \varphi\dot{q} + q\dot{\varphi}$$

Since the internal energy density $u(s, \{n_A\}, q)$ is a state function density, it is written as,

$$\dot{u} = \frac{\partial u}{\partial s}\dot{s} + \sum_{A=1}^{r} \frac{\partial u}{\partial n_A}\dot{n}_A + \frac{\partial u}{\partial q}\dot{q}$$

Taking into account the definitions of the intensive variables temperature T, chemical potential μ_A of substance A and electrostatic potential φ in equation (10.77), we have,

$$\dot{u} = T\dot{s} + \sum_{A=1}^{r} \mu_A \dot{n}_A + \varphi\dot{q}$$

Equating the two expressions for the internal energy density time derivative, we obtain an expression for the pressure time derivative,

$$\dot{p} = s\dot{T} + \sum_{A=1}^{r} n_A \dot{\mu}_A + q\dot{\varphi}$$

2. The internal energy density gradient u is expressed as,

$$\nabla u = s\nabla T + T\nabla s - \nabla p + \sum_{A=1}^{r} (n_A \nabla \mu_A + \mu_A \nabla n_A) + \varphi\nabla q + q\nabla\varphi$$

Since the internal energy density $u(s, \{n_A\}, q)$ is a state function density, it is written as,

$$\nabla u = \frac{\partial u}{\partial s}\nabla s + \sum_{A=1}^{r} \frac{\partial u}{\partial n_A}\nabla n_A + \frac{\partial u}{\partial q}\nabla q$$

Taking into account definitions (10.77), we can write,

$$\nabla u = T\nabla s + \sum_{A=1}^{r} \mu_A \nabla n_A + \varphi\nabla q$$

Equating the two expressions of the internal energy density gradient, we obtain an expression for the pressure gradient,

$$\nabla p = s\nabla T + \sum_{A=1}^{r} n_A \nabla \mu_A + q\nabla\varphi$$

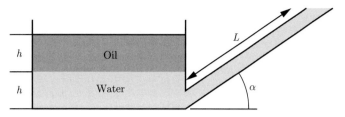

Figure 10.1   A container with equal heights of oil and water, with water filling the spout to its end.

10.3 Oil and Water Container

A container with a long spout is filled with oil and water in such a way that the spout is filled to its etop (Fig. 10.1). The water and the oil surfaces are in contact with the atmosphere. The height h of the water and oil are the same in the container. Determine the expression of the tilt angle α of the spout in terms of the mass densities m_w and m_o of water and oil.

Solution:

The hydrostatic pressure p at the bottom of the container is the sum of the atmospheric pressure p_0, the oil pressure $m_o\,g\,h$ and the water pressure $m_w\,g\,h$,

$$p = p_0 + m_o\,g\,h + m_w\,g\,h$$

The hydrostatic pressure at the bottom of the spout is equal to the hydrostatic pressure at the bottom of the container. It is the sum of the atmospheric pressure p_0 and the water pressure $m_w\,g\,L\,\sin\alpha$,

$$p = p_0 + m_w\,g\,L\,\sin\alpha$$

From these two equations, we find the expression of the tilt angle α,

$$\sin\alpha = \frac{m_w + m_o}{m_w}\,\frac{h}{L}$$

10.4 Floating Tub Stopper

A spherical weight of radius R is blocking a horizontal circular hole at the bottom of a container filled with a liquid (Fig. 10.2). The liquid of mass density m is at a height H above the hole ($H > R/2$) and the lowest point of the sphere is at a height h below it ($h < R/2$). The pressure above the liquid and below the sphere is the atmospheric pressure p_0. Determine Archimedes' force F_A exerted by the liquid on the spherical plug,

$$\boldsymbol{F}_A = - \int_S p\,d\boldsymbol{S}$$

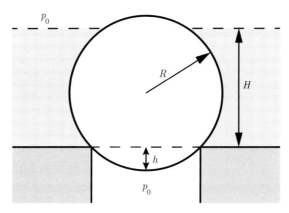

Figure 10.2 A spherical floater used as tub stopper.

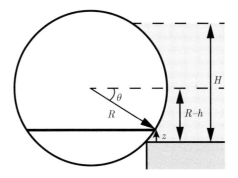

Figure 10.3 An infinitesimal spherical ring inside a floater.

Solution:

In order to find Archimedes' force $\boldsymbol{F}_A$ exerted by the liquid on the spherical plug, we first need to determine the infinitesimal force $d\boldsymbol{F}_A(z)$ exerted by the liquid on an infinitesimal spherical ring located at a height z above the bottom of the container (Fig. 10.3). According to the definition of Archimedes' force, we have,

$$d\boldsymbol{F}_A(z) = -p(z)\, d\boldsymbol{S}(z)$$

The pressure $p(z)$ exerted by the liquid at height z is given by,

$$p(z) = m\,g\,(H - z)$$

By symmetry, the vector $d\boldsymbol{S}(z)$ is oriented vertically downwards. Thus, the surface $d\boldsymbol{S}(z)$ of the infinitesimal spherical ring is,

$$d\boldsymbol{S}(z) = -R^2\,\sin\theta(z)\,d\theta \int_0^{2\pi} d\phi\,\hat{\boldsymbol{z}} = -2\pi\,R^2\,\sin\theta(z)\,d\theta\,\hat{\boldsymbol{z}}$$

where $\hat{\boldsymbol{z}}$ is the unit vector along the z-axis,

$$\sin\theta(z) = \frac{R - h - z}{R} \qquad\qquad \text{and} \qquad\qquad R\,d\theta = dz$$

Thus, the infinitesimal surface element becomes,

$$dS(z) = -2\pi (R - h - z) \, dz \, \hat{z}$$

and the infinitesimal force is recast in terms of the vertical coordinate z as,

$$dF_A(z) = 2\pi \, mg \, (H - z)(R - h - z) \, dz \, \hat{z}$$

The integral over the vertical coordinate z is written as,

$$F_A = 2\pi \, mg \int_0^H (H - z)(R - h - z) \, dz \, \hat{z}$$

or equivalently as,

$$F_A = 2\pi \, mg \int_0^H \left(H(R-h) - zH - z(R-h) + z^2 \right) dz \, \hat{z}$$

Thus, Archimedes' force is given by,

$$F_A = \pi \, mg \, H^2 \left(R - h - \frac{H}{3} \right) \hat{z}$$

10.5 Temperature Profile of the Earth's Atmosphere

Model the temperature profile $T(z)$ of the Earth's atmosphere as a function of height z. Ignore winds, clouds and the many effects due to the presence of moisture in the air and treat the air as an ideal gas. Assume that the Earth's atmosphere reaches an equilibrium mostly by matter transfer. Suppose that when an air mass moves up or down, it does so adiabatically because the air thermal conductivity is small.

a) Show that,

$$T(z) = T_0 - \frac{g}{c_p^*} \, z$$

     where c_p^* is the specific heat at constant pressure per unit mass.

b) Deduce from the temperature profile $T(z)$, the pressure profile $p(z)$ and the mass density profile $m(z)$ of the Earth atmosphere,

$$p(z) = p_0 \left(\frac{T(z)}{T_0} \right)^{c+1} \qquad \text{and} \qquad m(z) = m_0 \left(\frac{T(z)}{T_0} \right)^{c}$$

     where c is defined in equation (5.60).

Solution:

a) For an adiabatic process, the internal energy differential dU is only due to the infinitesimal work $\delta W = -p \, dV$ performed on the air,

$$dU = c \, NR \, dT = -p \, dV$$

Since the air is treated as an ideal gas,

$$U = c\,NR\,T = c\,p\,V$$

Here, the pressure p is a function of volume V, thus

$$\frac{dU}{dV} = c\,\frac{dp}{dV}\,V + c\,p = -p$$

which implies that,

$$\frac{dV}{V} = -\frac{c}{c+1}\frac{dp}{p}$$

The internal energy differential dU becomes,

$$dU = c\,NR\,dT = \frac{c\,V}{c+1}\,dp$$

The hydrostatic pressure $p\,(z)$ decreases linearly with altitude,

$$dp = -\,m\,g\,dz$$

Thus,

$$\frac{dT}{dz} = -\frac{m\,g\,V}{(c+1)\,NR}$$

Using the definition of the specific heat at constant pressure per unit mass,

$$c_p^* = \frac{C_p}{m\,V} = \frac{(c+1)\,NR}{m\,V}$$

the temperature gradient reduces to,

$$\frac{dT}{dz} = -\frac{g}{c_p^*} \qquad\qquad \text{thus} \qquad\qquad dT = -\frac{g}{c_p^*}\,dz$$

When integrating this relation from an initial position z_0 where the temperature is T_0 to a final position z where the temperature is $T\,(z)$, we find,

$$T\,(z) = T_0 - \frac{g}{c_p^*}\,z$$

b) According to property (5.83),

$$\frac{T^{c+1}}{p} = \text{const}$$

where $c + 1 = c\,\gamma$, we deduce the pressure profile from the temperature profile,

$$p\,(z) = p_0\left(\frac{T\,(z)}{T_0}\right)^{c+1}$$

According to the property,

$$m\,V = \text{const}$$

and to the ideal gas equation of state,

$$\frac{p\,V}{T} = c\,NR = \text{const} \qquad \text{thus} \qquad \frac{p}{m\,T} = \text{const}$$

we deduce the following relation,

$$\frac{m\,(z)}{m_0} = \frac{p\,(z)\,T_0}{p_0\,T\,(z)}$$

Using the pressure profile as a function of temperature, we obtain the mass profile as function of temperature,

$$m\,(z) = m_0 \left(\frac{T\,(z)}{T_0} \right)^{c}$$

10.6 Stratospheric Balloon

Model the rise of a balloon of mass M, which rises from the ground into the stratosphere.[18] At ground level, the balloon has a volume V_0, which is much smaller than the volume V_{max} it has when it is fully inflated. Nonetheless, the volume V_0 is sufficient to lift the payload. The ballom is filled with helium, which is considered an ideal gas. Use Archimedes' principle given in exercise 10.4.5 and the model of Earth's atmosphere established in exercise 10.5.

a) Find the maximum height z_{max} reached by the balloon.
b) Show that Archimedes' force F_A exerted on the balloon is uniform as long as the balloon is not fully inflated.

Solution:

a) The mass density profile established in exercise 10.5 is written as,

$$m\,(z_{max}) = m_0 \left(\frac{T\,(z_{max})}{T_0} \right)^{c} = m_0 \left(1 - \frac{g}{c_p^* T_0} z_{max} \right)^{c}$$

which implies that the maximum height z_{max} is given by,

$$z_{max} = \frac{c_p^* T_0}{g} \left(1 - \frac{m\,(z_{max})}{m_0} \right)^{1/c}$$

At the maximum height z_{max}, Archimedes' force, given in exercise 10.4.5, is equal and opposite to the weight. Thus, these forces have equal norms,

$$m\,(z_{max})\,V_{max}\,g = M g$$

---

[18] T. Yamagami, Y. Saito, Y. Matsuzuka, M. Namiki, M. Toriumi, R. Yokota, H. Hirosawa, K. Matsushima, *Development of the Highest Altitude Balloon*, Adv. Space Res. **33**, 1653–1659 (2004).

which implies that,

$$m\left(z_{\max}\right) = \frac{M}{V_{\max}}$$

Thus, the maximum height $z_{\max}$ is recast as,

$$z_{\max} = \frac{c_p^* T_0}{g}\left(1 - \frac{M}{m_0 V_{\max}}\right)^{1/c}$$

b) As the balloon rises, i.e. $V < V_{\max}$, the pressure inside the balloon is equal to the atmospheric pressure and the balloon is always at thermal equilibrium with the atmospheric air because it rises so slowly. At height z, Archimedes' force is given by,

$$\boldsymbol{F}_A\left(z\right) = m\left(z\right) V\left(z\right) g\,\hat{\boldsymbol{z}}$$

According to exercise 10.5,

$$m V = \text{const} \qquad \text{thus} \qquad m\left(z\right) V\left(z\right) = m_0 V_0$$

Hence, Archimedes' force is constant,

$$\boldsymbol{F}_A\left(z\right) = m_0 V_0 g\,\hat{\boldsymbol{z}} = \textbf{const}$$

10.7 Velocity Field Inside a Pipe

A fluid flows through a pipe which is shaped in such a way that the velocity field depends linearly on the position x along the pipe (Fig. 10.4). At the inlet ($x = 0$), the velocity is v_0. At the outlet ($x = L$), it is $3 v_0$. Determine the acceleration $a\left(x\right)$ of a small volume of fluid.

Numerical Application:

$v_0 = 3$ m/s, $L = 0.3$ m.

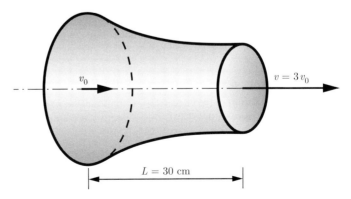

Figure 10.4 A pipe imposes a certain velocity field $v\left(x, t\right)$ to the fluid flowing through it.

Solution:

The velocity field $v(x, t)$ is a linear function of the spatial coordinate x and a function of time. At the initial time $t = 0$, the small element of fluid is at the inlet. At the final time $t = t_f$, it is at the outlet. The velocity field $v(x, t)$ has to satisfy the initial and final conditions,

$$v(0, 0) = v_0 \qquad \text{and} \qquad v(L, t_f) = 3\, v_0$$

Thus, the velocity field is given by,

$$v(x, t) = v_0 \left(1 + \frac{2x}{L} \right)$$

The acceleration field $a(x, t)$ is obtained by taking the unidimensional material derivative (10.18) of the velocity field,

$$a(x, t) = \dot{v}(x, t) = \partial_t v(x, t) + v(x, t)\, \partial_x v(x, t)$$

The velocity field is a stationary field because it does not depend explicitly on time, i.e. $\partial_t v(x, t) = 0$. Thus, the acceleration field $a(x, t)$ yields,

$$a(x, t) = v_0 \left(1 + \frac{2x}{L} \right) \frac{2\, v_0}{L} = \frac{2\, v_0^2}{L} \left(1 + \frac{2x}{L} \right)$$

which implies that,

$$a(0, 0) = \frac{2\, v_0^2}{L} = 60 \,\text{m/s} \qquad \text{and} \qquad a(L, t_f) = \frac{6\, v_0^2}{L} = 180 \,\text{m/s}$$

10.8 Divergence of a Velocity Field

Establish the continuity equation (10.34) for the mass density by working out the change in mass inside an infinitesimal cubic box centred at a position written in Cartesian coordinates as (x, y, z). The box has square faces orthogonal to the Cartesian axes and the dimensions of the edges of the box are dx, dy and dz. The velocity field is $\mathbf{v}(x, y, z)$.

Solution:

First, we consider the faces of the box that are orthogonal to the x-axis. The mass flow through the face located at position $x - dx/2$ is determined by the velocity $v_x(x - dx/2, y, z)$ and the mass flow through the face located at position $x + dx/2$ is determined by the velocity $v_x(x + dx/2, y, z)$. The infinitesimal mass variation dM_x inside the box during an infinitesimal time interval dt is due to the mass flowing through these two faces. Thus, the infinitesimal mass variation is written as,

$$dM_x = m\left(x - \frac{dx}{2}, y, z\right) v_x\left(x - \frac{dx}{2}, y, z\right) dy\,dz\,dt$$
$$- m\left(x + \frac{dx}{2}, y, z\right) v_x\left(x + \frac{dx}{2}, y, z\right) dy\,dz\,dt$$

where $m(x, y, z)$ is the mass density. The signs on the right hand side of this equation are due to the fact that the velocity $v_x(x - dx/2, y, z)$ is positive for an inflow of mass and the velocity $v_x(x + dx/2, y, z)$ is positive for an outflow of mass. The first-order series expansion of the mass densities $m(x \pm dx/2, y, z)$ and the velocities $v_x(x \pm dx/2, y, z)$ are given by,

$$m\left(x \pm \frac{dx}{2}, y, z\right) = m(x, y, z) \pm \frac{1}{2} \partial_x m(x, y, z)\, dx$$
$$v_x\left(x \pm \frac{dx}{2}, y, z\right) = v_x(x, y, z) \pm \frac{1}{2} \partial_x v_x(x, y, z)\, dx$$

With this, the expression for the infinitesimal mass variation becomes,

$$dM_x = \left(m - \frac{1}{2} \partial_x m\, dx\right)\left(v_x - \frac{1}{2} \partial_x v_x\, dx\right) dy\,dz\,dt$$
$$- \left(m + \frac{1}{2} \partial_x m\, dx\right)\left(v_x + \frac{1}{2} \partial_x v_x\, dx\right) dy\,dz\,dt$$

which reduces to,

$$dM_x = - \left(v_x\, \partial_x m + m\, \partial_x v_x\right) dx\,dy\,dz\,dt$$

In the same line of thought, the infinitesimal mass variation dM_y inside the box during an infinitesimal time interval dt due to the mass flowing through the two faces orthogonal to the y-axis is given by,

$$dM_y = - \left(v_y\, \partial_y m + m\, \partial_y v_y\right) dx\,dy\,dz\,dt$$

and the infinitesimal mass variation dM_z inside the box during an infinitesimal time interval dt due to the mass flowing through the two faces orthogonal to the z-axis is given by,

$$dM_z = - \left(v_z\, \partial_z m + m\, \partial_z v_z\right) dx\,dy\,dz\,dt$$

The partial time derivative of the mass density is defined as,

$$\partial_t m = \frac{dM_x + dM_y + dM_z}{dx\,dy\,dz\,dt}$$

which implies that,

$$\partial_t m = - \left(v_x\, \partial_x + v_y\, \partial_y + v_z\, \partial_z\right) m - m\left(\partial_x v_x + \partial_y v_y + \partial_z v_z\right)$$

Using the vectorial relations,

$$\boldsymbol{v} \cdot \boldsymbol{\nabla} = v_x\, \partial_x + v_y\, \partial_y + v_z\, \partial_z$$
$$\boldsymbol{\nabla} \cdot \boldsymbol{v} = \partial_x v_x + \partial_y v_y + \partial_z v_z$$

the partial time derivative of the mass density is recast as,

$$\partial_t\, m = -\, (\boldsymbol{v} \cdot \boldsymbol{\nabla})\, m - (\boldsymbol{\nabla} \cdot \boldsymbol{v})\, m$$

Using the definition (10.18) of material derivative,

$$\dot{m} = \partial_t\, m + (\boldsymbol{v} \cdot \boldsymbol{\nabla})\, m$$

we obtain the continuity equation (10.34) for the mass density,

$$\dot{m} + (\boldsymbol{\nabla} \cdot \boldsymbol{v})\, m = 0$$

Thermodynamics of Irreversible Processes

The exercises given in the last section of Chapter 11 are presented here with their solutions.

11.1 Heat Diffusion Equation

Show that the temperature profile (11.44),

$$T(x,t) = \frac{C}{\sqrt{t}} \exp\left(-\frac{x^2}{4\lambda t}\right)$$

where T is the temperature and x the spatial coordinate, is a solution of the heat diffusion equation (11.37).

Solution:

To show that the temperature profile $T(x,t)$ is a solution of the heat diffusion equation (11.37), we need to compute the partial derivatives of this function. The partial derivative of the temperature T with respect to t is given by,

$$\frac{\partial T}{\partial t} = \frac{\partial}{\partial t}\left(\frac{C}{\sqrt{t}}\right) \exp\left(-\frac{x^2}{4\lambda t}\right) + \frac{C}{\sqrt{t}} \frac{\partial}{\partial t}\left(\exp\left(-\frac{x^2}{4\lambda t}\right)\right)$$

$$= -\frac{C}{2\,t^{3/2}} \exp\left(-\frac{x^2}{4\lambda t}\right) + \frac{Cx^2}{4\lambda t^{5/2}} \exp\left(-\frac{x^2}{4\lambda t}\right)$$

The first-order partial derivative of the temperature T with respect to x yields,

$$\frac{\partial T}{\partial x} = \frac{C}{\sqrt{t}} \frac{\partial}{\partial x}\left(\exp\left(-\frac{x^2}{4\lambda t}\right)\right) = -\frac{Cx}{2\lambda t^{3/2}} \exp\left(-\frac{x^2}{4\lambda t}\right)$$

which implies that the product of λ and the second-order partial derivative of the temperature T with respect to x is given by,

$$\lambda \frac{\partial^2 T}{\partial x^2} = \lambda \frac{\partial}{\partial x}\left(\frac{\partial T}{\partial x}\right)$$

$$= -\lambda \frac{\partial}{\partial x}\left(\frac{Cx}{2\lambda t^{3/2}}\right) \exp\left(-\frac{x^2}{4\lambda t}\right) - \lambda \frac{Cx}{2\lambda t^{3/2}} \frac{\partial}{\partial x}\left(\exp\left(-\frac{x^2}{4\lambda t}\right)\right)$$

$$= -\frac{C}{2\,t^{3/2}} \exp\left(-\frac{x^2}{4\lambda t}\right) + \frac{Cx^2}{4\lambda t^{5/2}} \exp\left(-\frac{x^2}{4\lambda t}\right)$$

Thus, we find that the expressions for $\partial T/\partial t$ and $\lambda\,\partial^2 T/\partial x^2$ are identical, which establishes that the temperature profile $T(x, t)$ is a solution of the heat diffusion equation (11.37).

11.2 Thermal Dephasing

A long copper rod of thermal diffusivity λ is heated at one end with a flame passing underneath it periodically, while the other end is located so far away that it remains at room temperature T_0. Consider the rod as a one-dimensional system with a periodic temperature variation of amplitude ΔT at $x = 0$, given by,

$$T(0, t) = T_0 + \Delta T \cos(\omega t)$$

where x is the the spatial coordinate along the wire. Once the rod has reached a regime at which every point of the rod has a periodic temperature variation, show that the temperature profile is given by,

$$T(x, t) = T_0 + \Delta T \exp\left(-\frac{x}{d}\right) \cos\left(\omega t - \frac{x}{d}\right) \qquad \text{where} \qquad d = \sqrt{\frac{2\lambda}{\omega}}$$

The temperature oscillation at position x is dephased by an angle $-x/d$ with respect to the oscillation at position $x = 0$. The amplitude of oscillation is attenuated by a factor $\exp(-x/d)$.

Solution:

To show that the temperature profile $T(x, t)$ is a solution of the heat diffusion equation (11.37), we compute the partial derivatives of this function. The partial derivative of the temperature T with respect to t is given by,

$$\frac{\partial T}{\partial t} = \Delta T \exp\left(-\frac{x}{d}\right) \frac{\partial}{\partial t}\left(\cos\left(\omega t - \frac{x}{d}\right)\right)$$

$$= -\omega\,\Delta T \exp\left(-\frac{x}{d}\right) \sin\left(\omega t - \frac{x}{d}\right)$$

$$= -\frac{2\lambda\,\Delta T}{d^2} \exp\left(-\frac{x}{d}\right) \sin\left(\omega t - \frac{x}{d}\right)$$

The first-order partial derivative of the temperature T with respect to x yields,

$$\frac{\partial T}{\partial x} = \frac{\partial}{\partial x}\left(\Delta T \exp\left(-\frac{x}{d}\right) \cos\left(\omega t - \frac{x}{d}\right)\right)$$

$$= -\frac{\Delta T}{d} \exp\left(-\frac{x}{d}\right) \left(\cos\left(\omega t - \frac{x}{d}\right) - \sin\left(\omega t - \frac{x}{d}\right)\right)$$

which implies that the product of λ and the second-order partial derivative of the temperature T with respect to x is given by,

$$\lambda \frac{\partial^2 T}{\partial x^2} = \lambda \frac{\partial}{\partial x} \left(\frac{\partial T}{\partial x} \right)$$

$$= -\frac{\lambda \, \Delta T}{d} \frac{\partial}{\partial x} \left(\exp \left(-\frac{x}{d} \right) \left(\cos \left(\omega \, t - \frac{x}{d} \right) - \sin \left(\omega \, t - \frac{x}{d} \right) \right) \right)$$

$$= \frac{\lambda \, \Delta T}{d^2} \exp \left(-\frac{x}{d} \right) \left(\cos \left(\omega \, t - \frac{x}{d} \right) - \sin \left(\omega \, t - \frac{x}{d} \right) \right.$$

$$\left. - \sin \left(\omega \, t - \frac{x}{d} \right) - \cos \left(\omega \, t - \frac{x}{d} \right) \right)$$

$$= -\frac{2 \, \lambda \, \Delta T}{d^2} \exp \left(-\frac{x}{d} \right) \sin \left(\omega \, t - \frac{x}{d} \right)$$

Thus, we find that the expressions for $\partial T / \partial t$ and $\lambda \, \partial^2 T / \partial x^2$ are identical, which establishes that the temperature profile $T(x, t)$ is a solution of the heat diffusion equation.

11.3 Heat Equation with Heat Source

The heat diffusion equation was establish in § 11.4.2, in the absence of any source term due to a conductive electric current density, i.e. $j_q = q_e j_e = 0$. Show that for an electric conductor in the presence of a conductive electric current density j_q, the heat equation becomes,

$$\partial_t \, T = \lambda \, \boldsymbol{\nabla}^2 \, T - \frac{\tau}{c} j_q \cdot \boldsymbol{\nabla} \, T + \frac{j_q^2}{\sigma \, c}$$

where λ is the thermal diffusivity, σ is the electric conductivity, τ is the Thomson coefficient of the electric conductor and c is the specific heat density of the conduction electrons.

Solution:

In the frame of reference of the electric conductor, i.e. $v = 0$, and in the absence of any mechanical constraint on the metal, i.e. $\pi_u = 0$, the internal energy continuity equation (10.43) is reduced to,

$$\partial_t \, u + \boldsymbol{\nabla} \cdot j_u = 0$$

According to relation (11.33),

$$\partial_t \, u = c \, \partial_t \, T$$

Taking into account relation (11.102) and the definition (11.104) of the Thomson coefficient, relation (11.100) yields,

$$\boldsymbol{\nabla} \cdot j_u = -\kappa \, \boldsymbol{\nabla}^2 \, T + \tau \, j_q \cdot \boldsymbol{\nabla} \, T - \frac{j_q^2}{\sigma} \tag{11.1}$$

Thus, we obtain the heat equation,

$$\partial_t T = \lambda \nabla^2 T - \frac{T}{c} \boldsymbol{j}_q \cdot \nabla T + \frac{j_q^2}{\sigma c}$$

It contains a heat source, which consists of a Thomson and a Joule heating.

11.4 Joule Heating in a Wire

Estimate the temperature profile of a wire of length L and radius r when an electric current I is driven through it, from the left end to the right end, causing it to heat up. The wire has an electric conductivity σ and a thermal conductivity κ. The heat propagates down the wire to its ends, i.e. no heat is dissipated at the surface of the wire. The Thomson heating is negligible compared to the Joule heating. The left end and right end are kept at a constant temperature T_0. Determine the temperature profile $T(x)$ along the wire when it has reached a stationary state.

Solution:

In the frame of reference of the wire, i.e. $\boldsymbol{v} = \boldsymbol{0}$, in the absence of any mechanical constraint on the metal, i.e. $\pi_u = 0$, and in a stationary state, i.e. $\dot{u} = 0$, the internal energy continuity equation (10.43) yields the condition,

$$\nabla \cdot \boldsymbol{j}_u = 0$$

Neglecting the Thomson heating, i.e. $\nabla \varepsilon = \boldsymbol{0}$, relation (11.100) reduces to,

$$\kappa \nabla^2 T = -\frac{j_q^2}{\sigma}$$

where

$$\nabla^2 T = \partial_x^2 T \hat{\boldsymbol{x}} \qquad \text{and} \qquad \boldsymbol{j}_q = \frac{I}{\pi r^2} \hat{\boldsymbol{x}}$$

Thus,

$$\partial_x^2 T = -\frac{I^2}{\pi^2 r^4 \kappa \sigma}$$

The indefinite integral of this equation over x is,

$$\partial_x T = -\frac{I^2}{\pi^2 r^4 \kappa \sigma} x + A$$

where A is a constant. The indefinite integral of this equation over x is in turn,

$$T(x) = -\frac{1}{2} \frac{I^2}{\pi^2 r^4 \kappa \sigma} x^2 + Ax + B$$

where B is a constant. The constants are determined by the boundary conditions on the temperature,

$$T(0) = T(L) = T_0$$

which implies that,

$$A = \frac{1}{2} \frac{I^2 L}{\pi^2 r^4 \kappa \sigma} \qquad \text{and} \qquad B = T_0$$

Thus,

$$T(x) = \frac{1}{2} \frac{I^2}{\pi^2 r^4 \kappa \sigma} x(L - x) + T_0$$

Using the definition of the electric resistance $R(L)$ of the wire of resistivity $\rho = 1/\sigma$, length L and section $A = \pi r^2$,

$$R = \rho \frac{L}{A} = \frac{1}{\sigma} \frac{L}{\pi r^2}$$

the temperature profile can be written as,

$$T(x) = \frac{R I^2}{\kappa L A} \frac{1}{2} x(L - x) + T_0$$

which is maximal for $x = L/2$ in the middle of the wire.

11.5 Thomson Heating in a Wire

Estimate the temperature profile of a wire of length L and radius r when an electric current I is driven through it, from the left end to the right end, causing it to heat up. The wire has an electric conductivity σ and a Thomson coefficient τ. The heat is entirely carried by the wire to its ends. The heat dissipated at the surface of the wire and the Joule heating are assumed negligible. The left end is kept at a constant temperature T_0. Determine the temperature profile $T(x)$ along the wire when it has reached a stationary state. Also give an expression for the temperature at the right end in terms of the Thomson coefficient τ and the electric resistance R of the wire.

Solution:

In the frame of reference of the wire, i.e. $\mathbf{v} = \mathbf{0}$, in the absence of any mechanical constraint on the metal, i.e. $\pi_u = 0$, and in a stationary state, i.e. $\dot{u} = 0$, the internal energy continuity equation (10.43) yields the condition,

$$\boldsymbol{\nabla} \cdot \boldsymbol{j}_u = 0$$

Thus, the power density (11.103) is reduced to,

$$\boldsymbol{j}_q \cdot \left(\tau \boldsymbol{\nabla} T - \frac{\boldsymbol{j}_q}{\sigma} \right) = 0$$

which has to hold for any conductive electric current density $\boldsymbol{j}_q$. This implies that,

$$\boldsymbol{\nabla} T = \frac{\boldsymbol{j}_q}{\sigma \tau}$$

According to relation (11.22), the temperature gradient is given by,

$$\nabla T = \frac{T(x) - T_0}{x}\,\hat{x}$$

The conductive electric current density $\boldsymbol{j}_q$ is written as,

$$\boldsymbol{j}_q = \frac{I}{\pi r^2}\,\hat{x}$$

where I is the electric current. Thus,

$$T(x) = T_0 + \frac{I}{\pi r^2 \sigma \tau}\,x$$

At the right end, i.e. $x = L$, the temperature is,

$$T(L) = T_0 + \frac{LI}{\pi r^2 \sigma \tau}$$

The electric resistance $R(L)$ of the wire of resistivity $\rho = 1/\sigma$, length L and section $A = \pi r^2$ is given by,

$$R(L) = \rho\frac{L}{A} = \frac{1}{\sigma}\frac{L}{\pi r^2}$$

Therefore, the temperature on the right end can be written as,

$$T(L) = T_0 + \frac{R(L)I}{\tau}$$

11.6 Heat Exchanger

A heat exchanger is made up of two identical pipes separated by an impermeable diathermal wall of section area A, thickness h and thermal conductivity κ. In both pipes, a liquid flows at uniform velocities $\boldsymbol{v}_1 = v_1\,\hat{x}$ and $\boldsymbol{v}_2 = -v_2\,\hat{x}$, with $v_1 > 0$ and $v_2 > 0$, where $\hat{x}$ is the unit vector that is parallel to the liquid flow in pipe 1. The temperature T_1 of the liquid in pipe 1 is larger than the temperature T_2 of the liquid in pipe 2, i.e. $T_1 > T_2$. Thus, there is a heat current density $\boldsymbol{j}_Q = j_Q\hat{y}$, with $j_Q > 0$ going across the wall separating the pipes, where $\hat{y}$ is the unit vector orthogonal to the wall and oriented positively from pipe 1 to pipe 2. There is no liquid current density across the wall, i.e. $\boldsymbol{j}_C = 0$. Heat conductivity is considered negligible in the direction of the flow and yet large enough to ensure a homogeneous temperature across any section of both pipes. Consider that the heat exchanger has reached a stationary state.

a) Show that the temperature profiles in the fluids are given by the differential equations,

$$\partial_x T_1 = -\frac{\kappa}{h\ell c_1 v_1}\,(T_1 - T_2)$$
$$\partial_x T_2 = \frac{\kappa}{h\ell c_2 v_2}\,(T_1 - T_2)$$

where c_1 and c_2 are the specific heat densities of liquids 1 and 2, and κ is the thermal conductivity of the diathermal wall and ℓ is a characteristic length for the thermal transfer.

b) Show that the convective heat current density $j = c_1 v_1 T_1 + c_2 v_2 T_2$ is homogeneous.

c) Determine the temperature difference $\Delta T(x) = T_1(x) - T_2(x)$.

d) Determine the temperature profiles $T_1(x)$ and $T_2(x)$.

e) Show that on a distance that is short enough, i.e. $x/d \ll 1$,

$$T_1(x) = \frac{j + c_2 v_2 \Delta T(0)}{c_1 v_1 + c_2 v_2} - \frac{\kappa \Delta T(0)}{h \ell c_1 v_1} x$$

$$T_2(x) = \frac{j - c_1 v_1 \Delta T(0)}{c_1 v_1 + c_2 v_2} + \frac{\kappa \Delta T(0)}{h \ell c_2 v_2} x$$

Solution:

a) According to relation (10.101), since there is no liquid current density across the wall, i.e. $\boldsymbol{j}_C = \boldsymbol{0}$, the heat current density $\boldsymbol{j}_Q$ across the wall is equal to the internal energy current density $\boldsymbol{j}_{u2}$, which is the opposite of the internal energy current density $\boldsymbol{j}_{u1}$ due to energy conservation,

$$\boldsymbol{j}_Q = -\boldsymbol{j}_{u1} = \boldsymbol{j}_{u2}$$

Since the flow of liquid in both pipes is uniform, there is no fluid expansion, i.e. $\boldsymbol{\nabla} \cdot \boldsymbol{v}_1 = \boldsymbol{\nabla} \cdot \boldsymbol{v}_2 = 0$. Moreover, since there is no mechanical constraint on the liquids, i.e. $\pi_{u_1} = \pi_{u_2} = 0$, the internal energy continuity equations (10.43) for the liquid in pipes 1 and 2 can be recast as,

$$\dot{u}_1 = -\boldsymbol{\nabla} \cdot \boldsymbol{j}_{u1} = \boldsymbol{\nabla} \cdot \boldsymbol{j}_Q$$

$$\dot{u}_2 = -\boldsymbol{\nabla} \cdot \boldsymbol{j}_{u2} = -\boldsymbol{\nabla} \cdot \boldsymbol{j}_Q$$

The internal energy densities u_1 and u_2 are written in terms of the temperatures T_1 and T_2 of the liquids as,

$$u_1 = c_1 T_1 \quad \text{and} \quad u_2 = c_2 T_2$$

In a stationary state the time derivatives of the internal energy densities vanish, i.e. $\partial_t u_1 = \partial_t u_2 = 0$. Thus, by applying relation (10.18) to the internal energy densities u_1 and u_2 in the particular case of a uniform liquid flow at velocities $\boldsymbol{v}_1 = v_1 \hat{\boldsymbol{x}}$ and $\boldsymbol{v}_2 = -v_2 \hat{\boldsymbol{x}}$, we find,

$$\dot{u}_1 = \partial_t u_1 + \boldsymbol{v}_1 \cdot \boldsymbol{\nabla} u_1 = v_1 \partial_x u_1 = c_1 v_1 \partial_x T_1$$

$$\dot{u}_2 = \partial_t u_2 + \boldsymbol{v}_2 \cdot \boldsymbol{\nabla} u_2 = -v_2 \partial_x u_2 = -c_2 v_2 \partial_x T_2$$

According to equation (10.104) for an infinitesimal section of wall of volume dV, of infinitesimal cross section dA and of thickness h, the thermal power is given by,

$$P_Q = -dV \boldsymbol{\nabla} \cdot \boldsymbol{j}_Q = -h \, dA \, \boldsymbol{\nabla} \cdot \boldsymbol{j}_Q$$

In view of the discrete Fourier law (11.21), the divergence of the heat current density j_Q is written as,

$$\nabla \cdot j_Q = -\frac{P_Q}{h\, dA} = -\frac{\kappa}{h\ell}(T_1 - T_2)$$

Thus, the spatial derivatives of the temperatures are given by,

$$\partial_x T_1 = -\frac{\kappa}{h\ell\, c_1\, v_1}(T_1 - T_2)$$

$$\partial_x T_2 = \frac{\kappa}{h\ell\, c_2\, v_2}(T_1 - T_2)$$

b) In view of the spatial derivatives of the temperatures,

$$\partial_x j = \partial_x (c_1\, v_1\, T_1 + c_2\, v_2\, T_2) = c_1\, v_1\, \partial_x T_1 + c_2\, v_2\, \partial_x T_2 = 0$$

which implies that the convective heat current density j is homogeneous.

c) The difference between the spatial derivatives of the temperatures can be written as,

$$\partial_x (T_1 - T_2) = -\frac{1}{d}(T_1 - T_2)$$

where the decay length d is given by,

$$\frac{1}{d} = \frac{\kappa}{h\ell\, c_1\, v_1} + \frac{\kappa}{h\ell\, c_2\, v_2} = \frac{\kappa}{h\ell}\left(\frac{c_1\, v_1 + c_2\, v_2}{c_1\, v_1\, c_2\, v_2}\right)$$

Thus, the temperature difference $\Delta T(x) = T_1(x) - T_2(x)$ decays exponentially,

$$\Delta T(x) = \Delta T(0) \exp\left(-\frac{x}{d}\right)$$

d) The convective heat current density j can be recast as,

$$j = c_1\, v_1\, T_1(x) + c_2\, v_2\left(T_1(x) - \Delta T(0) \exp\left(-\frac{x}{d}\right)\right)$$

$$j = c_1\, v_1\left(T_2(x) + \Delta T(0) \exp\left(-\frac{x}{d}\right)\right) + c_2\, v_2\, T_2(x)$$

Thus, the temperature profiles are given by,

$$T_1(x) = \frac{1}{c_1\, v_1 + c_2\, v_2}\left(j + c_2\, v_2\, \Delta T(0) \exp\left(-\frac{x}{d}\right)\right)$$

$$T_2(x) = \frac{1}{c_1\, v_1 + c_2\, v_2}\left(j - c_1\, v_1\, \Delta T(0) \exp\left(-\frac{x}{d}\right)\right)$$

e) When the heat transfer occurs on a distance that is short enough, i.e. to first-order in x/d, the temperature profiles reduce to,

$$T_1(x) = \frac{j + c_2\, v_2\, \Delta T(0)}{c_1\, v_1 + c_2\, v_2} - \frac{\kappa\, \Delta T(0)}{h\ell\, c_1\, v_1}\, x$$

$$T_2(x) = \frac{j - c_1\, v_1\, \Delta T(0)}{c_1\, v_1 + c_2\, v_2} + \frac{\kappa\, \Delta T(0)}{h\ell\, c_2\, v_2}\, x$$

11.7 Harman Method

A rod is connected at both ends to electrodes. The electric wires that connect the rod to each electrode are strong enough to carry an electric current flowing through the rod and yet thin enough for the heat transfer to be negligible. The contact resistance and the heat radiated from the rod are negligible. In these experimental conditions, an adiabatic measurement of the resistivity of the material can be performed. As Harman suggested in his seminal paper,[19] experimental conditions can be found such that the Joule and Thomson heating have negligible effects. Use the empirical linear equations (11.92) to show that the adiabatic resistivity thus measured is given by,

$$\rho_{\text{ad}} = \rho \left(1 + \frac{\varepsilon^2}{\kappa \, \rho} T \right)$$

where $\rho = 1/\sigma$ is the isothermal resistivity, κ is the thermal conductivity and ε is the Seebeck coefficient of the rod material.

Solution:

In the experiment analysed here, the conductive electric current density is $\boldsymbol{j}_q = q_e \boldsymbol{j}_e$ and there is no chemical effect, i.e. $\boldsymbol{\nabla} \mu_e = 0$, which implies that $\boldsymbol{\nabla} \bar{\mu}_e = q_e \boldsymbol{\nabla} \varphi$ according to relation (11.93) since the electric charge q_e is a constant. Thus, the transport equations (11.95) are written as,

$$\begin{cases} \boldsymbol{j}_q = -\sigma \varepsilon \, \boldsymbol{\nabla} T - \sigma \, \boldsymbol{\nabla} \varphi \\ \boldsymbol{j}_Q = -\kappa \, \boldsymbol{\nabla} T + T \varepsilon \, \boldsymbol{j}_q \end{cases}$$

An adiabatic resistivity is measured in the absence of a heat current density, i.e. $\boldsymbol{j}_Q = \boldsymbol{0}$ (§ 11.4.9). Hence, the second transport equation implies that the temperature is given by,

$$\boldsymbol{\nabla} T = \frac{\varepsilon}{\kappa} T \boldsymbol{j}_q$$

Thus, according to the first transport equation, the conductive electric current density is written as,

$$\boldsymbol{j}_q = -\sigma \, \boldsymbol{\nabla} \varphi - \sigma \frac{\varepsilon^2}{\kappa} T \boldsymbol{j}_q$$

The isothermal conductivity σ is the inverse of the isothermal resistivity σ, i.e. $\sigma = 1/\rho$. Therefore, in view of definition (11.81) of the adiabatic resistivity ρ_{ad}, the electric potential gradient is given by,

$$\boldsymbol{\nabla} \varphi = -\rho \left(1 + \frac{\varepsilon^2}{\rho \, \kappa} T \right) \boldsymbol{j}_q = -\rho_{\text{ad}} \boldsymbol{j}_q$$

which implies that the adiabatic resistivity ρ_{ad} is written in terms of the isothermal resistivity as,

$$\rho_{\text{ad}} = \rho \left(1 + \frac{\varepsilon^2}{\rho \, \kappa} T \right)$$

[19] T. C. Harman, *Special Techniques for Measurement of Thermoelectric Properties*, J. App. Phys. **29**, 1373 (1958).

Harman suggests to determine the resistivity ρ in the same sample by using an alternative current of high enough frequency so that no temperature gradient has time to build up in each half period of the current. Then, the ratio $\left(\varepsilon^2/\rho\,\kappa\right)\,T$ can be deduced from the measurements of the resistivities ρ and ρ_{ad}. This ratio is called the ZT coefficient of the material. It is a figure of merit that characterises materials for thermoelectric power generation. As an alternative to a high frequency, a transient method was suggested,[20] and corrections for non-adiabatic conditions in the steady state measurements were also analysed.[21]

11.8 Peltier Generator

A Peltier generator is made up of two thermoelectric elements connected in series (Fig. 11.1). One side of the Peltier generator is maintained at temperature T^+ and the other temperature T^-. The electric current I generated by the Peltier generator flows through the thermoelectric materials labelled 1 and 2. The plate which is heated up to temperature T^+ connects electrically these two materials but is not electrically available to the user. Its electric potential is V^+. The other ends of the thermoelectric materials are on the cold side, at a temperature T^-. They are connected to the electric leads of the device. A load resistance R_0 is connected to these leads. The voltage V designates the electric potential difference between the leads.

Analyse the operation of this generator using the electric charge and heat transport equations,

$$\boldsymbol{j}_{q_1} = -\,\sigma_1\,\varepsilon_1\,\boldsymbol{\nabla}\,T_1 - \sigma_1\,\boldsymbol{\nabla}\,\varphi_1 \quad \text{and} \quad \boldsymbol{j}_{Q_1} = -\,\kappa_1\,\boldsymbol{\nabla}\,T_1 + T_1\,\varepsilon_1\boldsymbol{j}_{q_1}$$

$$\boldsymbol{j}_{q_2} = -\,\sigma_2\,\varepsilon_2\,\boldsymbol{\nabla}\,T_2 - \sigma_2\,\boldsymbol{\nabla}\,\varphi_2 \quad \text{and} \quad \boldsymbol{j}_{Q_2} = -\,\kappa_2\,\boldsymbol{\nabla}\,T_2 + T_2\,\varepsilon_2\boldsymbol{j}_{q_2}$$

The thermoelectric materials 1 and 2 have a length d and a cross-section surface area A, which can be written as,

$$d = \int_0^d dr \cdot \hat{\boldsymbol{x}} \qquad\qquad A = \int_S d\boldsymbol{S} \cdot \hat{\boldsymbol{x}}$$

where $\hat{\boldsymbol{x}}$ is a unit vector oriented clockwise along the electric current density $\boldsymbol{j}_q$, and the infinitesimal length and surface vectors $d\boldsymbol{r}$ and $d\boldsymbol{S}$ are oriented in the same direction. The temperature difference between the hot and cold ends is given by,

$$\Delta T = T^+ - T^- = \int_0^d d\boldsymbol{r} \cdot \boldsymbol{\nabla}\,T_1 = \int_0^d d\boldsymbol{r} \cdot \left(-\boldsymbol{\nabla}\,T_2\right)$$

[20] E. E. Castillo, C. L. Hapenciuc, and Th. Borca-Tasciuc, *Thermoelectric Characterization by Transient Harman Method under Non-Ideal Contact and Boundary Conditions*, Rev. Sci. Instruments **81**, 033902 (2010).
[21] I.-J. Roh, Y. G. Lee, M.-S. Kang, J.-U. Lee, S.-H. Baek, S. K. Kim, B.-K. Ju, D.-B. Hyun, J.-S. Kim, B. Kwon, *Harman Measurements for Thermoelectric Materials and Modules under Non-Adiabatic Conditions*, Sci. Rep., **6**, 39131 (2016).

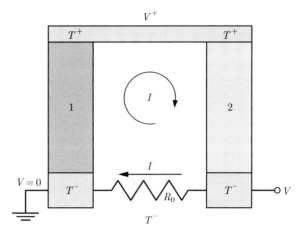

Figure 11.1 A Peltier generator has a load represented by the resistance R_0 connected to its leads. V is the voltage between the leads. The electric bridge at V^+ is not accessible to the user. The regions marked 1 and 2 represent the two thermoelectric materials. The regions marked T^+ and T^- are the hot and cold sides of the device.

Likewise, the electric potential differences $\Delta \varphi_1$ and $\Delta \varphi_2$ between the hot and cold ends are written as,

$$\Delta \varphi_1 = V^+ = \int_0^d dr \cdot \nabla \varphi_1$$

$$\Delta \varphi_2 = V^+ - V = \int_0^d dr \cdot (-\nabla \varphi_2)$$

The electric charge conservation implies that the electric current densities are the same in each material, i.e. $j_{q_1} = j_{q_2}$. The electric current I flowing through materials 1 and 2 is the integral of the electric current densities j_{q_1} and j_{q_2} over the cross-section area A,

$$I = \int_S j_{q_1} \cdot dS = \int_S j_{q_2} \cdot dS$$

According the relation (10.104), the thermal powers P_{Q_1} and P_{Q_2} are the integrals of the heat current densities j_{Q_1} and j_{Q_2} flowing through materials 1 and 2 over the cross-section area A,

$$P_{Q_1} = \int_S (-j_{Q_1}) \cdot dS \qquad P_{Q_2} = \int_S j_{Q_2} \cdot dS$$

Determine:

a) the thermal power P'_Q applied on the hot side of the device when no electric current flows through the device.

b) the effective electric resistance R of the two thermoelectric materials when the temperatures are equal, i.e. $T^+ = T^-$, and no electric current flows through the resistance R_0, i.e. when $R_0 = \infty$. Instead, an electric current flows through the thermoelectric materials.

c) the electric current I as a function of the temperature difference ΔT.

d) the thermodynamic efficiency of the generator defined as,

$$\eta = \frac{R_0\, I^2}{P_Q}$$

where here, P_Q is the thermal power at the hot side when the electric current is flowing through the device. Show that the optimum load resistance is given by

$$\frac{R_0}{R} = \sqrt{1+\zeta}$$

where ζ is a dimensionless parameter given by, [22]

$$\zeta = \frac{T^+\,(\varepsilon_1 - \varepsilon_2)^2}{(\kappa_1 + \kappa_2)\left(\dfrac{1}{\sigma_1} + \dfrac{1}{\sigma_2}\right)}$$

Solution:

a) When no electric current flows through the device, i.e. $\boldsymbol{j}_{q_1} = \boldsymbol{j}_{q_2} = \boldsymbol{0}$, in order to determine the thermal power P'_Q, we integrate the heat transport equations over the volume V. The integrals over the volume are the product of integrals over the cross-section area A times integrals over the length d of the thermoelectric materials,

$$\int_S (-\boldsymbol{j}'_{Q_1}) \cdot d\boldsymbol{S} \int_0^d d\boldsymbol{r} \cdot \hat{\boldsymbol{x}} = \kappa_1 \int_0^d d\boldsymbol{r} \cdot \boldsymbol{\nabla}\, T_1 \int_S d\boldsymbol{S} \cdot \hat{\boldsymbol{x}}$$

$$\int_S \boldsymbol{j}'_{Q_2} \cdot d\boldsymbol{S} \int_0^d d\boldsymbol{r} \cdot \hat{\boldsymbol{x}} = \kappa_2 \int_0^d d\boldsymbol{r} \cdot (-\boldsymbol{\nabla}\, T_2) \int_S d\boldsymbol{S} \cdot \hat{\boldsymbol{x}}$$

which reduce to,

$$P'_{Q_1} = \kappa_1 \frac{A}{d}\, \Delta T \qquad\qquad P'_{Q_2} = \kappa_2 \frac{A}{d}\, \Delta T$$

Thus, the total thermal power is given by,

$$P'_Q = P'_{Q_1} + P'_{Q_2} = (\kappa_1 + \kappa_2)\, \frac{A}{d}\, \Delta T$$

b) When the temperatures of the hot and cold sources are equal, i.e. $T^+ = T^-$, the temperature gradients vanish, i.e. $\boldsymbol{\nabla}\, T_1 = \boldsymbol{\nabla}\, T_2$, which implies that there is no thermoelectric effect. The integrals of the electric charge transport equations over the volume are the product of integrals over the cross-section area A times integrals over the length d of the thermoelectric materials,

$$\int_S (-\boldsymbol{j}_{q_1}) \cdot d\boldsymbol{S} \int_0^d d\boldsymbol{r} \cdot \hat{\boldsymbol{x}} = \sigma_1 \int_0^d d\boldsymbol{r} \cdot \boldsymbol{\nabla}\, \varphi_1 \int_S d\boldsymbol{S} \cdot \hat{\boldsymbol{x}}$$

$$\int_S (-\boldsymbol{j}_{q_2}) \cdot d\boldsymbol{S} \int_0^d d\boldsymbol{r} \cdot \hat{\boldsymbol{x}} = -\sigma_2 \int_0^d d\boldsymbol{r} \cdot (-\boldsymbol{\nabla}\, \varphi_2) \int_S d\boldsymbol{S} \cdot \hat{\boldsymbol{x}}$$

Since, the load resistance is infinite, i.e. $R_0 = \infty$, the electric current used for the measurement flows in the opposite direction, $I \to -I$ under the condition that $V^+ < V$. In this case, the electric charge transport equations integrated over the volume reduce to,

---

[22] H. J. Goldsmid, *Introduction to Thermoelectricity*, Springer (2010).

$$I = \sigma_1 \frac{A}{d} \Delta \varphi_1 = \sigma_1 \frac{A}{d} V^+$$

$$I = -\sigma_2 \frac{A}{d} \Delta \varphi_2 = -\sigma_2 \frac{A}{d} \left(V^+ - V \right)$$

The potential difference across the thermoelectric materials 1 and 2 connected in series is given by,

$$V = \Delta \varphi_1 - \Delta \varphi_2 = I \frac{d}{A} \left(\frac{1}{\sigma_1} + \frac{1}{\sigma_2} \right) = I \frac{2d}{A} \frac{1}{\sigma}$$

where σ is the effective conductivity of both thermoelectric materials. Since the effective electric resistivity ρ is the inverse of the effective electric conductivity ρ,

$$\rho = \frac{1}{\sigma} = \frac{1}{2} \left(\frac{1}{\sigma_1} + \frac{1}{\sigma_2} \right)$$

Thus, the potential difference across both thermoelectric materials is written as,

$$V = \rho \frac{2d}{A} I = R I$$

where $2d$ is the effective length of both materials of length d each, section area A, connected in series, and R is their effective resistance. Hence,

$$R = \rho \frac{2d}{A} = \frac{1}{2} \left(\frac{1}{\sigma_1} + \frac{1}{\sigma_2} \right) \frac{2d}{A}$$

c) The integrals of the electric charge transport equations over the volume are the product of integrals over the cross-section area A times integrals over the length d of the thermoelectric materials,

$$\int_S j_{q_1} \cdot dS \int_0^d dr \cdot \hat{x} = -\sigma_1 \varepsilon_1 \int_0^d dr \cdot \nabla T_1 \int_S dS \cdot \hat{x}$$

$$- \sigma_1 \int_0^d dr \cdot \nabla \varphi_1 \int_S dS \cdot \hat{x}$$

$$\int_S j_{q_2} \cdot dS \int_0^d dr \cdot \hat{x} = \sigma_2 \varepsilon_2 \int_0^d dr \cdot (- \nabla T_2) \int_S dS \cdot \hat{x}$$

$$+ \sigma_2 \int_0^d dr \cdot (- \nabla \varphi_2) \int_S dS \cdot \hat{x}$$

The electric charge transport equations integrated over the volume reduce to,

$$I = -\sigma_1 \varepsilon_1 \frac{A}{d} \Delta T - \sigma_1 \frac{A}{d} V^+$$

$$I = \sigma_2 \varepsilon_2 \frac{A}{d} \Delta T + \sigma_2 \frac{A}{d} \left(V^+ - V \right)$$

Ohm's law for the load resistance is written as,

$$V = R_0 I$$

In view of this relation that charaterises the electric properties of the load, the charge transport equations can be recast as,

$$V^+ = -\frac{1}{\sigma_1}\left(\frac{d}{A}I - \sigma_1\,\varepsilon_1\,\Delta T\right)$$

$$V^+ = \frac{1}{\sigma_2}\left(\left(\frac{d}{A} + \sigma_2\,R_0\right)I - \sigma_2\,\varepsilon_2\,\Delta T\right)$$

which implies that the electric current is given by,

$$I = \frac{\varepsilon_2 - \varepsilon_1}{\left(\dfrac{1}{\sigma_1} + \dfrac{1}{\sigma_2}\right)\dfrac{d}{A} + R_0}\,\Delta T = \frac{\varepsilon_2 - \varepsilon_1}{R + R_0}\,\Delta T$$

This result for the current I is consistent with the current obtained in the analysis of the Seebeck loop (§ 11.6.2), which is equivalent to a Peltier generator (Fig. 11.1) in which the load resistance is set at zero, i.e. $R_0 = 0$.

d) In order to determine the thermal power P_Q entering through the hot plate at temperature T^+, we integrate the heat transport equations over the volume V. The integrals over the volume are the product of integrals over the cross-section area A times integrals over the length d of the thermoelectric materials,

$$\int_S (-j_{Q_1}) \cdot dS \int_0^d dr \cdot \hat{x} = \kappa_1 \int_0^d dr \cdot \nabla T_1 \int_S dS \cdot \hat{x}$$
$$- T^+ \varepsilon_1 \int_S j_{q_1} \cdot dS \int_0^d dr \cdot \hat{x}$$

$$\int_S j_{Q_2} \cdot dS \int_0^d dr \cdot \hat{x} = \kappa_2 \int_0^d dr \cdot (-\nabla T_2) \int_S dS \cdot \hat{x}$$
$$+ T^+ \varepsilon_2 \int_S j_{q_2} \cdot dS \int_0^d dr \cdot \hat{x}$$

which reduce to,

$$P_{Q_1} = \kappa_1 \frac{A}{d}\Delta T - T^+\,\varepsilon_1\,I$$

$$P_{Q_2} = \kappa_2 \frac{A}{d}\Delta T + T^+\,\varepsilon_2\,I$$

and implies that,

$$P_Q = P_{Q_1} + P_{Q_2} = (\kappa_1 + \kappa_2)\frac{A}{d}\Delta T + T^+\,(\varepsilon_2 - \varepsilon_1)\,I$$

Therefore, the efficiency η for any load resistance R_0 is given by,

$$\eta = \frac{R_0\,I^2}{P_Q} = \frac{R_0\,\dfrac{(\varepsilon_2 - \varepsilon_1)^2\,\Delta T^2}{(R + R_0)^2}}{(\kappa_1 + \kappa_2)\dfrac{A}{d}\Delta T + T^+\,(\varepsilon_2 - \varepsilon_1)^2\,\dfrac{\Delta T}{R + R_0}}$$

which can be recast as,

$$\eta = \frac{\Delta T}{T^+} \frac{\dfrac{R_0}{R}}{\dfrac{(\kappa_1 + \kappa_2)}{T^+ (\varepsilon_2 - \varepsilon_1)^2} \left(\dfrac{1}{\sigma_1} + \dfrac{1}{\sigma_2}\right) \left(1 + \dfrac{R_0}{R}\right)^2 + \left(1 + \dfrac{R_0}{R}\right)}$$

Using the definition of the coefficient $\zeta > 0$, the temperature difference $\Delta T = T^+ - T^- > 0$ and the ratio $r = 1 + R_0/R > 0$, the efficiency is reduced to,

$$\eta = \left(1 - \frac{T^-}{T^+}\right) \frac{r - 1}{\zeta^{-1} r^2 + r}$$

To find the optimal load resistance, we have to optimise the efficiency η with respect to the ratio r,

$$\frac{d\eta}{dr} = \left(1 - \frac{T^-}{T^+}\right) \frac{\zeta^{-1} r^2 + r - (r - 1)\left(\zeta^{-1} r^2 + r\right)}{\left(\zeta^{-1} r^2 + r\right)^2} = 0$$

which implies that,

$$r^2 - 2r - \zeta = 0$$

Thus, the optimum ratio $r > 0$, is given by,

$$r = 1 + \sqrt{1 + \zeta}$$

Hence, for an optimal resistance load, the efficiency is given by,

$$\eta = \left(1 - \frac{T^-}{T^+}\right) \frac{\zeta\sqrt{1 + \zeta}}{\left(1 + \sqrt{1 + \zeta}\right)^2 + \zeta\left(1 + \sqrt{1 + \zeta}\right)} \le 1 - \frac{T^-}{T^+}$$

In the limit $\zeta \to \infty$, the efficiency of the Peltier generator η tends towards the Carnot efficiency η_C (7.46),

$$\lim_{\zeta \to \infty} \eta = 1 - \frac{T^-}{T^+} = \eta_C$$

11.9 ZT Coefficient of a Thermoelectric Material

The transport properties of a thermoelectric material of cross section area A and length L are defined by the transport equations,

$$\boldsymbol{j}_q = -\sigma \epsilon \boldsymbol{\nabla} T - \sigma \boldsymbol{\nabla} \varphi \qquad \text{and} \qquad \boldsymbol{j}_Q = -\kappa \boldsymbol{\nabla} T + T \varepsilon \boldsymbol{j}_q$$

in conformity with relations (11.92), where $\boldsymbol{\nabla} \mu_e = \boldsymbol{0}$, and (11.95). The efficiency η of the thermoelectric material is defined as,

$$\eta = -\frac{P_q}{P_Q}$$

where P_Q is the heat entering at the hot end and P_q is the electric power defined as,

$$P_q = \int_V \boldsymbol{j}_q \cdot (-\boldsymbol{\nabla}\,\varphi)\, dV$$

Write the efficiency η in terms of the ratio[23],

$$r = \frac{I}{\kappa}\frac{L}{A}\frac{1}{\Delta T}$$

where I is the electric current flowing through the thermoelectric material. In the limit where the thermoelectric effect is much smaller than the thermal power, i.e. $r\varepsilon \ll 1/T^+$, show that the optimal efficiency η is given by,

$$\eta = \left(1 - \frac{T^-}{T^+}\right)\frac{\sigma\,\varepsilon^2}{4\,\kappa}\,T^+$$

The coefficient $(\sigma\,\varepsilon^2/\kappa)\,T^+$ is called the 'ZT coefficient' of the thermoelectric material. The term in brackets is the Carnot efficiency.

Solution:

In order to determine the thermal power P_Q, we integrate the heat transport equation over the volume V. The integral over the volume is the product of an integral over the cross-section area A times an integral over the length L of the thermoelectric material,

$$\int_S \boldsymbol{j}_Q \cdot d\boldsymbol{S} \int_0^L d\boldsymbol{r} \cdot \hat{\boldsymbol{x}} = \kappa \int_0^L d\boldsymbol{r} \cdot (-\boldsymbol{\nabla}\,T)\int_S d\boldsymbol{S} \cdot \hat{\boldsymbol{x}} + T^+\varepsilon \int_S \boldsymbol{j}_q \cdot d\boldsymbol{S} \int_0^L d\boldsymbol{r} \cdot \hat{\boldsymbol{x}}$$

where $\hat{\boldsymbol{x}}$ is a unit vector in the direction of the current densities $\boldsymbol{j}_Q$ and $\boldsymbol{j}_q$, and the infinitesimal length and surface vectors $d\boldsymbol{r}$ and $d\boldsymbol{S}$ are oriented in the same direction. The thermal power P_Q and the electric current I are defined as,

$$P_Q = \int_S \boldsymbol{j}_Q \cdot d\boldsymbol{S} \qquad I = \int_S \boldsymbol{j}_q \cdot d\boldsymbol{S} \qquad \Delta T = \int_0^L d\boldsymbol{r} \cdot (-\boldsymbol{\nabla}\,T)$$

The cross-section surface area A and the length L can be written as,

$$A = \int_S d\boldsymbol{S} \cdot \hat{\boldsymbol{x}} \qquad L = \int_0^L d\boldsymbol{r} \cdot \hat{\boldsymbol{x}}$$

Thus, the thermal power P_Q is written as,

$$P_Q = \kappa\,\frac{A}{L}\,\Delta T + T^+\,\varepsilon\,I$$

Similarly, in order to determine the electric power P_q, we deduce $\boldsymbol{\nabla}\,\varphi$ from the electric charge transport equation and integrate the scalar product between $-\boldsymbol{\nabla}\,\varphi$ and the electric current density $\boldsymbol{j}_q$ over the volume V of the thermoelectric material,

$$\int_V \boldsymbol{j}_q \cdot (-\boldsymbol{\nabla}\,\varphi)\, dV = -\varepsilon \int_S \boldsymbol{j}_q \cdot d\boldsymbol{S} \int_0^L d\boldsymbol{r} \cdot (-\boldsymbol{\nabla}\,T) + \frac{1}{\sigma}\int_S \boldsymbol{j}_q \cdot d\boldsymbol{S} \int_0^L \boldsymbol{j}_q \cdot d\boldsymbol{r}$$

[23] G. J. Snyder, T. S. Ursell, *Thermoelectric Efficiency and Compatibility*, Phys. Rev. Lett. **91** (4), 138301 (2003).

The electric power P_q is defined as,

$$P_q = \int_V \boldsymbol{j}_q \cdot (-\boldsymbol{\nabla}\,\varphi)\, dV$$

and

$$I\frac{L}{A} = \int_0^L \boldsymbol{j}_q \cdot d\boldsymbol{r}$$

Thus, the electric power P_q is written as,

$$P_q = -\varepsilon I \Delta T + \frac{I^2}{\sigma}\frac{L}{A}$$

Hence, the efficiency η of the thermoelectric material is given by,

$$\eta = -\frac{P_q}{P_Q} = \frac{\varepsilon I \Delta T - \dfrac{I^2}{\sigma}\dfrac{L}{A}}{\kappa \dfrac{A}{L}\Delta T + T^+ \varepsilon I}$$

which can be recast as,

$$\eta = \frac{\Delta T}{T^+}\left(\frac{\varepsilon - \dfrac{I}{\sigma}\dfrac{L}{A}\dfrac{1}{\Delta T}}{\varepsilon + \dfrac{\kappa}{I}\dfrac{A}{L}\dfrac{\Delta T}{T^+}}\right)$$

Using the dimensionless ratio,

$$r = \frac{I}{\kappa}\frac{L}{A}\frac{1}{\Delta T}$$

the efficiency η becomes,

$$\eta = \frac{\Delta T}{T^+}\left(\frac{\varepsilon - \dfrac{\kappa}{\sigma}r}{\varepsilon + \dfrac{1}{r}\dfrac{1}{T^+}}\right) = \eta = \frac{\Delta T}{T^+}\left(\frac{r\left(\varepsilon - \dfrac{\kappa}{\sigma}r\right)}{r\varepsilon + \dfrac{1}{T^+}}\right)$$

Using the relation $\Delta T = T^+ - T^-$ in the limit $r\varepsilon \ll 1/T^+$, the efficiency η is reduced to,

$$\eta = \left(1 - \frac{T^-}{T^+}\right) T^+ r\left(\varepsilon - \frac{\kappa}{\sigma}r\right)$$

To find the optimum efficiency ratio r, we have to optimise the efficiency η with respect to the ratio r,

$$\frac{d\eta}{dr} = \left(1 - \frac{T^-}{T^+}\right) T^+ \left(\varepsilon - \frac{2\kappa}{\sigma}r\right) = 0$$

which implies that the optimal ratio is,

$$r = \frac{\sigma\varepsilon}{2\kappa}$$

Thus, the optimal efficiency of the thermoelectric material is,

$$\eta = \left(1 - \frac{T^-}{T^+}\right)\frac{\sigma\varepsilon^2}{4\kappa}T^+$$

which is a quarter of the product of the Carnot efficiency and the 'ZT coefficient' $\left(\sigma \varepsilon^2/\kappa\right) T^+$. In the limit where the thermoelectric effect is much smaller than the thermal power,

$$I \frac{\varepsilon}{\kappa} \frac{L}{A} \frac{T^+}{\Delta T} \ll 1$$

where according to Ohm's law and the Seebeck effect, the order of magnitude of the electric current I is given by,

$$I = \frac{\varepsilon \, \Delta T}{R} = \varepsilon \, \sigma \, \Delta T \frac{A}{L}$$

Thus, the condition is recast as,

$$\frac{\sigma \, \varepsilon^2}{\kappa} T^+ \ll 1$$

11.10 Transverse Transport Effects

A transport equation such as Ohm's law (11.74),

$$\boldsymbol{\nabla} \varphi = - \boldsymbol{\rho} \cdot \boldsymbol{j}_q$$

relates two vectors, which are the conductive electric current density $\boldsymbol{j}_q$ and electric potential gradient $\boldsymbol{\nabla} \varphi$, through a linear application, which is the electric resistivity $\boldsymbol{\rho}$. Mathematically, a vector is a rank-1 tensor and a linear application between two vectors is a rank-2 tensor.

a) Show that the electric resistivity $\boldsymbol{\rho}$ can be decomposed into the sum of a symmetric part $\boldsymbol{\rho}^s$ and an antisymmetric part $\boldsymbol{\rho}^a$.

b) Show that the antisymmetric part $\boldsymbol{\rho}^a$ has a contribution to the transport that can be written as,

$$\boldsymbol{\nabla}^a \varphi = - \rho^a \left(\hat{\boldsymbol{u}} \times \boldsymbol{j}_q\right)$$

where $\boldsymbol{\nabla}^a \varphi$ is the antisymmetric part of the electric potential gradient and $\hat{\boldsymbol{u}}$ is a unit axial vector.

The decomposition and the expression for the antisymmetric part of the electric potential gradient is a general result that applies for any empirical linear relation between a current density vector and a generalised force vector.

Solution:

a) The components of the symmetric rank-2 resistivity tensor are written as,

$$\rho_{ij}^s = \frac{1}{2} \left(\rho_{ij} + \rho_{ji}\right)$$

The components of the antisymmetric rank-2 resistivity tensor are given by,

$$\rho_{ij}^a = \frac{1}{2} \left(\rho_{ij} - \rho_{ji}\right)$$

The components of the rank-2 resistivity tensor are written as,

$$\rho_{ij} = \frac{1}{2}\left(\rho_{ij} + \rho_{ji}\right) + \frac{1}{2}\left(\rho_{ij} - \rho_{ji}\right)$$

which implies that,

$$\rho_{ij} = \rho_{ij}^s + \rho_{ij}^a$$

Thus, the conductivity tensor ρ is the sum of the symmetric rank-2 conductivity tensor ρ^s and the antisymmetric rank-2 conductivity tensor ρ^a,

$$\rho = \rho^s + \rho^a$$

a) The linear application $\rho^a \cdot \boldsymbol{j}_q$ can be written in components as,

$$\begin{pmatrix} 0 & \rho_{12}^a & \rho_{13}^a \\ -\rho_{12}^a & 0 & \rho_{23}^a \\ -\rho_{13}^a & -\rho_{23}^a & 0 \end{pmatrix} \begin{pmatrix} j_{q_1} \\ j_{q_2} \\ j_{q_3} \end{pmatrix} = \begin{pmatrix} \rho_{12}^a j_{q_2} + \rho_{13}^a j_{q3} \\ -\rho_{12}^a j_{q_1} + \rho_{23}^a j_{q3} \\ -\rho_{13}^a j_{q_1} - \rho_{23}^a j_{q_2} \end{pmatrix}$$

The vector product $\rho^a \left(\hat{\boldsymbol{u}} \times \boldsymbol{j}_q\right)$ is written in components as,

$$\begin{pmatrix} \rho^a \hat{u}_1 \\ \rho^a \hat{u}_2 \\ \rho^a \hat{u}_3 \end{pmatrix} \times \begin{pmatrix} j_{q_1} \\ j_{q_2} \\ j_{q3} \end{pmatrix} = \begin{pmatrix} \rho^a \hat{u}_2 j_{q3} - \rho^a \hat{u}_3 j_{q_2} \\ \rho^a \hat{u}_3 j_{q_1} - \rho^a \hat{u}_1 j_{q3} \\ \rho^a \hat{u}_1 j_{q_2} - \rho^a \hat{u}_2 j_{q_1} \end{pmatrix}$$

The identification of the components of these two vectors yields,

$$\hat{u}_1 = -\frac{\rho_{23}^a}{\rho^a} \qquad \hat{u}_2 = \frac{\rho_{13}^a}{\rho^a} \qquad \hat{u}_3 = -\frac{\rho_{12}^a}{\rho^a}$$

Since the vector $\hat{\boldsymbol{u}}$ is a unit axial vector,

$$\hat{\boldsymbol{u}}^2 = \hat{u}_1^2 + \hat{u}_2^2 + \hat{u}_3^2 = \frac{1}{(\rho^a)^2}\left((\rho_{23}^a)^2 + (\rho_{13}^a)^2 + (\rho_{13}^a)^2\right) = 1$$

which implies that,

$$\rho^a = \sqrt{(\rho_{23}^a)^2 + (\rho_{13}^a)^2 + (\rho_{12}^a)^2}$$

Thus, the antisymmetric part of the electric potential gradient is given by,

$$\boldsymbol{\nabla}^a \varphi = -\rho^a \cdot \boldsymbol{j}_q = -\rho^a \left(\hat{\boldsymbol{u}} \times \boldsymbol{j}_q\right)$$

where the unit axial vector $\hat{\boldsymbol{u}}$ is written in components as,

$$\hat{\boldsymbol{u}} = \frac{1}{\rho_a} \begin{pmatrix} -\rho_{23}^a \\ \rho_{13}^a \\ -\rho_{12}^a \end{pmatrix}$$

11.11 Hall Effect

An isotropic conductor is in the presence of a magnetic induction field $\boldsymbol{B}$. The electric resistivity rank-2 tensor is a function of the magnetic induction field $\boldsymbol{B}$ and Ohm's law is written as,

$$\boldsymbol{\nabla}\,\varphi = -\,\rho\,(\boldsymbol{B})\cdot\boldsymbol{j}_q$$

The reversibility of the dynamics at the microscopic scale, implies that the transpose of the electric resistivity tensor is obtained by reversing the orientation of the magnetic induction field $\boldsymbol{B}$.[24] Thus,

$$\rho^T\,(\boldsymbol{B}) = \rho\,(-\,\boldsymbol{B})$$

This result cannot be established in a thermodynamic framework but requires the use of a statistical physics. In a linear electromagnetic framework, when the magnetic induction field $\boldsymbol{B}$ is applied orthogonally to the conductive electric current density $\boldsymbol{j}_q$, show that Ohm's law can be written as,

$$\boldsymbol{\nabla}\,\varphi = -\,\rho\cdot\boldsymbol{j}_q - \mathcal{H}\boldsymbol{j}_q \times \boldsymbol{B}$$

where the first term is Ohm's law (11.74) in the absence of a magnetic induction field $\boldsymbol{B}$ and the second term is the Hall effect (11.75) in a direction that is orthogonal to the magnetic induction field $\boldsymbol{B}$ and to the conductive electric current density. Use the result established in § (11.10).

Solution:

We showed in § 11.10 that the electric resistivity tensor $\rho\,(\boldsymbol{B})$ can be expressed as the sum of a symmetric part $\rho^s\,(\boldsymbol{B})$ and an antisymmetric part $\rho^a\,(\boldsymbol{B})$. Thus, Ohm's law (11.74) is recast as,

$$\boldsymbol{\nabla}\,\varphi = -\,\rho^s\,(\boldsymbol{B})\cdot\boldsymbol{j}_q - \rho^a\,(\boldsymbol{B})\cdot\boldsymbol{j}_q$$

The electric resistivity tensor $\rho\,(\boldsymbol{B})$ is a linear function of the magnetic induction field $\boldsymbol{B}$ in a linear electromagnetic framework. According to the statistical relation based on the reversibility of the dynamics at the microscopic scale, the electric resistivity tensor is antisymmetric when a magnetic induction field $\boldsymbol{B}$ is present. Thus, the symmetric part of the resistivity tensor $\rho^s = \rho\,(\boldsymbol{0}) \equiv \rho$ is independent of the magnetic induction field since,

$$\rho^T\,(\boldsymbol{0}) = \rho\,(\boldsymbol{0})$$

According to the result established in § (11.10) for the antisymmetric part, Ohm's law is recast as,

$$\boldsymbol{\nabla}\,\varphi = -\,\rho\cdot\boldsymbol{j}_q - \rho^a\,(\boldsymbol{B})\,\hat{\boldsymbol{u}} \times \boldsymbol{j}_q$$

where $\rho^a\,(\boldsymbol{B})$ is a linear function of the magnetic induction field $\boldsymbol{B}$ and $\hat{\boldsymbol{u}}$ is a dimensionless unit vector that can be chosen orthogonal to the conductive electric current density $\boldsymbol{j}_q$

---

[24] L. D. Landau, E. M. Lifshitz, L.-P. Pitaevskii, *Electrodynamics of Continuous Media, Landau and Lifshitz Course of Theoretical Physics volume 8*, Pergamon Press, 3rd edition, (2000).

without imposing restrictions. The anisotropy unit vector $\hat{u}$ is due to the presence of the magnetic induction field $\boldsymbol{B}$ that breaks the isotropy of the conductor leading to off-diagonal terms in the electric resistivity tensor $\rho\,(\boldsymbol{B})$. Therefore, the anisotropy unit vector $\hat{u}$ is oriented along the magnetic induction field $\boldsymbol{B}$. When the magnetic induction field $\boldsymbol{B}$ is applied orthogonally to the conductive electric current density $\boldsymbol{j}_q$, the anisotropic term in Ohm's law can be recast as,

$$\rho^a\,(\boldsymbol{B})\,\hat{u}\times\boldsymbol{j}_q = -\,\mathcal{H}\,\boldsymbol{B}\times\boldsymbol{j}_q$$

where $\mathcal{H} = -\,\rho^a\,(\boldsymbol{B})\,/\|\boldsymbol{B}\|$ is a scalar coefficient. Thus, Ohm's law is recast as,

$$\boldsymbol{\nabla}\,\varphi = -\,\boldsymbol{\rho}\cdot\boldsymbol{j}_q - \mathcal{H}\boldsymbol{j}_q\times\boldsymbol{B}$$

where the first term is Ohm's law (11.74) and the second term is the Hall effect (11.75).

11.12 Heat Transport and Crystal Symmetry

A crystal consists of a honeycomb lattice. It is invariant under a rotation of angle $\theta = \pi/6$ in the horizontal plane around the vertical axis. This means that the physical properties of the crystal are the same after such a rotation. Show that the symmetric thermal conductivity tensor $\boldsymbol{\kappa}$ is written in components as,

$$\boldsymbol{\kappa} = \begin{pmatrix} \kappa_\perp & 0 & 0 \\ 0 & \kappa_\perp & 0 \\ 0 & 0 & \kappa_\| \end{pmatrix}$$

where $\kappa_\|$ is the thermal conductivity along the vertical rotation axis and $\kappa_\perp$ is the thermal conductivity in the horizontal plane of rotation.

Solution:

The symmetric thermal conductivity tensor $\boldsymbol{\kappa}$ can be written in components as,

$$\boldsymbol{\kappa} = \begin{pmatrix} \kappa_{11} & \kappa_{12} & \kappa_{13} \\ \kappa_{12} & \kappa_{22} & \kappa_{23} \\ \kappa_{13} & \kappa_{23} & \kappa_{33} \end{pmatrix}$$

The rotation matrix $\mathcal{R}_\theta$ describing a rotation of angle $\theta = \pi/6$ in the horizontal plane around the vertical axis that leaves the thermal conductivity tensor invariant and its inverse $\mathcal{R}_\theta^{-1}$ read,

$$\mathcal{R}_\theta = \frac{1}{2}\begin{pmatrix} 1 & -\sqrt{3} & 0 \\ \sqrt{3} & 1 & 0 \\ 0 & 0 & 1 \end{pmatrix} \quad\text{and}\quad \mathcal{R}_\theta^{-1} = \mathcal{R}_{-\theta} = \frac{1}{2}\begin{pmatrix} 1 & \sqrt{3} & 0 \\ -\sqrt{3} & 1 & 0 \\ 0 & 0 & 1 \end{pmatrix}$$

Since the physical properties of the crystal are invariant under the rotation $\mathcal{R}_\theta$, we rotate Fourier's law (11.26) by an angle θ,

$$\mathcal{R}_\theta\cdot\boldsymbol{j}_Q = -\,\mathcal{R}_\theta\cdot\boldsymbol{\kappa}\cdot\boldsymbol{\nabla}\,T = -\,\mathcal{R}_\theta\cdot\boldsymbol{\kappa}\cdot\mathcal{R}_{-\theta}\cdot\mathcal{R}_\theta\cdot\boldsymbol{\nabla}\,T$$

where $\mathcal{R}_{-\theta} \cdot \mathcal{R}_\theta = \mathbb{1}$. Since Fourier's law (11.26) is invariant under such a rotation,

$$\mathcal{R}_\theta \cdot j_Q = j_Q \qquad \text{and} \qquad \mathcal{R}_\theta \cdot \nabla T = \nabla T$$

and

$$\mathcal{R}_\theta \cdot \kappa \cdot \mathcal{R}_{-\theta} = \kappa \qquad \text{or} \qquad \mathcal{R}_\theta \cdot \kappa = \kappa \cdot \mathcal{R}_\theta$$

which is written in components as,

$$\begin{pmatrix} 1 & -\sqrt{3} & 0 \\ \sqrt{3} & 1 & 0 \\ 0 & 0 & 1 \end{pmatrix} \begin{pmatrix} \kappa_{11} & \kappa_{12} & \kappa_{13} \\ \kappa_{12} & \kappa_{22} & \kappa_{23} \\ \kappa_{13} & \kappa_{23} & \kappa_{33} \end{pmatrix} = \begin{pmatrix} \kappa_{11} & \kappa_{12} & \kappa_{13} \\ \kappa_{12} & \kappa_{22} & \kappa_{23} \\ \kappa_{13} & \kappa_{23} & \kappa_{33} \end{pmatrix} \begin{pmatrix} 1 & -\sqrt{3} & 0 \\ \sqrt{3} & 1 & 0 \\ 0 & 0 & 1 \end{pmatrix}$$

The solutions of this matrix system are,

$$\kappa_{12} = \kappa_{13} = \kappa_{23} = 0 \qquad \text{and} \qquad \kappa_{11} = \kappa_{22}$$

With the identifications,

$$\kappa_\parallel = \kappa_{33} \qquad \text{and} \qquad \kappa_\perp = \kappa_{11} = \kappa_{22}$$

the thermal conductivity tensor κ reduces to,

$$\kappa = \begin{pmatrix} \kappa_\perp & 0 & 0 \\ 0 & \kappa_\perp & 0 \\ 0 & 0 & \kappa_\parallel \end{pmatrix}$$

11.13 Planar Ettingshausen Effect

In this chapter, several examples of a current density in one direction inducing the gradient of an intensive quantity in another direction were shown. These effects are referred to by the name of their discoverers : Righi-Leduc (11.29), Hall (11.75), Nernst (11.85), Ettingshausen (11.80). The latter refers to a temperature gradient induced by an orthogonal electric charge current density. It was pointed out recently that this effect can occur in a crystal, which consists of two types of electric charge carriers and presents a strong crystalline anisotropy in the plane where the heat and electric charge transport take place. No magnetic induction field needs to be applied orthogonally in order to observe this effect.[25]

The material has two types of electric charge carriers, electrons (e) and holes (h). Assume that no 'chemical reaction' takes place between them. The thermoelectric properties are isotropic, i.e. the same in all directions. Therefore, the Seebeck tensors for the electrons and holes are given by,

$$\varepsilon_e = \begin{pmatrix} \varepsilon_e & 0 \\ 0 & \varepsilon_e \end{pmatrix} \qquad \text{and} \qquad \varepsilon_h = \begin{pmatrix} \varepsilon_h & 0 \\ 0 & \varepsilon_h \end{pmatrix}$$

However, the conductivities differ greatly in two orthogonal directions. Therefore, the conductivity tensors are given by,

---

[25] C. Zhou, S. Birner, Y. Tang, K. Heinselman, M. Grayson, *Driving Perpendicular Heat Flow : (p × n)-Type Transverse Thermoelectrics for Microscale and Cryogenic Peltier Cooling*, Phys. Rev. Lett. **110**, 227701 (2013).

$$\boldsymbol{\sigma}_e = \begin{pmatrix} \sigma_{e,aa} & 0 \\ 0 & \sigma_{e,bb} \end{pmatrix} \qquad \text{and} \qquad \boldsymbol{\sigma}_h = \begin{pmatrix} \sigma_{h,aa} & 0 \\ 0 & \sigma_{h,bb} \end{pmatrix}$$

where a and b label the a-axis and the b-axis that are orthogonal crystalline axes.

Consider an electric charge transport along the x-axis at an angle θ from the a-axis and show that this electric current density $\boldsymbol{j}_q$ induces a heat current density $\boldsymbol{j}_Q$ along the y-axis. This is the planar Ettingshausen effect. It can be understood by establishing the following facts :

a) Show that the Seebeck tensor for this crystal is given by,[26]

$$\boldsymbol{\varepsilon} = (\boldsymbol{\sigma}_e + \boldsymbol{\sigma}_h)^{-1} \cdot (\boldsymbol{\sigma}_e \cdot \boldsymbol{\varepsilon}_e + \boldsymbol{\sigma}_h \cdot \boldsymbol{\varepsilon}_h)$$

b) Show that the Seebeck tensor for this crystal is diagonal and written as,

$$\boldsymbol{\varepsilon} = \begin{pmatrix} \varepsilon_{aa} & 0 \\ 0 & \varepsilon_{bb} \end{pmatrix}$$

where the diagonal component ε_{aa} is different from ε_{bb} in general. The matrix is given here for a vector basis along the crystalline a-axis and b-axis.

c) Write the components of the Seebeck tensor with respect to the coordinate basis (x, y),

$$\boldsymbol{\varepsilon} = \begin{pmatrix} \varepsilon_{xx} & \varepsilon_{xy} \\ \varepsilon_{yx} & \varepsilon_{yy} \end{pmatrix}$$

in terms of the diagonal components ε_{aa} and ε_{bb} of the Seebeck tensor with respect to the coordinate basis (a, b).

d) The heat current density $\boldsymbol{j}_Q$ is related to the electric charge current density $\boldsymbol{j}_q$ by,

$$\boldsymbol{j}_Q = \boldsymbol{\Pi} \cdot \boldsymbol{j}_q$$

which is a local version of the Peltier effect (11.108). The Peltier tensor is related to the Seebeck tensor by,

$$\boldsymbol{\Pi} = T \boldsymbol{\varepsilon}$$

In particular, for an electric charge current density $\boldsymbol{j}_q = j_{q,x} \hat{\boldsymbol{x}}$, where $\hat{\boldsymbol{x}}$ is a unit vector along the x-axis, show that the component $j_{Q,y}$ along the y-axis of the heat current density $\boldsymbol{j}_Q = j_{Q,x} \hat{\boldsymbol{x}} + j_{Q,y} \hat{\boldsymbol{y}}$, where $\hat{\boldsymbol{y}}$ is a unit vector along the y-axis, is given by,

$$j_{Q,y} = \frac{1}{2} T (\varepsilon_{aa} - \varepsilon_{bb}) \sin(2\theta) j_{q,x}$$

Thus, the planar Ettingshausen effect is maximal for an angle $\theta = \pi/4$.

Solution:

a) The electric charge transport equations for the electrons and the holes are given by,

$$\boldsymbol{j}_{q,e} = -\boldsymbol{\sigma}_e \cdot \boldsymbol{\varepsilon}_e \cdot \boldsymbol{\nabla} T - \boldsymbol{\sigma}_e \cdot \boldsymbol{\nabla} \varphi$$

$$\boldsymbol{j}_{q,h} = -\boldsymbol{\sigma}_h \cdot \boldsymbol{\varepsilon}_h \cdot \boldsymbol{\nabla} T - \boldsymbol{\sigma}_h \cdot \boldsymbol{\nabla} \varphi$$

[26] S. D. Brechet et J.-Ph. Ansermet, *Heat-Driven Spin Currents on Large Scales.*, physica status solidi (RRL) **5** (12), 423–425 (2011).

The Seebeck effect is observed when the electric current density vanishes, i.e. $\boldsymbol{j}_q = \boldsymbol{j}_{q,e} + \boldsymbol{j}_{q,h} = \boldsymbol{0}$. Thus,

$$\boldsymbol{j}_q = \boldsymbol{j}_{q,e} + \boldsymbol{j}_{q,h} = -(\boldsymbol{\sigma}_e \cdot \boldsymbol{\varepsilon}_e + \boldsymbol{\sigma}_h \cdot \boldsymbol{\varepsilon}_h) \cdot \nabla T - (\boldsymbol{\sigma}_e + \boldsymbol{\sigma}_h) \cdot \nabla \varphi = \boldsymbol{0}$$

According to the Seebeck effect (11.83),

$$\nabla \varphi = -(\boldsymbol{\sigma}_e + \boldsymbol{\sigma}_h)^{-1} \cdot (\boldsymbol{\sigma}_e \cdot \boldsymbol{\varepsilon}_e + \boldsymbol{\sigma}_h \cdot \boldsymbol{\varepsilon}_h) \cdot \nabla T = -\boldsymbol{\varepsilon} \cdot \nabla T$$

Thus, the Seebeck tensor is given by,

$$\boldsymbol{\varepsilon} = (\boldsymbol{\sigma}_e + \boldsymbol{\sigma}_h)^{-1} \cdot (\boldsymbol{\sigma}_e \cdot \boldsymbol{\varepsilon}_e + \boldsymbol{\sigma}_h \cdot \boldsymbol{\varepsilon}_h)$$

b) The Seebeck tensor is written in components,

$$\boldsymbol{\varepsilon} = \begin{pmatrix} \dfrac{\sigma_{e,aa}\,\varepsilon_e + \sigma_{h,aa}\,\varepsilon_h}{\sigma_{e,aa} + \sigma_{h,aa}} & 0 \\ 0 & \dfrac{\sigma_{e,bb}\,\varepsilon_e + \sigma_{h,bb}\,\varepsilon_h}{\sigma_{e,bb} + \sigma_{h,bb}} \end{pmatrix}$$

where the diagonal components are given by,

$$\varepsilon_{aa} = \frac{\sigma_{e,aa}\,\varepsilon_e + \sigma_{h,aa}\,\varepsilon_h}{\sigma_{e,aa} + \sigma_{h,aa}} \quad \text{and} \quad \varepsilon_{bb} = \frac{\sigma_{e,bb}\,\varepsilon_e + \sigma_{h,bb}\,\varepsilon_h}{\sigma_{e,bb} + \sigma_{h,bb}}$$

c) The rotation matrix $\mathcal{R}_\theta$ that rotates the crystalline a-axis and b-axis by an angle θ in plane on the x-axis and y-axis, and its inverse $\mathcal{R}_{-\theta}$, are written in components as,

$$\mathcal{R}_\theta = \begin{pmatrix} \cos\theta & -\sin\theta \\ \sin\theta & \cos\theta \end{pmatrix} \quad \text{and} \quad \mathcal{R}_{-\theta} = \begin{pmatrix} \cos\theta & \sin\theta \\ -\sin\theta & \cos\theta \end{pmatrix}$$

where $\mathcal{R}_{-\theta} \cdot \mathcal{R}_\theta = \mathbb{1}$. The coordinates of the electric potential gradient $\nabla \varphi$ in the coordinate basis (x, y) are related to coordinates in the coordinate basis (a, b) by,

$$\begin{pmatrix} \partial_x \varphi \\ \partial_y \varphi \end{pmatrix} = \begin{pmatrix} \cos\theta & -\sin\theta \\ \sin\theta & \cos\theta \end{pmatrix} \begin{pmatrix} \partial_a \varphi \\ \partial_b \varphi \end{pmatrix}$$

The coordinates of the temperature gradient ∇T in the coordinate basis (x, y) are related to coordinates in the coordinate basis (a, b) by,

$$\begin{pmatrix} \partial_x T \\ \partial_y T \end{pmatrix} = \begin{pmatrix} \cos\theta & -\sin\theta \\ \sin\theta & \cos\theta \end{pmatrix} \begin{pmatrix} \partial_a T \\ \partial_b T \end{pmatrix}$$

Thus, in view of the Seebeck effect (11.83), the coordinates of the Seebeck tensor in the coordinate basis (x, y) are related to coordinates in the coordinate basis (a, b) by,

$$\begin{pmatrix} \varepsilon_{xx} & \varepsilon_{xy} \\ \varepsilon_{yx} & \varepsilon_{yy} \end{pmatrix} = \begin{pmatrix} \cos\theta & -\sin\theta \\ \sin\theta & \cos\theta \end{pmatrix} \begin{pmatrix} \varepsilon_{aa} & 0 \\ 0 & \varepsilon_{bb} \end{pmatrix} \begin{pmatrix} \cos\theta & \sin\theta \\ -\sin\theta & \cos\theta \end{pmatrix}$$

which implies that,

$$\begin{pmatrix} \varepsilon_{xx} & \varepsilon_{xy} \\ \varepsilon_{yx} & \varepsilon_{yy} \end{pmatrix} = \begin{pmatrix} \varepsilon_{aa}\cos^2\theta + \varepsilon_{bb}\sin^2\theta & (\varepsilon_{aa} - \varepsilon_{bb})\sin\theta\cos\theta \\ (\varepsilon_{aa} - \varepsilon_{bb})\sin\theta\cos\theta & \varepsilon_{aa}\sin^2\theta + \varepsilon_{bb}\cos^2\theta \end{pmatrix}$$

d) The Peltier effect is given by,

$$j_Q = T\varepsilon \cdot j_q$$

For a heat current density $j_Q = j_{Q,x}\hat{x} + j_{Q,y}\hat{y}$ and an electric charge current density $j_q = j_{q,x}\hat{x}$, this effect is written in components as,

$$\begin{pmatrix} j_{Q,x} \\ j_{Q,y} \end{pmatrix} = T \begin{pmatrix} \varepsilon_{aa}\cos^2\theta + \varepsilon_{bb}\sin^2\theta & (\varepsilon_{aa} - \varepsilon_{bb})\sin\theta\cos\theta \\ (\varepsilon_{aa} - \varepsilon_{bb})\sin\theta\cos\theta & \varepsilon_{aa}\sin^2\theta + \varepsilon_{bb}\cos^2\theta \end{pmatrix} \begin{pmatrix} j_{q,x} \\ 0 \end{pmatrix}$$

Thus, the planar Ettingshausen effect is written as,

$$j_{Q,y} = \frac{1}{2} T(\varepsilon_{aa} - \varepsilon_{bb})\sin(2\theta)\, j_{q,x}$$

since $\sin(2\theta) = 2\sin\theta\cos\theta$.

11.14 Turing Patterns

A biological medium consists of two substances 1 and 2 of densities n_1 and n_2. This medium is generating both substances by processes characterised by the matter source densities $\pi_1(n_1, n_2)$ and $\pi_2(n_1, n_2)$. The substances 1 and 2 can diffuse inside this medium. The matter current densities j_1 and j_2 follow Fick's law (11.51),

$$j_1 = -D_1 \nabla n_1 \qquad \text{and} \qquad j_2 = -D_2 \nabla n_2$$

where $D_1 > 0$ and $D_2 > 0$ are the homogeneous diffusion constants of substances 1 and 2. The medium has a fixed volume, which means that its expansion rate vanishes, i.e. $\nabla \cdot v = 0$. Thus, the matter continuity equations for substances 1 and 2 are given by,

$$\dot{n}_1 + \nabla \cdot j_1 = \pi_1(n_1, n_2) \qquad \text{and} \qquad \dot{n}_2 + \nabla \cdot j_2 = \pi_2(n_1, n_2)$$

At equilibrium, the system is assumed to be homogeneous and characterised by the densities n_{01} and n_{02} of substances 1 and 2. In the neighbourhood of the equilibrium, the matter source densities $\pi_1(n_1, n_2)$ and $\pi_2(n_1, n_2)$ are given to first-order in terms of the density perturbations $\Delta n_1 = n_1 - n_{01}$ and $\Delta n_2 = n_2 - n_{02}$ by,

$$\pi_1(n_1, n_2) = \Omega_{11}\,\Delta n_1 + \Omega_{12}\,\Delta n_2$$
$$\pi_2(n_1, n_2) = \Omega_{21}\,\Delta n_1 + \Omega_{22}\,\Delta n_2$$

where the coefficients $\Omega_{11}, \Omega_{12}, \Omega_{21}, \Omega_{22}$ are given by,

$$\Omega_{11} = \frac{\partial \pi_1}{\partial n_1} \qquad \Omega_{12} = \frac{\partial \pi_1}{\partial n_2} \qquad \Omega_{21} = \frac{\partial \pi_2}{\partial n_1} \qquad \Omega_{22} = \frac{\partial \pi_2}{\partial n_2}$$

Assume that the processes to generate substances 1 and 2 are the two chemical reactions $1 \overset{a}{\longrightarrow} 2$ and $2 \overset{b}{\longrightarrow} 1$ described by the stoichiometric coefficients $\nu_{a1} = -1$, $\nu_{a2} = 1$, $\nu_{b1} = 1$, $\nu_{b2} = -1$ and the reaction rate densities ω_a and ω_b. Assume that the temperature T and the chemical potentials μ_1 and μ_2 are homogeneous, i.e. $\nabla T = \mathbf{0}$ and $\nabla \mu_1 =$

$\nabla \mu_2 = \mathbf{0}$. Analyse the evolution of the density perturbations Δn_1 and Δn_2 by using the following instructions :

a) Express the coefficients Ω_{11}, Ω_{12}, Ω_{21}, Ω_{22} in terms of the total density $n = n_1 + n_2$, the density perturbations Δn_1 and Δn_2, the temperature T and a scalar $W \geq 0$, which is a linear combination of Onsager matrix elements L_{aa}, L_{ab}, L_{ba} and L_{bb}. Begin by using the second law, i.e. $\pi_s \geq 0$, and the relation (8.68) for a mixture of ideal gas.

b) Determine the coupled time evolution equations for the density perturbations Δn_1 and Δn_2.

c) Show that under the condition imposed in $a)$ the relation,

$$\begin{pmatrix} \Delta n_1 \,(t) \\ \Delta n_2 \,(t) \end{pmatrix} = e^{\lambda t} \, \cos \left(\mathbf{k} \cdot \mathbf{r} + \varphi \right) \begin{pmatrix} \Delta n_1 \,(0) \\ \Delta n_2 \,(0) \end{pmatrix}$$

is a solution of the coupled time evolution equations with $\lambda < 0$.

Solution:

a) Using the definition (10.25) for the matter source densities $\pi_1 \, (n_1, n_2)$ and $\pi_2 \, (n_1, n_2)$ for the stoichiometric coefficients $\nu_{a1} = -1$, $\nu_{a2} = 1$, $\nu_{b1} = 1$, $\nu_{b2} = -1$, we can write that,

$$\pi_1 \, (n_1, n_2) = \Omega_{11} \, \Delta n_1 + \Omega_{12} \, \Delta n_2 = \nu_{a1} \, \omega_a + \nu_{b1} \, \omega_b = - \left(\omega_a - \omega_b \right)$$
$$\pi_2 \, (n_1, n_2) = \Omega_{21} \, \Delta n_1 + \Omega_{22} \, \Delta n_2 = \nu_{a2} \, \omega_a + \nu_{b2} \, \omega_b = \omega_a - \omega_b$$

Using the definition (8.18) for the chemical affinities $\mathcal{A}_a$ and $\mathcal{A}_b$,

$$\mathcal{A}_a = - \mu_1 \, \nu_{a1} - \mu_2 \, \nu_{a2} = \mu_1 - \mu_2$$
$$\mathcal{A}_b = - \mu_1 \, \nu_{b1} - \mu_2 \, \nu_{b2} = - \left(\mu_1 - \mu_2 \right) = - \mathcal{A}_a$$

Since there is no expansion of the system, i.e. $\nabla \cdot \mathbf{v} = 0$, the scalar linear empirical relations (11.6) reduce to,

$$\omega_a = L_{aa} \, \mathcal{A}_a + L_{ab} \, \mathcal{A}_b = \left(L_{aa} - L_{ab} \right) \mathcal{A}_a = \left(L_{aa} - L_{ab} \right) \left(\mu_1 - \mu_2 \right)$$
$$\omega_b = L_{ba} \, \mathcal{A}_a + L_{bb} \, \mathcal{A}_b = \left(L_{ba} - L_{bb} \right) \mathcal{A}_a = \left(L_{ba} - L_{bb} \right) \left(\mu_1 - \mu_2 \right)$$

Since the temperature and the chemical potentials are homogeneous, i.e. $\nabla T = \mathbf{0}$ and $\nabla \mu_1 = \nabla \mu_2 = \mathbf{0}$, the second law (10.88) reduces to,

$$\pi_s = \frac{1}{T} \left(\omega_a \, \mathcal{A}_a + \omega_b \, \mathcal{A}_b \right) = \frac{\mathcal{A}_a}{T} \left(\omega_a - \omega_b \right)$$
$$= \frac{\mathcal{A}_a^2}{T} \left(L_{aa} - L_{ab} - L_{ba} + L_{bb} \right) = \frac{W \, \mathcal{A}_a^2}{T} \geq 0$$

which implies that $W \geq 0$ since $T > 0$ and $\mathcal{A}_a^2 \geq 0$. Thus,

$$\Omega_{11} \, \Delta n_1 + \Omega_{12} \, \Delta n_2 = - \left(\omega_a - \omega_b \right) = - W \left(\mu_1 - \mu_2 \right)$$
$$\Omega_{21} \, \Delta n_1 + \Omega_{22} \, \Delta n_2 = \omega_a - \omega_b = W \left(\mu_1 - \mu_2 \right)$$

Now, we have to express the chemical potentials μ_1 and μ_2 in terms of the concentration n_1/n and n_2/n of substances 1 and 2 using the relation (8.68) for a mixture of ideal gases,

$$\mu_1\left(T, n_1, n\right) = \mu_1\left(T, n_1\right) + RT\ln\left(\frac{n_1}{n}\right)$$

$$= \mu_1\left(T, n_1\right) + RT\ln\left(\frac{n_{01} + \Delta n_1}{n}\right)$$

$$\mu_2\left(T, n_2, n\right) = \mu_2\left(T, n_2\right) + RT\ln\left(\frac{n_2}{n}\right)$$

$$= \mu_2\left(T, n_2\right) + RT\ln\left(\frac{n_{02} + \Delta n_2}{n}\right)$$

which can be recast as,

$$\mu_1\left(T, n_1, n\right) = \mu_1^0\left(T, n_1, n\right) + RT\ln\left(1 + \frac{\Delta n_1}{n}\right)$$

$$\mu_2\left(T, n_2, n\right) = \mu_2^0\left(T, n_1, n\right) + RT\ln\left(1 + \frac{\Delta n_2}{n}\right)$$

where the chemical potentials at equilibrium are,

$$\mu_1^0\left(T, n_1, n\right) = \mu_1\left(T, n_1\right) + RT\ln\left(\frac{n_{01}}{n}\right)$$

$$\mu_2^0\left(T, n_2, n\right) = \mu_2\left(T, n_2\right) + RT\ln\left(\frac{n_{02}}{n}\right)$$

At equilibrium the chemical potentials are equal,

$$\mu_1^0\left(T, n_1, n\right) = \mu_2^0\left(T, n_2, n\right) \equiv \mu^0\left(T, n_1, n_2\right)$$

Thus, for small density perturbations, i.e. $\Delta n_1 \ll 1$ and $\Delta n_2 \ll 1$, the chemical potentials become,

$$\mu_1\left(T, n_1, n\right) = \mu^0\left(T, n_1, n_2\right) + RT\frac{\Delta n_1}{n}$$

$$\mu_2\left(T, n_1, n\right) = \mu^0\left(T, n_1, n_2\right) + RT\frac{\Delta n_2}{n}$$

Hence,

$$\Omega_{11}\,\Delta n_1 + \Omega_{12}\,\Delta n_2 = -W\left(\mu_1 - \mu_2\right) = -\frac{RTW}{n}\left(\Delta n_1 - \Delta n_2\right)$$

$$\Omega_{21}\,\Delta n_1 + \Omega_{22}\,\Delta n_2 = W\left(\mu_1 - \mu_2\right) = \frac{RTW}{n}\left(\Delta n_1 - \Delta n_2\right)$$

which implies that the coefficients are given by,

$$\Omega_{11} = \Omega_{22} = -\frac{RTW}{n} \leq 0 \qquad \text{and} \qquad \Omega_{12} = \Omega_{21} = \frac{RTW}{n} \geq 0$$

b) Using the matter current densities $\boldsymbol{j}_1$ and $\boldsymbol{j}_2$, the matter continuity equations can be recast as,

$$\dot{n}_1 = D_1\,\boldsymbol{\nabla}^2\,n_1 + \pi_1\left(n_1, n_2\right) \qquad \text{and} \qquad \dot{n}_2 = D_2\,\boldsymbol{\nabla}^2\,n_2 + \pi_2\left(n_1, n_2\right)$$

where the Laplacian $\nabla^2 = \nabla \cdot \nabla$ is a scalar operator. Introducing the scalar $\Omega \equiv \Omega_{12} = \Omega_{21} = -\Omega_{11} = -\Omega_{22} \geq 0$ and using the relations for the matter source densities $\pi_1(n_1, n_2)$ and $\pi_2(n_1, n_2)$, the matter continuity equations become,

$$\dot{n}_1 = D_1 \nabla^2 n_1 - \Omega \Delta n_1 + \Omega \Delta n_2$$
$$\dot{n}_2 = D_2 \nabla^2 n_2 + \Omega \Delta n_1 - \Omega \Delta n_2$$

Since $\Delta n_1 = n_1 - n_{01}$ and $\Delta n_2 = n_2 - n_{02}$ where n_{01} and n_{02} are constants,

$$\dot{n}_1 = \Delta \dot{n}_1 \qquad \dot{n}_2 = \Delta \dot{n}_2 \qquad \nabla^2 n_1 = \nabla^2(\Delta n_1) \qquad \nabla^2 n_2 = \nabla^2(\Delta n_2)$$

Thus, the coupled time evolution equations for the density perturbations Δn_1 and Δn_2 yield,

$$\Delta \dot{n}_1 = D_1 \nabla^2(\Delta n_1) - \Omega \Delta n_1 + \Omega \Delta n_2$$
$$\Delta \dot{n}_2 = D_2 \nabla^2(\Delta n_2) + \Omega \Delta n_1 - \Omega \Delta n_2$$

c) The coupled time evolution equations can be written as a matrix system,

$$\begin{pmatrix} \Delta \dot{n}_1 \\ \Delta \dot{n}_2 \end{pmatrix} = \begin{pmatrix} -\Omega + D_1 \nabla^2 & \Omega \\ \Omega & -\Omega + D_2 \nabla^2 \end{pmatrix} \begin{pmatrix} \Delta n_1 \\ \Delta n_2 \end{pmatrix}$$

Replacing the solution in the coupled time evolution equations and then using the relations,

$$\Delta \dot{n}_1 = \lambda \Delta n_1 \qquad \text{and} \qquad \nabla^2(\Delta n_1) = -k^2 \Delta n_1$$
$$\Delta \dot{n}_2 = \lambda \Delta n_2 \qquad \text{and} \qquad \nabla^2(\Delta n_2) = -k^2 \Delta n_2$$

the matrix system can be recast as,

$$\begin{pmatrix} -\Omega - D_1 k^2 - \lambda & \Omega \\ \Omega & -\Omega - D_2 k^2 - \lambda \end{pmatrix} \begin{pmatrix} \Delta n_1 \\ \Delta n_2 \end{pmatrix} = 0$$

For non-trivial solutions, the determinant of this matrix vanishes,

$$(\Omega + D_1 k^2 + \lambda)(\Omega + D_2 k^2 + \lambda) - \Omega^2 = 0$$

which is recast as,

$$\lambda^2 + 2\omega\lambda + \alpha = 0$$

where,

$$\omega \equiv \frac{1}{2}\left(2\Omega + (D_1 + D_2) k^2\right) \geq 0$$
$$\alpha \equiv (\Omega + D_1 k^2)(\Omega + D_2 k^2) - \Omega^2 \geq 0$$

which implies that $\omega^2 - \alpha \geq 0$. The solutions of the quadratic equation in λ are,

$$\lambda_1 = -\omega + \sqrt{\omega^2 - \alpha}$$
$$\lambda_2 = -\omega - \sqrt{\omega^2 - \alpha}$$

These solutions are called the **Lyapunov exponents** of the system. Under the assumption of a closed system undergoing chemical reactions transforming substance 1 into substance 2 and vice versa, the Lyapunov exponents are negative, i.e. $\lambda_1 < 0$ and $\lambda_2 < 0$, which corresponds to stable solutions. For the formation of instabilities, where the density perturbations grow exponentially, at least one of the Lyapunov exponents has to be positive, i.e. $\lambda_1 > 0$ or $\lambda_2 > 0$. Thus, in a closed system, density perturbations cannot grow. In order to see the rise and growth of instabilities, which can lead to the formation of patterns called **Turing patterns**, there has to be a source for the substances 1 and 2, which is the environment.[27]

11.15 Ultramicroelectrodes

In electrochemistry, the observed current is generally determined by ion diffusion in the electrolyte. It was found that diffusion-limited currents can be avoided by using very small electrodes, known as '**ultramicroelectrodes**'.[28] [29] [30] In order to capture how conductive current densities (also called diffusion current densities) vary with the size of the electrode, consider a spherical electrode and a conductive matter current density with spherical symmetry, $\boldsymbol{j}_A = j_{Ar}\hat{\boldsymbol{r}} \equiv j_r\hat{\boldsymbol{r}}$. Show that when the system reaches a stationary state, the conductive matter current density is non-zero. The analysis of the transient behaviour would show that the stationary state is reached faster when the electrode is smaller.[31] In spherical coordinates (r, θ, ϕ), taking into account the spherical symmetry of the conductive matter current density, i.e. $\partial/\partial\theta = 0$ and $\partial/\partial\phi = 0$, the matter diffusion equation (11.54) for a solute of concentration $c(r, t)$ reads,

$$\frac{\partial c(r, t)}{\partial t} = D\left(\frac{\partial^2 c(r, t)}{\partial r^2} + \frac{2}{r}\frac{\partial c(r, t)}{\partial r}\right)$$

The boundary conditions are,

$$c(r > r_0, t = 0) = c^* \qquad \text{and} \qquad \lim_{r \to \infty} c(r, t) = c^*$$

where c^* is the concentration very far away from the electrode and r_0 is the radius of the electrode. According to relation (11.51), the conductive matter current density scalar j_r that characterises this electrode is,

$$j_r(r_0, t) = -D\left.\frac{\partial c(r, t)}{\partial r}\right|_{r=r_0}$$

Establish the following results :

a) The diffusion equation recast in terms of the function $w(r, t) = r\,c(r, t)$ has the structure of a diffusion equation where the spherical coordinate r plays an analogous role to a Cartesian coordinate.

[27] R. Phillips, *Physical Biology of the Cell*, Taylor & Francis, New York, 2nd edition (2012).
[28] K. Aoki, K. Akimoto, K. Tokuda, H. Matsuda, J. Osteryoung, *Linear Sweep Voltammetry at Very Small Stationary Disk Electrodes*, J. Electroanal. Chem. **171**, 219–230 (1984).
[29] M. Fleschmann, S. Pons, *The Behavior of Microdisk and Microring Electrodes*, J. Electroanal. Chem. **222**, 107–115 (1987).
[30] A. M. Bond, K. B. Oldham, C. G. Zoski, *Steady-State Voltammetry*, Analytica Chimica Acta, **216**, 177–230 (1989).
[31] J. Heinze, *Ultramicroelectrodes in Electrochemistry*, Angew. Chem. Int. Ed. Engl. **32**, 1268–1288 (1993).

b) The diffusion equation,

$$\frac{\partial w(r,t)}{\partial t} = D \frac{\partial^2 w(r,t)}{\partial r^2}$$

admits the solution,

$$w(r,t) = B \int_{\nu_0}^{\nu} \exp\left(-\nu'^2\right) d\nu' \qquad \text{where} \qquad \nu = \frac{r}{2\sqrt{Dt}}$$

First, write $w(r,t) = f(\eta)$ where the variable η is a dimensionless function of r and t given by,

$$\eta(r,t) = \frac{r^2}{Dt}$$

c) In the limit where the radius of the electrode is negligible, i.e. $r = 0$, the scalar conductive matter current density is given by,

$$j_r(0,t) = \frac{B}{8\sqrt{D}t^{3/2}}$$

d) After a transient behaviour, the scalar conductive matter current density reaches a stationary value,

$$j_r(r_0, \infty) = -\frac{Dc^*}{r_0}$$

Solution:

a) To show that the function $w(r,t) = rc(r,t)$ satisfies a diffusion equation for the spatial variable r, we compute the partial derivatives taking into account the fact that the variables r and t are independent. Using the diffusion equation for the matter concentration $c(r,t)$, the partial time derivative of $w(r,t)$ can be written as,

$$\frac{\partial w(r,t)}{\partial t} = \frac{r \partial c(r,t)}{\partial t} = D\left(r \frac{\partial^2 c(r,t)}{\partial r^2} + 2 \frac{\partial c(r,t)}{\partial r}\right)$$

The second-order partial spatial derivative is given by,

$$\frac{\partial^2 w(r,t)}{\partial r^2} = \frac{\partial}{\partial r}\left(\frac{\partial}{\partial r}(rc(r,t))\right) = \frac{\partial c(r,t)}{\partial r} + \frac{\partial}{\partial r}\left(r \frac{\partial c(r,t)}{\partial r}\right)$$

$$= r \frac{\partial^2 c(r,t)}{\partial r^2} + 2 \frac{\partial c(r,t)}{\partial r}$$

Thus,

$$\frac{\partial w(r,t)}{\partial t} = D \frac{\partial^2 w(r,t)}{\partial r^2}$$

b) The partial derivatives of the function $w(r,t)$ have to be recast in terms of the partial derivatives of the function $f(\eta)$ where $\eta(r,t) = r^2/Dt$.[32] The partial derivatives of the function $\eta(r,t) = r^2/Dt$ are,

$$\frac{\partial \eta}{\partial t} = -\frac{r^2}{Dt^2} = -\frac{\eta}{t} \qquad \text{and} \qquad \frac{\partial \eta}{\partial r} = \frac{2r}{Dt} = \frac{2\eta}{r}$$

[32] K. F. Riley, M. P. Hobson, S. J. Bence, *Mathematical Methods for Physics and Engineering*, Cambridge University Press, 3rd edition (2006), § 20.5.

Since $f(\eta) = w(r, t)$, the first-order partial derivatives of the function $w(r, t)$ are recast in terms of the first-order derivative of the function $f(\eta)$ as,

$$\frac{\partial w}{\partial t} = \frac{df}{d\eta}\frac{\partial \eta}{\partial t} = -\frac{r^2}{Dt^2}\frac{df}{d\eta} = -\frac{\eta}{t}\frac{df}{d\eta}$$

$$\frac{\partial w}{\partial r} = \frac{df}{d\eta}\frac{\partial \eta}{\partial r} = \frac{2r}{Dt}\frac{df}{d\eta} = \frac{2\eta}{r}\frac{df}{d\eta}$$

The second-order partial derivative of the function $w(r, t)$ is recast in terms of the second-order derivative of the function $f(\eta)$ as,

$$\frac{\partial^2 w}{\partial r^2} = \frac{\partial}{\partial r}\left(\frac{2r}{Dt}\frac{df}{d\eta}\right) = \frac{2}{Dt}\frac{df}{d\eta} + \frac{2r}{Dt}\frac{d^2 f}{d\eta^2}\frac{\partial \eta}{\partial r} = \frac{2\eta}{r^2}\frac{df}{d\eta} + \frac{4\eta^2}{r^2}\frac{d^2 f}{d\eta^2}$$

Thus, the diffusion equation becomes,

$$-\frac{\eta}{t}\frac{df}{d\eta} = \frac{2D\eta}{r^2}\frac{df}{d\eta} + \frac{4D\eta^2}{r^2}\frac{d^2 f}{d\eta^2}$$

Using the definition of the dimensionless function η, this differential equation is recast as,

$$4\eta\frac{d^2 f}{d\eta^2} + (\eta + 2)\frac{df}{d\eta} = 0$$

Here, we introduce a function $g(\eta)$ as the derivative of $f(\eta)$ with respect to η,

$$g(\eta) = \frac{df}{d\eta}$$

Thus, the differential equation becomes,

$$4\eta\frac{dg(\eta)}{d\eta} + (\eta + 2)g(\eta) = 0$$

which is recast as,

$$\frac{dg(\eta)}{g(\eta)} = -\left(\frac{1}{4} + \frac{1}{2\eta}\right)d\eta$$

When integrating from η_0 to η we find,

$$\int_{g_0}^{g(\eta)}\frac{dg'(\eta')}{g'(\eta')} = -\int_{\eta_0}^{\eta}\left(\frac{1}{4} + \frac{1}{2\eta'}\right)d\eta'$$

where $g_0 = g(\eta_0)$. The solution is,

$$\ln\left(\frac{g(\eta)}{g_0}\right) = -\frac{1}{2}\ln\left(\frac{\eta}{\eta_0}\right) - \frac{1}{4}(\eta - \eta_0)$$

which is recast as,

$$\ln\left(\frac{g(\eta)\eta^{1/2}}{g_0\eta_0^{1/2}}\right) = -\frac{1}{4}(\eta - \eta_0)$$

and implies that,

$$g(\eta) = \frac{A}{\eta^{1/2}} \exp\left(-\frac{\eta}{4}\right)$$

where the constant $A = g_0 \, \eta_0^{1/2} \exp(\eta_0/4)$. When integrating from η_0 to η we find,

$$f(\eta) = \int_{\eta_0}^{\eta} g'(\eta') \, d\eta' = A \int_{\eta_0}^{\eta} \frac{1}{\eta'^{1/2}} \exp\left(-\frac{\eta'}{4}\right) d\eta'$$

Using the change of variable,

$$\nu = \frac{\eta^{1/2}}{2} = \frac{r}{2\sqrt{Dt}} \qquad \text{then} \qquad d\nu = \frac{d\eta}{4\eta^{1/2}}$$

and taking into account the definition $h(\nu) = f(\eta) = w(r,t)$, we can recast the solution as,

$$w(r,t) = h(\nu) = B \int_{\nu_0}^{\nu} \exp\left(-\nu'^2\right) d\nu'$$

where the constant $B = 4A$. For $\nu_0 = 0$, the solution is the error function $h(\nu) = \mathrm{erf}(\nu)$ multiplied by a constant.

c) The matter concentration $c(r,t)$ is given by,

$$c(r,t) = \frac{w(r,t)}{r} = \frac{B}{r} \int_{\nu_0}^{\nu} \exp\left(-\nu'^2\right) d\nu'$$

where $\nu(r,t)$ and we choose $\nu_0(r,t) = \nu(r_0,t)$. Thus, the conductive matter current density scalar at the electrode of radius r_0 is written as,

$$j_r(r_0,t) = -D \left.\frac{\partial c(r,t)}{\partial r}\right|_{r=r_0}$$
$$= \frac{BD}{r_0^2} \int_{\nu_0}^{\nu} \exp\left(-\nu'^2\right) d\nu' \Bigg|_{r=r_0} - \frac{BD}{r_0^2} \exp\left(-\nu^2\right) \left.\frac{\partial \nu}{\partial r}\right|_{r=r_0}$$

where we used the fact that the upper integration bound $\nu(r,t)$ is a function of r. In this relation, the integral vanishes since ν evaluated at r_0 is ν_0, which means that the upper and lower integration bounds are equal. Taking into account that,

$$\exp\left(-\nu^2\right) \left.\frac{\partial \nu}{\partial r}\right|_{r=r_0} = \exp\left(-\frac{r^2}{4Dt}\right) \left.\frac{\partial}{\partial r}\left(\frac{r}{2\sqrt{Dt}}\right)\right|_{r=r_0}$$
$$= \frac{1}{2\sqrt{Dt}} \exp\left(-\frac{r_0^2}{4Dt}\right)$$

we find,

$$j_r(r_0,t) = -\frac{B}{2r_0^2} \sqrt{\frac{D}{t}} \exp\left(-\frac{r_0^2}{4Dt}\right)$$

In the limit where the radius of the electrode is negligible, i.e. $r_0 = 0$, the scalar conductive matter current density is given by,

$$j_r(0,t) = \lim_{r_0 \to 0} j_r(r_0,t) = \lim_{r_0 \to 0}\left(-\frac{B}{2r_0^2}\sqrt{\frac{D}{t}}\left(1 - \frac{r_0^2}{4Dt}\right)\right) = \frac{B}{8\sqrt{D}t^{3/2}}$$

d) The stationary state is reached in the limit where $t \to \infty$. In the stationary limit,

$$\lim_{t \to \infty} v\left(r_0, t\right) = \lim_{t \to \infty} \frac{r_0}{2\sqrt{Dt}} = 0$$

and initially,

$$v_0 = \lim_{t \to 0} v\left(r_0, t\right) = \lim_{t \to 0} \frac{r_0}{2\sqrt{Dt}} = \infty$$

Thus, in the stationary state, the general expression for the scalar conductive matter current density obtained in c) reduces to,

$$j_r\left(r_0, \infty\right) = -D \left. \frac{\partial c\left(r, \infty\right)}{\partial r} \right|_{r=r_0} = -\frac{BD}{r_0^2} \int_0^\infty \exp\left(-v^2\right) dv$$

The error function erf (x) is defined as,

$$\text{erf}\left(x\right) = \frac{2}{\sqrt{\pi}} \int_0^x \exp\left(-v^2\right) dv$$

and erf $(\infty) = 1$. Thus, the scalar conductive matter current density is recast as,

$$j_r\left(r_0, \infty\right) = -\frac{\sqrt{\pi} BD}{2\,r_0^2} \, \text{erf}\left(\infty\right) = -\frac{\sqrt{\pi} BD}{2\,r_0^2} = -\frac{D\,c^*}{r_0}$$

where

$$B = \frac{2\,r_0\,c^*}{\sqrt{\pi}}$$

11.16 Effusivity

Two long blocks, made up of different homogeneous materials, are at temperatures T_1 and T_2 when they are brought into contact with one another. The interface quickly reaches a temperature T_0 given by,

$$T_0 = \frac{E_1\,T_1 + E_2\,T_2}{E_1 + E_2}$$

where $E_1 = \sqrt{\kappa_1\,c_1} > 0$ and $E_2 = \sqrt{\kappa_2\,c_2} > 0$ are called the **effusivities** of materials 1 and 2, where κ_1 and κ_2 are the thermal conductivities and c_1 and c_2 the specific heat per unit volume of both materials. If material 1 is very hot, but it has a low thermal conductivity κ_1 and specific heat c_1, and to the contrary material 2 has large thermal conductivity κ_2 and specific heat c_2, then the temperature of the interface T_0 is almost T_2, i.e. material 2 does not 'feel the heat' of material 1. Establish this result by using the following instructions :

a) Consider an x-axis normal to the interface with $x = 0$ at the interface, $x < 0$ in material 1 and $x > 0$ in material 2. Let $T_1\left(x, t\right)$ and $T_2\left(x, t\right)$ be the solutions of the heat diffusion equation (11.35) in materials 1 and 2. Determine the boundary conditions on $T_1\left(x, t\right)$ and $T_2\left(x, t\right)$ at the interface.

b) Using an approach that is analogous to the one presented in § 11.4.3, show that the general solutions for the temperature profiles $T_1(x,t)$ and $T_2(x,t)$ are given by,

$$T_1(x,t) = C_1 + D_1 \operatorname{erf}\left(\frac{x}{2\sqrt{\lambda_1 t}}\right) \qquad \text{where} \qquad x \le 0$$

$$T_2(x,t) = C_2 + D_2 \operatorname{erf}\left(\frac{x}{2\sqrt{\lambda_2 t}}\right) \qquad \text{where} \qquad x \ge 0$$

where $\operatorname{erf}(\nu)$ is the error function defined as,

$$\operatorname{erf}(\nu) = \frac{2}{\sqrt{\pi}} \int_0^\nu \exp\left(-s^2\right) ds$$

and C_1, C_2, D_1 and D_2 are constant coefficients.

c) Use the boundary conditions to determine the coefficients in terms of the temperatures T_0, T_1 and T_2. Show that the temperature T_0 is given by the effusivity relation just after the two blocks have reached a common temperature at the interface.

Solution:

a) At the interface, one boundary condition is that the temperatures of both materials are equal,

$$T_1(0,t) = T_2(0,t)$$

The other boundary condition is that the heat current densities $\boldsymbol{j}_{Q_1} = -\kappa_1\, \partial_x\, T_1\, \hat{\boldsymbol{x}}$ and $\boldsymbol{j}_{Q_2} = -\kappa_2\, \partial_x\, T_2\, \hat{\boldsymbol{x}}$ are equal as well,

$$\kappa_1 \frac{\partial T_1(0,t)}{\partial x} = \kappa_2 \frac{\partial T_2(0,t)}{\partial x}$$

where $\hat{\boldsymbol{x}}$ is the unit vector along the x-axis.

b) The heat diffusion equation in block 1 reads,

$$\frac{\partial T_1(x,t)}{\partial t} = \lambda_1 \frac{\partial^2 T_1(x,t)}{\partial x^2}$$

The partial derivatives of the function $T_1(x,t)$ have to be recast in terms of the partial derivatives of the function $f_1(\eta_1)$ where $\eta_1(x,t) = x^2/\lambda_1 t$.[33] The partial derivatives of the function $\eta_1(x,t) = x^2/\lambda_1 t$ are,

$$\frac{\partial \eta_1}{\partial t} = -\frac{x^2}{\lambda_1 t^2} = -\frac{\eta_1}{t} \qquad \text{and} \qquad \frac{\partial \eta_1}{\partial x} = \frac{2x}{\lambda_1 t} = \frac{2\eta_1}{x}$$

Since $f_1(\eta_1) = T_1(x,t)$, the first-order partial derivatives of the function $T_1(x,t)$ are recast in terms of the first-order derivative of the function $f_1(\eta_1)$ as,

$$\frac{\partial T_1}{\partial t} = \frac{df_1}{d\eta_1} \frac{\partial \eta_1}{\partial t} = -\frac{x^2}{\lambda_1 t^2} \frac{df_1}{d\eta_1} = -\frac{\eta_1}{t} \frac{df_1}{d\eta_1}$$

$$\frac{\partial T_1}{\partial x} = \frac{df_1}{d\eta_1} \frac{\partial \eta_1}{\partial x} = \frac{2x}{\lambda_1 t} \frac{df_1}{d\eta_1} = \frac{2\eta_1}{x} \frac{df_1}{d\eta_1}$$

[33] K. F. Riley, M. P. Hobson, S. J. Bence, *Mathematical Methods for Physics and Engineering*, Cambridge University Press, 3rd edition (2006), § 20.5.

The second-order partial derivative of the function $T_1(x,t)$ is recast in terms of the second-order derivative of the function $f_1(\eta_1)$ as,

$$\frac{\partial^2 T_1}{\partial x^2} = \frac{\partial}{\partial x}\left(\frac{2x}{\lambda_1 t}\frac{df_1}{d\eta_1}\right) = \frac{2}{\lambda_1 t}\frac{df_1}{d\eta_1} + \frac{2x}{\lambda_1 t}\frac{d^2 f_1}{d\eta_1^2}\frac{\partial\eta_1}{\partial x} = \frac{2\eta_1}{x^2}\frac{df_1}{d\eta_1} + \frac{4\eta_1^2}{x^2}\frac{d^2 f_1}{d\eta_1^2}$$

Thus, the heat diffusion equation becomes,

$$-\frac{\eta_1}{t}\frac{df_1}{d\eta_1} = \frac{2\lambda_1\eta_1}{x^2}\frac{df_1}{d\eta_1} + \frac{4\lambda_1\eta_1^2}{x^2}\frac{d^2 f_1}{d\eta_1^2}$$

Using the definition of the dimensionless function η_1, this differential equation is recast as,

$$4\eta_1\frac{d^2 f_1}{d\eta_1^2} + (\eta_1 + 2)\frac{df_1}{d\eta_1} = 0$$

Here, we introduce a function $g_1(\eta_1)$ as the derivative of $f_1(\eta_1)$ with respect to η_1,

$$g_1(\eta_1) = \frac{df_1}{d\eta_1}$$

Thus, the differential equation becomes,

$$4\eta_1\frac{dg_1(\eta_1)}{d\eta_1} + (\eta_1 + 2)g_1(\eta_1) = 0$$

which is recast as,

$$\frac{dg_1(\eta_1)}{g_1(\eta_1)} = -\left(\frac{1}{4} + \frac{1}{2\eta_1}\right)d\eta_1$$

When integrating from η_0 to η_1 we find,

$$\int_{g_0}^{g_1(\eta_1)}\frac{dg_1'(\eta_1')}{g_1'(\eta_1')} = -\int_{\eta_0}^{\eta_1}\left(\frac{1}{4} + \frac{1}{2\eta_1'}\right)d\eta_1'$$

where $g_0 = g_1(\eta_0)$. The solution is,

$$\ln\left(\frac{g_1(\eta_1)}{g_0}\right) = -\frac{1}{2}\ln\left(\frac{\eta_1}{\eta_0}\right) - \frac{1}{4}(\eta_1 - \eta_0)$$

which is recast as,

$$\ln\left(\frac{g_1(\eta_1)\eta_1^{1/2}}{g_0\eta_0^{1/2}}\right) = -\frac{1}{4}(\eta_1 - \eta_0)$$

and implies that,

$$g_1(\eta_1) = \frac{A_1}{\eta^{1/2}}\exp\left(-\frac{\eta_1}{4}\right)$$

where the constant $A_1 = g_0\eta_0^{1/2}\exp(\eta_0/4)$. When integrating from η_0 to η_1 we find,

$$f_1(\eta_1) = \int_{\eta_0}^{\eta_1}g_1'(\eta_1')d\eta_1' = A_1\int_{\eta_0}^{\eta_1}\frac{1}{\eta_1'^{1/2}}\exp\left(-\frac{\eta_1'}{4}\right)d\eta_1'$$

Using the change of variable,

$$\nu_1 = \frac{\eta_1^{1/2}}{2} = \frac{x}{2\sqrt{\lambda_1 t}} \qquad \text{then} \qquad d\nu_1 = \frac{d\eta_1}{4\eta_1^{1/2}}$$

and taking into account the definition $h_1(\nu_1) = f_1(\eta_1)$, we can recast the solution as,

$$h_1(\nu_1) = B_1 \int_{\nu_0}^{\nu_1} \exp\left(-\nu_1'^2\right) d\nu_1'$$

where the constant $B_1 = 4A_1$, which can be rewritten as,

$$h_1(\nu_1) = C_1 + B_1 \int_0^{\nu_1} \exp\left(-\nu_1'^2\right) d\nu_1'$$

where the constant C_1 is given by,

$$C_1 = -\int_0^{\nu_0} \exp\left(-\nu_1'^2\right) d\nu_1'$$

The error function $\operatorname{erf}(\nu)$, defined as,

$$\operatorname{erf}(\nu) = \frac{2}{\sqrt{\pi}} \int_0^{\nu} \exp\left(-s^2\right) ds$$

is an odd function, i.e. $\operatorname{erf}(-\nu) = -\operatorname{erf}(\nu)$ such that $\operatorname{erf}(0) = 0$ and $\operatorname{erf}(\infty) = 1$. The derivative of the error function $\operatorname{erf}(\nu)$ is given by,

$$\frac{d\operatorname{erf}(\nu)}{d\nu} = \frac{2}{\sqrt{\pi}} \exp\left(-\nu^2\right)$$

Thus, using the error function we find,

$$h_1(\nu_1) = C_1 + D_1 \operatorname{erf}(\nu_1)$$

where $D_1 = (\sqrt{\pi}/2) B_1$. Since $h_1(\nu_1) = T_1(x,t)$ and $\nu_1 = x/2\sqrt{\lambda_1 t}$, we have,

$$T_1(x,t) = C_1 + D_1 \operatorname{erf}\left(\frac{x}{2\sqrt{\lambda_1 t}}\right) \qquad \text{where} \qquad x \le 0$$

Likewise, we obtain the temperature profile in block 2,

$$T_2(x,t) = C_2 + D_2 \operatorname{erf}\left(\frac{x}{2\sqrt{\lambda_2 t}}\right) \qquad \text{where} \qquad x \ge 0$$

c) At the interface, i.e. $x = 0$, the first boundary condition, $T_1(0,t) = T_2(0,t) = T_0$, is satisfied provided that,

$$C_1 = C_2 = T_0$$

The blocks are long enough so that the temperatures at the end of each block is, at all time, equal to its initial temperatures. This is written as,

$$T_1 = T_1(-\infty, t) = T_0 + D_1 \operatorname{erf}(-\infty) = T_0 - D_1$$
$$T_2 = T_2(\infty, t) = T_0 + D_2 \operatorname{erf}(\infty) = T_0 + D_2$$

Hence, the temperature profiles can be recast as,

$$T_1(x,t) = T_0 + (T_0 - T_1)\,\text{erf}\left(\frac{x}{2\sqrt{\lambda_1 t}}\right) \qquad \text{and} \qquad x \leq 0$$

$$T_2(x,t) = T_0 + (T_2 - T_0)\,\text{erf}\left(\frac{x}{2\sqrt{\lambda_2 t}}\right) \qquad \text{and} \qquad x \geq 0$$

The spatial derivative of the temperature profiles are given by,

$$\frac{\partial T_1(0,t)}{\partial x} = (T_0 - T_1)\,\frac{d}{d\nu}\Big(\text{erf}(\nu)\Big)\Big|_{\nu=0}\,\frac{\partial}{\partial x}\left(\frac{x}{2\sqrt{\lambda_1 t}}\right)\Big|_{x=0}$$

$$= \frac{T_0 - T_1}{\sqrt{\pi\,\lambda_1 t}}$$

$$\frac{\partial T_2(0,t)}{\partial x} = (T_2 - T_0)\,\frac{d}{d\nu}\Big(\text{erf}(\nu)\Big)\Big|_{\nu=0}\,\frac{\partial}{\partial x}\left(\frac{x}{2\sqrt{\lambda_2 t}}\right)\Big|_{x=0}$$

$$= \frac{T_2 - T_0}{\sqrt{\pi\,\lambda_2 t}}$$

The second boundary condition, i.e. $\kappa_1\,\partial_x T_1(0,t) = \kappa_2\,\partial_x T_2(0,t)$, is recast as,

$$\kappa_1\,\frac{T_0 - T_1}{\sqrt{\pi\,\lambda_1 t}} = \kappa_2\,\frac{T_2 - T_0}{\sqrt{\pi\,\lambda_2 t}}$$

It has to hold for all times t after the interface has reached temperature T_0. According to relation (11.36),

$$\lambda_1 = \frac{\kappa_1}{c_1} \qquad \text{and} \qquad \lambda_2 = \frac{\kappa_2}{c_2}$$

Thus,

$$\frac{T_1 - T_0}{T_0 - T_2} = \frac{\kappa_2}{\kappa_1}\sqrt{\frac{\lambda_1}{\lambda_2}} = \frac{\kappa_2}{\kappa_1}\sqrt{\frac{\kappa_1\,c_2}{\kappa_2\,c_1}} = \sqrt{\frac{\kappa_2\,c_2}{\kappa_1\,c_1}} = \frac{E_2}{E_1}$$

which implies that,

$$T_0 = \frac{E_1\,T_1 + E_2\,T_2}{E_1 + E_2}$$

References

1. M. Séguin-Ainé, *De l'influence des Chemins de Fer et de l'art de les tracer et de les conduire*, Imprimerie Pitrat Ainé (1887).

2. J. G. Crowther, *British Scientists of the XIXth Century*, vol. 1, Pelican Books, London (1940).

3. J. R. von Mayer, *Bemerkungen über die Kräfte der unbelebten Natur*, Annalen der Chemie (1842), cited by K. Simonyi, *Kulturgeschichte der Physik*, Harri Deutsch Verlag (2004), § 4.5.8.

4. J. P. Joule, *On the Existence of an Equivalent Relation between Heat and the Ordinary Forms of Mechanical Power*, Phil. Mag. London, Edinburgh, and Dublin, Series **3**, 27 (179), 205–207 (1845).

5. I. Müller, *A History of Thermodynamics*, Springer, Berlin-Heidelberg (2007).

6. Ch. Gruber, Ph.-A. Martin, *De l'atome antique à l'atome quantique*, Presses Polytechniques et Universitaires Romandes (2013).

7. H. von Helmholtz, *Über die Erhaltung der Kraft, eine physikalische Abhandlung*, Druck und Verlag von G. Reimer (1847).

8. K. F. Riley, M. P. Hobson, S. J. Bence, *Mathematical Methods for Physics and Engineering*, Cambridge University Press (2006), § 5.2, § 5.5, § 5.7.

9. J.-Ph. Ansermet, *Mécanique*, Traité de physique, Presses polytechniques et Universitaires Romandes (2013), § 4.3.2.

10. Ch. Gruber, S. D. Brechet, *Lagrange Equations Coupled to a Thermal Equation: Mechanics as Consequence of Thermodynamics*, Entropy, **13**, 367–378 (2011).

11. J.-Ph. Ansermet, *Mécanique*, Traité de physique, Presses Polytechniques et Universitaires Romandes (2013), § 1.17.3.

12. J.-Ph. Ansermet, *Mécanique*, Traité de physique, Presses Polytechniques et Universitaires Romandes (2013), § 3.20.

13. J.-Ph. Ansermet, *Mécanique*, Traité de physique, Presses Polytechniques et Universitaires Romandes (2013), § 5.31.

14. E. C. G. Stückelberg von Breidenbach, P. B. Scheurer, *Thermocinétique phénoménologique galiléenne*, Birhäuser Verlag, Basel and Stuttgart (1974).

15. J.-Ph. Ansermet, *Mécanique*, Traité de physique, Presses Polytechniques et Universitaires Romandes (2013), § 1.3.2.

16. E. A. Guggenheim, *Thermodynamics, an Advanced Treatment for Chemists and Physicists*, North-Holland Pub. Co., Amsterdam (1949).

17. B. Pascal, *Traité de l'équilibre des liqueurs et de la pesanteur de la masse de l'air*, Guillaume Desprez, Paris (1654).

18. E. H. Lieb, J. Yngvason, *The Physics and Mathematics of the Second Law of Thermodynamics*, Phys. Rep. **310**, 1–96 (1999).

19. E. H. Lieb, J. Yngvason, *A Fresh Look at Entropy and the Second Law of Thermodynamics*, Phys. Today **54**, 32–37 (2000).

20. H. U. Fuchs, *The Dynamics of Heat*, Springer, Berlin-Heidelberg (2010).

21. J. W. Gibbs, *Graphical Methods in the Thermodynamics of Fluids*, Transactions of the Connecticut Academy, II, 309–342, April–May (1873).

22. S. W. Hawking, G. F. R. Ellis, *The Large Scale Structure of Space-Time*, Cambridge University Press (1973).

23. J. Botsis, M. Deville, *Mécanique des milieux continus*, Traité de physique, Presses Polytechniques et Universitaires Romandes (2015).

24. A. Einstein, *Physics and Reality* (1936), cited in Albert Einstein, *Oeuvres choisies*, vol. 5, Seuil (1971).

25. A. Einstein, cited by E. H. Lieb and J. Yngvason in *A Fresh Look at Entropy and the Second Law of Thermodynamics*, Phys. Today **54**, 32–37 (2000).

26. I. Müller, *A History of Thermodynamics: The Doctrine of Energy and Entropy.* Springer, Berlin, Heidelberg (2007).

27. G. Jaumann, *Geschlossenes System physikalischer und chemischer Diffrentialgestze*, Sitzungbericht Akademie der Wissenschaften, Wien, 12 IIa (1911).

28. E. Lohr, *Entropie und geschlossenes Gleichungssystem*. Denkschrift der Akademie der Wissenshaften 93 (1926).

29. J.-L.-M. Poiseuille, *Le mouvement des liquides dans les tubes de petits diamètres*, Paris (1844).

30. Ch. Ferrari, Ch. Gruber, *Friction Force: From Mechanics to Thermodynamics*, Eur. J. P. **31**, 1159–1175 (2010).

31. G. Carrington, *Basic Thermodynamics*, Oxford Science Publications, New York (1994).

32. M. Goupil, *Du Flou au Clair, Histoire de l'affinité chimique*, Editions du Comité des Travaux historiques et scientifiques, Paris (1991), § 4.

33. H. B. Callen, *Thermodynamics and an introduction to Thermostatistics*, John Wiley & Sons, Inc. (1985), § 5.4.

34. H. B. Callen, *Thermodynamics and an Introduction to Thermostatistics*, John Wiley & Sons, Inc. (1985), p. 288.

35. H. von Helmholtz, *Die Thermodynamik chemischer Vorgänge*, Wissenschaftliche Abhandlungen von Hermann Helmholtz, vol. 2 (1883), p. 958–978.

36. I. K. Howard, *H Is for Enthalpy, Thanks to Heike Kamerlingh Onnes and Alfred W. Porter*, J. Chem. Educ. **79** (6), 697 (2002).

37. H. B. Callen, *Thermodynamics and an Introduction to Thermostatistics*, John Wiley & Sons Inc. (1985).

38. G. Lebon, D. Jou, J. Casas-Vazquez, *Understanding Non-Equilibrium Thermodynamics*, Springer Verlag, Berlin (2008), p. 22–23.

39. G. Lebon, D. Jou, J. Casas-Vazquez, *Understanding Non-Equilibrium Thermodynamics*, Springer Verlag, Berlin (2008), p. 23.

40. H. B. Callen, *Thermodynamics*, John Wiley & Sons Inc., New York (1960), § 7.

41. G. Bruhat, *Thermodynamique*, Masson & Cie, 6th edition, Paris (1968), § 63.

42. A. L. de Lavoisier, P.-S. de Laplace, *Mémoire sur la chaleur*, Mémoires de l'Académie des Sciences (1787).

43. R. Boyle, *De la nature de l'air*, Etienne Michallet, Paris (1669).

44. E. Mariotte, *A Continuation of New Experiments Physico-Mechanical, Touching the Spring and Weight of the Air and Their Effects*, Henry Hall, Oxford (1679).

45. J. Charles (1787) mentioned by L. Gay-Lussac in *Recherches sur la dilatation des gaz et des vapeurs*, Annales de chimie **43**, 157 (1802).

46. L. Gay Lussac, *Mémoire sur la combinaison des substances gazeuses, les unes avec les autres*, Mémoires de la Société d'Arceuil **2**, 207–234 (1809).

47. A. Avogadro, *Essai d'une manière de déterminer les masses relatives des molécules élémentaires des corps*, Journal de Physique **73**, 58–76 (1810).

48. G. Bruhat, *Thermodynamique*, Masson & Cie, 6th edition, (1968), p. 124.

49. J. A. Deluc, *Recherche sur les modifications de l'atmosphère*, vol. 2, Genève, (1772).

50. C. Truesdell, *The Tragicomical History of Thermodynamics 1822–1854*, Springer-Verlag, New York (1980).

51. R. König, *Frontiers in Refrigeration and Cooling: How to Obtain and Sustain Ultra-Low Temperatures beyond Nature's Ambience*, Inter. J. of Refrig. **23**, 577–587 (2000).

52. L. Galgani, A. Scotti, *Remarks on Convexity of Thermodynamic Functions*, Physica **40**, 150–152 (1968).

53. H. B. Callen, *Thermodynamics and an Introduction to Thermostatistics*, Wiley, 2nd edition (1985), § 8.1.1.

54. M. Le Bellac, F. Mortessagne, G. G. Batrouni, *Equilibrium and Non-Equilibrium Statistical Thermodynamics*, Cambridge University Press (2004).

55. H. B. Callen, *Thermodynamics and an Introduction to Thermostatistics*, Wiley, 2nd edition (1985).

56. S. J. and K. M. Blundell, *Concepts in Thermal Physics*, Oxford University Press (2009), § 28.7.

57. M. Elenius, M. Dzugutov, *Evidence for a Liquid-Solid Critical Point in a Simple Monatomic System*, J. Chem. Phys. **131**, 104502 (2009).

58. S. Han, M. Y. Choi, P. Kumar, H. E. Stanley, *Phase Transitions in Confined Water Nanofilms*, Nat. Phys., **6**, 685 (2010).

59. P. Atkins, Julio de Paula, *Atkins' Physical Chemistry*, Oxford University Press, 7th edition (2002), p. 176, 181.

60. J. D. van der Waals, PhD thesis, *Over de continuiteit van den gas en vloeistoftestand* (1873).

61. J. C. Maxwell, *On the Dynamical Evidence of the Molecular Constitution of Bodies*, Nature, **11**, 357–359 (1875).

62. M. Legault, L. Blum, *The Coexistence Line in Mean Field Theories*, Fluid Phase Equilibria, **91**, 55–66 (1993).

63. Peter Atkins, Julio de Paula, *Atkins' Physical Chemistry*, Oxford University Press, 7th edition (2002), p. 186.

64. C. Johnston, *Advances in Thermodynamics of the van der Waals Fluid*, Morgan & Claypool Publishers (2014).

65. H. B. Callen, *Thermodynamics and an Introduction to Thermostatistics*, Wiley, 2nd edition (1985), § 8.2.2.

66. P. R. Mashaei, M. Shahryari, S. Madani, *Analytical Study of Multiple Evaporator Heat Pipe with Nanofluid: A Smart Material for Satellite Equipment Cooling Application*, Aerospace Science and Technology **59**, 112–121 (2016).

67. J.-P. Borel, A. Chatelain, *Surface Stress and Surface Tension: Equilibrium and Pressure in Small Particles*, Surf. Sci. **156**, 572–579 (1985).

68. Ph. Buffat, J.-P. Borel, *Size Effect on the Melting Temperature of Gold Particles*, Phys. Rev. A **13** (6), 2287 (1976).

69. R. W. Siegel, *Cluster-Assembled Nanophase Materials*, Annu. Rev. Mater. Sci. **21**, 559–578 (1991).

70. S. Carnot, *Reflections on the Motive Power of heat and on Machines Fitted to Develop That Power*, John Wiley & Sons (1897).

71. Ph. Depont, *L'entropie et tout ça, le roman de la thermodynamique*, Cassini (2001).

72. R. Feynman, R. B. Leighton, M. Sands, *The Feynman Lectures on Physics*, Adison-Wesley (1963).

73. F. L. Curzon, B. Alborn, *Efficiency of a Carnot Engine at Maximum Power Output*, American Journal of Physics **43**, 22 (1975).

74. G. Bruhat, *Thermodynamique*, Masson & Cie, 6th edition (1968), p. 173.

75. D. Kondepudi, I. Prigogine, *Modern Thermodynamics*, John Wiley & Sons Ltd (1998).

76. M. Goupil, *Du Flou au Clair, Histoire de l'Affinité de Cardan à Prigogine*, Editions du Comité des Travaux historiques et scientifiques, Paris (1991).

77. C. K. W. Friedli, *Chimie générale pour ingénieur*, Presses Polytechniques et Universitaires Romandes (2010).

78. P. Infelta, *Introductory Thermodynamics*, Brown Walker Press, Boca Raton Florida (2004).

79. E. A. Guggenheim, *Thermodynamics, an Advanced Treatment for Chemists and Physicists*, North-Holland (1977).

80. A. Marchand, A. Facoult, *La thermodynamique mot à mot*, De Boek Université (1995).

81. H. B. Callen, *Thermodynamics*, 1st edition, John Wiley & Sons, New York (1960), § D6.

82. B. Dreyfus, A. Lacaze, *Cours de thermodynamique*, Dunod (1971).

83. H. Girault, *Electrochimie physique et analytique*, Presses Polytechniques et Universitaires Romandes (2012), p. 235.

84. H. Reiss, *Methods of Thermodynamics*, Dover, Mineola, New York (1996), § 5.39 and following.

85. R. A. Horn, C. R. Johnson, *Matrix Analysis*, Cambridge University Press (1990).

86. S. R. de Groot, P. Mazur, *Non-Equilibrum Thermodynamics*, Dover, New York (1984), p. 371–375.

87. Y. V. Kuzminskii, V. A. Zasukha, G. Y. Kuzminskaya, *Thermoelectric Effects in Electrochemical Systems: Nonconventional Thermogalvanic Cells*, J. Power Sources **52**, 231–242 (1994).

88. S. W. Lee, Y. Yang, H.-W. Lee, H. Ghasemi, D. Kraemer, G. Chen, Y. Cui, *An Electrochemical System for Efficiently Harvesting Low–Grade Heat Energy*, Nat. Commun. **5**, 3942 (2014).

89. D. C. Mattis, *The Theory of Magnetism Made Simple*, World Scientific, New Jersey (2006).

90. W. Gilbert, P. Fleuy Mottelay, *De Magnete*, Dover Publication Inc., New York (1958).

91. S. Kohout, J. Roos, H. Keller, *Novel Sensor Design for Torque Magnetometry*, Rev. Sci. Instrum. **78**, 013903 (2007).

92. W. Känzig, *History of Ferroelectricity 1938–1955*, Ferroelectrics. **74**, 285–291 (1987).

93. H. A. Pohl, *The Motion and Precipitation of Suspensoids in Divergent Electric Fields*, J. Appl. Phys. **22** (7), 869–871, (1951) and H. A. Pohl, *Some Effects of Nonuniform Fields on Dielectrics*, J. Appl. Phys. **29** (8), 1182–1188 (1958).

94. A. Ashkin, J. M. Dziedzic, J. E. Bjorkholm, S. Chu, *Observation of a Single-Beam Gradient Force Optical Trap for Dielectric Particles*, Opt. Lett. **11** (5) 288–290 (1986).

95. J. C. Maxwell, *On the Physical Lines of Force*, Philos. Mag. **4**, 161 (1861).

96. J. H. Poynting, *On the Transfer of Energy in the Electromagnetic Field*, Philos. Trans. Royal Soc. **175**, 343–361 (1884).

97. P. Debye, *Polar Molecules*, Dover, New York (1945).

98. R. Clausisus, *Die mechanische U'gretheorie* (1879).

99. O. F. Mossotti, Mem. di mathem. e fisica in Modena **24**, 49 (1850).

100. J. Larmor, *A Dynamical Theory of the Electric and Luminiferous Medium. Part III. Relations with Material Media*, Philos. Trans. Royal Soc. **190**, 205–493 (1897).

101. F. Bloch, *Nuclear Induction*, Phys. Rev. **70**, 460–473 (1946).

102. F. Reuse, *Electrodynamique*, Traité de physique, Presses Polytechniques et Universitaires Romandes (2012), p. 43, 111, 174, 175.

103. F. Reuse, *Electrodynamique*, Traité de physique, Presses Polytechniques et Universitaires Romandes (2012), p. 119.

104. F. Reuse, *Electrodynamique*, Traité de physique, Presses Polytechniques et Universitaires Romandes (2012), p. 181–182.

105. F. Reuse, *Electrodynamique*, Traité de physique, Presses Polytechniques et Universitaires Romandes (2012), p. 117.

106. F. Reuse, *Electrodynamique*, Traité de physique, Presses Polytechniques et Universitaires Romandes (2012), p. 175.

107. S. D. Brechet, A. Roulet, J.-Ph. Ansermet, *Magnetoelectric Ponderomotive Force*, Mod. Phys. Lett. B **27** (21), 1350150 (2013).

108. F. Reuse, *Electrodynamique*, Traité de physique, Presses Polytechniques et Universitaires Romandes (2012), p. 303, 305.

109. S. D. Brechet, F. A. Reuse, J.-Ph. Ansermet, *Thermodynamics of Continuous Media with Electromagnetic Fields*, Eur. Phys. J. B **85**, 412 (2012).

110. F. Reuse, *Electrodynamique*, Traité de physique, Presses Polytechniques et Universitaires Romandes (2012), p. 118.

111. F. Reuse, *Electrodynamique*, Traité de physique, Presses Polytechniques et Universitaires Romandes (2012), p. 174.

112. F. Reuse, *Electrodynamique*, Traité de physique, Presses Polytechniques et Universitaires Romandes (2012), p. 120, 177.

113. F. Reuse, *Electrodynamique*, Traité de physique, Presses Polytechniques et Universitaires Romandes (2012), p. 58.

114. F. Reuse, *Electrodynamique*, Traité de physique, Presses Polytechniques et Universitaires Romandes (2012), p. 86.

115. F. Reuse, *Electrodynamique*, Traité de physique, Presses et Polytechniques Universitaires Romandes (2012), p. 88.

116. F. Reuse, *Electrodynamique*, Traité de physique, Presses et Polytechniques Universitaires Romandes (2012), p. 93.

117. F. Reuse, *Electrodynamique*, Traité de physique, Presses Polytechniques et Universitaires Romandes (2012), p. 82, 86.

118. F. Reuse, *Electrodynamique*, Traité de physique, Presses Polytechniques et Universitaires Romandes (2012), p. 28–29.

119. F. Reuse, *Electrodynamique*, Traité de physique, Presses Polytechniques et Universitaires Romandes (2012), p. 45.

120. F. Reuse, *Electrodynamique*, Traité de physique, Presses Polytechniques et Universitaires Romandes (2012), p. 103.

121. F. Reuse, *Electrodynamique*, Traité de physique, Presses Polytechniques et Universitaires Romandes (2012), p. 186.

122. F. Reuse, *Electrodynamique*, Traité de physique, Presses Polytechniques et Universitaires Romandes (2012), p. 154.

123. F. Reuse, *Electrodynamique*, Traité de physique, Presses Polytechniques et Universitaires Romandes (2012), p. 219.

124. F. Reuse, *Electrodynamique*, Traité de physique, Presses Polytechniques et Universitaires Romandes (2012), p. 222.

125. F. Reuse, *Electrodynamique*, Traité de physique, Presses Polytechniques et Universitaires Romandes (2012), p. 220.

126. S. D. Brechet, J.-Ph. Ansermet *Thermodynamics of a Continuous Medium with Electric and Magnetic Dipoles*, Eur. Phys. J. B **86**, 318 (2013).

127. P. Debye, *Einige Bermerkungen zur Magnetisierung bei tiefer Temperatur*, Ann. Phys. **81**, 1154 (1926).

128. D. Liu, *Origin and Tuning of the Magnetocaloric Effect in the Magnetic Refrigerant $Mn_{1.1}Fe_{0.9}(P_{0.8}Ge_{0.2})$*, Phys. Rev. B **79**, 014435, (2009).

129. A. Kitanovski, P. W. Golf, *Innovative Ideas for Future Research on Magnetocaloric Technologies*, Int. J. Refrig. **33**, 449–464 (2010).

130. C. Kittel, *Introduction to Solid State Physics*, John Wiley & Sons Inc., Hoboken, 5th edition (1976).

131. C. P. Slichter, *Principles of Magnetic Resonance*, Springer-Verlag, Berlin-Heidelberg (1990), § 6.3.

132. S. Ozeki, J. Miyamoto, S. Ono, C. Wakai, T. Watanabe, *Water-Solid Interactions under Steady Magnetic Fields: Magnetic Field-Induced Adsorption and Desorption of Water*, J. Phys. Chem. **100**, 4205–4212 (1996).

133. T. M. Squires, S. R. Quake, *Microfluidics: Fluid Physics at the Nanoliter Scale*, Rev. Mod. Phys. **77**, 977 (2005).

134. A. M. Tishin, Y. I. Spichkin, *Recent Progress in Magnetocaloric Effect: Mechanisms and Potential Applications*, Int. J. Refrig. **37**, 223–229 (2014).

135. C. Hagmann, D. J. Benfod, P. L. Richards, *Paramagnetic Salt Pill Design for Magnetic Refrigerators used in Space Applications*, Cryogencis, **34** (3), 213–219 (1994).

136. X. Bohigas, E. Molins, A. Roig, J. Tejada, X. X. Zhang, *Room-Temperature Magnetic Refrigerator using Permanent Magnets*, IEEE Trans. Mag. **36** (3), 538–544 (2000).

137. S. Carnot, *Réflexions sur la puissance motrice du feu et sur les machines propres à développer cette puissance.*, Bachelier, Paris (1924).

138. Ch. Ferrari, Ch. Gruber, *Friction Force: From Mechanics to Thermodynamics*, Eur. J. Phys. **31**, 1159 (2010).

139. C. Eckart, *The Thermodynamics of Irreversible Processes. I. The Simple Fluid*, Phys. Rev. **58**, 267–269 (1940).

140. C. Eckart, *The Thermodynamics of Irreversible Processes. II. Fluid Mixtures*, Phys. Rev. **58**, 269–275 (1940).

141. J. Botsis, M. Deville, *Mécanique des milieux continus*, Presses Polytechniques et Universitaires Romandes (2006), p. 49–51.

142. I. L. Ryhming, *Dynamique des fluides*, Presses Polytechniques et Universitaires Romandes (2009), p. 27.

143. F. Reuse, *Electrodynamique*, Traité de physique, Presses Polytechniques et Universitaires Romandes (2012), equation (2.57).

144. J. Botsis, M. Deville, *Mécanique des milieux continus*, Traité de physique, Presses Polytechniques et Universitaires Romandes (2015), § 2.4.2.

145. J.-Ph. Ansermet, *Mécanique*, Traité de physique, Presses Polytechniques et Universitaires Romandes (2013) § 3.21.

146. E. C. G. Stückelberg von Breidenbach, P. B. Scheurer, *Thermocinétique phénoménologique galiléenne*, Birkhauser Verlag, Basel and Stuttgart, (1974) p. 177.

147. G. Gremaud, *Théorie eulérienne des milieux déformables*, Presses Polytechniques et Universitaires Romandes (2013), p. 11.

148. S. D. Brechet, J.-Ph. Ansermet, *Thermodynamics of Continuous Media with Intrinsic Rotation and Magnetoelectric Coupling*, Continuum Mech. Therm. **26** (2), 115–142 (2014).

149. T. Yamagami, Y. Saito, Y. Matsuzuka, M. Namiki, M. Toriumi, R. Yokota, H. Hirosawa, K. Matsushima, *Development of the Highest Altitude Balloon*, Adv. Space Res. **33**, 1653–1659 (2004).

150. L. Dufour, *Sur une variation de température qui accompagne la diffusion des gaz à travers une cloison de terre poreuse*, Archives des Sciences Physiques et naturelles, Genève **49**, (1874).

151. L. Dufour, *Diffusion des gaz à travers les parois poreuses*, Archives des sciences physiques et naturelles **45**, 9–12 (1872).

152. L. Dufour, *Über die Diffusion der Gase durch poröse Wände und die sie begleitenden Temperaturveränderungen*, Annalen der Physik **28**, 490 (1873)

153. M. Reichl, M. Herzog, A. Goetz, D. Braun, *Why Charged Molecules Move across a Temperature Gradient: The Role of Electric Fields*, Phys. Rev. Lett. **112**, 198101 (2014).

154. L. A. Belfiore, *Transport Phenomena for Chemical Reactor Design*, Wiley, Hoboken, New Jersey (2003) p. 700–702.

155. L. A. Belfiore, M. N. Karim, C. J. Belfiore, *Tubular Bioreactor Models That Include Onsager-Curie Scalar Cross-Phenomena to Describe Stress-Dependent Rates of Cell Proliferation*, Biophys. Chem. 135 (1–3), 41–50 (2008).

156. K. D. Andrews, P. Feugier, R. A. Black, J. A. Hunt, *Vascular Prostheses Performance Related To Cell-Shear Responses*, Journal of Surgical Research **149** (1), 39–46 (2007).

157. L. A. Belfiore, *Soret Diffusion and Non-Ideal Dufour Conduction in Macroporous Catalysts with Exothermic Chemical Reaction at Large Intrapellet Damköhler Numbers*, Canadian J. Chem. Eng. **85**, 268–279 (2007).

158. L. Gravier, S. Serrano-Guisan, F. Reuse and J.-Ph. Ansermet, *Thermodynamic Description of Heat and Spin Transport in Magnetic Nanostructures*, Phys. Rev. B **73**, 024419 (2006).

159. J-Ph. Ansermet, *Thermodynamic Description of Spin Mixing in Spin-Dependent Transport*, IEEE Trans. Mag. 329–335 (2008).

160. C. Eckart, *The Thermodynamics of Irreversible Processes. I. The Simple Fluid.* Phys. Rev. **58**, 267 (1940).

161. C. Eckart, *The Thermodynamics of Irreversible Processes. II. Fluid Mixtures.* Phys. Rev. **58**, 269 (1940).

162. L. Onsager, *Reciprocal Relations in Irreversible Processes. I*, Phys. Rev. **37**, 405 (1931).

163. L. Onsager, *Reciprocal Relations in Irreversible Processes. II*, Phys. Rev. **38**, 2265 (1931).

164. H. B. G. Casimir, *On Onsager's Principle of Microscopic Reversibility*, Rev. Mod. Phys. **17**, 343 (1945).

165. I. Prigogine, *Etude thermodynamique des phénomènes irréversibles*, Desoer, Liège, (1947).

166. I. Prigogine, I. Stengers, *La Nouvelle Alliance*, Editions Gallimard (1991).

167. I. Prigogine, I. Stengers, *Order out of Chaos: Man's New Dialogue with Nature*, Flamingo, New-York City (1984).

168. P. Curie, *Sur la symétrie dans les phénomènes physiques, symétrie d'un champ électrique et d'un champ magnétique*, J. Phys. Théor. Appl. **3**, 393 (1894).

169. H. B. Callen, *Thermodynamics and an Introduction to Thermostatistics*, 2nd edition, Wiley, New York (1985).

170. J. Fourier, *Théorie de la chaleur*, Firmin Didot, Paris (1822).

171. A. Righi, *Rotazione delle linee isotermiche del bismuto posto in un campo magnetico*, Atti della Reale Accademia dei Lincei, Rendiconti **4**, 284 (1887).

172. A. Leduc, *Sur la conductibilité calorifique du bismuth dans un champ magnétique et la déviation des lignes isothermes*, Journal de Physique **6**, 378 (1887).

173. A. Fick, *Über diffusion*, Poggendorff's Annalen der Physik **94**, 59–86 (1855).

174. L. Dufour, *Uber die Diffusion der Gase durch poröse Wände und die sie begleitenden Temperaturveränderungen*, Annalen der Physik **28**, 490 (1873).

175. C. Soret, *Sur l'état d'équilibre que prend, au point de vue de sa concentration, une dissolution saline primitivement homogène, dont deux parties sont portées à des températures différentes*, Archives des Sciences Physiques et Naturelles, Genève, **2**, 48–61 (1879).

176. S. D. Brechet et, J.-Ph. Ansermet, *Heat-Driven Spin Currents on Large Scales*, physica status solidi (RRL) **5** (12), 423–425 (2011).

177. G. S. Ohm, *Die galvanische Kette*, T. H. Riemann, Berlin (1827).

178. E. H. Hall, *On a New Action of the Magnet on Electric Currents*, Am. J. Mathemat., **2** (3), 287–292 (1879).

179. A. von Ettinghausen, Walther Nernst, *Über das Auftreten electromotorischer Kräfte in Metallplatten, welche von einem Wärmestrome durchflossen werden und sich im magnetischen Felde befinden*, Annalen der Physik, **265** (10), 343–347 (1886).

180. T. J. Seebeck, *Magnetische Polarisation der Metalle und Erze durch Temperatur-Differenz*, Abh. Akad. Wiss. Berlin, 289–346 (1822).

181. W. Thomson, *On a Mechanical Theory of Thermoelectric Currents*, Proc. Royal Soc. Edinburgh 91–98 (1851).

182. J. P. Joule, *On the Effects of Magnetism upon the Dimensions of Iron and Steel Bars*, Phil. Mag., London, Edinburgh, and Dublin, **30**, 225–241 (1847).

183. J. C. A. Peltier, *Nouvelles expériences sur la caloricité des courants électriques*, Ann. Chim. Phys. **56**, 371 (1834).

184. L. Gravier, S. Serrano-Guisan, F. Reuse, J.-Ph. Ansermet, *Spin-Dependent Peltier Effect of Perpendicular Currents in Multilayered Nanowires*, Phys. Rev. B **73**, 052410 (2006).

185. J.-Ph. Ansermet, *Thermodynamic Description of Spin Mixing in Spin-Dependent Transport*, IEEE Trans. Magn. **44** (3), 329 (2008).

186. T. Valet, A. Fert, *Theory of the Perpendicular Magnetoresistance in Magnetic Multilayers*, Phys. Rev. B **48**, 7099 (1993).

187. T. C. Harman, *Special Techniques for Measurement of Thermoelectric Properties*, J. App. Phys. **29**, 1373 (1958).

188. H. J. Goldsmid, *Introduction to Thermoelectricity*, Springer, Berlin-Heidelberg (2010).

189. G. J. Snyder, T. S. Ursell, *Thermoelectric Efficiency and Compatibility*, Phys. Rev. Lett. **91** (4) 138301 (2003).

190. L. D. Landau, E. M. Lifshitz, L.-P. Pitaevskii, *Electrodynamics of Continuous Media, Landau and Lifshitz Course of Theoretical Physics,* volume 8, Pergamon Press, Oxford, 3rd edition (2000).

191. C. Zhou, S. Birner, Y. Tang, K. Heinselman, M. Grayson, *Driving Perpendicular Heat Flow : $(p \times n)$-Type Transverse Thermoelectrics for Microscale and Cryogenic Peltier Cooling*, Phys. Rev. Lett. **110**, 227701 (2013).

192. K. Aoki, K. Akimoto, K. Tokuda, H. Matsuda, J. Osteryoung, *Linear Sweep Voltammetry at Very Small Stationary Disk Electrodes*, J. Electroanal. Chem. **171**, 219–230 (1984).

193. M. Fleschmann, S. Pons, *The Behavior of Microdisk and Microring Electrodes*, J. Electroanal. Chem. **222**, 107–115 (1987).

194. A. M. Bond, K. B. Oldham, C. G. Zoski, *Steady-State Voltammetry*, Anal. Chim. Acta. **216**, 177–230 (1989).

195. J. Heinze, *Ultramicroelectrodes in Electrochemistry*, Angew. Chem. Int. Ed. Engl. **32**, 1268–1288 (1993).

Index